Air Quality

While the field of air quality is well established, it is evolving in response to new regulations driven by climate change. The seventh edition of this essential textbook presents the most current information, incorporating updates across all sections to reflect the latest technologies, regulations, and air quality management strategies. All chapters are thoroughly reviewed for the latest updates and the chapters on regulations, modeling, monitoring, and control technologies are revised. New sections are added on topics such as the co-benefits of climate change, air pollution, and hydrogen vehicles. The section on indoor air quality is thoroughly updated as it remains a growing concern in society.

New in this edition:

- Includes the latest updates on air quality legislation in the United States.
- Presents the latest information on the state of air quality monitoring.
- Provides updates on the effects of air pollution and CO_2 on climate change.
- Explains the impact of recent changes in energy production on pollutant emissions and the associated co-benefits from climate change mitigation.
- Incorporates broadened coverage of air pollutant emissions and quality in other countries with an increased emphasis on global air quality.
- Includes a new section on ozone deposition.
- Provides new emerging pollutants such as per- and polyfluoroalkyl substances (PFAS).

This textbook is for students taking introductory courses in air quality, air pollution, and climate change. The end of each chapter includes a list of recommended readings as well as questions to check reading comprehension. Professors can take advantage of an arsenal of included tools, such as lecture slides, a test bank for each chapter, and answer guides.

Air Quality

Seventh Edition

Joshua S. Fu and Wayne T. Davis

CRC Press
Taylor & Francis Group
Boca Raton London New York

CRC Press is an imprint of the
Taylor & Francis Group, an **informa** business

Designed cover image: Top Image © Shutterstock; Bottom Image © Getty Images

Seventh edition published 2026
by CRC Press
2385 NW Executive Center Drive, Suite 320, Boca Raton FL 33431

and by CRC Press
4 Park Square, Milton Park, Abingdon, Oxon, OX14 4RN

CRC Press is an imprint of Taylor & Francis Group, LLC

First edition published by CRC Press 1985
Sixth edition published by CRC Press 2021

ISBN: 978-1-041-12312-5 (hbk)
ISBN: 978-1-041-12313-2 (pbk)
ISBN: 978-1-003-66410-9 (ebk)

DOI: 10.1201/9781003664109

Typeset in Times
by Apex CoVantage, LLC

Access the Instructor and Student Resources: www.routledge.com/cw/fu

To the memory of Ed Lewis, publisher and friend, and Thad Godish, author and friend.

This book is your continuing legacy.

Contents

Preface

This seventh edition of *Air Quality* has been written (as have previous editions) to provide readers (students, instructors, consultants, government personnel, and others) with a comprehensive overview of air quality, the science that continues to provide a better understanding of atmospheric chemistry and its effects on public health and the environment, and the regulatory and technological management practices employed in achieving air quality goals.

It has been over four decades since Dr. Godish began to pen the words that would comprise the first edition. Much has changed. The internet, that enormous library of information that was so much a part of the writing of this edition, was virtually unknown outside the military and academe (and even there it was still a foundling). The word processor had not quite replaced the typewriter (when "cutting and pasting" was literally that). How long ago, technologically, the early to mid-1980s now seems.

In the United States, we are now entering the 75th year of significant regulatory efforts to "protect and enhance" the quality of the nation's air. We accepted the challenge of expanding our concerns to protect the Earth's atmosphere from the effects of ozone-destroying chemicals that have threatened to destroy the ozone layer. We are currently moving toward accepting the challenges that the planet faces, including greenhouse gas–associated increases in global surface temperatures and the environmental changes predicted to occur and already occurring across the globe.

In the 1970s and 1980s, acidic deposition was identified as a major environmental concern in North America, and we responded by enacting and then implementing major acidic deposition control measures. In conducting research studies on acidic deposition, it became increasingly apparent that it was only a part of a much larger environmental concern: atmospheric deposition, which includes mercury, nitrate nitrogen, and organochlorine compounds such as pesticides, polychlorinated biphenyls, dioxins, and furans. In the 2020s and beyond, we will need to focus more directly on the factors affecting climate change and their interrelated effects on air quality worldwide.

The health protection issues that dominate air quality management in the United States continue to evolve as more powerful statistical procedures increasingly demonstrate that pollutant exposures, at levels previously considered safe, could cause adverse health effects, with a resultant need for more stringent regulatory requirements. This has been particularly the case for ozone and $PM_{2.5}$ (particulate matter with an aerodynamic diameter of ≤ 2.5 µm).

Although we Americans and other citizens of the planet want clean air, regulatory efforts impose real costs on the regulated community. While we expressed concern in our previous edition of *Air Quality*, we are even more concerned as we enter the mid-2020s that efforts need to be increased to provide improvements in air quality and reductions in greenhouse gas emissions to avoid potential significant health effects and potentially irreversible changes to our planet. We also recognize that this is not a challenge for the United States alone, but rather it is a global challenge. An extensive glossary is included at the end of the book, as well as a list of *Readings* and *Questions* at the end of each chapter to help pique further interest and to facilitate discussion in the classroom. This edition includes significant revisions and updates on the regulatory aspects related to air quality, emissions of pollutants, and particularly the area of greenhouse gas emissions.

The reader is encouraged to utilize the excellent resources and updates that are available on the US EPA website *http://www.epa.gov* and its related subdirectories. The subdirectory*/technical-air-pollution-resources* contains extensive information on Air Emissions; Air Modeling; Analyzing Air Pollution Impacts and Controls; and National Ambient Air Quality Standards (NAAQS). The subdirectory*/environmental-topics/air-topics/*contains information on Pollution and Air Quality; Air Pollutants (Criteria Air Pollutants, Greenhouse Gas Emissions, Hazardous Air Pollutants); Indoor Air; Data (air pollution trends by pollutant); Research; and Air Quality Management. The sites have search functions that allow the reader to find the latest fact sheets summarizing the current status and history of air quality standards, including the status of greenhouse gas legislation. The site is an invaluable source of information for both students and practitioners in the field of air quality, although a bit wieldy due to the vast amount of information on the site.

Air Quality continues to be a very readable text for advanced-level undergraduate and beginning-level graduate students in environmental science; environmental management technical programs at the junior college level; and programs serving public health, industrial hygiene, and engineering needs. It is useful as a supplement to engineering curricula, where the primary focus is the design and operation of pollution control equipment. It is also written for a variety of nonuniversity and noncollege readers who have a professional or personal interest in the field.

Acknowledgments

As co-authors to the fifth, sixth, and seventh editions of *Air Quality*, Dr. Davis and I acknowledge Dr. Godish's substantial input to *Air Quality*, as he was the original author of the first four editions. Dr. Godish initiated the original edition in 1984 with the late Ed Lewis, the publisher of Lewis Publishers, Inc. Like Dr. Godish, Dr. Davis had the pleasure of working with Ed on a number of activities. Through the work of Dr. Godish and Mr. Lewis, *Air Quality* was conceived and has been a valuable book used by many colleges and universities in the area of environmental science for the last 41 years.

Dr. Godish was a professor of natural resources and environmental management at Ball State University in Muncie, Indiana. He earned his PhD at Penn State in 1969. He had just agreed to begin working on the fifth edition when, unfortunately, he died on June 20, 2009, at the age of 67. We knew Dr. Godish and always had a great respect for his zeal in developing and publishing textbooks related to both indoor and ambient air quality. I am honored to have been asked by Taylor & Francis Group/CRC to prepare the seventh edition and to ensure the continuation of the textbook that was developed by our friend and colleague, Dr. Godish.

We continue to be indebted to the many colleagues in colleges and universities in the United States and other parts of the world who have chosen to use *Air Quality* in their courses and who continue to share valuable feedback. Through your support, you have enabled us to expand our classrooms to a far broader audience.

Authors

Joshua S. Fu currently holds multiple positions at the University of Tennessee, Knoxville (UTK). He is Chancellor's Professor in the Department of Civil and Environmental Engineering, and concurrently serves as the John D. Tickle Professor and James G. Gibson Professor in the Tickle College of Engineering. In addition to his UTK roles, he is the inaugural professor of the UT-Oak Ridge Innovation Institute's Bredesen Center for Interdisciplinary Research and Graduate Education and holds a joint appointment with Oak Ridge National Laboratory as a Joint Faculty Professor in the Integrated Computational Earth Sciences Group in the Computational Sciences and Engineering Division, fostering strong university–national lab collaborations. Prior to joining UTK in 2000, Dr. Fu worked as a senior scientific applications analyst and software engineer at Lockheed Martin/EPA. He earned his BS in environmental engineering from National Cheng Kung University, an MS in environmental engineering and water resources from UCLA, and a PhD in civil engineering from North Carolina State University.

Dr. Fu's research focuses on air quality, climate change, energy, and human health. His work also emphasizes the applications of artificial intelligence (AI) and machine learning (ML) to air quality, climate change, human health, global atmospheric deposition mapping, and air quality and weather forecasting. At UTK, he has taught courses on air quality management, climate change, and pollution control, authored more than 220 peer-reviewed journal articles, and has served in multiple editorial roles, including co-editor of *Atmospheric Chemistry and Physics*, editor of *Aerosol and Air Quality Research*, and served as ad hoc editor-in-chief and associate editor of the *Journal of the Air & Waste Management Association*. One of his air quality studies was recognized by Elsevier as among the most cited papers in *Atmospheric Environment* (2007–2012). Dr. Fu has served as principal investigator and co-investigator on numerous projects funded by federal agencies such as the CDC, DHS, DOE, US DOT, LLNL, NASA, NIH, NOAA, NSF, ORNL, US EPA, and USDA, as well as by state agencies and industry partners. His contributions have earned him numerous honors, including Fellow status in both the American Association for the Advancement of Science (AAAS) and the Air & Waste Management Association (AWMA), Board-Certified Environmental Engineer Member (BCEEM) recognition from the American Academy of Environmental Engineers and Scientists (AAEES), and multiple distinguished and endowed professorships. He is also a recipient of AWMA's Lyman Ripperton Environmental Education Award. In addition to his academic contributions, Dr. Fu serves as Vice Chair of the WMO's MMF-GTAD Initiative Steering Committee and WMO Research Board's Task Team on AI for Weather (AI4Wx), contributes to the UN ECE Task Force on Hemispheric Transport of Air Pollution, leads the Model Intercomparison Study in Asia, and served as an expert member of the Arctic Council's AMAP SLCFs. He currently chairs the Association of Environmental Engineering and Science Professors (AEESP)'s Lecturers Committee and the Climate Change Division of the AWMA. Locally, he served as a member of the Knox County Air Pollution Control Board in Tennessee and currently serves on the State of Tennessee Air Pollution Control Board.

Wayne T. Davis is emeritus Chancellor and emeritus Dean of the Tickle College of Engineering at the University of Tennessee, Knoxville (UTK), and retired in 2019. He served as associate dean for research and technology in the college from 2003 to 2008. He is also an emeritus professor of civil and environmental engineering. He earned his AB in physics from Pfeiffer University (1969), an MS in physics from Clemson (1971), and an MS in environmental engineering and PhD in civil engineering from UTK (1973 and 1975, respectively). He has conducted research and teaching in the area of air quality management and pollution control for more than 45 years at UTK. He is the author/co-author/editor of numerous research publications, including the *Air Pollution Control Engineering Manual* (published by John Wiley Publishers) and the graduate textbook *Air Pollution: Its Origin and Control* (published by Elsevier). He has been involved in numerous projects funded by the US EPA, DOE, ORNL, NSF, DOT, and various state agencies and industrial companies, particularly as related to the monitoring and control of sulfur dioxide, ozone/precursors, and particulate matter. Dr. Davis is a recipient of the Lyman Ripperton Outstanding Professor Award presented by the International Air and Waste Management Association (AWMA), where he is a fellow member; he also received a Lifetime Achievement Award from the Institute of Professional and Environmental Practice (Pittsburgh, Pennsylvania) in 2007. He is also a Board-Certified Environmental Engineering Member (BCEEM) of the American Academy of Environmental Engineers and Scientists. Dr. Davis served as chair of the Knox County (Tennessee) Air Pollution Control Board for over 22 years and served on the State of Tennessee's Air Pollution Control Board for nine years.

1 Atmosphere

1.1 CHEMICAL COMPOSITION

1.1.1 General Composition

The Earth's atmosphere is a mixture of gases and particulate-phase substances. The most abundant of these, nitrogen (N_2) and oxygen (O_2), comprise approximately 78% and 21%, respectively, of atmospheric mass and volume. A number of trace gases make up the remaining 1%. Average concentrations (with the exception of stratospheric ozone, O_3) are reported in Table 1.1. These include gases present in essentially constant concentrations: N_2, O_2, argon (Ar), neon (Ne), helium (He), krypton (Kr), hydrogen (H_2), and xenon (Xe). Others vary temporally and spatially. These include water vapor (H_2O), carbon dioxide (CO_2), carbon monoxide (CO), methane (CH_4), O_3, the nitrogen oxides (nitrous oxide [N_2O], nitric oxide [NO], and nitrogen dioxide [NO_2]), ammonia (NH_3), formaldehyde (HCHO), sulfur dioxide (SO_2), a number of reduced sulfur compounds (dimethyl sulfide [$[CH_3]_2S$], carbon disulfide [CS_2], carbonyl sulfide [COS], and hydrogen sulfide [H_2S]), and odd hydrogen species (hydroxyl radical [OH], hydroperoxyl radical [HO_2], and hydrogen peroxide [H_2O_2]). In addition to these gas-phase substances, the atmosphere contains trace quantities of particulate nitrate $\left(NO_3^-\right)$, ammonium $\left(NH_4^+\right)$, and sulfate $\left(SO_3^{2-}\right)$.

Although N_2 is the most abundant constituent of the atmosphere, it has a relatively limited direct role in atmospheric and life processes. It serves as a precursor molecule for the formation of NO_3^-, from which plant processes synthesize amino acids, proteins, chlorophyll, and nucleic acids (organic molecules that are directly or indirectly essential to all living things). The conversion of N_2 to NO_3^- occurs as a result of atmospheric and symbiotic biological processes.

Nitrogen reacts with O_2 to produce nitrogen oxides (NO_x), which include N_2O, NO, NO_2, gas-phase nitric acid (HNO_3), and short-lived substances such as dinitrogen pentoxide (N_2O_5) and nitrate radical (NO_3). Concentrations of these compounds or substances, unlike their precursors (N_2 and O_2), vary significantly in time and space. Nitrous oxide (N_2O), a relatively inert gas commonly referred to as "laughing gas," was, until several decades ago, thought to be present in the atmosphere at constant levels. It is one of a number of substances whose concentrations are increasing as a result of human activities.

The evolution of free atmospheric O_2 at elevated concentrations set the stage for the evolution of oxidative metabolism, the series of energy-transferring chemical reactions that sustain most life-forms. Oxygen, as a consequence, is vital to almost all living things. It is a precursor for the production of O_3 and the development of the O_3 layer, the stratospheric region that absorbs high-energy ultraviolet (UV) light radiation streaming into the Earth's atmosphere from the sun. By absorbing most of the UV radiation incident on the Earth and its atmosphere, the O_3 layer shields most organic materials and living things from UV's destructive energy.

On average, background O_3 levels in the troposphere are approximately 20 parts per billion by volume (ppbv). Ozone concentrations vary, and there is evidence that average levels are increasing as a result of human activities. At heights of 10 to 50 km (6.2–31 mi), O_3 concentrations increase dramatically, with peak mixing ratio (parts per million by volume [ppmv]) concentrations (defined in Chapter 7) observed at 35 km (21.7 mi.; Figure 1.1); concentrations based on partial pressure or molecular density peak at 25 km (15.5 mi). At approximately 15 km (9.3 mi), O_3 concentrations are ~0.1 to 0.5 ppmv; at 20 km (12.4 mi), 3 ppmv; at 35 km, 8 to 10 ppmv; and at 50 km, 2 ppmv.

In addition to these altitude-based differences, stratospheric O_3 concentrations vary from day to day, from one season to another, and latitudinally. There are also significant variations that result

DOI: 10.1201/9781003664109-1

TABLE 1.1
Atmospheric Gases

Chemical Species	Symbol	Concentration (%, ppmv, ppbv, pptv)
Nitrogen	N_2	78.084%
Oxygen	O_2	20.948
Argon	Ar	0.934
Water vapor	H_2O	0.1–30,000 ppmv
Carbon dioxide	CO_2	~420 ppmv
Neon	Ne	18.18
Ozone (stratosphere)	O_3	0.5–10
Helium	He	5.24
Methane	CH_4	~1.87
Krypton	Kr	1.14
Hydrogen	H_2	0.50
Xenon	Xe	0.09
Nitrous oxide	N_2O	~0.33
Carbon monoxide	CO	110 ppbv (50–200)
Ozone	O_3	20
Ammonia	NH_3	4
Formaldehyde	HCOH	0.1–1
Sulfur dioxide	SO_2	~1
Nitrogen dioxide	NO_2	~1
Carbonyl sulfide	COS	500 pptv
Carbon disulfide	CS_2	1–300
Dimethyl sulfide	$(CH_3)_2S$	10–100
Hydrogen sulfide	H_2S	~50
Nitric oxide	NO	~50
Hydroxyl radical	OH	0.1–10

from quasi-biennial oscillations (air movement in the stratosphere), the 11-year sunspot cycle, and volcanic eruptions.

Ozone concentrations are also measured and reported as total column O_3 (vertical sum, reported as Dobson units). Mean monthly column O_3 concentrations are illustrated as a function of latitude and time of year in Figure 1.2. Column O_3 concentrations are highest at latitudes where O_3 production is relatively low. Highest O_3 values are found at high latitudes in winter and early spring; the lowest values are found in the tropics. Most O_3 production occurs near the equator where high levels of solar radiation are received; the observed stratospheric O_3 distribution reflects a strong poleward transport during winter.

In contrast to N_2 and O_2, the atmospheric concentration of CO_2 is relatively low, approximately 0.042% or 420 ppmv. Carbon dioxide is enormously important. It is one of the two principal raw materials from which green plants (during photosynthesis) make food molecules on which most living things depend. Life is carbon based, and CO_2 is the source of that carbon. Carbon dioxide is also a major greenhouse gas (GHG) and, because of its thermal absorptivity, is responsible in good measure for maintaining a favorable global heat balance. However, carbon dioxide is also increasing at a rate of approximately 2.5 ppmv/year and is also a major contributor to global warming. Other greenhouse gases are often referred to with respect to their global warming potential (GWP) when compared to CO_2 which, by definition, has a value of 1.0. The GWP is the ratio of a specific gas's ability to trap heat when compared to the same mass of carbon dioxide. The latest summary of the

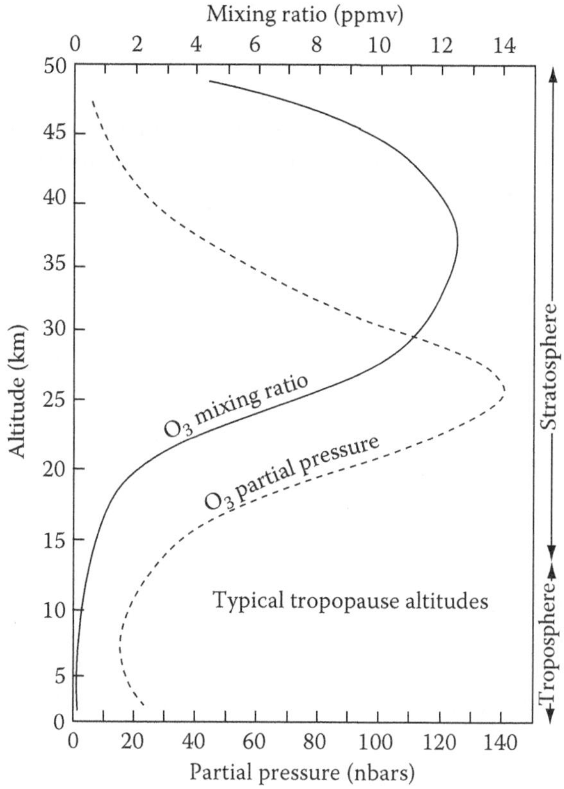

FIGURE 1.1 Atmospheric O_3 concentrations expressed as mixing ratios (ppmv) and partial pressure (nbars) as a function of height. (From National Research Council, *Rethinking the Ozone Problem in Urban and Regional Air Pollution*, National Academy Press, Washington, DC, 1991. With permission.)

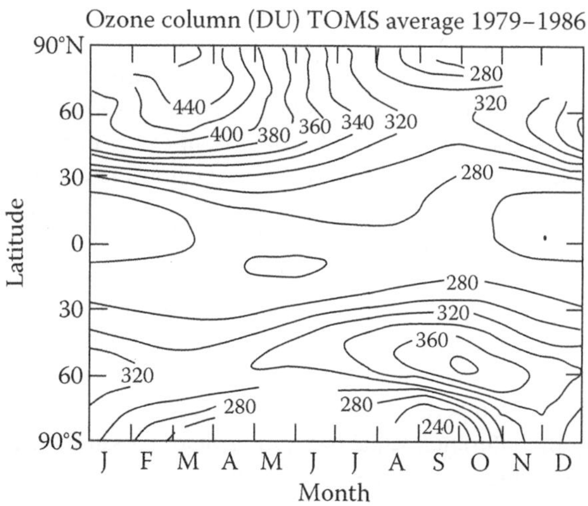

FIGURE 1.2 Total column O_3 (Dobson units) averaged over the period 1979 to 1986 reflecting latitudinal and seasonal differences. (From WMO, *Report of the International Ozone Trends Panel: 1988*, World Meteorological Organization, Geneva, 1990.)

GWP of various atmospheric gases is contained in the Intergovernmental Panel on Climate Change (IPCC) Sixth Assessment Report (AR6) that was published in 2021 and contains GWP values for more than 80 gases, including several of those reported in Table 1.1.

Water vapor is the atmospheric constituent with the highest degree of variability (0.1–30,000 ppmv). Like CO_2, it is a major greenhouse gas, absorbing thermal energy radiated from the Earth's surface. Water vapor is significant in the atmosphere because it readily changes phases. Upon cooling, it condenses to form tiny droplets of liquid H_2O.

Helium, Ar, Ne, Kr, and Xe are noble gases. They are inert and, as a consequence, do not seem to have any major effect on, or role in, the atmosphere.

Many of the variable trace gases are produced either biogenically or geogenically. Most important among these are NH_3, CH_4, H_2S, CO, and SO_2. Ammonia, CH_4, and H_2S are produced primarily by biological decomposition. Methane (CH_4) is a thermal absorber and serves as a greenhouse gas and while its global average concentration is only about 1.92 ppmv as of 2024, its concentration is increasing at about 10 ppbv/year. It is significant that methane has a GWP value of 28. While its concentration in the environment is substantially lower than that of CO_2, one kg of methane in the atmosphere has the GWP of 28 kg of CO_2. Another way of expressing this is to state that 1 kg of methane is equal to 28 kg of carbon dioxide equivalency or 28 kg of CO_2e.

Other gas-phase species may be found in the atmosphere (in addition to those listed in Table 1.1). These include organic substances released by plants, microorganisms, and, to a lesser extent, animals. During the growing season, plants emit large quantities of volatile substances such as isoprenes and pinenes. These biogenic substances are rapidly removed by sink processes. Therefore, they are not described as being "normal" constituents of the atmosphere. Another gas, often used as a tracer gas in atmospheric dispersion studies due to its chemical stability and ability to be measured at very small concentrations, is sulfur hexaflouride (SF_6). The concentration of SF_6 in the atmosphere, based on NOAA data (2024), is approximately 11.5 pptv, but it is currently increasing at approximately 0.40 pptv/year and has a GWP of 25,200.

1.1.2 Water

Water in its liquid phase covers approximately 70% of the Earth's surface; in its solid phase, it forms the polar ice caps, glaciers, episodic snow and ice of our seasons, and ice crystals of cold and very cold clouds. In its vapor phase, it is a significant component of the atmosphere, varying from trace levels up to approximately 3%.

Water is a unique constituent of the atmosphere in that it occurs under ambient conditions as a solid, liquid, and gas. In some circumstances, all three phases are present at the same time. Phase changes are significant in atmospheric processes that produce "weather."

In its vapor phase, H_2O represents a variable component of the mixture of gases that comprise the atmosphere. The combined pressure of these gases (Dalton's law) describes atmospheric pressure. The partial pressure of H_2O vapor in an atmosphere is described as vapor pressure. Vapor pressure increases and decreases with increasing or decreasing concentrations of H_2O vapor. Water vapor concentrations may be reported as absolute or relative humidity. Absolute humidity may be expressed as density (mass/volume) or vapor pressure (millimeters of mercury).

Relative humidity is expressed as a percentage; it is the percentage of H_2O vapor a volume of air holds at a given temperature. Because air can hold more H_2O vapor at higher temperatures, relative humidity values decrease as temperature increases. As air is cooled, relative humidity increases until it reaches 100%, or the saturation value (saturation vapor pressure). If cooled below the saturation value, condensation occurs. This is described as the dew point, and the temperature at which the dew point occurs is called the dew point temperature.

The relationships between temperature and relative humidity, relative humidity and dew point temperature, and relative and absolute humidity can be seen in the psychrometric chart in Figure 1.3. As warm, moist air rises, it cools below the dew point. Subsequent condensation produces clouds with

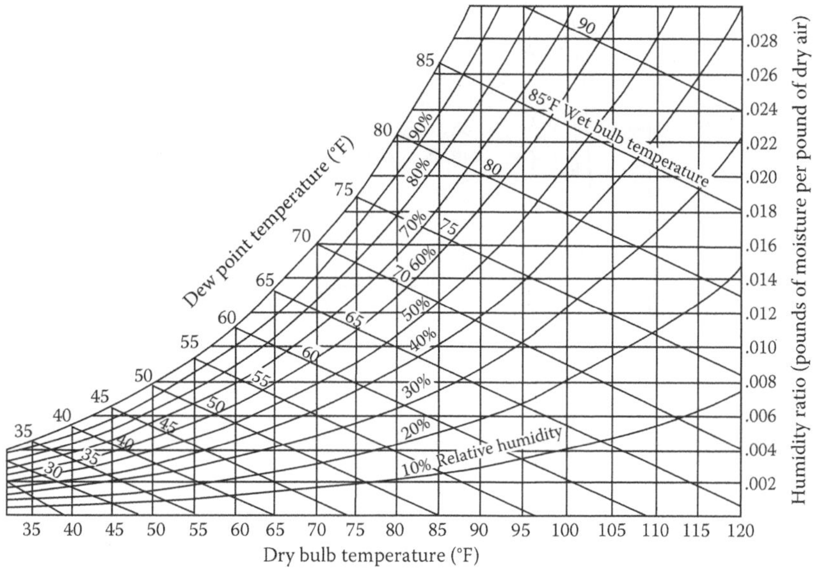

FIGURE 1.3 Psychrometric chart.

droplet sizes one millionth the size of a rain droplet. Clouds are masses of air with enormous numbers of condensed H_2O droplets or ice particles that initially are too small to appreciably fall through the atmosphere. As these droplets grow by a millionfold, precipitation occurs. Clouds have very important physical effects. They reflect sunlight back to space and are major contributors to the Earth's albedo, that is, its ability to reflect sunlight. They also absorb heat, retarding its flow back to space.

1.2 PHYSICAL PHENOMENA

The atmosphere is characterized by its constituent gas- and particulate-phase substances and the physical forces and phenomena that act upon and within it. These include solar radiation, thermal energy, gravity, air density, and associated pressure, and the movement of individual air molecules and the atmosphere itself.

1.2.1 Solar Radiation

The sun radiating energy approximately as a blackbody (perfect absorber and emitter of radiation at all wavelengths) at an effective temperature of 6,000 K, "showers" the atmosphere and the Earth's surfaces with an enormous quantity of electromagnetic energy. The total amount of energy emitted by the sun and received at the extremity of the Earth's atmosphere is constant, 1,370 $W/m^2/s$. That received per unit area of the Earth's surface is 343 $W/m^2/s$.

Spectral characteristics of solar radiation, both external to the Earth's atmosphere and at the ground, can be seen in Figure 1.4. More than 99% of the energy flux from the sun is in the spectral region of 0.15 to 4 μm, with approximately 50% in the visible light region of 0.4 to 0.7 μm. The solar spectrum peaks at 0.49 μm, the green portion of visible light. Because of the dominance of visible light in the solar spectrum, it is not surprising that life processes such as photosynthesis and photoperiodism (responses to changing day length) are dependent on specific visible wavelength bands.

Incoming solar radiation is absorbed by atmospheric gases such as O_2, O_3, CO_2, and H_2O vapor. UV lights, at wavelengths of less than 0.18 μm (180 nm), are strongly absorbed by O_2 at altitudes above 100 km (62 mi). Ozone below 60 km (37.2 mi) absorbs most UV between 0.2 and 0.3 μm (200–300 nm).

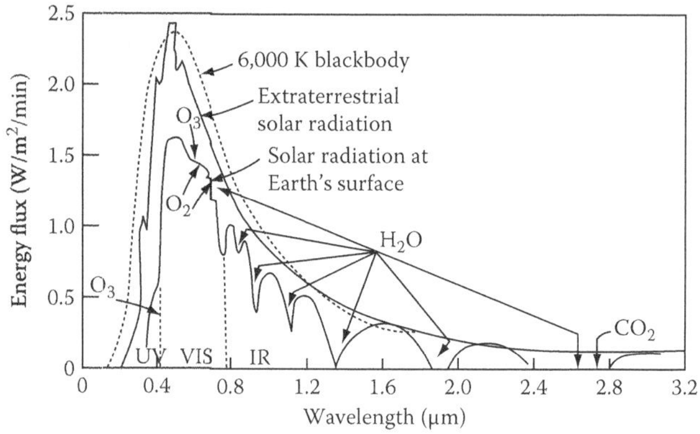

FIGURE 1.4 Solar spectra. (From Krondratjev, *K.Y., Radiation in the Atmosphere,* Academic Press, New York, 1969. With permission.)

Atmospheric absorption above 40 km (24.8 mi) results in the attenuation of approximately 5% of the incoming solar radiation. Under clear-sky conditions, another 10% to 15% is absorbed by the lower atmosphere or is scattered back to space, with 80% to 85% reaching the ground.

The Earth, however, has variable but significant cloud cover so that under average conditions of cloudiness, a significant portion of the sun's energy incident on the atmosphere does not reach the Earth's surface. Cloud droplets and other atmospheric aerosols, as well as the Earth's surfaces, scatter sunlight back to space. This backscatter, as seen from the moon, makes the Earth a relatively bright planet. This phenomenon describes the Earth's albedo. The albedo varies considerably, with an average value estimated to be approximately 30% to 35%; 30% being the most widely referenced value. Approximately 55% of the Earth's albedo is due to clouds, 23% to cloud-free atmosphere, and about 22% to the Earth's surface.

As can be seen in Figure 1.4, a significant portion of solar radiation incident on the atmosphere is not received on the ground. In addition to reflection by clouds and atmospheric aerosols, considerable attenuation of solar energy results from the absorption of different portions of the solar spectrum by atmospheric constituents such as H_2O vapor, CO_2, O_3, and O_2. Carbon dioxide and H_2O vapor are primarily absorbed in the infrared region, O_3 in the UV and portions of the visible region, and O_2 in portions of the visible and infrared regions.

1.2.2 Terrestrial Thermal Radiation

Because the Earth's temperature is lower than that of the sun, it absorbs and reradiates energy at longer wavelengths than it receives. The Earth and its atmosphere radiate energy nearly as a blackbody, with an effective temperature of 290 K. This results in an emission spectrum primarily in the infrared region of 1 to 30 μm, with a peak at ~11 μm. A nighttime thermal emission spectrum is illustrated in Figure 1.5 (note the significant absorption due to CO_2 and H_2O vapor). The atmosphere is particularly transparent to wavelengths in the infrared spectral region of 7 to 13 μm (described as the atmospheric window). Nearly 80% of thermal energy emitted by the Earth's atmospheric window escapes to space. It is notable that most non-CO_2/H_2O vapor greenhouse gases (i.e., O_3, CH_4, N_2O, and chlorofluorocarbons [CFCs]) have strong absorption bands in this spectral region. Water vapor is the most significant of all greenhouse gases, absorbing five times the amount of thermal radiation as all other greenhouse gases combined.

1.2.3 Energy Balance

The Earth's overall energy balance is illustrated in Figure 1.6. Incoming solar radiation produces an energy input of 343 W/m^2/s. Of this, approximately 103 W/m^2/s is reflected back to space. The

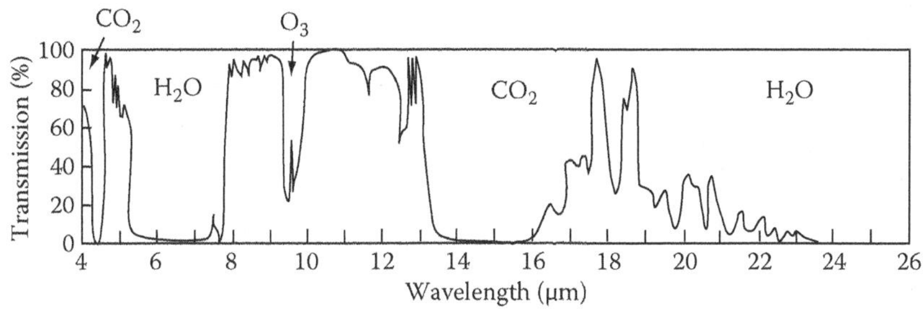

FIGURE 1.5 Nighttime Earth atmospheric emission spectrum. (From Gates, D.M., *Energy Exchange in the Atmosphere,* Harper and Row Monographs, New York, 1962. With permission.)

fraction absorbed by the Earth-atmosphere system is approximately 240 W/m^2/s. This corresponds to the net outgoing radiation flux of 240 W/m^2/s (Figure 1.6). The radiative flux emitted at the Earth's surface is 390 W/m^2/s, which substantially exceeds the outgoing flux of 240 W/m^2/s. The difference reflects the significant effect that clouds, H_2O vapor, and greenhouse gases have in absorbing and emitting thermal energy. Absorption of thermal energy by the atmosphere provides the Earth with a relatively moderate average global surface temperature of approximately 15°C (59°F).

1.2.4 Surface Temperatures

The quantity of solar radiation received at any location on the Earth's surface is a direct function of the sun's angle of inclination. As a consequence, considerably more energy is received and absorbed in equatorial than polar regions. In the tropics, average temperatures exceed 22°C (71.6°F). Average temperature differences between the tropics and poles are approximately 35°C (95°F).

As a result of these differences, thermal energy is transported poleward by air and ocean currents. Warm currents move toward the poles; cold currents move toward the equator. Considerable energy transfer also occurs by evaporation and condensation.

The unequal distribution of solar energy on the Earth's surface and heat transfer mechanisms result in observed differences in regional climate. Changes in the Earth's orientation to the sun, which occur over the course of the year, result in the temporal and climatic temperature differences associated with the seasons, most notably in middle and high latitudes.

1.2.5 Vertical Temperature Gradients

The atmosphere is characterized by significant temperature gradients in the vertical dimension. These temperature gradients define the various layers or zones of the atmosphere (Figure 1.7).

The troposphere, the lowest layer of the atmosphere, extends upward to the tropopause, which begins at altitudes of 8 to 18 km (4.96–11.2 mi) depending on the latitude and time of year. The troposphere is at its maximum height over the equator and its lowest height over the poles. Temperature decreases steadily from an average of approximately 15°C (59°F) to −60°C (−76°F) at 15 km (9.3 mi). On average, temperature decreases at a rate of −6.5°C/km. This progressive temperature decline is due to the increasing distance from the sun-warmed Earth.

Because of the intense movement of thermal energy associated with the heating of the Earth's surface and subsequent transfer to the atmosphere, the troposphere is relatively unstable. This instability results in changes in day-to-day temperature, cloud cover, and precipitation—a phenomenon we call weather. The troposphere is where we humans live and all biological activity occurs.

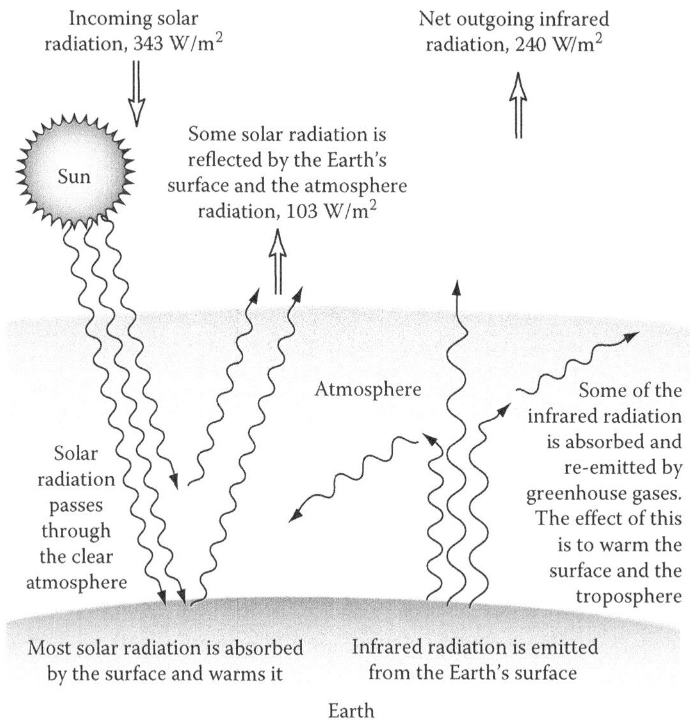

FIGURE 1.6 Earth's energy balance. (From Seinfeld, J.H. and Pandis, S.N., *Atmospheric Chemistry and Physics: Air Pollution to Climate Change*, John Wiley & Sons, New York, 1998. With permission.)

The troposphere can be divided into two regions, that is, the planetary boundary layer (PBL), which extends upward to 1 km (0.62 mi) or more, and the free troposphere, which extends up to the tropopause.

The tropopause, like other pauses, is an atmospheric zone where temperatures are isothermal, that is, no temperature change occurs with height. It is a relatively stable region forming a boundary between the troposphere, where temperature decreases with height, and the stratosphere, where temperature increases with height. The height of the tropopause is often represented as extending from a few kilometers to approximately 10 km (~1 to 6 mi; Figure 1.7).

Above the tropopause, temperature increases to an altitude of 45 to 55 km (27.9–34.1 mi), where the temperature is approximately −2°C (28.4°F). As such, stratospheric temperatures are inverted relative to the troposphere. Consequently, the stratosphere is relatively stable, with little vertical convective mixing. The increase of temperature with height in the stratosphere results from the absorption of UV by O_3. The highest temperatures occur at the top of the stratosphere because of high-efficiency UV absorption.

Few clouds (although stratospheric cloudiness is increasing) are present in the stratosphere, and air movement is primarily horizontal. There is little manifestation of weather-type processes.

The stratosphere and the atmospheric layers above it make up the upper atmosphere, the troposphere being the lower atmosphere. Above the stratopause, temperature decreases with altitude to −90°C (−130°F) at 85 km (52.7 mi), the coldest location in the atmosphere. This region, the mesosphere, is characterized by rapid vertical mixing.

The thermosphere begins above the mesopause altitudes of 90 km (55.8 mi) and extends outward to the atmosphere's extremity (~1,000 km, 620 mi). The high temperatures that characterize

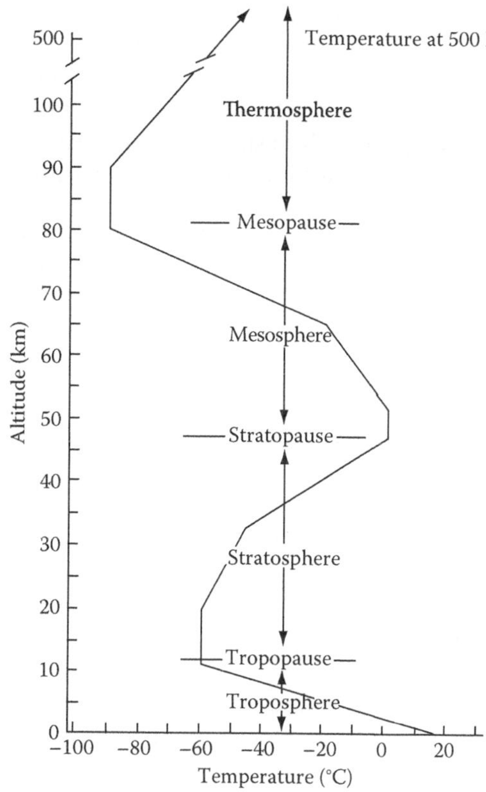

FIGURE 1.7 Vertical temperature profile and associated atmospheric zones.

it are produced as a result of absorption of short-wavelength radiation by molecules of N_2 and O_2. Temperatures reach approximately 1,200°C (2,128°F) at 500 km (310 mi). These are, however, thermodynamic temperatures, which are a measure of the energy of molecules. Because relatively few molecules are present, standard thermometer temperatures are lower because fewer air molecules collide with the thermometer.

Molecules of N_2 and O_2 are photoionized in the thermosphere region of 90 to 400 km (55.8–248 mi). These molecules and atoms lose one or more electrons and thus become positively charged ions. Electrons are set free and travel as electrical currents. The principal chemical species found in the thermosphere are N_2, O_2, N, O, and the ions N_2^+, O_2^+, N^+, and O^+.

This ionized layer of the thermosphere is described as the ionosphere. It is responsible for the formation of the aurora borealis (northern lights) and aurora australis (southern lights). Auroras are closely related in time with solar flare activity and occur at the Earth's magnetic poles. As masses of protons and electrons approach the Earth from the sun, they are captured by its magnetic field, which moves them toward the poles. As these ions move, O_2 and N_2 are energized, emitting light characteristic of auroral displays.

The electrical nature of the ionosphere varies as a result of changes in ion density. Because ion production requires direct solar radiation, the concentration of charged particles decreases at night. The ionosphere reflects AM radio signals back to Earth during nighttime hours.

1.2.6 Gravity

Atmospheric gases are held close to the Earth's surface by gravity. According to Newton's law of universal gravitation, all bodies in the universe attract each other with a force equal to

$$F = Gm_1m_2 / r^2 \tag{1.1}$$

where m_1 and m_2 are the masses of two bodies, G is the universal gravity constant (6.67×10^{-11} nm^2/kg^2), and r is the distance between the two bodies. It can be seen from Equation 1.1 that the force (F) of gravity decreases as an inverse square of the distance between two bodies.

In addition to holding air molecules close to the Earth's surface, gravity plays a significant role in atmospheric processes. As air cools, it becomes more dense and sinks. On warming, it becomes less dense and rises. These mass/density changes result in vertical air motions and contribute to planetary air circulation. Gravity also affects the atmospheric residence time of atmospheric aerosols.

1.2.7 Atmospheric Density

Although mixing ratios of atmospheric gases remain relatively constant up to an altitude of 80 km (49.6 mi), the absolute concentration or numbers of molecules decreases exponentially with height. As molecules become fewer, their density (mass per unit volume) decreases (Figure 1.8). Approximately 80% to 90% of atmospheric mass is below an altitude of 12 km (7.4 mi); 99% is below 33 km (20.5 mi). The total mass of the atmosphere is estimated to be approximately 5×10^{18} kg (11×10^{18} lb). The molar mass or molecular weight of the lower atmosphere is 28.96 g/mol.

1.2.8 Atmospheric Pressure

Pressure, defined as force per unit area, is produced when constantly moving air molecules strike an object, rebound, and transfer momentum to it. Momentum transferred is a function of the average kinetic energy associated with air molecules and is proportional to the absolute temperature. At higher temperatures, molecules have more kinetic energy, and thus atmospheric pressure increases.

Although molecules in the atmosphere are in constant motion, they are, nevertheless, pulled toward the Earth's surface by gravity. As such, the atmosphere exerts pressure on all surfaces.

Objects (including humans) support a column of air directly above them. This column exerts a force on such objects equal to its weight or density. This force is, on average, maximal at sea level, where atmospheric pressure is equal to a standard atmosphere, 760 mmHg or 760 torr. In the international system of units (SI), the unit of pressure is newtons per square meter, called a pascal (Pa). The pressure of the standard atmosphere in SI units is 1.01325×10^5 Pa. Many meteorologists describe atmospheric pressure in millibars (1 mbar = 100 Pa). At sea level, atmospheric pressure is 1,013.25 mbar. In US units, it is 29.92 in Hg and 14.7 lb/in^2 (psi).

As atmospheric density decreases with height, atmospheric pressure decreases as well. Because pressure is influenced by temperature, this decrease is not quite the smooth exponential curve observed for atmospheric density in Figure 1.8.

The mean surface pressure at sea level, as indicated, is 1,013 mbar. The average global surface pressure, which includes pressure values over oceans and land, is estimated to be 984 mbar (29 in Hg).

Small horizontal and vertical pressure differences occur in the troposphere as a result of thermal conditions. As air warms, atmospheric pressure increases, it expands and becomes less dense. As it cools, it becomes more dense. In the first case, atmospheric pressure decreases; in the second, it increases. Thermally related pressure differences characterize large-scale air movements (low-pressure and high-pressure systems are described later in this chapter).

1.3 ATMOSPHERIC MOTION

The atmosphere is characterized by a continuous exchange of energy, momentum, and moisture, with the ocean and land beneath it. Although the amount of energy entering and leaving the Earth's atmosphere is essentially equal, it is not equal over every part of the Earth. This unequal distribution

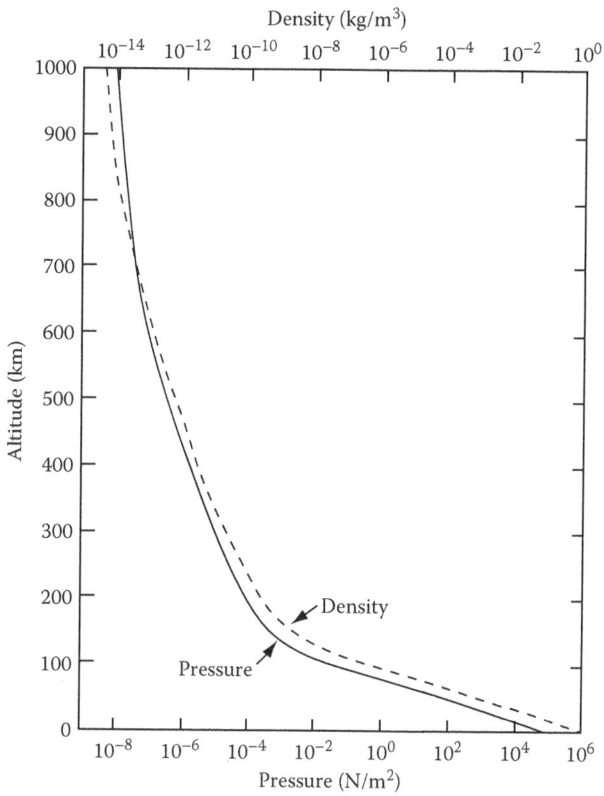

FIGURE 1.8 Changes in atmospheric density and pressure with altitude. (From National Atmospheric and Oceanographic Administration, *U.S. Standard Atmosphere*, NOAA S/T 76-1562, Washington, DC, 1976.)

of solar energy resulting from differences in latitude, cloud cover, and absorptivity of the Earth's surface is responsible for local, regional, and large-scale air motions.

1.3.1 Winds

Wind is the term commonly used to describe air movement in the horizontal dimension. Winds result from differences in air pressure that are caused by unequal heating of the Earth's surface. In the absence of friction and the Earth's rotation, air would flow from areas of high to low pressure. The direction of airflow is controlled by a combination of the pressure gradient force, Coriolis effect, and friction.

1.3.2 Pressure Gradient Force

Pressure differences in the lower atmosphere occur along a gradient in the horizontal dimension. The force associated with these differences is called the pressure gradient force (PGF). The nature of PGFs can be seen from weather maps on which lines of equal pressure (isobars) are plotted (Figure 1.9). The spacing of isobars describes pressure gradients. Isobars relatively close together indicate a strong or steep pressure gradient with high associated winds. If these isobars were contour intervals, they would describe steep or mountainous terrain. Isobars relatively far apart are characterized by small pressure gradients and light winds. The PGF is directed perpendicular to isobars from high to low pressure.

FIGURE 1.9 Surface weather map over North America.

1.3.3 Coriolis Effect

As air flows from either the tropics or the high (55°) middle latitudes, it seems to be deflected from its expected path northward or southward. This apparent deflection to the east (in the northern hemisphere) in poleward flow and to the west in flows toward the equator is due to the Earth's rotation. As the Earth moves from west to east, air seems to lag behind the Earth. This effect of the Earth's rotation is described as the Coriolis effect (CE). Because of the CE, wind does not move perpendicular across isobars as it would if only the PGF were acting on it. The observed apparent deflection depends on latitude and wind velocity. The CE is greatest at high latitudes and is, for the most part, nonexistent at the equator. Deflection associated with the CE increases with wind speed because faster winds cover a greater distance than slower winds in the same period.

1.3.4 Friction

Below a height of several kilometers, friction opposes PGF and thus moderates it. Friction reaches its maximum potential near the Earth's surface. It reduces wind velocity and, as a consequence, "deflection" associated with the CE. Surface winds turn toward low pressure until friction and the CE balance the PGF. Friction turns the wind in the direction of low pressure, decreasing upward.

The extent to which the gradient wind is reduced and its direction changed depends on the roughness of the terrain (surface roughness and temperature gradients in the frictional layer). As friction decreases with increasing height, wind speeds increase. As convective airflows diminish at night,

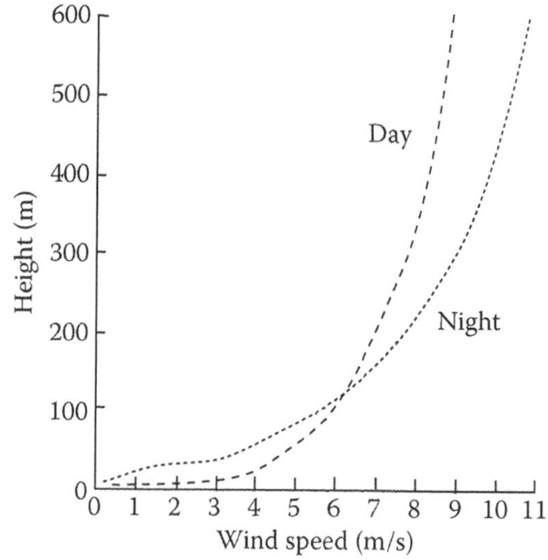

FIGURE 1.10 Surface day–night winds speeds as a function of elevation. (From Turner, D.B., EPA Publication AP-26, US EPA, Washington, DC, 1969.)

frictional effects are less. This can be seen in the day–night differences in wind speed as a function of height (Figure 1.10).

Above the PBL (~1 km thick) the effects of friction are relatively small. The CE causes wind to turn so that when the PGF and CE are in balance, wind flows parallel to isobars at constant speed. Under these idealized conditions, airflow is said to be in geostropic (turned by the Earth) balance. Geostropic winds flow parallel to isobars with velocities proportional to PGFs. Strong winds are associated with steep pressure gradients, light winds with weak gradients. The geostropic model approximates airflows aloft, although they are not purely geostropic.

1.3.5 Cyclones and Anticyclones

On weather maps, isobars form circular or roughly circular cells of high or low pressure. As such, airflow in these cells follows a curved path that tends to be parallel to isobars. These winds blowing at a constant speed are called gradient winds. They are produced as a result of the combined effects of the PGF and CE.

In the northern hemisphere, the CE deflects airflow to the right; the combined effects of CE and PGF create deflection to the left in low-pressure systems. Consequently, airflow is counterclockwise (cyclonic). Air flowing out from the center of the cell (directed by the PGF) in a high-pressure system is opposed by the CE directed inward; therefore, air flows clockwise.

Because airflow in low-pressure systems is cyclonic, centers of low pressure are called cyclones. Cyclonic flow is counterclockwise in the northern hemisphere and clockwise in the southern hemisphere. Airflow in a high-pressure system is anticyclonic (opposite the Earth's rotation). It is clockwise in the northern hemisphere and counterclockwise in the southern hemisphere.

When the force of friction is added to the PGF and CE, air moves across the isobars at various angles depending on roughness of the terrain, but always moving from higher to lower pressure.

In low-pressure systems, pressure decreases inward and friction causes net airflow (convergence) toward their centers. In high-pressure systems, pressure decreases outward and friction causes air to flow away (divergence) from their centers.

Both low-pressure and high-pressure systems are characterized by airflows in the vertical dimension. Near the Earth's surface, air spirals inward in low-pressure systems, reducing the area

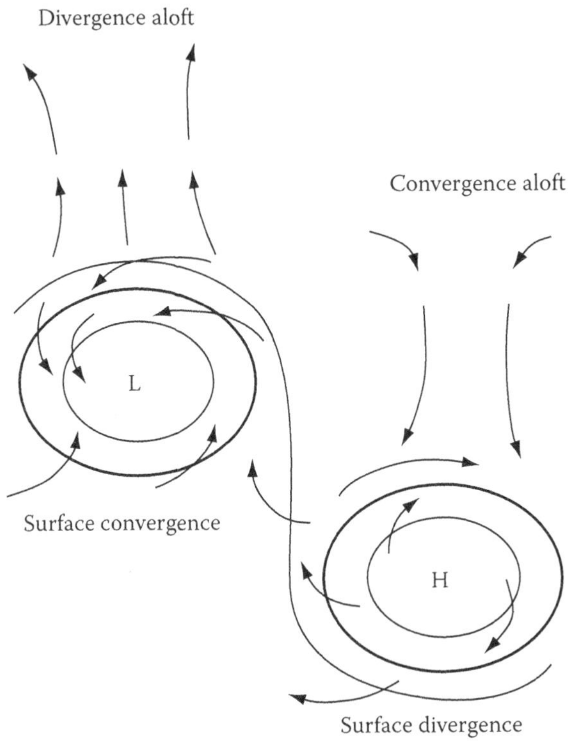

FIGURE 1.11 Airflows associated with low-pressure and high-pressure systems.

occupied by the air mass. As this convergence occurs, air must flow upward to compensate. As such, surface convergence results in divergence aloft (Figure 1.11). As rising air results in cloud formation and precipitation, low-pressure systems are characterized by unstable atmospheric conditions and stormy weather.

In anticyclones, the outflow of air near the Earth's surface is accompanied by convergence from aloft and subsidence (sinking) of air toward the surface (Figure 1.11). As air descends, it compresses and warms air beneath it (see Chapter 3). As a consequence, cloud formation and precipitation generally do not occur. High-pressure systems are characterized by fair, sunny weather during warm months, and cold, sunny weather in cold months.

Most cyclones and anticyclones have cross-sectional diameters of 100 to 1,000 km (62–620 mi). They are typically migratory, moving from one area of the Earth to another. Cyclones of very large scale are called hurricanes or typhoons.

Many migratory cyclones and anticyclones form in temperate regions as a consequence of movements of tropical and polar air. They move from west to east at approximately 800 km/day (500 mi/day) and typically have a life span of a few weeks.

1.3.6 General Circulation of the Atmosphere

In the eighteenth century, Sir George Hadley attempted to explain global air circulation using a simple model in which sensible heat would be transported from the equatorial region to the poles by two large, simple convective cells. Hadley theorized that intense solar heating of land and water surfaces at the equator would produce warm air that would rise to the top of the troposphere where it would lose its thermal energy, sink to the Earth at the poles, and return along the surface to the

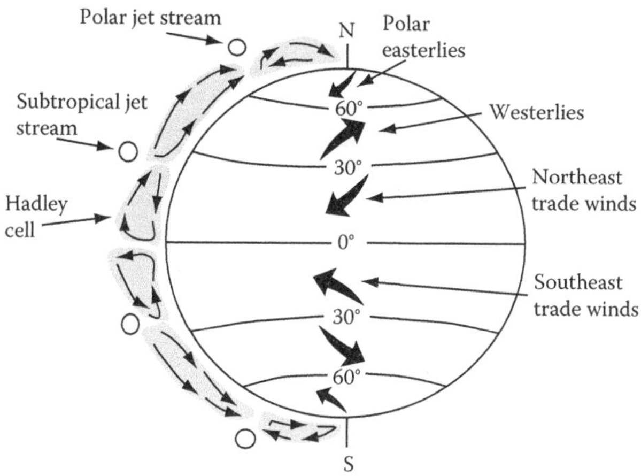

FIGURE 1.12 Three-zone model describing planetary airflow.

equator where it would be reheated. Because of factors such as the CE and friction, Hadley's model does not adequately describe global air circulation.

Global air circulation is better described by the three-zone model illustrated in Figure 1.12. Hadley-type circulation cells are formed between the equator and 30°N and 30°S latitude. Thermal air circulation on both sides of the equator is characterized by the northward or southward flow of warm tropical air aloft and cool southerly or northerly flows along the surface toward the equator. Airflows to the south (in the northern hemisphere) are easterly, that is, they appear to be moving from an easterly direction. These are the trade winds. Their flow is determined in part by the CE.

Hadley-type airflows also occur in polar regions. Warm air from temperate regions moves poleward at relatively high tropospheric altitudes, cools by radiation, and descends at the poles. Cool or cold polar air then moves southward (in the northern hemisphere). These winds appear to be coming from the east and are described as the polar easterlies.

Due to the intrusion of air from the tropics and polar regions, atmospheric circulation in temperate regions (40° to 55° latitude) is highly variable. Consequently, airflow patterns are less of the Hadley-type characteristic of tropical and polar zones. In temperate regions of the northern hemisphere, surface winds generally flow from the southwest.

Areas of relative calm occur at approximately 30°N and 30°S latitude. These areas are characterized by the subsidence of relatively cool air, commonly associated with semipermanent marine high-pressure systems. They have sunny skies, low precipitation, and low winds. These systems are centered over oceans, with their positions shifting only slightly in the summer and winter. Deserts are common on the west coasts of continents where such semipermanent marine high-pressure systems occur.

In the upper troposphere (7.5–12 km, 4.7–7.4 mi), west to east airflows occur in a wavelike fashion (in contrast to the relatively closed high-pressure and low-pressure systems that occur near the Earth's surface). These streams or rivers of westerly flowing air (jet streams) meander for thousands of kilometers around the Earth. They have widths that vary from 100 to 500 km (62–310 mi) and are usually only a few kilometers deep. Wind speeds are relatively high, with peak speeds of more than 200 kph (124 mi/h).

Jet streams form along zones called fronts (in the discontinuities associated with circulation cells in the three-zone model). A polar jet stream in middle latitudes is associated with a polar front that is situated between the cool polar easterlies and the warm westerlies. These relatively fast-moving upper air winds are produced as a result of strong temperature differences at the surface, which produce strong pressure gradients aloft.

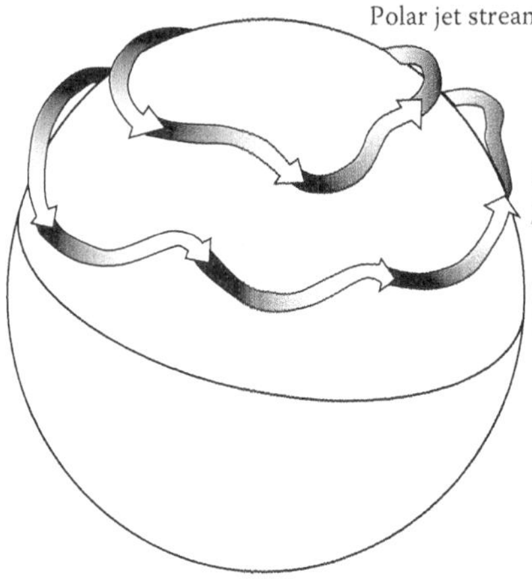

FIGURE 1.13 Generalized flow patterns of northern hemispheric jet streams.

As the polar front moves, so does the polar jet stream. As such, it meanders and sometimes splits. It travels at average speeds of approximately 125 kph (77.5 mi/h) in the winter and approximately 60 kph (37.2 mi/h) in the summer; the difference is a consequence of seasonal differences in temperatures and pressure gradients. During winter, the northern hemispheric polar jet stream may extend as far south as 30°N. In summer, its position moves much farther north, with an average position of 50°N in midsummer. It typically migrates between 30° and 70°N latitude and is often referred to as the midlatitude jet stream.

A semipermanent jet stream also occurs over the tropics in winter. This subtropical jet stream is somewhat slower than the midlatitude jet stream and is centered at 25°N latitude at an altitude of approximately 13 km (8.1 mi). It sometimes merges with the midlatitude jet stream. The generalized flow patterns of these northern hemispheric jet streams can be seen in Figure 1.13.

Jet streams significantly affect surface airflows. When they accelerate, divergence of air occurs at the altitude of the jet stream; this promotes convergence near the surface and induction of cyclonic motion. Jet streams supply energy to surface storms and direct their path. Jet streams also cause a convergence of air aloft and subsidence near the surface, resulting in intensification of high-pressure systems. As such, jet streams are often described as weather makers.

1.4 EVOLUTION OF THE ATMOSPHERE

Our solar system (including the Earth) is widely held by Earth scientists to have evolved from the condensation of an interstellar cloud of gases and dust about 4.6 billion years ago. The Earth's core consisted of iron and iron alloys; its mantle and crust were made up of oxides and silicates of metals.

The very early or primeval atmosphere is believed to have been a mixture of CO_2, CO, N_2, and H_2. Although large amounts of gases were lost to space, other gases were brought to the Earth's surface through volcanism. These gases (CO_2, CO, N_2, and H_2) are emitted into the present day's atmosphere by volcanoes.

As volcanoes brought gases to the surface, chemical reactions made possible a variety of new gas-phase species and increases in the levels of others. Water and CO were formed by reactions catalyzed by metal oxides in the Earth's crust.

$$4H_2 + CO_2 \rightarrow CH_4 + 2H_2O \quad (1.2)$$

$$H_2 + CO_2 \rightarrow CO + H_2O \quad (1.3)$$

As an enormous amount of H_2O vapor was produced over time, the Earth's atmosphere underwent significant changes. As H_2O vapor condensed, the world's oceans were formed. Because of the solubility of CO_2 in H_2O, large quantities of CO_2 dissolved in the newly formed oceans. Dissolved CO_2 was precipitated (first by chemical and then biological processes) to form the Earth's large deposits of sedimentary carbonate rock (limestone, dolomite, etc.).

Because N_2 is relatively inert, insoluble in water, and noncondensable, it accumulated to become the most abundant gas in the atmosphere. Although O_2 was present in combined forms such as metal oxides, silicates, and CO_2, there was no free O_2 present in early Earth's history. Production of free O_2 required the evolution of life-forms, which began approximately 4 billion years ago.

The evolution of life was preceded by the formation of molecular species such as hydrogen cyanide (HCN), NH_3, and H_2S. Small organic molecules such as methanol (CH_3OH) and HCHO were produced from reactions with CO_2, CO, and H_2 in the presence of UV light and catalysts.

$$CO_2 + 2H_2 \rightarrow CH_3OH \tag{1.4}$$

$$CO + H_2 \rightarrow HCHO \tag{1.5}$$

These small molecules served as precursors for more complex molecules such as amino acids and simple peptides.

Very early life-forms, which likely evolved in the Earth's oceans, lived in an environment without free O_2. Approximately 3.5 billion years ago, some organisms developed the ability to photosynthetically produce organic molecules and free O_2.

$$CO_2 + H_2O \rightarrow CH_2O + O_2 \tag{1.6}$$

Most of this early O_2 reacted with crustal materials and was removed from the evolving atmosphere. About 2 billion years ago, free O_2 began to accumulate, reaching current levels about 400 million years ago. Present-day steady-state O_2 concentrations are maintained by photosynthesis and biological removal processes.

The abundance of free O_2 led to the formation of O_3 and the development of the O_3 layer, which shielded the Earth from biologically destructive UV light and made possible the development of terrestrial life.

As life evolved, CO_2 continued to be removed from the atmosphere by sedimentary activity and formation of large deposits of coal, oil, and natural gas. As a consequence, CO_2 became a minor or trace gas. In the last half to 1 billion years, average levels of the major atmospheric constituents have remained relatively constant.

Noble gases are characterized by their lack of chemical reactivity. Noble gases such as Ar and He were present in the Earth's early atmosphere and continue to be there. Both originated as products of crustal radioactive decay and have increased in concentration over time. However, because of its heavier atomic mass, Ar tends to be retained in the atmosphere, whereas He (because of its low mass) tends to be slowly lost to space.

Hydrogen also has a low mass and, as such, has a greater potential to be lost to space.

Both He and H_2 were once believed to have been left over from the interstellar cloud that formed the Earth. It seems, however, that present-day He and H_2 levels are maintained by geogenic processes. As indicated previously, the atmosphere consists of a variety of trace gases that comprise less than 1% of the atmosphere. Concentrations of some of these trace gases (e.g., CO_2, CH_4, and N_2O) have increased significantly over the last two centuries. Increases have occurred in tropospheric O_3 and decreases in stratospheric O_3; the atmospheric chlorine (Cl) level has increased fivefold.

The atmosphere has evolved considerably over the past 4.6 billion years. For most of that time, its evolution was shaped entirely by natural forces. For the past two centuries, atmospheric evolution has come under the influence of industrial, agricultural, and cultural activities of humans. The atmosphere continues to evolve. Humans now have a significant influence on the nature and pace of this evolution. As such, it is a new day in the Earth's history.

READINGS

Boubel, R.W., Fox, D.L., Turner, D.B., and Stern, A.C., *Fundamentals of Air Pollution*, 3rd ed., Academic Press, Orlando, FL, 1994.

Brasseur, G.P., Orlando, J.J., and Tyndall, G.S., Eds., *Atmospheric Chemistry and Global Change*, Oxford University Press, Oxford, 1999.

Bristlecombe, P., *Air Composition and Chemistry*, 2nd ed., Cambridge University Press, Cambridge, UK, 1996.

Finlayson-Pitts, B.J., and Pitts, J.N., Jr., *Chemistry of the Upper and Lower Atmosphere: Theory, Experiments and Applications*, Academic Press, Orlando, FL, 2000.

IPCC, Summary for policymakers, in *Climate Change 2021: The Physical Science Basis*, Contribution of Working Group I to the Sixth Assessment Report of the Intergovernmental Panel on Climate Change, Masson-Delmotte, V., Zhai, P., Pirani, A., Connors, S.L., Péan, C., Berger, S., Caud, N., Chen, Y., Goldfarb, L., Gomis, M.I., Huang, M., Leitzell, K., Lonnoy, E., Matthews, J.B.R., Maycock, T.K., Waterfield, T., Yelekçi, O., Yu, R., and Zhou, B., Eds., Cambridge University Press, Cambridge, UK and New York, NY, USA, 2021, pp. 3–32. https://doi.org/10.1017/9781009157896.

Liou, K.N., *Radiation and Cloud Processes in the Atmosphere*, Oxford University Press, Oxford, 1992.

Lutgens, F.K., Tarbuck, E.G., and Tasa, D.G., *The Atmosphere: An Introduction to Meteorology*, 13th ed., Pearson/Prentice Hall, Upper Saddle River, NJ, 2016.

National Oceanic and Atmospheric Administration, *U.S. Standard Atmosphere*, NOAA S/T 76-1562, Washington, DC, 1976.

Seinfeld, J.H., and Pandis, S.N., *Atmospheric Chemistry and Physics: From Air Pollution to Climate Change*, 3rd ed., John Wiley & Sons, New York, 2016.

Turkekian, K.L., *Global Environmental Change: Past, Present and Future*, Prentice Hall, Saddlebrook, NJ, 1996.

US EPA, *Understanding Global Warming Potentials*, 2020. www.epa.gov/ghgemissions/understanding-global-warming-potentials.

Warneck, P., *Chemistry of the Natural Atmosphere*, 2nd ed., Academic Press, Orlando, FL, 1999.

QUESTIONS

1. What are the major nonvarying constituents of the atmosphere?
2. What is the significance of atmospheric O_2?
3. Why are concentrations of water vapor and CO_2 so variable?
4. What gases absorb incoming solar radiation?
5. What gases absorb outgoing thermal radiation?
6. Why don't atmospheric gases simply flow away from the Earth's surface and be lost to space?
7. With one exception, the relative concentrations of atmospheric gases do not change with height. Indicate this exception, its concentration patterns, and reasons why this is the case.
8. There is a considerable difference between solar energy received at the extremity of the Earth's atmosphere and that received at the surface. Describe these differences and the factors responsible.
9. Significant temperature changes occur with height out to a distance of 500 km. Describe these differences and the factors that contribute to them.
10. Thermal emissions from the Earth and its atmosphere are characterized by electromagnetic energy in which spectral range?
11. What is the atmospheric window? What is its significance?
12. What is atmospheric pressure? What causes it?
13. Why is water vapor such an important atmospheric constituent?
14. The atmosphere is in constant motion. Why?
15. Why does air flow counterclockwise in a low-pressure system in the northern hemisphere?

16. Air flowing from the equator in the middle latitudes in the northern hemisphere appears to flow from the southwest to the northeast. Why?
17. The three-zone model is used to describe what type of air motion?
18. Why does airflow aloft differ from that near the surface?
19. What causes the Earth to have an albedo? What is the significance of the Earth's albedo?
20. How, where, and why are jet streams formed?
21. How did O_2 originate in the atmosphere?
22. How have atmospheric CO_2 levels changed over time?
23. What is the origin of H_2 and He in the atmosphere?
24. The composition of the Earth's atmosphere has changed in the past 200 years. What were/are these changes?
25. What does a GWP of 10 mean?

2 Atmospheric Pollution and Pollutants

As seen in Chapter 1, the atmosphere is a mixture of gaseous substances produced over the Earth's long history by biogenic, geogenic, and atmospheric processes. As humans evolved, the atmosphere began to be affected by human activities. Initially, these effects were small and insignificant. With the enormous growth in human numbers, resource use, and technological advancements, the effect of humans on the atmosphere has been and continues to be significant.

We generally view (in the absence of human activity) the atmosphere and other environmental media, such as water and land, to be "clean." Upon contamination with our "wastes," they become "polluted." The concept of pollution includes a sense of degradation, a loss of quality, a departure from purity, and adverse environmental effects. Air becomes polluted when it is changed by the introduction of gas-phase or particulate-phase substances or energy forms (heat, noise, radioactivity) so that locally, regionally, or globally altered atmosphere poses harm to humans, biological systems, materials, or the atmosphere itself.

Air pollution concerns, historically, have focused on the freely moving air of the outdoor environment described as ambient air. It is the pollution of ambient air that is the primary focus of regulatory programs, as well as of this book. However, it is but one of several air pollution or air quality concerns. Significant air pollutant exposures occur in industrial workplaces and the built environments of our homes, offices, and industrial buildings and from personal habits such as smoking and vaping.

2.1 NATURAL AIR POLLUTION

Contamination or pollution of the atmosphere occurs as a consequence of natural processes as well as human activity (the latter are called anthropogenic). Although regulatory programs focus exclusively on anthropogenic air pollution, it is important to understand that nature contributes to atmospheric pollution and, in some instances, causes significant air quality problems. Volcanic activity in the nineteenth century (e.g., Tambora and Krakatoa) and the more recent eruptions of Mount Saint Helens in Washington State (1980), El Chinchon in Mexico (1982), Mount Pinatubo in the Philippines (1991), and Kilauea in Hawaii (1983–2018), highlight the significance of natural sources of air pollution. Similarly, major wildfire events including forest, brush, and bush fires in Yellowstone National Park (1988), the western United States (2019–2022), Australia (2019–2020), and Canada (2023–2025), as well as dust events in Illinois (2023) and the transport of Saharan dust (2024) to parts of the US, particularly Florida and Texas, further demonstrate the impact of naturally occurring pollutants on both regional and global air quality. While the more recent forest and bush fires have increasingly been linked directly to climate change, scientists have warned that climate change contributes indirectly by creating hotter, drier, drought conditions that increase both frequency and intensity of such fires. These fires can be ignited by various sources, including lightning, sparks from electrical lines during high winds, campfires, discarded cigarettes, and, in some cases, arsonists. While emissions from these fires are often classified as natural air pollution, some may arguably be considered anthropogenic, resulting from inadequate forest management practices. The consequences of uncontrolled fires include widespread habitat destruction and the release of CO_2, unburned hydrocarbons, and particulate matter (smoke), all of which can significantly impact the health of animals and mammals, including humans, while also contributing to further climate change.

DOI: 10.1201/9781003664109-2

In addition to volcanoes, dust storms, and forest fires, natural air pollution occurs from a variety of sources: plant and animal decomposition, pollen and spores, volatile hydrocarbons (HCs) emitted by vegetation, ocean spray, soil erosion and mineral weathering by wind, emission of gas-phase substances from soil and water surfaces, ozone (O_3) and nitrogen oxides (NO_x) from electrical storms, and O_3 from stratospheric intrusion and photochemical reactions. Emissions to the atmosphere associated with geochemical processes are described as geogenic; those associated with biological processes are biogenic.

Natural pollutants pose air quality concerns when generated in significant quantities near human settlements. With the exception of major events such as dust storms, forest fires, and volcanoes, natural air pollution has not been a major societal concern. On a mass basis, nature pollutes more than humans do, as vocal opponents of contemporary air pollution control efforts maintain. However, such pollution sometimes has relatively lower significance in causing health and welfare effects because (1) levels of contaminants associated with natural air pollution are typically very low; (2) large distances often separate sources of natural pollution and large human populations; and (3) major sources of natural pollution, such as forest fires, dust storms, and volcanoes, are episodic and transient.

A major potential exception to this relegation of natural air pollution to a status of relative insignificance is the biogenic emission of photochemically active nonmethane hydrocarbons (NMHCs) such as isoprene and α-pinene from vegetation. Natural NMHCs play a significant role in photochemical oxidant production in some nonurban and urban areas.

2.2 ANTHROPOGENIC AIR POLLUTION

Anthropogenic air pollution has been and continues to be viewed as a serious environmental and public health problem. Its seriousness lies in the fact that elevated pollutant levels are produced in environments where harm to human health and welfare is more likely. It is this potential that makes anthropogenic air pollution the significant concern that it is.

In a historical sense, air pollution became a serious problem when humans discovered the utility of fire. Smoke generated in the incomplete combustion of wood plagued early humans and generations to come. For civilizations in western Europe, the decimation of forests and a switch to use of soft coal as a fuel source in the fourteenth century added a new dimension to air pollution problems. The emission of pollutants from coal combustion was intense, and coal smoke became a major environmental concern. The industrial revolution that began in the early nineteenth century and the technological developments of the nineteenth through twenty-first centuries (e.g., motor vehicles and power plants fueled by fossil-based fuels) are what have given rise to the atmospheric pollution characteristics of our times. This is in large part due to society's consumption of energy worldwide and the resultant emissions of unwanted pollutants emitted during consumption. As a result of the concerns over global warming, the anthropogenic emission of greenhouse gases has also come under scrutiny in recent years and is, in large part, a result of the combustion of fossil and biofuels for energy production. Thus, an understanding of the amount of energy consumption from the wide variety of energy sources used is critical to addressing ways to reduce the anthropogenic emission of pollutants into the atmosphere, as well as greenhouse gases.

One of the best and timely sources of information on energy consumption in the US is the report of the US Energy Information Administration found at www.eia.gov and updated annually. The latest report showed an energy consumption of 101.3 quadrillion Btu in 2018. Globally energy consumption is often reported in millions of tonnes of energy equivalent (Mtoe). Mtoe is the amount of energy created by burning one metric tonne (1000 kg) of oil. For comparison purposes, among different energy sources (coal, oil, natural gas, etc.), one Mtoe is equivalent to 3.968×10^7 Btu. On that basis the US consumption as reported by the US EIA would be approximately 2,550 Mtoe. A private organization, Enerdata (www.enerdata.net), provides estimates of annual energy consumption on a worldwide basis. Based on their findings and as reported in their Global Energy Statistical

Yearbook 2019 (which summarizes energy consumption in 2018), China was the country with the largest consumption of energy at 3,164 Mtoe, with the United States being second at 2,258 Mtoe and India third at 929 Mtoe. Total global consumption of energy was estimated to be approximately 14,000 Mtoe. The differences between the estimation of consumption by different groups will vary due to the large number of individual sources of data and assumptions needed to arrive at both US and global estimates. The consumption in the United States accounts for approximately 16%–18% of total global consumption. It is worth noting that the US consumption of energy has been relatively flat since the year 2000 (increasing by about 3.5% in 2018), whereas China's and India's consumption of energy have increased by about 180% and 110%, respectively, in that same time period, due to rapid growth in transportation, industry, and the use of energy for electricity. This rapid growth has also led to increasing challenges with air pollution and the emission of greenhouse gases. On a per capita basis, the United States consumed approximately three times the energy of China and ten times the energy of India.

Figure 2.1 illustrates the complexity of the energy sources and the end-use sectors for 2023, using the United States as an example. Overall, the sources of energy in descending order in 2023 were petroleum (38%), natural gas (36%), coal (9%), renewable energy (hydro, wind, solar, etc.) (9%), and nuclear power (9%). While overall energy production in the United States has slightly decreased since 2000, production from nuclear and petroleum and petroleum sources has remained relatively unchanged. In contrast, major shifts have occurred in coal, nature gas, and renewable energy, reflecting concerns over air pollution, global warming, and economic conditions. Coal decreased by 181%, whereas natural gas and renewables increased by 42% and 20%, respectively. Approximately 35% of the total consumption of energy in 2023 was for the production of electricity, which is highlighted in the center of the figure to illustrate the various end-use sectors that used

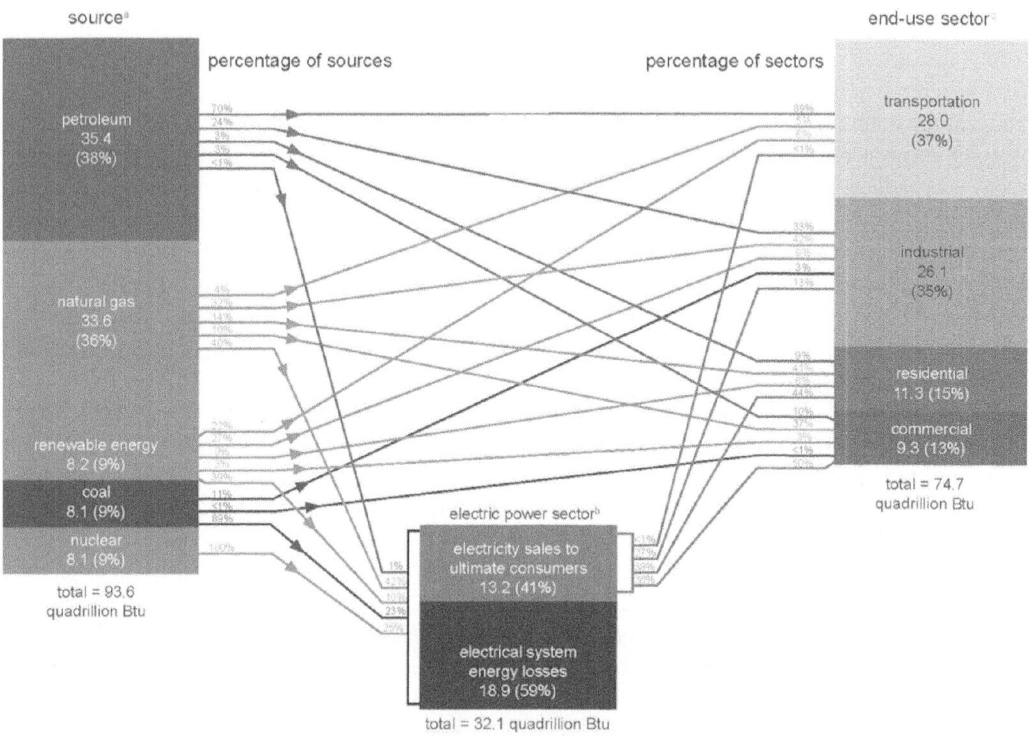

FIGURE 2.1 US energy consumption by source and sector, 2023 (quadrillion Btu). (Courtesy of United States Energy Information Administration [www.eia.gov].)

electricity. Notably, the electric power sector is only 41% efficient (13.2 out of 32.1 quadrillion Btu) due in part to the combination of the thermal efficiency of fuel combustion systems and electrical system energy losses from the transmission of electricity. Coal-fired power plants have typical thermal efficiencies of 37%, whereas newer combined cycle gas-fired power plants can be as high as 60%–70%. Energy losses for other sectors, such as transportation, industry, and nonelectric residential/commercial heating and air conditioning, are not shown explicitly in the figure; however, they are included in the numbers associated with the end-use sectors. The end-users of energy in descending order were transportation (37%), industrial (35%), residential (15%), and commercial (13%). Transportation is the primary user of petroleum at 70% of total consumption. As will be discussed in more detail in Chapter 9, the thermal efficiency of gasoline and diesel engines ranges from 20% to 50% (compared to slightly higher energy efficiency of electricity production). Recent developments in the efficiency of modern gasoline engines suggest that thermal efficiencies of 50% might be achievable. Improvements directed toward reducing energy losses associated with both electricity production and automotive engines result in the reduction of both air pollutants and greenhouse gas emissions into the atmosphere. The recent significant focus on electric vehicles has the potential to shift the focus away from the efficiency of the gasoline or diesel engine to one focused on the efficiency of the energy sources providing electricity. While coal and natural gas each provide about 16% and 43%, respectively, and renewables provide now about 21% of the electricity produced in 2023, the US EIA's Annual Energy Outlook 2025 predicted that renewables (including solar, wind, and hydroelectric power) will become the largest producer of electricity by 2050. While this could significantly reduce the emission of greenhouse gases, the political and economic climate could affect these types of predictions.

2.2.1 Smog

Historically, in London and other British industrial cities, smoke from coal combustion episodically mixed with fog from the North Sea. These occurrences were marked by an extreme reduction in visibility, the stench of sulfurous emissions and, in severe episodes, illness and death. Such pollution fogs were experienced in London for more than a century. The term *smog,* used to describe severe ambient air pollution conditions, was derived from the words *smoke* and *fog.* Such smog was gray in color and was referred to as a "gray smog" or "London-type" smog. The term smog as it is used today is broadly applied to atmospheric pollution conditions characterized by a significant reduction in visibility. It generally involves photochemical reactions in the atmosphere resulting in a slightly brownish-yellow character, as compared to a gray color. The term is often applied without consideration of pollutant types, sources, or smog-forming processes. Figure 2.2 is an example of smog experienced by a colleague in 2021 taken at the Park Hyatt Shanghai in Shanghai, China, showing the significant degradation of visibility with a characteristic brownish-yellow cast, although not shown in the black-and-white photo.

The best-known smog problem in the United States is associated with the Los Angeles Basin of southern California. Although this Los Angeles–type smog is characterized by a significant reduction in visibility, the conditions that produce smog in southern California are very different from those that produced smog in London and other northern European cities. The climate of the Los Angeles Basin is semiarid, with extended periods of sunny weather. Additionally, Los Angeles is situated at approximately 30°N latitude, where the subsidence of air from a tropical Hadley cell (see Chapter 1) produces semipermanent marine high-pressure systems with cell centers over Hawaii. This subsidence of air produces elevated inversions that are closest to the surface at the eastern edge, or continental side, of the marine high, that is, over the southern coast of California. These relatively low subsidence inversions significantly reduce pollution dispersion in the vertical dimension. In addition, mountains to the north and east serve as barriers, inhibiting air movement in northerly and easterly directions and further reducing dispersion. Poor dispersion conditions in the Los Angeles Basin contribute to smog formation. Other factors also play a role. The

FIGURE 2.2 Smog in Shanghai, China, 2021. (Photo by Jia Xing.)

major source of pollutants in the basin is light-duty motor vehicles (with approximately 8 million registered in Los Angeles County) that emit exhaust and evaporative gases into the atmosphere. Motor vehicle–generated pollutants, abundant sunlight, poor dispersion conditions resulting from inversions (see Chapter 3), and topographical barriers contribute to the development of chemically complex photochemical smog. Because of the presence of nitrogen dioxide (NO_2), Los Angeles–type smog is brownish yellow.

Los Angeles and London smogs reflect relatively unique conditions of smog development. Smogs that occasionally form over large cities around the world often exhibit conditions that are somewhat intermediate between the two. These cities have, or have had, significant emissions of pollutants from both transportation and fuel combustion.

In recent years, changing weather patterns particularly in the western United States have contributed to an increase in the frequency and intensity of wildfires. These wildfires are a significant source of air pollution and play a major role in the formation of smog. When vegetation burns, it releases large quantities of smoke containing fine particulate matter ($PM_{2.5}$), carbon monoxide (CO), nitrogen oxides (NO_x), and volatile organic compounds (VOCs). These pollutants can react in the atmosphere, especially under sunlight, to form ground-level ozone, a key component of smog. The resulting smog can travel far beyond the fire zone, reducing visibility and posing serious health risks to both humans and animals, particularly those with respiratory conditions. As climate change and land management practices continue to drive more frequent and intense wildfires, their contribution to regional and even cross-border smog events is expected to grow.

2.2.2 Haze

The term *haze* is often used in the nomenclature of air quality. Although haze and smog are closely related, both representing atmospheric conditions characterized by visibility reduction, they differ in both intensity and geography. Haze typically refers to the large-scale visibility reduction that occurs in the lower atmosphere near the ground that is a result of the presence of submicron suspended particulate matter that scatters visible light. It can be observed in cities across the world, particularly in summer months. Both haze and photochemical smog can be present at the same time, depending on whether the conditions exist for atmospheric photochemical reactions to occur.

As discussed in the previous section, wildfires and dust storms are major contributors to haze, which significantly reduces visibility and degrades air quality. Wildfires emit smoke containing fine particulate matter and various gases, while dust storms lift large quantities of soil and sand into the atmosphere. These airborne particles scatter and absorb sunlight, forming a dense, murky layer in the air. When combined, smoke and dust can create widespread haze that adversely affects human health, disrupts transportation, and harms the environment across vast regions.

2.2.3 Nontraditional Air Pollutants

In traditional treatments of the subject of air pollution, pollutants are characterized as being particulate-phase or gaseous-phase substances. Air quality can, however, be degraded by other contaminants. These include noise, heat, ionizing radiation, and electromagnetic fields associated with the transmission and use of electricity. With the exception of noise, discussed extensively in Chapter 12, nontraditional air pollutants are considered to be relatively minor environmental concerns, and thus their treatment in this book is limited.

2.2.4 Gases and Particles

Because of effects on human health, the atmosphere, vegetation, and materials, pollutant gases and particles have received considerable research and regulatory attention. Although the terms *gases* and *particles* are used, pollutants exist in all three phases of matter. Particles include both solid and liquid substances. When dispersed in the atmosphere, solid and liquid particles form aerosols. Aerosols may reduce visibility, soil materials, and affect human health.

Fume aerosols are produced by the condensation of metal vapors, dust aerosols from the fragmentation of matter, mists from the atomization of liquids or condensation of vapors, and smoke from the incomplete combustion of organic materials. The photochemical aerosol of Los Angeles–type smog, as well as haze, is produced, in part, by condensation of vapors resulting from photochemical reactions in the atmosphere. Acid aerosol, which contributes to the formation of haze and the acidification of precipitation, is, for the most part, a result of secondary reactions that occur in the atmosphere due to absorption of acid gases into water vapor or particulate matter where the vapor has been condensed.

2.2.4.1 Sources

Gaseous pollutants are emitted from a variety of identifiable sources, including those associated with transportation, stationary fuel combustion, industrial processes, waste disposal, and so on. They are also produced in the atmosphere as a result of chemical reactions. The former are classified as primary pollutants; the latter, secondary. Sources of primary pollutants may be classified as mobile or stationary, combustion or non-combustion, area or point, direct or indirect. The use of these classifications reflects both regulatory and administrative approaches to implementing air pollution control programs at federal, state, and local levels.

Mobile sources include automobiles, both traditional and electric, along with trains, airplanes, and similar forms of transportation, while other sources are stationary. A point source is a stationary source whose emissions degrade air quality. An area source is composed of a number of small stationary and mobile sources that individually have only a small effect on air quality. Viewed collectively over an area, they may have an enormous effect. Area sources include emissions from motor vehicles, aircraft, and trains, open burning, wildfires, and a variety of small miscellaneous sources. These include indirect sources such as shopping centers and athletic stadiums that contribute to elevated pollutant levels as a result of attracting motor vehicle traffic.

TABLE 2.1
National Emission Estimates of Primary Pollutants (10^3 t/year) for the United States in 1970

Source Category	CO	PM_{10}	$PM_{2.5}$	NO_x	SO_2	VOC
Fuel combustion—electric utilities	237	1,775	NA	4,900	17,398	30
Fuel combustion—industrial	770	641	NA	4,325	4,568	150
Fuel combustion—other	3,625	455	NA	836	1,490	541
Chemical and allied product manuf.	3,397	235	NA	271	591	1,341
Metals processing	3,644	1,316	NA	77	4,775	394
Petroleum and related industries	2,179	286	NA	240	881	1,194
Other industrial processes	620	5,832	NA	187	846	270
Solvent utilization	NA	NA	NA	0	NA	7,174
Storage and transport	NA	NA	NA	0	NA	1,954
Waste disposal and recycling	7,059	999	NA	440	8	1,984
Highway vehicles	163,231	480	NA	12,624	273	16,910
Off-highway vehicles	11,371	164	NA	2,652	278	1,616
Miscellaneous	7,909	NA	NA	330	110	1,101
Total	204,042	NA	NA	26,882	31,218	34,659
Wildfires	6,766	NA	NA	NA	NA	917
Total without wildfires	197,276	NA	NA	26,882	31,218	33,742
Miscellaneous without wildfires	NA	NA	NA	NA	110	184
Stationary fuel combustion	4,632	NA	NA	10,061	23,456	721
Industrial and other processes	16,899	NA	NA	1,215	7,101	14,311
Transportation	174,602	NA	NA	15,276	551	18,526
Miscellaneous	7,909	839	NA	330	110	1,101
Total	204,042	13,022	NA	26,882	31,218	34,659

Does not include biogenic emissions.

Source: National Annual Emissions Trend—Criteria pollutants National Tier 1 for 1970–2024, https://www.epa.gov/air-emissions-inventories/air-pollutant-emissions-trends-data.

2.2.4.2 National Emission Estimates

National emission estimates for five regulated primary pollutants and major source categories are presented in Tables 2.1 and 2.2. Emission estimates in Table 2.1 are for 1970, when the United States began a major regulatory effort to control ambient air pollution. Emission estimates reflecting the status of this nation relative to the same pollutants and source categories almost five decades later (2018) are summarized in Table 2.2.

Transportation sources include motor vehicles, aircraft, trains, ships, boats, and a variety of off-road vehicles. Major stationary fuel combustion sources include fossil fuel–fired electricity-generating plants, industrial and institutional boilers, and home space heaters. Industrial process emissions include pollutants produced in a broad range of industrial activities, including mineral ore smelting, petroleum refining, oil and gas production and marketing, chemical production, paint application, industrial organic solvent use, food processing, mineral rock crushing, and others. Emissions from waste disposal result from on-site and municipal incineration and open burning. Miscellaneous sources include forest fires, agricultural burning, coal refuse and structural fires, and a variety of organic solvent uses. In addition to the five pollutant categories indicated in Tables 2.1 and 2.2, the United States has major regulatory programs for lead (Pb) and O_3. Ozone is not inventoried because it is a secondary pollutant produced as a result of atmospheric reactions.

TABLE 2.2
National Emission Estimates of Primary Pollutants (10^3 t/year) for the United States in 2024

Source Category	CO	PM_{10}	$PM_{2.5}$	NO_x	SO_2	VOC
Fuel combustion—electric utilities	409	94	81	967	649	27
Fuel combustion—industrial	875	285	243	545	219	116
Fuel combustion—other	3,446	520	514	31	29	486
Chemical and allied product manuf.	128	18	13	57	72	69
Metals processing	345	42	30	705	53	22
Petroleum and related industries	750	28	23	271	335	2,899
Other industrial processes	461	738	347	1	126	386
Solvent utilization	1	5	4	3	0	2,777
Storage and transport	6	34	13	82	1	605
Waste disposal and recycling	1,477	253	227	1,745	34	192
Highway vehicles	12,322	189	67	1,578	10	896
Off-highway vehicles	11,983	99	93	204	19	990
Miscellaneous	12,733	14,114	3,433	967	127	3,317
Total	44,936	16,419	5,090	6,940	1,675	12,783
Wildfires	3,692	770	504	40	36	974
Total without wildfires	41,244	15,650	4,585	6,900	1,639	11,809
Miscellaneous without wildfires	9,042	13,344	2,929	164	91	2,343
Stationary fuel combustion	4,730	899	838	2,263	897	630
Industrial and other processes	3,167	1,118	659	1,150	622	6,950
Transportation	24,305	288	160	3,323	30	1,886
Miscellaneous	12,733	14,114	3,433	204	127	3,317
Total	44,936	16,419	5,090	6,940	1,675	12,783

Does not include biogenic emissions.

Source National Annual Emissions Trend-Criteria pollutants National Tier 1 for 1970–2024, https://www.epa.gov/air-emissions-inventories/air-pollutant-emissions-trends-data.

Emissions of Pb to the atmosphere in 1970 and 2020 are summarized in Table 2.3 by source category and showed a 99.8% reduction in emissions to 1,000 tons per year. By 2014, the reduction had increased to 99.7% (730 tons/year) as a result of emission standards and the establishment of Pb air quality standards in 1978 and a further increase in the stringency of the standard in 2020 to its current value. Emission reductions have been particularly significant in the transportation sector with the phaseout of Pb use in gasoline. However, Pb continues to be an environmental concern as air emissions from the past were deposited into the environment and still represent a risk.

National emission estimates are published periodically by the US Environmental Protection Agency (US EPA). Since 1970, the process has been modified several times. Nonmethane hydrocarbon emissions are now reported as volatile organic compound (VOC) emissions, and particulate matter emissions as PM_{10} (particulate matter with an aerodynamic diameter of $\leq$ 10 μm) and $PM_{2.5}$ (aerodynamic diameter of $\leq$ 2.5 μm). By definition, $PM_{2.5}$ is a subset of PM_{10} so these reported values in Table 2.2 cannot be summed. Emission estimates presented in Tables 2.1 and 2.2 do not include fugitive particulate matter emissions or biogenic emissions. In addition, some of the PM values in Table 2.1 were listed by the authors as Not Available (NA) since these were either not backcalculated or reported since the size specific PM standards were established later or the methodology changed, preventing comparison between the two tables.

TABLE 2.3
National Lead Emission Estimates for the United States in 1970 and 2020 (10^3 t/year)

Source	1970	2020	% Reduction
Transportation	180.30	0.43	99.8
Stationary source fuel combustion	10.62	0.06	99.5
Industrial processes	26.40	0.12	99.6
Solid waste disposal	2.20	—	—
Total	219.47	0.61	99.8

Source: https://gispub.epa.gov/air/trendsreport/2024/#naaqs_trends.

2.2.4.3 National Emissions Picture

Table 2.1 provides a variety of insights into the nature of the air quality problem the United States faced in 1970 as it began to enact and implement tough air pollution control legislation and subsequent regulation. As can be seen, (1) transportation was responsible for approximately 80% of the CO emissions and 50% of the NO_x and VOC emissions; (2) fuel combustion was the major source of sulfur dioxide (SO_2) at 75% and a major contributor to NO_x at 37%; and (3) industrial processes were the largest source of PM_{10} and second-largest source of SO_2. Motor vehicles were also the major source of atmospheric Pb. Of the source categories for which major control programs have been initiated, transportation, particularly motor vehicles, had been the major direct or indirect source for CO, VOCs, NO_x, Pb, and O_3. Consequently, motor vehicles have justifiably received enormous regulatory attention and investment in pollution control systems.

Stationary fuel combustion sources have been the largest source of SO_2 and the second largest source of NO_x emissions. Sulfur oxides emitted by fossil fuel (especially coal) fired electrical power-generating stations have been of particular concern, and since 1990, these facilities have been required to significantly reduce emissions of SO_x and NO_x for purposes of controlling acidic deposition (see Chapters 4 and 8).

The relative overall effectiveness of pollution control efforts over the past five decades can be seen when Tables 2.1 and 2.2 are compared. Emission reduction percentages associated with specific pollutants include CO (78%), NO_x (75%), SO_2 (95%), and VOC (63%), although the percentages vary by source category due to the fact that reductions in specific categories are dependent on specific regulations that are category specific and based on the difficulty of controlling emissions from each category. Particulate matter emissions have also been reduced significantly, although they are more difficult to assess from Tables 2.2 and 2.3 due to changing definitions of the particulate matter standards since 1970. However, a review of reductions by source category shows a range from about 25% up to 95% reduction.

The relatively modest reductions in emissions over the last five decades somewhat masks the fact that individual sources (e.g., an automobile or an electric utility power plant) have made substantial reductions in emissions. Taking transportation as an example, and although it has been intensely regulated, emission reductions may seem modest. This has been due, in part, to (1) an increase in the number of motor vehicles since 1970, (2) a significant increase in miles traveled per vehicle, and (3) a relatively small number (~10%) of primarily older motor vehicles that disproportionately contributes to motor vehicle emissions. Similarly, the relatively modest reductions associated with fuel combustion emissions of NO_x from electric utilities (81%) reflect an increased number of sources due to increased population and demand for electricity.

Decreases in CO and VOC emissions are a result of the application of catalytic converters and VOC control technology on light-duty motor vehicles. Decreases in PM_{10} can be attributed to the installation of pollution control equipment on industrial sources, more efficient particulate control technologies on combustion sources, changes in fuel use practices in the direction of cleaner-burning fuels, the banning of open burning of solid wastes, and limitations on combustion practices used for solid waste incineration. The observed decrease in SO_x emissions has been due to the use of lower sulfur (S) fuels, the installation of flue gas desulfurization systems on coal-fired plants, and the more recent retirement of a number of coal-fired power plants with their replacement by natural gas–fired plants and renewable energy sources such as solar and wind.

Although the United States has made significant progress in reducing emissions of Pb and PM_{10} and modest overall progress in reducing emissions of other primary pollutants and enhancing the quality of ambient air, it is apparent that continuing efforts to limit emissions are required to achieve air quality goals. Many air quality control regions in the country still do not have acceptable air quality. Significant reduction in emissions of atmospheric pollutants can be expected as (1) the population of motor vehicles with stricter emission limits increases, (2) toxic pollutant (most of which are VOCs) reductions continue as a result of new regulatory requirements, and (3) the current move toward more electric vehicles is implemented. A significant reduction of SO_2 from 2012 levels occurred with the implementation of Phase II of the acidic deposition reduction requirements.

The emission estimates in Tables 2.1, 2.2, and 2.3 only include primary pollutants for which air quality standards have been promulgated. They do not include O_3-destroying chemicals or greenhouse gases such as carbon dioxide (CO_2) and methane (CH_4).

2.3 GAS-PHASE POLLUTANTS

A large variety of gas-phase substances are emitted to the atmosphere from both anthropogenic and natural sources. Hundreds of different chemicals may be present in the polluted air of the many metropolitan areas that exist on the planet. More than 400 different gas-phase species have been identified in automobile exhaust alone, and many additional substances are produced as a result of atmospheric chemistry.

Only a relatively small number of the many hundreds of gaseous substances found in the atmosphere are either toxic enough or present in sufficient concentrations to directly or indirectly cause harm to humans or their environment. Major gas-phase substances identified as having potentially significant effects on humans or our environment are discussed in the following pages in the context of their nature, sources, background and urban–suburban levels, atmospheric chemistry, and sink processes.

2.3.1 Carbon Oxides

Billions of metric tons of the carbon oxides, CO and CO_2, are emitted by natural and anthropogenic sources. Because of its mammalian toxicity, CO has been a major pollutant and exposure concern for decades. Carbon dioxide, on the other hand, is relatively nontoxic. Although many atmospheric scientists began to express concern about increasing CO_2 levels and their potential effect on global climate decades ago, only since the late twentieth century have CO_2 and other greenhouse gas emissions and global warming become major public policy issues. The twenty-first century and particularly the last several years have seen an even greater escalation of the concerns related to global warming.

2.3.1.1 Carbon Dioxide

Carbon dioxide is a relatively abundant and variable constituent of the atmosphere. It is produced and emitted naturally to the atmosphere by aerobic biological processes, combustion, and the weathering of carbonates in rock and soil. It is taken up by plants and used in the process of photosynthesis. Anthropogenic sources are fossil fuel combustion and land use conversion. The latter case includes clearing of forest lands with subsequent biomass burning and use for agricultural production, residential/commercial purposes, and industrial development.

Enormous changes in atmospheric CO_2 levels have occurred over the course of the Earth's history. This has included the removal of carbon (C) from its storage (in vast quantities) in the Earth's crust as coal, oil, and natural gas. Because of their availability and ease of use, these fossil energy sources have made the industrial revolution and our modern technological world possible. As these are consumed at a prodigious rate, it is not surprising that global CO_2 levels have increased. Based on ice core studies, the atmospheric concentration of CO_2 for the 800,000 years prior to the industrial revolution varied from 200 to 300 ppmv with cycles varying from 25,000 to 100,000 years. It increased to 370 ppmv at the turn of the millennium to approximately 430 ppmv in May 2025. Current estimates indicate that global CO_2 levels are increasing at approximately 3.6 ppmv per year. The pattern of this increase can be seen from data collected at Mauna Loa Observatory in Hawaii in the period 1960 to 2024 (Figure 2.3). Note the strong upward trend, as well as pronounced seasonal variations. The latter (with a variability of ~8 ppmv) is associated with photosynthetic consumption during spring and summer, with an excess of plant respiratory emissions during autumn and winter. The seasonal cycle ranges from approximately 1 ppmv at the South Pole, where the amplitude remains quite small, to 15–16 ppmv at Point Barrow, Alaska, occasionally reaching ~24 ppmv in extreme years.

The cyclic movement of C and other substances between organisms and their environment, as well as various components of the environment, is addressed in most biological, ecological, and environmental texts. As CO_2 has become an increasing environmental concern, the traditional C cycle has been modified in recent years to include reservoirs (sinks) and C fluxes associated with anthropogenic emissions. The most recent global carbon cycle reported in the IPCC's *Climate Change 2021: The Physical Science Basis* is shown in Figure 2.4. The values in the figure indicate carbon reservoirs in pentagrams of carbon (PgC) and fluxes in PgC/year. For easy reference to other data in the textbook, 1 PgC is equivalent to 1 gigatonne (Gt C) or 10^9 tonnes (metric tons) or 10^{15} grams. The black numbers in each carbon reservoir (indicated by boxes in Figure 2.4) represent the carbon mass prior to the industrial era (about 1750), and the black arrows with black numbers indicate the estimated carbon fluxes for that same period. Prior to the industrial era, the fluxes into and from the atmosphere were ostensibly in balance with about 589 PgC in the atmosphere. The gray-shaded numbers in each reservoir represent the net accumulation (+) or decrease (−) in carbon associated with anthropogenic activity from about 1750 to 2011. For example, both the atmosphere and the intermediate and deep sea show accumulations, whereas the fossil fuel reserves show a depletion of about 365 PgC. The gray arrows and associated numbers show the annual fluxes of carbon averaged over the period of 2000–2009 for various sources with both fossil fuels (7.8 PgC/year) and land use changes (1.1 PgC) being the largest anthropogenic sources of flux into the atmosphere. The average increase into the atmosphere was estimated to be about 4 PgC/year with the result of increasing the CO_2 concentration as illustrated in Figure 2.3.

Although CO_2 is relatively soluble in water (H_2O) and is taken up by plants, the timescale for C cycling in the terrestrial and ocean spheres is on the order of decades to thousands of years. Consequently, CO_2 has a relatively long atmospheric lifetime (~100 years).

A more recent global CO_2 budget presented in Table 2.4 quantifies the CO_2 emissions over the last six decades from human activities and the sink processes that remove them. The total anthropogenic flux of 11.5 Gt C/year shown for 2018 is small when compared with natural fluxes from the ocean and terrestrial ecosystems illustrated in Figure 2.4; however, the consistent increase is of major concern. CO_2 emissions and fluxes are generally reported in either PgC or Gt C/year, illustrating the carbon component. The actual emissions of CO_2 (if reported as CO_2) would be approximately 3.67 times higher due to the ratio of the molecular weight of CO_2 to C. Global anthropogenic emissions continue to increase and have nearly doubled in the last three decades. Although sink processes are relatively slow, they are nevertheless significant.

The US EPA reported that the nation's total anthropogenic emissions of greenhouse gases in 2022 were 6,343 million metric tons of CO_2 equivalent gases consisting of 80% CO_2, 11% CH_4, 6% N_2O, and 3% fluorinated gases. The CO_2 component would have been about 5.08 Gt per year of CO_2 or 1.39 Gt C/year. Figure 2.5 illustrates the global and select country anthropogenic emissions

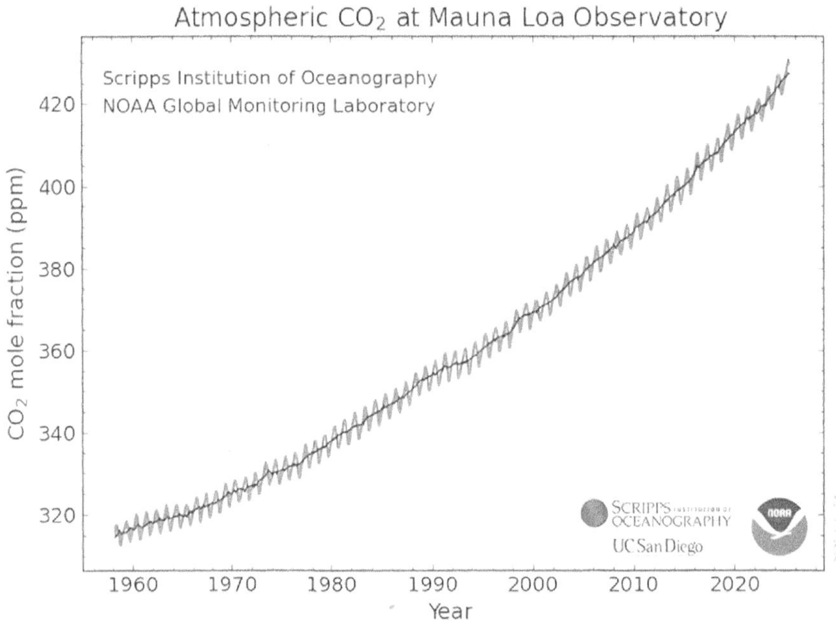

FIGURE 2.3 Annual trends and seasonal variation in CO_2 levels measured at Mauna Loa Observatory, HI, from 1960 to 2023. (From National Oceanic and Atmospheric Administration Global Greenhouse Gas Network, www.esrl.noaa.gov/gmd/ccgg/trends/, 2023.)

of the CO_2 component of greenhouse gases from 1990 to 2024 for fossil fuels (91% of total GHG emissions) in Gt C/year. Emissions by country show China to be the highest and India to be fourth highest and continuing to grow while the United States (second highest) and the European Union (EU28) are declining. While the energy consumption in the United States has been relatively flat since 2000, the downward trend in CO_2 emissions since about 2007 is a result of multiple factors including a significant shift from coal to less carbon-intensive natural gas with its higher thermal efficiency, the increased use of renewable energy such as wind and solar, and forest management practices. Based on the information shown in Figure 2.5, the United States emits approximately 13% of the global emissions of CO_2, a value that is only slightly less than its percentage of total global energy consumption, while China is the largest contributor contributing approximately 31%. On a per capita basis, the United States emits more than twice the CO_2 emissions of the next highest country, a value that reflects its greater per capita energy consumption than other countries. Transportation and electricity production are the two largest and almost equal sources of CO_2 emissions in the United States, each contributing approximately one-third of the emissions. The recent increased worldwide interest in improving the efficiency of engines, the increased usage of hybrid vehicles, and anticipated future adoption of electric vehicles will likely result in electricity becoming the largest source of CO_2 emissions within the next decade.

2.3.1.2 Carbon Monoxide

Carbon monoxide is a colorless, odorless, and tasteless gas. It is emitted from a number of anthropogenic and natural sources and produced in the atmosphere from oxidation of CH_4 and other NMHCs (Table 2.5). Major sources of anthropogenic emissions are technological processes (combustion and industrial) and biomass burning.

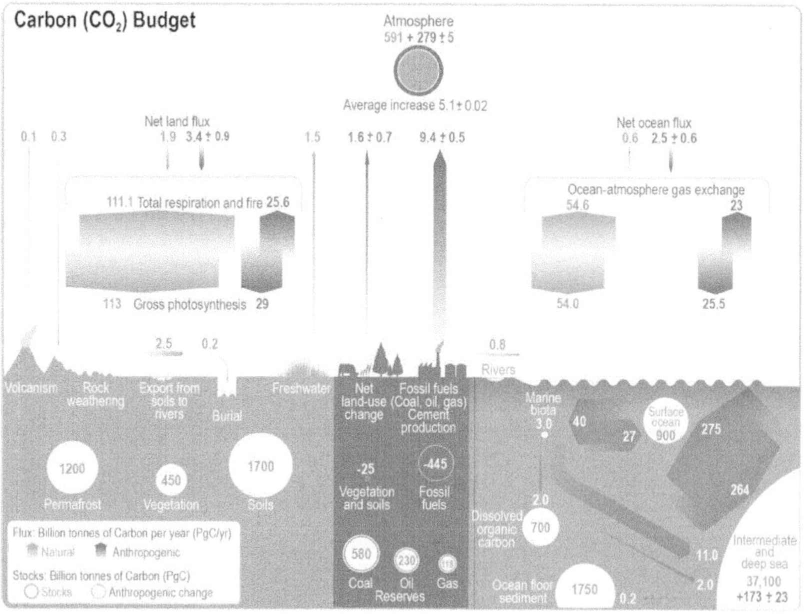

FIGURE 2.4 Global carbon cycle indicating reservoirs and fluxes. (Figure 5.12 in IPCC, 2021: Chapter 5. In: *Climate Change 2021: The Physical Science Basis. Contribution of Working Group I to the Sixth Assessment Report of the Intergovernmental Panel on Climate Change* [Canadell, J.G., P.M.S. Monteiro, M.H. Costa, L. Cotrim da Cunha, P.M. Cox, A.V. Eliseev, S. Henson, M. Ishii, S. Jaccard, C. Koven, A. Lohila, P.K. Patra, S. Piao, J. Rogelj, S. Syampungani, S. Zaehle, and K. Zickfeld], 2021: Global Carbon and Other Biogeochemical Cycles and Feedbacks. In: Climate Change 2021: The Physical Science Basis. Contribution of Working Group I to the Sixth Assessment Report of the Intergovernmental Panel on Climate Change [Masson-Delmotte, V., P. Zhai, A. Pirani, S.L. Connors, C. Péan, S. Berger, N. Caud, Y. Chen, L. Goldfarb, M.I. Gomis, M. Huang, K. Leitzell, E. Lonnoy, J.B.R. Matthews, T.K. Maycock, T. Waterfield, O. Yelekçi, R. Yu, and B. Zhou (eds.)]. Cambridge University Press, Cambridge, United Kingdom and New York, NY, USA, pp. 673–816, https://doi.org/10.1017/9781009157896.007. With permission.)

TABLE 2.4
Global Carbon Budget for CO_2 Showing Sources and Partitioning/Sinks Associated with Human Activities

	Mean (Gt C/year)						
Total emissions	**1960–1969**	**1970–1979**	**1980–1989**	**1990–1999**	**2000–2009**	**2014–2023**	**2023**
Fossil CO_2	3.0±0.2	4.7±0.2	5.5±0.3	6.4±0.3	7.8±0.4	9.7±0.5	10.1±0.5
Land use change	1.6±0.7	1.4±0.7	1.4±0.7	1.6±0.7	1.4±0.7	1.1±0.7	1.0±0.7
Total emissions	4.6±0.7	6.1±0.7	6.9±0.8	7.9±0.8	9.2±0.8	10.8±0.9	11.1±0.9
Partitioning							
Growth in CO_2 conc.	1.7±0.07	2.8±0.07	3.4±0.02	3.1±0.02	4.0±0.02	5.2±0.02	5.9±0.2
Ocean sink	1.2±0.4	1.5±0.4	1.9±0.4	2.1±0.4	2.3±0.4	2.9±0.4	2.9±0.4
Terrestrial sink	1.2±0.5	2.0±0.8	1.8±0.8	2.5±0.6	2.8±0.7	3.2±0.9	2.3±1.0
Budget imbalance	0.5	−0.1	−0.2	0.1	0.0	−0.4	0.0

Source: Global Carbon Budget 2024, *Earth System Science Data*, Vol. 17, Issue 3, 965–1039, 2025, https://doi.org/10.5194/essd-17-965-2025, 2025.

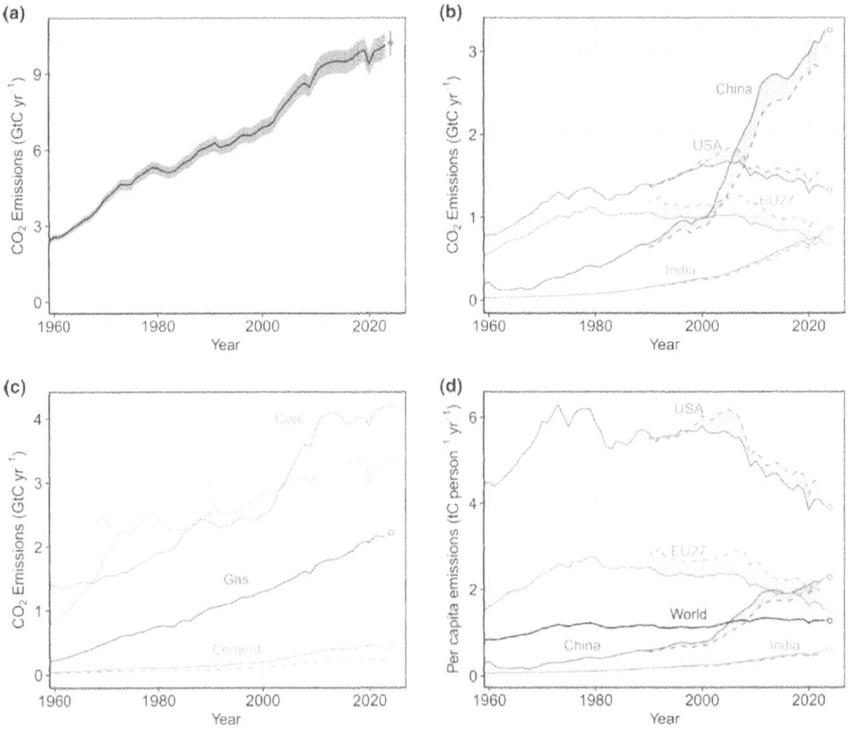

FIGURE 2.5 Fossil CO_2 emissions for (a) the globe, including an uncertainty of ±5% (gray shading) and a projection through the year 2024 (red dot and uncertainty range); (b) territorial (solid lines) and consumption (dashed lines) emissions for the top three country emitters (USA, China, India) and for the European Union (EU27); (c) global emissions by fuel type, including coal, oil, gas, cement, and cement minus cement carbonation (dashed); and (d) per capita emissions for the world and for the large emitters as in panel (b). (From *Earth System Science Data*, Vol. 17, Issue 3, 965–1039, 2025, https://doi.org/10.5194/essd-17-965-2025, 2025. With permission.)

Carbon monoxide is an intermediate product in combustion oxidation of fuels and biomass:

$$2C + O_2 \rightarrow 2CO \tag{2.1}$$

$$2CO + O_2 \rightarrow 2CO_2 \tag{2.2}$$

If insufficient O_2 is present, combustion is incomplete and significant quantities of CO are produced.

Direct anthropogenic emissions from combustion and industrial processes and photolytic oxidation of anthropogenic NMHCs account for approximately 25% to 30% of CO emissions and production in the northern hemisphere. Carbon monoxide emissions from transportation sources are particularly significant in North America and Europe.

2.3.1.2.1 *Concentrations*

Background CO levels vary spatially and temporally, with a concentration range of 40 to 200 ppbv and an average of 120 ppbv at 45°N latitude. Background concentrations have been increasing (because of emissions and production associated with human activity) at a rate of approximately 1% per year, with most of the increase occurring in the northern middle latitudes. The highest concentrations are observed in winter when sink processes are less efficient.

TABLE 2.5
Estimated Sources and Sinks of Carbon Monoxide

	Range (10^8 t CO/year)
Sources	
Technological	3.30–6.06
Biomass burning	3.30–7.71
Biogenic	0.07–1.76
Oceans	0.22–2.20
Methane oxidation	4.40–11.01
NMHC oxidation	2.20–6.60
Total sources	19.80–29.73
Sinks	
Reaction with OH	15.42–28.63
Soil uptake	2.75–7.05
Stratospheric loss	~1.10
Total sink losses	23.13–33.04

Source: Seinfeld, J.H. and Pandis, S.N.: *Atmospheric Chemistry and Physics: Air Pollution to Climate Change*. 1998. Copyright Wiley-VCH Verlag GmbH & Co. KGaA. Reproduced with permission.

Ambient CO concentrations in metropolitan areas are orders of magnitude higher than background concentrations. These levels are primarily associated with transportation emissions and are closely related to traffic density and meteorological conditions. The highest concentrations occur along major traffic arteries during morning and evening rush hours and decrease relatively rapidly with distance from roadways. While the United States has both 1-hour (35 ppmv) and 8-hour average (9 ppmv) National Ambient Air Quality Standards [NAAQS]), all areas have met the standards since 2010. Figure 2.6 shows the trend for average 8-hour concentrations which was less than 2 ppmv as of 2023. The shaded areas in the figure and similar figures reported later in this chapter represent the 10th and 90th percentile values. Elevated CO levels are still prevalent in some cities and countries that have high traffic densities and/or are located at relatively high altitudes such as high-altitude Mexico City, China, and other South Asian countries. These are generally a result of emissions from rapidly increasing overall traffic volumes that outpaced the reductions achieved on a per vehicle basis as a result of applying and implementing new vehicle emission standards.

2.3.1.2.2 Sink Processes

When sink processes are considered, CO becomes significantly more important than regulatory concerns would indicate. Carbon monoxide affects the tropospheric concentrations of both hydroxyl radical (OH·) and O_3 and, as a consequence, the oxidizing potential of the atmosphere. Because CO and CH_4 (a major greenhouse gas) compete for OH·, CO indirectly affects tropospheric concentrations of CH_4 and stratospheric H_2O vapor (derived from CH_4 oxidation).

Major sink processes include OH·-related photochemistry and uptake by soil. In Table 2.5, these account for an estimated 14.02 to 26.03 × 10^8 and 2.5 to 6.4 × 10^8 metric t/year (15.42–28.63 × 10^8 and 2.75–7.05 × 10^8 t/year) of CO, respectively. Carbon monoxide is oxidized in the atmosphere to CO_2 upon reaction with OH·:

$$CO + OH\cdot \rightarrow CO_2 + H\cdot \tag{2.3}$$

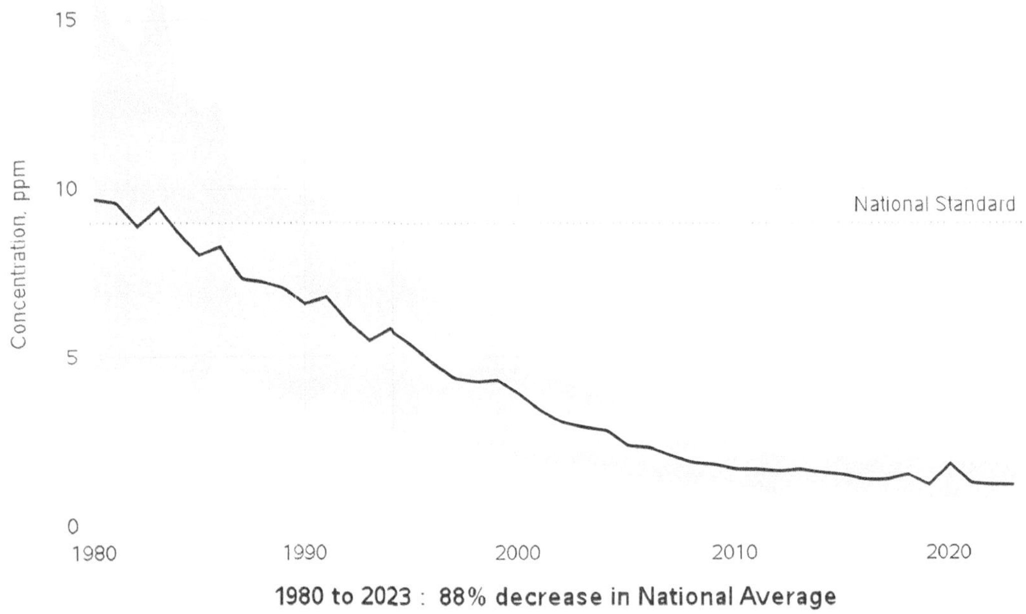

FIGURE 2.6 CO air quality (annual 2nd maximum 8-hour average concentrations observed at 32 sites in the United States from 1980 to 2023). (From US EPA, https://www.epa.gov/air-trends/carbon-monoxide-trends.)

Subsequent reactions of H· and oxygen (O_2) produce the highly reactive hydroperoxyl radical, HO_2·:

$$H\cdot + O_2 + M \rightarrow HO_2\cdot + M \tag{2.4}$$

where M is an energy-absorbing molecule, for example, nitrogen (N_2) or O_2. Summarizing,

$$2CO + H\cdot + 2O_2 \rightarrow 2CO_2 + HO_2\cdot \tag{2.5}$$

Nitrogen dioxide is produced upon reaction with HO_2 in the presence of significant concentrations of nitric oxide (NO):

$$HO_2\cdot + NO \rightarrow NO_2 + OH\cdot \tag{2.6}$$

Nitrogen dioxide is subsequently photolyzed to produce O_3:

$$NO_2 + hv \rightarrow NO + O(^3P) \tag{2.7}$$

$$O(^3P) + O_2 + M \rightarrow O_3 + M \tag{2.8}$$

$$CO + 2O_2 + hv \rightarrow CO_2 + O_3 \tag{2.9}$$

where $O(^3P)$ is ground-state atomic oxygen and hv is a photon of light energy.

As a result of these reactions, CO is converted to CO_2, with one molecule of O_3 produced for each CO molecule oxidized. As such, CO oxidation in the troposphere is an important (but low-level) source of O_3.

As indicated previously, CO competes with CH_4 for OH·. Methane oxidation, as seen in Table 2.5, is also a major source of CO, accounting for approximately 15% to 20% of its background atmospheric concentration. The atmospheric residence time of CO is approximately 1 month in the tropics to 4 months at midlatitudes in the spring and autumn.

2.3.2 Sulfur Compounds

Sulfur compounds are emitted from a variety of natural and anthropogenic sources and produced as a result of atmospheric chemical processes. Major atmospheric S compounds include sulfur dioxide (SO_2) and reduced S compounds such as hydrogen sulfide (H_2S), dimethyl sulfide ($(CH_3)_2S$), carbon disulfide (CS_2), and carbonyl sulfide (COS).

2.3.2.1 Sulfur Oxides

Sulfur oxides are emitted to the atmosphere by volcanoes and a number of anthropogenic sources. They are also formed in the atmosphere as a result of oxidation of naturally or anthropogenically produced reduced S compounds.

The sulfur oxides, which include sulfur trioxide (SO_3) and SO_2, are produced when fuels and biomass are combusted and S-containing metal ores are smelted. Because of its high affinity for H_2O vapor, the atmospheric lifetime of SO_3 is relatively short (seconds). The formation of SO_2, SO_3, and sulfuric acid (H_2SO_4) by relatively slow, direct oxidation processes is described by the following equations:

$$S + O_2 \rightarrow SO_2 \quad (2.10)$$

$$2\ SO_2 + O_2 \rightarrow 2\ SO_3 \quad (2.11)$$

$$SO_3 + H_2O + M \rightarrow H_2SO_4 + M \quad (2.12)$$

Sulfur dioxide is a colorless gas with a sulfury odor that can be easily detected at concentrations of 0.38 to 1.15 ppmv. It has a pungent, irritating odor above 3.0 ppmv.

2.3.2.1.1 Concentrations

Background levels of SO_2 vary from approximately 20 pptv over marine surfaces to 160 pptv over relatively clean areas of the North American continent. Average annual concentrations in urban areas range from 1.5 ppbv in Auckland, New Zealand, to more than 90 ppbv in cities in China. Shorter-term, 24-hour concentrations may vary from 5 ppbv in Auckland to 300 to 400 ppbv in Chinese cities. Maximum 1-hour concentrations of several hundred parts per billion by volume (ppbv) were once common in US cities, such as Chicago and Pittsburgh, and are still common in industrialized cities in developing countries. The highest maximum 1-hour concentrations of several parts per million by volume (ppmv) have been reported downwind of non-ferrous metal smelters in the past. Figure 2.7 illustrates concentrations of SO_2 in the United States (37 sites). The values reported are the 99th percentile of the daily maximum 1-hour averages (NAAQS for SO_2). The ambient concentrations have decreased significantly and are below the NAAQS as a result primarily of implementation of control technologies on sulfur-emitting combustion sources and smelters and the reduced use of coal. Based on information available in the US EPA Green Book (*www.epa.gov/green-book*), there were still 28 counties, cities or parishes in the United States that were nonattainment for the SO_2 1-hour standard of 75 ppbv in June 2025.

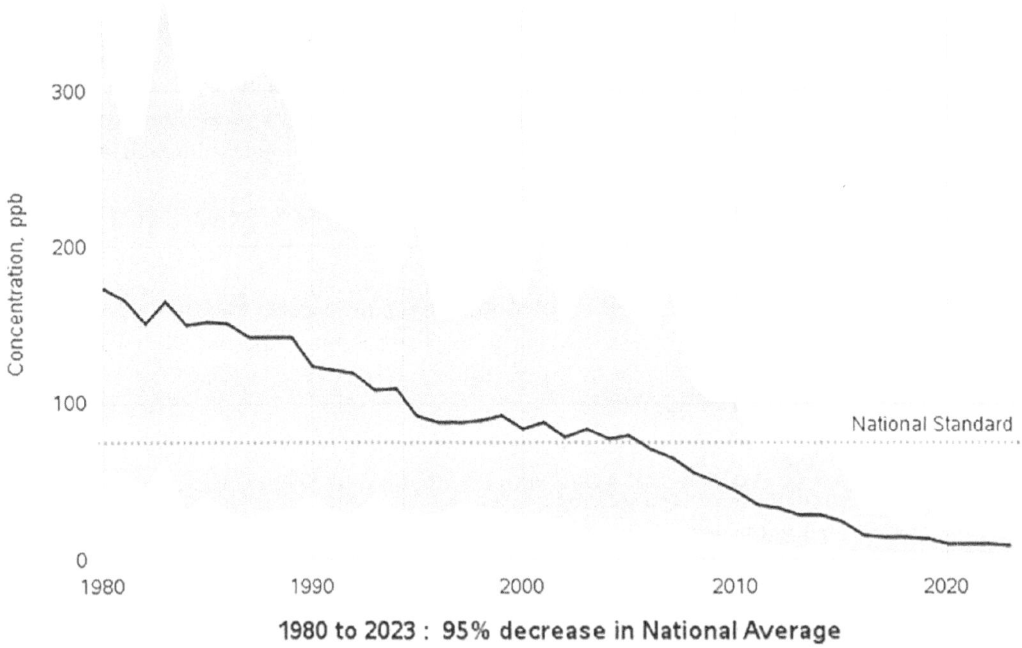

FIGURE 2.7 SO_2 Air Quality (annual 99th percentile of the daily maximum 1-hour average concentrations observed at 32 sites in the United States from 1980 to 2023). (From US EPA, https://www.epa.gov/air-trends/sulfur-dioxide-trends)

2.3.2.1.2 Sink Processes

After emission to the atmosphere, SO_2 is oxidized by various gas-phase and aqueous-phase reactions. Oxidation may be direct (Equations 2.11 and 2.12), photochemical, or catalytic. Gas-phase oxidation may occur as a result of reactions with OH·, O_3, $O(^3P)$, and peroxy radicals (HO_2·, RO_2·). The dominant sink mechanism for SO_2 is reaction with OH·:

$$SO_2 + OH \cdot \rightarrow HOSO_2 \tag{2.13}$$

$$HOSO_2 + O_2 \rightarrow HO_2 \cdot + SO_3 \tag{2.14}$$

$$SO_3 + H_2O \rightarrow H_2SO_4 \tag{2.15}$$

The reaction of SO_3 with H_2O vapor produces H_2SO_4, which forms an aerosol by nucleation and condensation processes. Sulfuric acid may react with ammonia (NH_3) to form ammonium salts. The lifetime of SO_2 (based on reaction with OH·) is approximately 1 week.

Sulfur dioxide can also be oxidized to form dilute sulfurous acid (H_2SO_3) in fog, cloud, and rain droplets, as well as hygroscopic aerosols:

$$SO_2 + H_2O \rightarrow SO_2 \cdot H_2O \tag{2.16}$$

$$SO_2 \cdot H_2O \rightarrow H_2SO_3 \tag{2.17}$$

In the aqueous phase, SO_2 can be oxidized to H_2SO_4 by reaction with nitrous acid (HNO_2), O_3, hydrogen peroxide (H_2O_2), organic peroxides, and catalysis by iron (Fe) and manganese (Mn).

Both SO_2 and its oxidation products are removed from the atmosphere by wet and dry deposition processes. Dry deposition to plant and other surfaces is relatively efficient. At a dry deposition velocity of 1 cm/s, the atmospheric lifetime in the 1-km-thick planetary boundary layer has been estimated to be approximately 1 day. This removal is enhanced when clouds are present.

2.3.2.2 Reduced Sulfur Compounds

Carbonyl sulfide is the S gas present at highest levels in the background global atmosphere, with an average concentration of 500 pptv. It has low chemical reactivity, with an atmospheric lifetime of approximately 44 years. Carbon disulfide is produced from sources similar to COS. However, it is more photochemically reactive, with a lifetime of approximately 12 days. Global background levels are variable, ranging from 15 to 190 pptv. Dimethyl sulfide is released from the oceans in large quantities. It has a short atmospheric lifetime (0.6 days) and is rapidly converted to SO_2.

Hydrogen sulfide is the only reduced S compound that poses environmental concerns. Although it is a relatively toxic gas, atmospheric concentrations are usually too low to pose a threat to human health. It has a characteristic rotten egg odor, detectable by humans at concentrations as low as 500 pptv. It is an air quality concern because of the malodor it produces. Other reduced S malodorants include methyl and ethyl mercaptans, which smell like rotting cabbage.

Anthropogenic sources of H_2S include oil and gas extraction, petroleum refining, coke ovens, and kraft paper mills. Although anthropogenic sources produce less than 5% of global emissions, the highest ambient concentrations (and, as a consequence, malodor problems) occur near anthropogenic emission sources. Concentrations in industrial and surrounding ambient environments are usually above the odor threshold of 500 pptv. Background concentrations are in the 30 to 100 pptv range.

2.3.3 Nitrogen Compounds

A number of gas-phase and particulate-phase nitrogen (N) compounds are found in the atmosphere. These include N_2, nitrous oxide (N_2O), NO, NO_2, nitrate radical ($NO_3\cdot$), dinitrogen pentoxide (N_2O_5), HNO_2, nitric acid (HNO_3), organic nitrates such as peroxyacyl nitrate ($CH_3COO_2NO_2$), that is, PAN, and reduced N compounds such as NH_3 and hydrogen cyanide (HCN). The ionic species, NO_2^-, NO_3^-, and NH_4^+, are found in the aqueous phase.

Nitrogen is the most abundant gas in the atmosphere. Under ambient conditions, it is relatively unreactive and plays no direct role in atmospheric chemistry. However, as a result of combustion reactions and biological processes, it serves as a precursor molecule for the production of NO and NO_2, which play significant roles in tropospheric and stratospheric chemistry. Because they are readily converted from one to the other, NO and NO_2 concentrations are expressed as NO_x.

Reactive nitrogen (NO_y) includes NO_x compounds as well as their atmospheric oxidation products, for example, HNO_2, HNO_3, $NO_3\cdot$, N_2O_5, peroxynitric acid (HNO_4), PAN and its homologues, alkyl nitrates ($RONO_2$) and peroxyalkyl nitrates ($ROONO_2$). Total NO_y is a measure of the total oxidized N content of the atmosphere.

Despite relatively low concentrations, NO, NO_2, organic nitrates, gas-phase and particulate-phase N acids, and NH_3 play significant roles in atmospheric chemistry (particularly in polluted environments). Nitric oxide and NO_2 are of major environmental concern because atmospheric concentrations increase significantly as a result of human activities, and because they serve as precursor molecules for many important atmospheric reactions.

2.3.3.1 Nitrous Oxide

Nitrous oxide is a colorless, slightly sweet, relatively nontoxic gas. It is widely used as an anesthetic in medicine and dentistry. Human exposure to elevated concentrations produces a kind of hysteria, and as such, it is often referred to as laughing gas.

Nitrous oxide concentration in the atmosphere varied from 200 to 300 ppbv in cycles in preindustrial times but has now risen to about 332 ppbv. Estimated annual emissions to the atmosphere are 17.9 Tg (tetragrams) N/year, with approximately 60% being produced by nitrification and denitrification processes in undisturbed terrestrial environments and in the world's oceans. About 7 Tg N/year, or approximately 38%, is associated with anthropogenic emissions including agricultural tillage, fertilizer use, and animal wastes.

Nitrous oxide has no known tropospheric sink. As a result, it has a very long atmospheric lifetime (~130 years). Its photolysis and subsequent oxidation by singlet oxygen ($O(^1D)$) in the stratosphere is the primary sink process. As such, N_2O is the major natural source of NO_x in the stratosphere, where it plays a significant role in stratospheric O_3 chemistry.

Increased atmospheric N_2O levels pose two major environmental concerns: stratospheric O_3 depletion and, because of its thermal absorptivity, global warming (see Chapter 4).

2.3.3.2 Nitric Oxide

Nitric oxide is a colorless, odorless, tasteless, relatively nontoxic gas. It is produced naturally in soil through biological nitrification and denitrification processes and as a result of biomass burning, lightning, and oxidation of NH_3 by photochemical processes. It is also transported from the stratosphere into the troposphere.

Nitric oxide is produced anthropogenically in high-temperature combustion processes:

$$N_2 + O_2 \rightarrow 2NO \tag{2.18}$$

Because this reaction is endothermic, equilibrium moves to the right at high temperatures and to the left at lower temperatures. If cooling is rapid, equilibrium is not maintained and significant NO emissions occur. High combustion temperatures, rapid cooling associated with gas flows from combustion chambers, and instantaneous dilution contribute to high NO emissions.

Major anthropogenic sources include gasoline-powered and diesel-powered vehicles, fossil fuel–fired electricity-generating stations, industrial boilers, municipal incinerators, and home space heating.

Nitric oxide concentrations are not generally reported separately from those of NO_2. However, early studies indicate that NO concentrations peak during the early morning rush hour (6–9 a.m.) and then are rapidly depleted by oxidation to NO_2 and various scavenging processes (Figure 2.8).

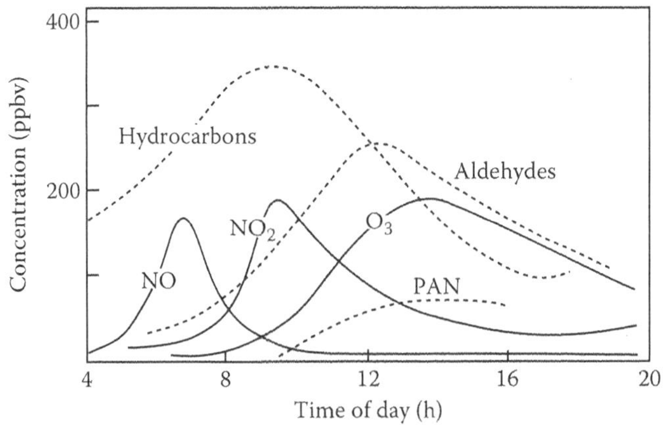

FIGURE 2.8 Time-dependent variation of NO, NO_2, NMHC, and O_3 in an urban atmosphere on a smoggy day. (From Goody, R., *Principles of Atmospheric Physics and Chemistry*, Oxford University Press, Oxford, 1995. With permission.)

2.3.3.3 Nitrogen Dioxide

Nitrogen dioxide is a colored gas that varies from yellow to brown depending on its concentration in the atmosphere. It has a pungent, irritating odor and, because of its high oxidation rate, is relatively toxic and corrosive.

Nitrogen dioxide can be produced in a number of atmospheric processes. These include the relatively slow, direct oxidation reaction:

$$2NO + O_2 \rightarrow 2NO_2 \quad (2.19)$$

and more rapid photochemical reactions involving O_3, RO_2, and other hydrogen species ($OH\cdot$, $HO_2\cdot$, H_2O_2, etc.):

$$NO + O_3 \rightarrow NO_2 + O_2 \quad (2.20)$$

$$RO_2 \cdot + NO \rightarrow NO_2 + RO \cdot \quad (2.21)$$

$$HO_2 \cdot + NO \rightarrow NO_2 + OH \cdot \quad (2.22)$$

Because it is primarily produced by photochemical oxidation of NO, peak NO_2 levels in urban–suburban areas typically occur in midmorning (Figure 2.8).

2.3.3.4 Nitrogen Oxides

2.3.3.4.1 Concentrations

Average NO_x concentrations (based on the sum of NO and NO_2) measured in remote locations have been reported to range from 0.02 to 0.04 ppbv in marine environments and 0.02 to 10 ppbv in tropical forests. In rural US locations, NO_x concentrations are reported to range from 0.02 to 10 ppbv; in urban–suburban areas, from 10 to 1000 ppbv, historically. The NAAQS is based on the concentration of NO_2 since it tends to be representative of the trend of NO_x concentrations. Figure 2.9 shows that the ambient concentrations have steadily decreased from well above 100 ppbv in 1980 to well below the standard of 100 ppbv in 2023. The last nonattainment area achieved attainment status in 1998.

2.3.3.4.2 Sink Processes

Sink processes for NO and NO_2 involve chemical reactions that convert NO to NO_2 and NO_2 to HNO_3. Reaction of NO_2 with $OH\cdot$ is the major sink process:

$$NO_2 + OH\cdot + M \rightarrow HNO_3 + M \quad (2.23)$$

Nitrogen dioxide is also converted to HNO_3 by nighttime reactions involving O_3:

$$NO_2 + O_3 \rightarrow NO_3 \cdot + O_2 \quad (2.24)$$

$$NO_2 + NO_3 \cdot \rightarrow N_2O_5 \quad (2.25)$$

$$N_2O_5 + H_2O \rightarrow 2HNO_3 \quad (2.26)$$

Reactions between O_3 and NO_2 produce $NO_3\cdot$ that subsequently reacts with NO_2 to produce N_2O_5 (which reacts rapidly with H_2O vapor to produce HNO_3). Nitric acid is also produced in reactions involving $NO_3\cdot$ and formaldehyde (HCHO) or hydrocarbon radicals ($RH\cdot$):

$$NO_3 \cdot + HCHO \rightarrow HNO_3 + CHO \cdot \quad (2.27)$$

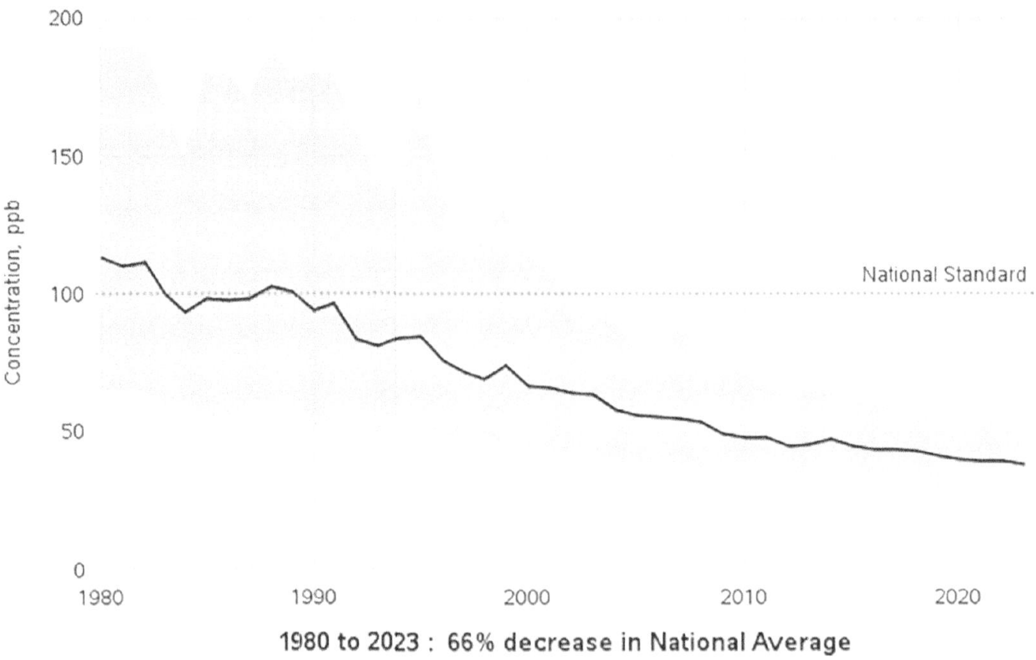

FIGURE 2.9 NO_2 air quality (annual 98th percentile of the daily maximum 1-hour average concentrations observed at 19 sites in the United States from 1980 to 2023). (From US EPA, https://www.epa.gov/air-trends/nitrogen-dioxide-trends.)

$$NO_3\cdot + RH\cdot \rightarrow HNO_3 + R\cdot \tag{2.28}$$

Nitrate radicals play a significant role in nighttime oxidation of biogenically produced VOCs such as isoprene and the pinenes—compounds that have a major role in tropospheric O_3 chemistry. Nitrate radicals are rapidly photolyzed at sunrise by a shift in the equilibrium between NO_2, $NO_3\cdot$, and N_2O_5 toward NO_2.

Nitric acid remains in the gas phase until it reacts with NH_3 to form ammonium salts:

$$NH_3 + HNO_3 \rightarrow NH_4NO_3 \tag{2.29}$$

Nitric acid and its salts are removed from the atmosphere by wet and dry deposition processes. In the lower troposphere, the atmospheric lifetime of nitrates is approximately 1 week.

Because of their low solubility, NO and NO_2 are readily transported to the upper troposphere, with the subsequent production of HNO_3 upon reaction with OH·. Nitric acid produced in the upper troposphere is not subject to dry deposition, and generally not to wet deposition. It can be transformed to NO_2 by photolysis:

$$HNO_3 + h\nu \rightarrow NO_2 + OH\cdot \tag{2.30}$$

and, to a lesser extent, oxidation by OH·. Because these reactions are relatively slow, HNO_3 serves as a reservoir species and carrier for NO_x in atmospheric transport.

2.3.3.5 Ammonia

Ammonia is the most abundant basic chemical substance in the atmosphere and the third most abundant N compound. Major sources include anaerobic decomposition of organic matter, animals and their wastes, biomass burning, soil humus formation, and application of anhydrous NH_3 to cropland. Other sources include coal combustion and industrial emissions. Background concentrations have been reported to vary from 100 pptv over remote oceans to 1 to 10 ppbv over continents. Because it is readily absorbed by H_2O and soil surfaces, NH_3's atmospheric lifetime is relatively short (~10 days).

Ammonia reacts rapidly with strong acids (e.g., H_2SO_4 and HNO_3) to produce ammonium salts. As a consequence, NH_3 plays an important role in removing SO_x and NO_x from the atmosphere. Ammonia can also be oxidized upon reacting with OH·.

2.3.3.6 Hydrogen Cyanide

Hydrogen cyanide (HCN) occurs in the atmosphere at low background levels (~160 pptv). Although it reacts relatively slowly with OH·, little is known about its atmospheric chemistry. It is typically emitted into the atmosphere as a result of burning of biomass such as forest fires, and its presence has often been used as a marker in identifying the presence of biomass-related smoke. Cyanide poisoning, due to inhalation of HCN, has also been well documented in residential and commercial building fires where HCN is formed due to the incomplete combustion of materials, such as wood, plastics, vinyl, wool, and silk, that contain nitrogen.

2.3.3.7 Organic Nitrate Compounds

Organic nitrate compounds are produced as a result of photochemical reactions involving NO_x and NMHCs. Peroxy radicals are formed from the photochemical oxidation of short-chained NMHCs such as acetaldehyde (CH_3CHO). Acetaldehyde, upon reaction with OH·, produces an acetyl radical that upon reaction with O_2 produces a peroxyacetyl radical. Subsequent reaction with NO_2 produces PAN:

$$CH_3CHO + OH\cdot \rightarrow CH_3CO + H_2O \tag{2.31}$$

$$CH_3CO + O_2 \rightarrow CH_3COO_2\cdot \tag{2.32}$$

$$CH_3COO_2\cdot + NO_2 + M \rightarrow CH_3COO_2NO_2 + M \tag{2.33}$$

Reactions involving propionylaldehyde and butrylaldehyde produce peroxypropionyl nitrate (PPN) and peroxybutryl nitrate (PBN).

Peroxy nitrates (PAN) were first identified as components of smog in the Los Angeles Basin in the 1950s. For most of the past 40+ years, PAN was thought to be associated only with polluted urban atmospheres with high photochemical activity. It is now known to be ubiquitously present in the atmosphere, being found in urban, rural, and remote environments. It is a potent eye irritant and the cause of "smog injury" on sensitive agricultural crops.

Peroxyacetyl nitrate and its homologues are relatively inert photochemically and insoluble in water. As such, they have relatively long atmospheric lifetimes. The principal sink process is thermal degradation to acetylperoxy radicals and NO_2. In the upper troposphere, PAN is quite stable and transported over long distances. It is an important reservoir and carrier of NO_x (similar to HNO_3, described previously). Because its decomposition increases with temperature, it can be formed in cooler regions, transported, and then decompose to release NO_2 in warmer regions. Peak concentrations of PAN in urban–suburban areas have been reported in the range of 10 to 50 ppbv.

2.3.4 Hydrocarbon Compounds

2.3.4.1 Hydrocarbon Chemistry

Hydrocarbon compounds include a large number of chemicals that exist under ambient environmental conditions as gases, volatile liquids, semivolatile substances, and solids. They include compounds that contain only hydrogen (H) and C (HCs), and HC derivatives that contain substances such as O, N, S, and various halogens.

All HC species are characterized by covalent chemical bonds between C and H, C and C, or C and other atomic species such as O. The covalently bonded atoms share electron pairs to attain chemical stability. A single covalent bond is formed when an electron pair is shared. In the simplest HC, CH_4, each H is bonded to C by a single bond. Hydrocarbon compounds that contain only single bonds are called alkanes or paraffins. Larger molecules contain two or more carbons covalently bonded to each other. The paraffins include compounds that are straight chained (e.g., CH_4 and ethane [C_2H_6]), branched (e.g., isobutane), and cyclic (e.g., cyclopropane). The alkanes or paraffins are relatively unreactive and therefore chemically stable; for example, *n*-butane has a relatively long atmospheric lifetime (~5 days).

Carbon atoms can share one to three electron pairs. A double bond is formed when two pairs of electrons are shared. Hydrocarbons that contain one or more double bonds are called alkenes or olefins; the simplest is ethylene (C_2H_4). The olefins also include a number of longer-chained compounds (e.g., propylene and butylene). Because double bonds are somewhat unstable, olefinic compounds are very reactive. Olefins and other compounds with double bonds serve as primary reactants in the photochemical processes that produce haze and smog. Their atmospheric lifetimes are on the order of several hours.

If two H atoms were removed from ethylene, the two carbons would share three electron pairs, forming acetylene (C_2H_2), which has a triple bond. Unlike the olefins that have atmospheric lifetimes of hours, alkynes such as C_2H_2 are long-lived, with an estimated lifetime of 14 days.

Hydrocarbons described previously, that is, the paraffins, olefins, and alkynes, are classified as aliphatic HCs because of their basic chain structure. Hydrocarbons, whose structure is based on the ring structure of benzene (C_6H_6), are called aromatic HCs. Aromatic HCs other than C_6H_6 include (among others) ethylbenzene, toluene, and the xylenes. This group of four aromatic compounds was formulated to produce the fuel additive, BTEX, which increases the octane rating of unleaded gasolines and was added to gasoline to maintain high octane levels as lead was phased out of gasoline. Due to its known health effects, the US EPA reduced the volume of BTEX in gasoline and ethanol was increased (i.e., E10, which has an ethanol content of 10%) as an octane booster. Most refiners today create a sub-octane gasoline such as 84 octane followed by blending with 10% ethanol to reach the minimum octane requirements at the pump of 87 octane.

Because of their widespread use as solvents and fuel additives, aromatic HCs are commonly found in ambient air. Although they contain at least three double bonds, aromatic HCs are generally less reactive than olefins. They vary in their reactivity and therefore estimated atmospheric lifetimes (C_6H_6, ~12 days; toluene, ~2 days; *m*-xylene, ~7.4 hours). Some reactive HCs (e.g., *m*-xylene) are likely to be quickly consumed in urban photochemistry, whereas others (e.g., C_6H_6) are even more stable than some of the paraffins.

The aromatic HCs described earlier are simple compounds based on a single C_6H_6 ring. Benzene rings also serve as the basic structural unit of larger, more complex compounds, including naphthalene (two C_6H_6 rings) and a family of polycyclic aromatic hydrocarbons (PAHs), which are solids under ambient atmospheric conditions. The best-known compound in this group is benzo-α-pyrene, which contains five C_6H_6 rings. The PAHs are produced in the combustion chemistry of organic fuel and materials and are commonly found in the elemental carbon (EC) fraction of atmospheric aerosol. Most PAHs are potential human carcinogens.

Hydrocarbons with only single bonds are described as being saturated; those with one or more double or triple bonds are unsaturated. Compounds that contain 1 to 4 C atoms are gases under ambient conditions; those with 5 to 12 C atoms, volatile liquids. Those with higher C numbers tend to be semivolatile liquids or solids, or nonvolatile solids.

Hydrocarbon compounds and their derivatives are classified based on their boiling points and volatility. Very volatile organic compounds (VVOCs) have boiling points that range from 0°C to 50°C to 100°C (32°F to 122°F to 212°F); VOCs, from 50°C to 100°C to 240°C to 260°C (464°F–500°F); semivolatile organic compounds (SVOCs), from 240°C to 260°C to 380°C to 400°C (716°F–752°F); and solid organic compounds, more than 400°C.

Hydrocarbon emissions are regulated by the US EPA. Because of the low reactivity of CH_4, HC concentrations in the atmosphere and national emissions estimates are reported as NMHCs. The

US EPA has recently begun to report emission estimates as VOCs (which include both VOCs and VVOCs).

A variety of substances are derived from the basic HC structure. These include, among others, oxygenated and halogenated HCs, both of which represent major atmospheric pollution concerns. Because they are subject to unique regulatory requirements, halogenated HCs are discussed apart from other NMHCs in this chapter.

2.3.4.2 Oxyhydrocarbons

A variety of HC compounds can be formed as a result of reaction with O_2 or O-containing compounds. These include alcohols, aldehydes, ketones, ethers, organic acids, and esters.

Alcohols, for example, methanol (CH_3OH) and ethanol (C_2H_5OH), are characterized by having an −OH moiety attached to one of their carbon atoms. The substitution of an −OH in the structure of CH_4 produces CH_3OH. Short-chain alcohols are liquid at ambient conditions and are commonly used as solvents. Methanol is used as an automotive fuel and C_2H_5OH as a fuel additive. Alcohols with two or more −OH moieties are called polyols. Low-volatility polyols such as ethylene glycol are used as motor vehicle coolants and in water-based paint and varnish products. Because of their widespread use, alcohols and polyols are frequently emitted to the atmosphere.

Aldehydes are characterized by a carbonyl moiety (C = O) on a terminal C atom. The simplest aldehyde is HCHO. It is widely used in industry as a chemical feedstock for the production of adhesives and other polymers. It is also a common by-product of combustion processes, as are the aldehydes CH_3CHO and acrolein. Significant aldehyde emissions are associated with light-duty motor vehicle operation. Aldehydes are also produced in photochemical reactions, with peak atmospheric levels occurring at solar noon. Both HCHO and CH_3CHO are important reactants in a variety of photochemical processes. Acetaldehyde has an estimated atmospheric lifetime of 11 hours, whereas HCHO's lifetime is approximately 1.2 days.

The ketones contain a carbonyl attached to a nonterminal C. Ketones are widely used as solvents and, because of their high volatility, are quickly released into the atmosphere. Acetone, one of the most widely used ketones, enters the atmosphere from both biogenic and anthropogenic sources and oxidation of paraffinic and olefinic compounds. Because of its long atmospheric lifetime (~100 days), acetone is found in the free troposphere at concentrations of 1 ppbv.

Ethers are VOCs used as solvents. They are characterized by a moiety in which O is bonded to two C atoms. They are not significant atmospheric pollutants. However, several ethers have been used as a source of O in reformulated automotive fuels. These include methyl-*t*-butyl ether (MTBE), ethyl-*t*-butyl ether (ETBE), and tertiary-amyl-methyl ether (TAME).

Organic acids are characterized by one or more carboxyl groups (–COOH). Organic acids commonly used in industry include acetic and formic acids. Although emissions are limited, organic acids are common in the atmosphere in the gas phase. Acetic acid is found in urban areas in the low ppbv range. A variety of dicarboxylic acids (two −COOH groups) have been observed in urban areas. These include oxalic, succinic, and malonic acids. They are produced from the photochemical oxidation of organic compounds in the atmosphere. Because of their low vapor pressures, dicarboxylic acids and other organic acids are predominantly found in or on aerosol particles. Organic acids are removed from the atmosphere by reaction with OH and by wet and dry deposition processes.

Esters are organic compounds produced by the reaction of alcohols with organic acids. They are widely used as solvents and industrial feedstock. Emissions to the atmosphere are relatively limited, and esters have negligible atmospheric significance.

2.3.4.3 Nonmethane Hydrocarbons

2.3.4.3.1 Sources and Emissions

Because of their role in atmospheric photochemistry, emissions of NMHCs or NMVOCs are regulated in many developed countries. Major anthropogenic sources include mobile and stationary source fuel usage and combustion; petroleum refining; petrochemical manufacturing; industrial, commercial, and individual solvent use; gas and oil production; and biomass burning. Emissions of NMHCs have been of particular concern in urban areas. A study for Beijing, China, conducted in March 2021 estimated that gasoline-related emissions contributed 22.35%, diesel vehicles 14.64%, LPG 26.97%, and other sources 36.04% (Li et al. 2023). On a global basis, it was reported in 2020 that global anthropogenic emissions of NMVOCs increased from about 115 Mt/year in 1970 to 165 Mt/year in 2012, even though both Europe and the United States had decreased by over 50% in the same period (Huang et al. 2017). Global emissions of NMVOCs increased by 40% between 1970 and 2017, driven primarily by the on-road transport, residential, energy, industrial, and solvent use sectors (McDuffie et al. 2020). On a global basis, source categories included industry (6%), transportation (18%), residential heating/cooking (19%), international shipping (2%), fuel production (36%), waste (2%), and solvents (17%).

Emissions associated with gasoline usage and petroleum-related industries include a variety of paraffinic, olefinic, and aromatic HCs; natural gas and light paraffins; and solvents, higher paraffinic, and aromatic HCs. Emissions from biogenic sources, which include trees and grasslands, and emissions from soils and oceans are approximately an order of magnitude higher on a global basis compared with anthropogenic emissions. Emissions from trees consist primarily of isoprene and monoterpenes, with some paraffins and olefins; from grasslands, light paraffins and higher HCs; from soils, mainly C_2H_6; and from oceans, light paraffins, olefins, and C_9 to C_{28} paraffins.

The US EPA estimated NMVOC emissions to be 1.28×10^7 t/year (1.16×10^7 metric t/year or 11.6 Mt/year) in 2024 as shown earlier in Table 2.2. Of these, 55% resulted from industrial and other processes; 15% from transportation sources (7.9% non-road/7.1% on-road); 34% from wildfires and miscellaneous sources; < 1% from electricity-producing fuel sources; and 5% from other fuel sources. A 63% decrease in total NMVOC emissions occurred between 1970 and 2024. On-road mobile source emissions decreased by 95%, reflecting the ability of technology to decrease individual source emissions in a situation where the total number of individual sources (i.e., on-road vehicles including electric vehicles) was growing. Unfortunately, this success was partially countered by specific industrial, non-road, and other source emissions that remained constant or increased during the same time period even though control technologies were implemented.

2.3.4.3.2 Identification

Many investigators have conducted studies to identify the many NMHC species and their derivatives that may be present in the atmosphere. Because of the complexities and expense associated with NMHC sampling and analysis, the atmosphere has not been well characterized, particularly in rural and remote areas. Many NMHC species produced as a result of photochemical reactions are short-lived or present in trace quantities and, in many cases, below the limit of detection of sampling systems utilized. Consequently, the presence of many NMHC compounds must be inferred from smog chamber studies and analysis of motor vehicle exhaust. In the latter case, more than 400 NMHCs and oxyhydrocarbon derivatives have been identified. Oxyhydrocarbons, which include aldehydes, ketones, organic acids, alcohols, ethers, esters, and phenol, have been reported to comprise 5% to 10% of the total NMHC concentration in auto exhaust. The aldehyde fraction is dominated by HCHO and CH_3CHO.

A variety of NMHC species have been found, or are believed to be present, in the atmosphere. These include straight, branch-chained, and cyclic paraffinic compounds; olefinic compounds with one or two double bonds; acetylene-type compounds; and C_6H_6 and its derivatives. Compounds with multiple C_6H_6 rings (PAHs) are found condensed on aerosol particles.

2.3.4.3.3 Concentrations

Most concentration data are based on measurement of total NMHCs, usually averaged from 6 a.m. to 9 a.m. (corresponding to early morning commuting traffic in urban areas). Concentrations are determined from flame ionization detector measurements and reported in CH_4 or C equivalents. The

TABLE 2.6
Major Hydrocarbon Compounds Measured in South Coast Air Basin (SoCAB) in Summer, CA, in 2021

Species	Concentration (ppbC)	Species	Concentration (ppbC)
i-Pentane	0.424	Isoprene	0.589
n-Butane	0.514	Benzene	0.446
n-Heptane	0.146	Toluene	0.381
n-Pentane	0.123	Ethylbenzene	0.06
Propene	0.575	*o*-Xylene	0.056
1-Butene+Isobutylene	0.273	*M*,p-Xylene	0.073
n-Hexane	1.593	*i*-Butane	0.301

Source: Wu, S. et al., AE, 2024, https://doi.org/10.1016/j.atmosenv.2024.120847.

air quality standard for this 3-hour period is 0.24 ppmv, with most urban concentrations at levels less than this. Substantially higher levels were reported in the Los Angeles Basin in the 1950s and 1960s, in the period prior to emission control requirements on motor vehicles.

Concentrations of individual NMHC species based on C equivalents (ppbc) are presented for 14 major HC species identified in samples collected over South Coast Air Basin, CA (Table 2.6). These include a large number of short-chained and long-chained paraffins, a relatively limited number of olefins, and a variety of aromatic HCs. Paracymene and isoprene are biogenic in origin. Acetylene is an indicator of motor vehicle emissions.

2.3.4.3.4 Sink Processes

The primary sink process for NMHCs and their oxygenated derivatives is the oxidative production of alkylperoxy radicals (ROO·) upon reaction with OH· or O_3. In the presence of NO, ROO· is converted to alkoxy radical (RO), which reacts with O_2 to produce aldehydes or, in the case of longer-chained NMHCs, butane, aldehydes, and ketones. Oxidation of C_2H_6, a relatively unreactive substance, is summarized in the following series of equations:

$$C_2H_6 + OH \cdot \rightarrow H_2O + C_2H_5 \cdot \tag{2.34}$$

$$C_2H_5 \cdot + O_2 \rightarrow C_2H_5OO \cdot \tag{2.35}$$

$$C_2H_5OO \cdot + NO \rightarrow C_2H_5O \cdot + NO_2 \tag{2.36}$$

$$4C_2H_5O \cdot O_2 \rightarrow 4CH_3CHO + 2H_2O \tag{2.37}$$

Ethane oxidation produces CH_3CHO, which, like other aldehydes, reacts more readily with OH· than its NMHC precursor. As a consequence, CH_3CHO reacts with OH· and, by a series of reactions, produces HCHO, which undergoes further oxidation or photodecomposition.

Formaldehyde, CH_3CHO, and acetone can photodecompose upon the absorption of ultraviolet (UV) light in the wavelength range of 330 to 350 nm. Photodecomposition of HCHO proceeds via either of two pathways. In both cases, photodecomposition produces CO. These reactions are summarized as follows:

$$HCHO + hv \rightarrow HCO + H \tag{2.38}$$

$$H + O_2 + M \rightarrow HO_2 \cdot + M \tag{2.39}$$

$$HCO + O_2 \rightarrow HO_2 \cdot + CO \quad (2.40)$$

$$HO_2 \cdot + NO \rightarrow NO_2 + OH \cdot \quad (2.41)$$

In the first pathway, photodecomposition of HCHO leads to the production of $HO_2\cdot$, which, upon reaction with NO, generates OH·, which then becomes available for NMHC oxidation. In another pathway (Equation 2.42), H + HCO are rapidly converted to H_2CO:

$$H + HCO \rightarrow H_2 + CO \quad (2.42)$$

Aldehydes and ketones produced from NMHC oxidation are removed from the atmosphere by scavenging processes involving dry and wet deposition. Oxidation of longer-chain aliphatic and aromatic HCs may lead to the formation of condensable products such as dicarboxylic acids, which enter the particulate phase and are removed by wet and dry deposition processes.

All NMHCs react relatively rapidly with OH·. However, NMHCs in the atmosphere vary considerably in their OH· reactivity, with unsaturated HCs, including olefinic and aromatic HCs, being the most reactive. Many unsaturated NMHCs react with O_3 at rates that are competitive with OH·. As indicated previously, the lifetime of olefins and biogenically produced terpenes is on the order of hours, whereas for less reactive paraffinic species, the atmospheric lifetime may be days. Because the most reactive compounds are removed at a faster rate, the abundance of NMHCs changes in the direction of less reactive species such as C_2H_6 in urban areas as well as those downwind.

2.3.4.3.5 Photochemical Precursors

Oxidation products produced in NMHC sink processes, for example, ROO·, RO·, $HO_2\cdot$, and CO, serve as major reactants in the production of photochemical smog and associated elevated tropospheric levels of O_3 and other oxidants (the formation of which will be discussed in the following section). In addition to oxidants, photochemical reactions produce a large variety of species that may comprise 95% of atmospheric NMHCs during smog episodes.

2.3.4.4 Methane

Methane has low reactivity with OH· relative to olefins, aromatics, and even other paraffins, and as a consequence, it has limited significance in the photochemical reactions responsible for producing elevated O_3 levels in urban areas. Methane and other low-reactivity paraffins such as C_2H_6 become important when polluted air masses (depleted of reactive NMHCs) travel downwind of urban sources. In addition, methane is a thermal absorber and, as a result of its increasing atmospheric levels, an important greenhouse gas.

Methane concentrations based on the analysis of ice cores and air trapped in Antarctic and Greenland ice show that the global concentration varied cyclically between 350 to as high as 800 ppbv over 800,000 years in cycles that appear to the eye to range from 25,000 to more than 100,000 years. However, the value has been increasing steadily above the historic cycles since the beginning of the industrial era. Figure 2.10 shows CH_4 concentrations from 1980 to the present (2025) as reported by NOAA and now updated monthly at esrl.noaa.gov/gmd/ccgg/trends/. In the 1980s to mid-1990s, the concentration was increasing at a rate of about 11.5 ppbv/year with a flattening during the period of about 1997–2007. In recent years, the rate of increase has accelerated notably. Between 2019 and 2023, the average annual rise in global methane concentrations was approximately 13.2 ppbv per year—a marked jump compared to earlier periods—compared with about 5.1 ppbv per year during 2009–2013 and 9.1 ppbv per year during 2014–2018. Specifically, the growth in 2022 reached 13.2 ± 0.8 ppb, and in 2023, it was 9.9 ± 0.6 ppb. Additionally, data from May 2025 places the global monthly mean at about 1,933.5 ppb—up from 1,925.7 ppb in May 2024.

While CH_4's atmospheric concentration (~1.93 ppm or ~1,933 ppb) remains much lower than that of CO_2 (~410 ppm), its significant global warming potential (GWP) makes it highly consequential.

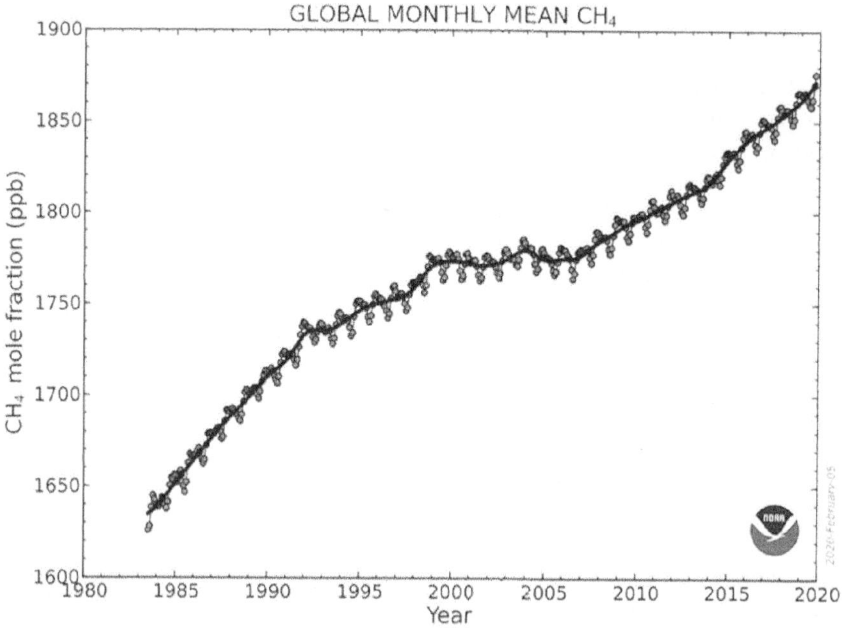

FIGURE 2.10 Recent CH_4 concentrations and trends. (From National Oceanic and Atmospheric Administration Global Greenhouse Gas Network, www.esrl.noaa.gov/gmd/ccgg/trends/, 2025.)

Over a 100-year time horizon, methane's GWP is around 28–30, meaning that each part per million of CH_4 has roughly the warming power of 28–30 ppmv of CO_2.

Emission of CH_4 to the atmosphere occurs from natural sources, natural sources influenced by human activity, and anthropogenic sources. Biogenic emissions result from anaerobic decomposition of organic matter in the sediments of swamps, lakes, rice paddies, and sewage wastes and from the digestion of cellulose by livestock (and other ruminants) and termites. Methane emissions also occur from coal and lignite mining, oil and natural gas extraction, petroleum refining, leakage from natural gas transmission lines, and automobile exhaust; CH_4 accounts for 15% of all motor vehicle exhaust HCs.

The major sink process for CH_4 is reaction with OH:

$$OH\cdot + CH_4 \rightarrow CH_3\cdot + H_2O \tag{2.43}$$

Subsequent reactions produce HCHO, CO, and ultimately, CO_2. The stratosphere is another important sink, with CH_4 photodecomposition being the major source of stratospheric H_2O vapor. The OH sink removes approximately 10 times more CH_4 than the stratosphere. One of the major factors contributing to increased atmospheric CH_4 levels may be CO depletion of OH· (Equation 2.3). The atmospheric lifetime of CH_4 is estimated to be 10 years.

2.3.4.5 Halogenated Hydrocarbons

A variety of halogenated HCs are emitted to the atmosphere from natural and anthropogenic sources. These include both volatile and semivolatile substances that contain one or more atoms of chlorine (Cl), bromine (Br), and fluorine (F). Due to their persistence, halogenated compounds are relatively unique atmospheric contaminants. They include volatile substances such as methyl chloride (CH_3Cl), methyl bromide (CH_3Br), methyl chloroform (CH_3CCl_3), trichloroethylene (CH_2CCl_3), perchloroethylene (C_2Cl_4), and carbon tetrachloride (CCl_4). Methyl chloride and CH_3Br are produced naturally;

other volatile halogenated HCs have been used industrially and commercially. A variety of semivolatile halogenated HCs are land, air, water, and biological contaminants. These include chlorinated HCs such as DDT, chlordane, dieldrin, aldrin, and others; polychlorinated biphenyls (PCBs) used as solvents and electrical transformer insulators; polybrominated biphenyls (PBBs) used as fire retardants; and dioxins and furans that are by-products of pesticide manufacture, bleaching of wood fiber to make paper, and incineration of paper and plastics.

Halogenated HCs that include Cl and F in their chemical structure are called chlorofluorocarbons (CFCs). The most commonly used CFCs have been trichlorofluoromethane (CCl_3F), difluorodichloromethane (CF_2Cl_2), and trichlorotrifluoroethane ($C_2Cl_3F_3$). These are generally described as CFC-11, CFC-12, and CFC-13, respectively. As a result of their important industrial and commercial properties (including low reactivity, low toxicity, thermal absorption, and solvent properties), CFCs have been used as aerosol propellants, refrigerants, degreasing solvents, and foaming agents.

Because of their low chemical reactivity, halogenated HCs break down slowly and, consequently, are characterized by long atmospheric lifetimes, estimated at approximately 1.5 years for CH_3Cl, 6 years for CH_3CCl_3, approximately 40 years for CCl_4, approximately 50 years for $CFCl_3$, and approximately 102 years for CF_2Cl_2.

Although halogenated compounds such as CH_3Cl and CH_3CCl_3 have tropospheric sinks, CFCs do not. As a consequence, concentrations in both the troposphere and stratosphere had been increasing until the early 1990s. Tropospheric concentrations of CFCs and other halocarbons are reported in the pptv range. Average approximate global concentrations reported for 1995 were CF_2Cl_2, 503 pptv; CH_3Cl, 600 pptv; CCl_3F, 268 pptv; CH_3CCl_3, 160 pptv; CCl_4, 132 pptv; $C_2Cl_3F_3$, 82 pptv; and CH_3Br, 12 pptv. Concentrations of CFCs and other halogenated HCs vary spatially over the surface of the Earth, with the highest concentrations observed in source regions over the northern hemisphere.

Total Cl concentrations in the troposphere peaked around 1994, with an average concentration of approximately 3.7 ppbv, slowly decreasing thereafter. Tropospheric Br concentrations, however, were still increasing. Both stratospheric Cl and Br peaked around the year 2000.

Brominated HCs are of environmental interest because Br atoms (as do Cl atoms) catalytically destroy O_3. Methyl bromide is the most abundant Br compound in the atmosphere, with average atmospheric concentrations in the range of 9 to 13 pptv. Its major sources include marine phytoplankton, agricultural pesticide use, and biomass burning. Other atmospheric Br sources include CF_3Br (Halon 1301) and CF_2BrCl (Halon 1211), which are used as fire control agents.

Various halocarbons can be characterized by their O_3-depleting potential. These include $CFCl_3$ (1.0), CF_2Cl_2 (0.82), $C_2Cl_3F_3$ (0.90), CCl_4 (1.2), CH_3CCl_3 (0.12), CH_3Br (0.64), CF_3Br (12.0), and CH_3Cl (0.02). Thus, CF_2Cl_2 is approximately 18% less effective in destroying O_3 than $CFCl_3$, and CCl_4 is 20% more effective than $CFCl_3$.

2.3.5 Photochemical Oxidants

Photochemical oxidants are produced in the atmosphere as a consequence of chemical reactions involving sunlight, NO_x, O_2, and a variety of NMHCs. Photochemical oxidants produced from these reactions include O_3, NO_2, PAN, odd hydrogen compounds, and RO_2. Because of the significant environmental effects of elevated levels associated with human activities, tropospheric O_3 has received major scientific and regulatory attention.

As indicated in Chapter 1, O_3 is a normal constituent of the atmosphere, with peak concentrations in the middle stratosphere. As a consequence of anthropogenic activities, O_3 represents two major and distinct environmental concerns. In the stratosphere, it is a depletion issue (Chapter 4); in the troposphere, it is one of the levels elevated significantly above background. Tropospheric O_3 production is the focus of this discussion.

Ozone is formed in the atmosphere when molecular O_2 reacts with $O(^3P)$:

$$O_2 + O(^3P) + M \rightarrow O_3 + M \tag{2.44}$$

In the troposphere, photodissociation of NO_2 at wavelengths of 280 to 430 nm is the primary source of $O(^3P)$:

$$NO_2 + hv \rightarrow NO + O(^3P) \tag{2.45}$$

The reaction of $O(^3P)$ with O_2 produces O_3, which reacts rapidly with NO to regenerate NO_2:

$$NO + O_3 \rightarrow NO_2 + O_2 \tag{2.46}$$

This O_3-producing process is represented diagrammatically in Figure 2.11.

Reactions shown in Equations 2.44 through 2.46 proceed rapidly, producing a small steady-state concentration of approximately 20 ppbv under solar noon conditions in midlatitudes at an atmospheric NO_2/NO concentration ratio of 1.

Ozone concentrations in both urban and nonurban atmospheres are often higher than those associated with NO_2 photolysis. Elevated O_3 levels occur as a result of chemical reactions that convert NO to NO_2 without consuming O_3. In very polluted and even lightly polluted atmospheres, shifts in O_3 chemistry occur in the presence of $RO_2\cdot$, produced by the oxidation of NMHCs (Equations 2.34 and 2.35). Peroxy radicals then react with NO:

$$RO_2 \cdot NO \rightarrow NO_2 + RO\cdot \tag{2.47}$$

$$NO_2 + hv \rightarrow NO + O(^3P) \tag{2.48}$$

$$O(^3P) + O_2 + M \rightarrow O_3 + M \tag{2.49}$$

$$\text{Net}: RO_2\bullet + O_2 + hv \rightarrow RO\bullet + O_3 \tag{2.50}$$

These reactions are represented diagrammatically in Figure 2.12.

The rate of O_3 formation is closely related to the concentration of $RO_2\cdot$. Peroxy radicals are produced when $OH\cdot$ and other odd hydrogen species react with vapor-phase NMHCs. Odd hydrogen species such as $OH\cdot$ are produced by reactions involving the photodissociation of O_3, aldehydes, and HNO_2. Ozone photodissociation and the formation of $OH\cdot$ are shown in Equations 2.51 and 2.52:

$$O_3 + hv \rightarrow O(^1D) + O_2 \tag{2.51}$$

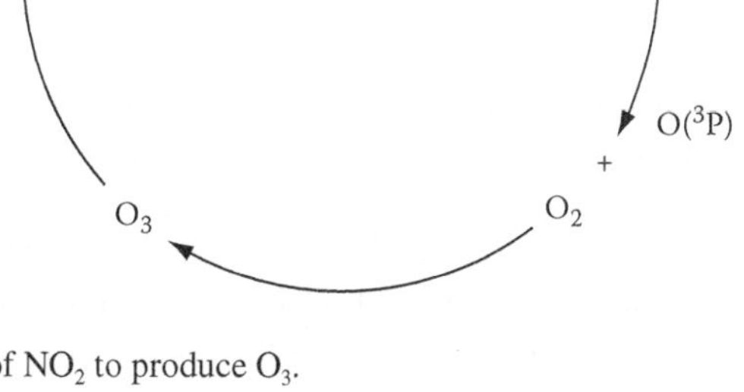

FIGURE 2.11 Photolysis of NO_2 to produce O_3.

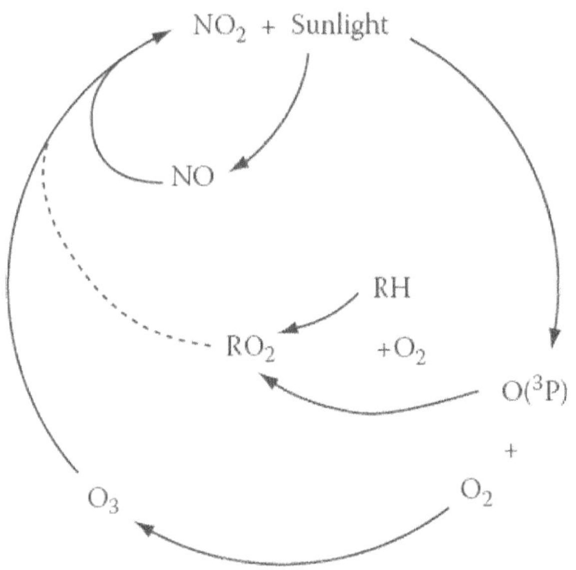

FIGURE 2.12 Photochemical reactions between NO_x and hydrocarbon radicals resulting in the production of O_3.

$$O(^1D) + H_2O \rightarrow 2OH\cdot \quad (2.52)$$

In very polluted atmospheres, O_3 concentrations at any time and place depend on the intensity of sunlight, NO_2/NO ratios, reactive NMHCs and their concentrations, and other pollutants, such as aldehydes and CO, which react photochemically to produce $RO_2\cdot$. The change in NO_2/NO ratios caused by atmospheric reactions involving $RO_2\cdot$ may result in elevated tropospheric O_3 levels.

2.3.5.1 Tropospheric Ozone Concentrations

In the relatively clean atmospheres of remote sites, ground-level O_3 concentrations are estimated to range from 20 to 50 ppbv during the warmer months of the year. Major sources of background O_3 are photochemical processes and the movement of stratospheric O_3 in the region of so-called tropopause folds into the lower atmosphere, particularly during the spring months. There is general agreement among atmospheric scientists that the former is a major source of background O_3. Depending on location, O_3 associated with processes uninfluenced by human activities may account for 20% to 50% of the observed rural monthly average concentrations, particularly during warmer months.

Ozone concentrations over large, economically developed landmasses such as North America, Europe, eastern Australia, and Southeast Asia are often significantly higher than those found in remote and rural areas. These relatively high O_3 levels are associated with anthropogenic emissions of NMHCs and NO_x, biogenic emissions of reactive NMHCs, and subsequent photochemical reactions summarized in Equations 2.47 through 2.50.

Our understanding of atmospheric chemistry associated with the formation of elevated tropospheric O_3 levels has evolved considerably from the 1950s and early 1960s, when it was understood to be an urban problem associated with cities such as Los Angeles. At that time, peak levels were observed at solar noon, with a subsequent rapid destruction associated with NO_x-scavenging processes (Figure 2.8). In the 1970s, elevated O_3 levels were observed in many rural areas far from the urban centers that were the focus of O_3 concerns. Rural O_3 levels elevated above background were observed to be associated with (1) long-range transport of O_3 isolated from normal scavenging processes by ground-based inversions, and (2) transport of O_3 precursors, such as NO_x, and less-reactive NMHCs.

In the mountain areas of eastern United States, elevated O_3 is observed to both be persistent and occur at night, contradicting the previous paradigm of O_3 production and destruction that was characteristic of urban areas with large motor vehicle populations. This is primarily a result of transport of ozone aloft which impacts directly on the higher altitude sites throughout the entire day and night. The ground-level ozone in the urban areas (usually at lower altitude) is scavenged throughout the night within the nighttime stable boundary layer where it reacts with surfaces (trees, buildings, etc.), and it is not replenished due to transport as the air upwind is undergoing the same scavenging process. The urban area typically experiences an increase in the ozone the next morning when the nighttime inversion breaks up due to thermal heating by the sun and brings ozone downward from above as well as from photochemically produced ozone that begins to occur.

In the United States, noontime peak concentrations were historically at their highest in the precursor-rich and sunlight-rich Los Angeles Basin, with 1-hour average concentrations ranging as high as 0.40 ppmv during severe smog episodes and concentrations of 0.15 to 0.30 ppmv not uncommonly reported. Daily maximum 8-hour average concentrations are summarized for selected US cities in Table 2.7 for the year 2012. The ozone levels of most cities were above the then-ambient air quality standard of 0.075 ppmv (maximum 8-hour average). The number of US cities exceeding the 0.075 ppmv air quality standard varied from year to year, with a hundred cities or more exceeding the standard in very hot summers. The standard was lowered to 0.070 ppmv (annual fourth-highest daily maximum 8-hour concentration—averaged over three years) in 2015. Figure 2.13 shows the progress toward meeting the most recent standard based on 134 sites. As of 2023, approximately 240 US counties and partial counties (including parishes and tribal areas) had not yet met the standard.

Although elevated urban O_3 levels have been a concern in the United States for decades, the problem is not limited to the United States. In Mexico City, hourly O_3 still frequently breaches the national 95 ppb standard. Recent analyses show that between 2020 and 2022 the highest 1-hour concentrations ranged from 0.152 to 0.158 ppmv, and days with 8-hour means above 0.110 ppmv represented 7%–10% of the year (Cortez-Huerta et al., 2025). Trend analyses of the full SIMAT

TABLE 2.7
Ozone Levels Measured in Selected US Metropolitan Areas in 2024

Metropolitan Area	O_3 (ppmv)[a]
Bakersfield, CA	0.088
Baltimore, MD	0.073
Birmingham, AL	0.068
Bridgeport, CT	0.082
Dallas, TX	0.089
Los Angeles, CA	0.107
Louisville, KY	0.080
New London, CT	0.073
Orange County, CA	0.080
Philadelphia, PA	0.075
Phoenix, AZ	0.083
Riverside, CA	0.099
St. Louis, MO	0.081

Source: US EPA, Air Quality Statistics Report, 2024.
[a] *Maximum 8-hour average ozone concentration.*

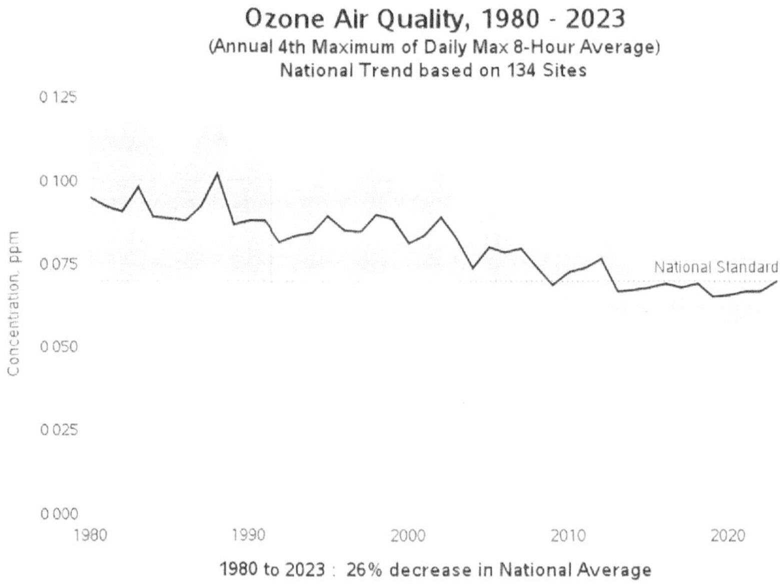

FIGURE 2.13 Ozone air quality (annual fourth-highest daily maximum 8-hour concentrations, averaged over three years) observed at 134 sites in the United States from 1980 to 2023. (From US EPA, https://www.epa.gov/air-trends/ozone-trends.)

network indicate a small but significant upward trend of 0.39 ppb year^{-1} since 2010, with daily MDA8 peaks reaching 0.180 ppmv in 2022 (Velasco et al., 2025). Across Northern Europe, recent heat-wave events now dominate exceedances. During the July 2022 UK heat dome, near-surface O_3 in central London reached 0.102 ppmv (1-hour) and sustained 0.050–0.075 ppmv through the afternoon, causing 18 exceedance days of the EU 60 ppbv MDA8 target (Ryan et al., 2023). Germany's 2022 national assessment recorded 1,257 hours above the 0.090 ppmv information threshold and a national maximum of 0.107 ppmv (German Environment Agency, 2023).

2.3.5.2 Ozone Sinks

The two major sink processes for O_3 are surface destruction or deposition and photochemical reactions. Surface deposition includes reaction with plants, bare land, ice and snow, and human-made structures. Deposition of O_3 is at its greatest over forests and croplands during daylight hours.

The primary sink for O_3 is its photodissociation upon the absorption of UV light and subsequent formation of OH· (Equations 2.51 and 2.52). In polluted atmospheres, O_3 reacts with NO to produce $NO_2 + O_2$ (Equation 2.46). During nighttime hours, it reacts with NO_2 to produce HNO_3 (Equations 2.24–2.26).

2.4 ATMOSPHERIC PARTICLES

Atmospheric particles include both solid-phase and liquid-phase substances that vary in size and density. As such, larger and denser particles settle out in a matter of minutes, whereas smaller, less dense particles remain suspended in the atmosphere for periods of days to weeks. Individually, particles differ in size, mass, density, morphology, chemical composition, and various chemical and physical properties.

Particles may be emitted from natural and anthropogenic sources. They may also be produced secondarily as a result of chemical reactions involving gas-phase pollutants.

Of special concern are particles that are less than 20 μm in aerodynamic diameter because they may remain suspended in the atmosphere and, depending on their size, settle out slowly. Such suspended particles characterize atmospheric aerosol.

The direct emission of particles from anthropogenic sources (and, to a lesser degree, natural sources) and the formation of secondary particles as a result of atmospheric chemistry pose major environmental concerns. These include (1) toxic effects of particle exposure on humans; (2) the reduction of visibility due to the light-scattering ability of particles; (3) local, regional, and potential global climatic effects associated with the ability of particles to scatter light back into space and to absorb incoming solar radiation and outgoing thermal radiation; and (4) nuisance effects associated with the soiling potential of deposited particles.

2.4.1 Classes and Sources

Atmospheric particles can be described as being primary or secondary based on their origin and the processes by which they form. Primary particles are emitted directly into the atmosphere from a variety of natural and anthropogenic sources. In the former case, these include volcanoes, forest fires, ocean spray, biological sources (mold, pollen, bacteria, and plant parts), and meteoric debris; the latter includes transportation, fuel combustion in stationary sources, a variety of industrial processes, solid waste disposal, and miscellaneous sources such as agricultural activities and fugitive emissions from roadways.

Secondary particles are produced from both naturally emitted gas-phase substances and anthropogenic sources as a result of chemical processes involving gases, aerosol particles, and H_2O and H_2O vapor. Secondary aerosol particles include sulfates, nitrates, and oxyhydrocarbons produced by direct, catalytic, and photochemical oxidation of S, N, and volatile NMHCs. Approximately 30% of atmospheric aerosol in the particle size range of less than 5 μm may be produced as a consequence of chemical reactions involving gases and vapors from natural sources.

2.4.2 Size

Particles that comprise atmospheric aerosol range in size over many orders of magnitude, from less than 0.005 nm to several hundred micrometers. The size ranges for a variety of particles, as well as their physical properties, are summarized in Figure 2.14.

Particle sizes (Figure 2.14) are based on what are described as aerodynamic equivalent diameters (AEDs). The AED refers to a spherical particle of unit density (1 g/cm^3) that settles at a standard velocity. It can also be expressed as the Stokes equivalent diameter (defined as the diameter of a sphere that has the same density and settling velocity as the particle). Stokes diameters are based on settling velocities, whereas the AED includes a standardized density of 1.

Size is an important property of particles because it affects their atmospheric history, which includes growth, accumulation of substances on their surface, concentrations, atmospheric lifetime, light scattering, and deposition in the human respiratory tract. Most atmospheric particles are very small (< 0.1 μm), whereas most aerosol mass is associated with particles more than 0.1 μm.

The size distribution of aerosol particles can be expressed in several different ways. These include particle number, mass, volume, and surface area in different size ranges (plotted logarithmically). Important characteristics of urban aerosols emerge when urban particle data are plotted as number, volume, surface, or mass distributions. In particle number plots, a large peak is observed at approximately 0.02 μm and a slight peak around 0.1 μm. Particle concentration distributions based on volume show two peaks, one in the 0.1 to 1.0 μm range and the second in the 1 to 2 μm range. In surface area distribution plots, a major peak is observed near 0.1 μm, with smaller peaks between 0.01 and 0.1 μm, and between 1 and 10 μm. These particle distribution plots indicate that urban aerosol concentrations are multimodal. Until recently, it was widely accepted that atmospheric aerosols consisted of three distinct particle modes. With increasing scientific attention being given to

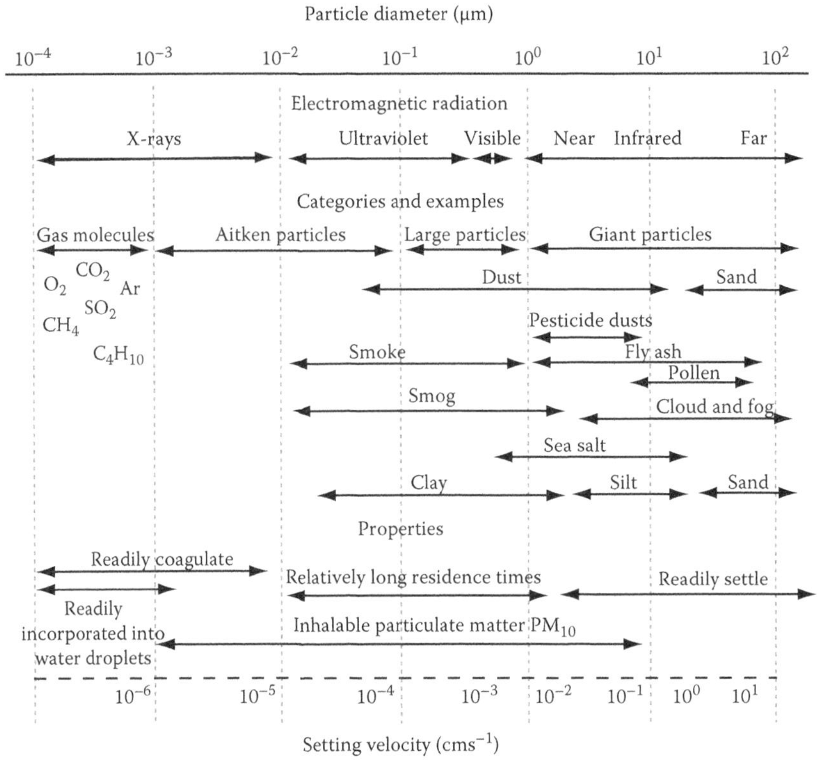

FIGURE 2.14 Size distribution and physical properties of common aerosol particles. (From Van Loon, G.W., and Duffy, S.J., *Environmental Chemistry: A Global Perspective,* Oxford University Press, Oxford, 2000. With permission.)

ultrafine particles (UFPs; diameters of < 0.01 μm), atmospheric particles are now being described in terms of the four particle modes illustrated in Figure 2.15. Although no vertical axis label is shown, it could represent particle distribution based on number, mass, surface area, or volume distribution. Figure 2.15 also shows major sources and particle removal processes for each mode. The UFP mode and the two peaks sometimes observed in the accumulation mode are illustrated by dashed lines.

As can be seen in Figure 2.15, atmospheric aerosols are divided into two major fractions based on their size, origin, and physical properties. Coarse particles have diameters of more than 2.5 μm, whereas fine particles are less than 2.5 μm. These particle modes or fractions are distinct relative to (1) their origin, (2) how they are affected by atmospheric processes, (3) chemical composition, (4) removal processes, (5) optical properties and light-scattering ability, and (6) deposition in the human respiratory tract.

The fine particle fraction includes three of the four modes illustrated; these are the ultrafine, Aitken nuclei, and accumulation modes. The particle diameter ranges of these modes are less than 0.01, 0.01 to 0.08, and approximately 0.10 to 1.2 μm, respectively. Most atmospheric particles are in the fine fraction. It comprises one-third of the particulate mass found in nonurban areas and one-half in urban areas.

Identification and definition of the UFP mode has been a relatively new development. It is sometimes referred to as the nucleation mode. UFPs are generated by poorly understood gas-to-particle conversion processes. Although they contribute little to particulate mass, they are often present in large numbers. Because of their human respiratory tract depositional and toxicological properties, UFPs are increasingly becoming a major public health concern (see Chapter 5).

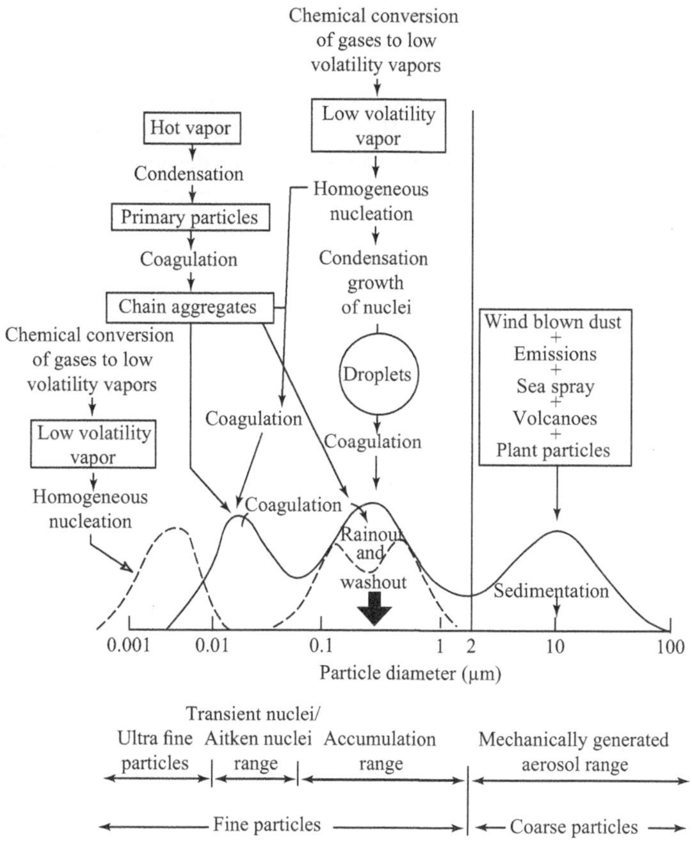

FIGURE 2.15 Particle size characteristics of atmospheric aerosols. (From Finlayson-Pitts, B.J., and Pitts, J.N., Jr., *Chemistry of the Upper and Lower Atmosphere: Theory, Experiments and Applications*, Academic Press, Orlando, FL, 2000. With permission.)

Aitken nuclei are produced in the ambient atmosphere by gas-to-particle conversion, as well as by combustion processes in which hot, supersaturated vapors are formed and undergo condensation. They also serve as nuclei for the condensation of low vapor pressure gas-phase substances. Because of their small size, they account for only a minor percentage of the mass of airborne particles. Their numbers rapidly diminish as they coagulate with each other to form larger particles.

Accumulation mode particles form from the coagulation of Aitken nuclei particles and condensation of vapors on existing particles. Their rate of growth depends on the number of particles present and their velocity and surface area. Because of growth-limiting physical factors, they do not grow into coarse particles. The term *accumulation mode* is used because removal processes are inefficient in this size range, with particles "accumulating" as a result. Although representing only a small fraction of particle numbers (~5%), they are a significant portion of aerosol mass (~50%). The accumulation mode seems to be bimodal in relatively polluted urban areas, with peaks at approximately 0.2 and 0.7 μm.

Coarse particles are usually produced by mechanical processes (fragmentation of matter and atomization of liquids). Most coarse particles are in the 2.5 to 10 μm size range, but they may be as large as 30 μm or more. Both fine and coarse particles up to a size range of 10 μm tend to remain suspended in the atmosphere for varying periods and are therefore described as suspended particles. Particles with diameters of more than 100 μm tend to settle out rapidly and are described as settleable particles.

In urban areas, particulate aerosols are generally evenly divided (on a mass basis) between fine and coarse particles. Under calm atmospheric conditions, the fine particle mass begins to exceed the coarse mass with increasing downwind distance from urban areas. Because of low settling velocities, fine particles are transported relatively long distances from source regions.

Particle size classes have been identified that reflect particle penetration and deposition in the human respiratory system, and sampling procedures have been designed to reflect the health-affecting potential of atmospheric aerosols. Size ranges of health significance are described and illustrated in Chapter 5.

A variety of sampling devices have been used to measure particle concentrations to determine compliance with air quality standards. These have included the high-volume sampler (Hi-Vol) and PM_{10} and $PM_{2.5}$ samplers. These reflect size-selective sampling considerations. They include the collection of particles below or within a specified aerodynamic size range defined by the upper 50% cutpoint size (where 50% of the particles at the specified diameter are collected). The upper cutoff size for the Hi-Vol sampler (used for regulatory purposes until 1987) varied from 25 to 40 μm, depending on the wind speed and direction. Hi-Vol concentrations were reported as total suspended particulates (TSPs). PM_{10} samplers have an upper 50% cutoff diameter of 10 μm, $PM_{2.5}$ samplers of 2.5 μm. PM_{10} includes most particles less than 10 μm and some particles greater than 10 μm. $PM_{2.5}$ includes most particles less than 2.5 μm and some particles greater than 2.5 μm.

PM_{10} samplers were designed to collect particles that can enter the thoracic region of the human respiratory system (see Chapter 5); $PM_{2.5}$ samplers were designed to collect particles that have a high probability of being deposited in lung tissue. However, because $PM_{2.5}$ samples include a fraction of coarse mode particles, $PM_{2.5}$ is equivalent to neither the fine particle mode nor the respirable particle mode described in Chapter 5.

2.4.3 Particle Shape

Particles vary in shape from simple spheres or crystalline cubes to the many that are irregular in their morphology. Some are flat and layered; others are globular. Shape is often determined by a particle's chemical composition and the processes by which it was formed.

Particles occur singly or form aggregates that may appear as single compact particles. Carbon particles produced in combustion processes often form aggregates with irregular shapes. Some are characterized as small clumps, whereas others are chain-like. Particle surfaces vary as well. Some are smooth, whereas others are uneven and porous.

2.4.4 Chemical Composition

Particles in atmospheric aerosol vary in their chemical composition. This reflects their origin, how they were initially produced, and their subsequent atmospheric history. Because of the large number of sources of primary particles, the formation of secondary particles by atmospheric chemistry, growth of some atmospheric particles, and sorption of gas-phase substances by others, particles suspended in the atmosphere contain hundreds of different chemical species. Notable because of their relatively high concentrations in PM samples are the following chemical groups: sulfates, nitrates, EC, condensed organic compounds, organic carbons (OCs), and metals.

Chemical composition varies as a function of particle size. Chemical substances are not equally distributed across the particle size spectrum. They tend to be in specific size ranges characteristic of sources and particle development. The smallest UFPs are produced by homogeneous nucleation and, as a result, contain secondary species such as sulfate and organic compounds. Those in the Aitken range are produced by combustion, coagulation of smaller particles, and condensation of products of gas-phase reactions. Such particles, as well as accumulation mode particles, tend to contain EC, metals characteristic of combustion, sulfates, nitrates, and polar organic compounds. Coarse particles tend to contain chemical species characteristic of materials that were mechanically fragmented or atomized. These include dusts from minerals processing and soil erosion. The latter contain common crustal (derived from the Earth's crust) elements such as O, silicon (Si), aluminum (Al), magnesium (Mg), calcium (Ca), and others. Nitrates and sulfates in coarse particles come primarily from the reaction of these gas-phase substances with preexisting coarse particles or their deposition.

Sulfate compounds comprise a major portion of the fine particle mass collected in ambient air in the United States: 36% of the fine particle mass at mixed rural, suburban, and urban sites in the eastern United States, and 44% at rural sites in the eastern United States (Table 2.8). At rural western US sites, particulate sulfate levels are considerably lower, representing about one-fourth of the composition of the relatively low-mass $PM_{2.5}$ concentrations.

Sulfates in $PM_{2.5}$ samples include ammonium sulfate ($(NH_4)_2SO_4$), ammonium acid sulfate (NH_4HSO_4), H_2SO_4, a variety of metal salts, and calcium sulfate ($CaSO_4$). Of these, $(NH_4)_2SO_4$ is the most common and abundant sulfate species present. Sulfuric acid may also comprise a significant fraction of atmospheric sulfate. In the global context, sulfate aerosol concentrations are in the range of 1 to 2 μg/m^3 in remote areas, more than 10 μg/m^3 in nonurban continental areas, and less than 10 μg/m^3 in areas under urban and anthropogenic influence.

EC, on average, represents a much smaller fraction of the United States' $PM_{2.5}$ mass, with 4%–6% average composition at rural and mixed eastern sites, increasing to as high as 9% in mixed western sites. The major source of EC is fuel combustion. Particles containing EC are relatively small, with average aerodynamic diameters of 0.01 μm. EC can, however, form larger aggregates. Concentrations of EC, as a percentage of collected $PM_{2.5}$, as high as 25% have been reported for Phoenix, AZ.

The OC fraction comprises more than 40% of $PM_{2.5}$ mass in the United States, with the highest percentage levels in western states at all site locations (Table 2.8). In cities such as Los Angeles, the OC fraction may be as high as 45%. $PM_{2.5}$ OC is derived from biogenic as well as anthropogenic sources. The composition of this fraction is complex even in remote areas. Nonurban aerosols contain a wide variety of organic compounds, including paraffins, olefins, aromatic HCs, fatty acids, alcohols, and organic bases. Paraffins in the C_{15} to C_{35} range are common.

Organic compounds found in or on particles associated with human activities and in "aged" air masses are more complex than in remote regions because there are more opportunities for oxidation reactions to take place. As a consequence, such particles contain, in addition to organic compounds from biogenic emissions, complex compounds emitted directly from anthropogenic sources or formed as a result of atmospheric chemical reactions. For example, direct emissions from automobiles

TABLE 2.8
Chemical Composition of $PM_{2.5}$ Particles

	Rural Measurements[a]		Supersites[b]	
	Eastern US	Western US	Eastern US	Western US
Constituent	% Contribution		% Contribution	
SO_4	44	25	36	15
EC	4	4	6	9
OC	31	41	26	44
NO_3	12	9	14	18
Crustal	9	18	4	8
$PM_{2.5}$ concentration μg/m^3)	7.9	3.8	14.2	11.4

[a] *IMPROVE:* the Interagency Monitoring of Protected Visual Environments network.
[b] *STN:* US EPA's Speciated Trends Network.

and heavy-duty diesel trucks are known to be sources of particulate-phase *n*-alkanes, *n*-alkanoic acids, aromatic aldehydes and acids, PAHs and their oxidized derivatives, steranes, pentacyclictriterpones, and azanaphthalenes. In secondary aerosols, difunction substitute derivatives of the form $X–(CH_2)_n–X$ (with n = 1–5) have been reported in the Los Angeles area. These include dicarboxylic acids, dialdehydes, glycols, esters, and organic nitrite and nitrate compounds. In urban areas of central Japan, dicarboxylic acids are reported to comprise 30% to 50% of total OCs in PM samples.

Inorganic nitrates are a major component of the fine particle aerosol mass, varying from 9% to 12% of $PM_{2.5}$ collected at rural sites to higher percentages of 14% to 18% at mixed, western US sites. Under severe smog conditions, inorganic nitrate levels as high as 22.6 μg/m^3 have been reported in Los Angeles, accounting for 25% of $PM_{2.5}$ mass. Ammonium nitrate is the major nitrate species in urban aerosol.

The crustal fraction represents mineral fragments from the Earth's crust. These become airborne as a result of soil erosion, mineral weathering processes, and dust storms. The crustal component of atmospheric aerosol is considerably higher in both rural and mixed sites in western states in the United States than in eastern states.

Metals as a group are also major constituents of atmospheric aerosol. Indeed, atmospheric aerosol is a major transport mechanism for a variety of metals around the planet, in some cases (e.g., mercury, Hg) resulting in ecosystem effects on a global scale.

Metals in crustal dust include iron (Fe), manganese (Mn), zinc (Zn), Pb, vanadium (V), chromium (Cr), nickel (Ni), copper (Cu), cobalt (Co), Hg, and cadmium (Cd). Biogenic particle emissions include low concentrations of Fe, Mn, Zn, Pb, V, Cr, Ni, Cu, Co, Hg, Cd, arsenic (Ar), and antimony (Sb).

Metals are commonly associated with aerosol particles of anthropogenic origin. Major sources include power plants, smelters, incinerators, cement kilns, home heating systems, and motor vehicles. The major source of metals from coal- and oil-fueled power plants is fly ash, the inorganic residue remaining after combustion. Fly ash is characterized by high concentrations of Fe, Al, silicon oxides, and carbonaceous compounds that have undergone various degrees of oxidation. In addition, fly ash from coal-fired power plants contains high concentrations of Mn, Zn, Pb, V, Cr, Ni, Cu, Co, As, and Sb. Oil fly ash can contain not only significant quantities of V and Ni but also abundant quantities of Mn, Zn, Pb, Cr, Cu, Co, As, and Sb. Metals in oil are often unique to particular oil fields, and as such, they may be used as tracers to determine the origin of atmospheric aerosols of concern (e.g., those in Arctic haze). Incineration of municipal waste is also an important source of atmospheric metals, including Hg.

Metals in atmospheric aerosol have been assumed to be present as insoluble oxides. However, in the case of oil fly ash, most metals may exist in soluble forms. Significant metal solubilization in atmospheric H_2O occurs by photoreduction processes that involve sulfate. Metals also condense on the surface of fly ash particles, thereby affecting their mobility and bioavailability in the environment.

Metal pollution is increasingly becoming a major environmental and public health concern. Metals released even in what has been believed to be small quantities can accumulate over time and biomagnify in food chains. There is increasing evidence that Hg emissions from coal-fired power plants, and mercury's subsequent accumulation and mobility in the environment, pose a new threat to aquatic and terrestrial ecosystems as well as to humans. Increasing attention is being focused on the metal components of atmospheric aerosol because of their catalytic properties and effects on biological functions. These potential effects of metals on the environment are described in greater detail in Chapters 5 and 6.

2.4.5 Source Apportionment

Because some sources emit particles with distinct chemical profiles, one should be able to calculate source contributions to PM samples. Such source apportionment can be done using models based on chemical mass balance, factor analysis, principal component analyses, multiple linear regression, and Lagrangian techniques. Source apportionment modeling of PM samples (using a chemical mass balance model) from sampling conducted in downtown Los Angeles (1982) indicated that diesel vehicle emissions, gasoline vehicle emissions, road dust, burning vegetation, secondary sulfate, and secondary nitrate accounted for approximately 18.8%, 5.7%, 12.4%, 9.6%, 20.9%, and 7.4% of collected PM mass, respectively. In Philadelphia, diesel emissions, road dust, and secondary sulfate accounted for 8.5%, 5.1%, and 88.5% of collected PM mass, respectively.

2.4.6 Behavior

Aerosol particles have their own unique atmospheric history. It may begin with the emission of primary particles from various natural or anthropogenic sources or by nucleation of condensable vapors produced as a consequence of chemical reactions between low-molecular-weight gas-phase chemical species. Depending on various chemical and physical characteristics, as well as the ambient environment, particles may change size, sorb and release vapor-phase molecules, collide, coalesce, or adhere to each other, sorb and condense H_2O vapor, undergo deliquescence (rapid uptake of H_2O), undergo changes in electrical condition, and be removed from the atmosphere by depositional processes.

Fine particles produced by gas-to-particle conversion processes commonly undergo changes in size. Homogeneous gas-phase chemical reactions produce condensable species such as H_2SO_4, which can nucleate to form a new particle (nucleation) or condense on the surface of an existing particle (condensation). The relative amount of nucleation compared with condensation depends on the rate of formation of the condensable species and surface area of existing particles. In urban areas, new particle formation seems to occur only near major sources of nuclei such as freeways. In many cases, the available surface area associated with aerosol particles rapidly scavenges newly formed condensable species.

Particle growth may occur as a result of condensation of homogeneous phase reaction products on particle surfaces, condensation of H_2O vapor, collisions between particles and subsequent coalescence or adherence, surface adsorption, and heterogeneous chemical reactions. In the last case, both gas-phase and particulate-phase substances participate in chemical reactions (as when SO_2 is oxidized to sulfite in an aqueous droplet). The rate of particle growth by this mechanism is limited by chemical reactions at the surface of or within the particle. Collisions with other particles that result in coalescence or adherence also result in particle growth. Associated with this phenomenon is a decrease in particle number and total surface area, whereas the average volume per particle increases. Coagulation produces chain agglomerates in soot and some metal-based particles.

Although fine particles are usually thought of as having the potential to increase in size as a result of the processes described earlier, in theory, they also have the potential to decrease in size. Aerosol species such as PAHs are semivolatile and thus can exist in the gas and particulate phases, depending on vapor pressure, particle surface area and composition, and atmospheric temperature. Daily temperature fluctuations can result in their desorption and absorption, with subsequent changes in particle mass as well as chemical composition.

Water is a very important component of many atmospheric particles. The behavior of atmospheric particles in association with changes in ambient humidity is of considerable importance in the global hydrologic cycle and energy budget, as well as in atmospheric chemistry and optical phenomena. Atmospheric sulfates are both hygroscopic (take up H_2O) and H_2O soluble. Their hygroscopicity results in physical and chemical changes that affect their size, shape, pH, reactivity, and refractive index. Crystalline solid particles such as $(NH_4)_2SO_4$ undergo a phase change with increasing relative humidity (RH) to become aqueous solution particles. This phase transition is abrupt, with a sudden uptake of H_2O taking place at an RH above the deliquescence point (usually ~80% RH).

Particles may undergo a change in their electrical state as they change size. A particle's charge depends on the state of its surface, the value of its dielectric constant, and its size. In general, particles greater than 3 μm carry a negative charge, and particles less than 0.01 μm carry a positive charge. Electrical charges can affect coagulation and the rate of dry deposition.

Particle behavior also includes their ability to scatter visible light and absorb and reradiate thermal energy. This behavior is affected by particle size, geometry, and chemical composition (see Chapter 4).

Like other components of the atmosphere, particles move in both horizontal and vertical dimensions. They are carried downwind by advective airflows and transported both upward and downward by convective air motion. Depending on their size, particles in the atmosphere are affected to different degrees by gravity. In the absence of significant convection, particles (particularly large particles) are drawn toward the surface of the Earth by gravity. Gravitational forces are opposed by a frictional force, which is a function of air viscosity, particle velocity, and particle diameter. As a result, larger particles "settle" more rapidly than smaller ones (see Chapter 3). As particle size decreases, the effect of gravity is reduced and particles settle out slowly. Small particles (< 1 μm) are subject to random movement-induced collisions with air molecules. This random movement of both air molecules and small particles is called Brownian motion or diffusion. Such movement results in the migration of particles from areas of high to low concentration. In a still atmosphere, particles that are less than 0.1 μm are primarily transported by Brownian motion. It is largely responsible for collisions between Aitken nuclei to form larger accumulation mode particles. It is also responsible for collisions that result in the adsorption of gas-phase substances onto particle surfaces.

2.4.7 Concentrations

Particulate matter measurements have been routinely done in the United States at more than 1,000 sites since the early 1970s. Initially, these were based on the use of a Hi-Vol sampler and were reported as TSP. After the NAAQS was changed to a PM_{10} standard (50 μg/m^3 annual average, and a 150 μg/m^3 24-hour average not to be exceeded more than once per year on average over a three-year period) in 1987, measurements were made with a sampler with a 50% cutoff diameter of 10 μm. In addition, measurements of fine particles ($PM_{2.5}$) were made in anticipation of the implementation of a $PM_{2.5}$ NAAQS (promulgated in 1997). In 2024, the US EPA strengthened the nation's air quality standards for fine particle pollution to improve public health protection by revising the primary annual $PM_{2.5}$ standard to 9.0 μg/m^3 (averaged over three years). It also retained the 24-hour $PM_{2.5}$ standard of 35 μg/m^3 (98th percentile, averaged over 3 years) and the 150 μg/m^3 24-hour PM_{10} standard. As such, considerable air quality data have been and continue to be collected on PM_{10}, and $PM_{2.5}$ concentrations at various locations in the United States. Figure 2.16 shows the trend of $PM_{2.5}$ from the 356 sites showing substantial reductions in ambient concentrations since 2000.

As of the US EPA's 2024 revision to the annual $PM_{2.5}$ standard (9 μg/m^3), initial assessments from 2020–2022 air quality data indicate that approximately 119 US counties may now fall short of this stricter limit, potentially placing them in Moderate nonattainment.

In the western US, counties expected to be highlighted include Shoshone County (ID), along with regions in Alaska (Fairbanks North Star Borough), Colorado (Denver County), Montana (Lincoln and Missoula Counties), Nevada (Clark, Douglas, Washoe Counties), New Mexico (Bernalillo County), and Utah (Salt Lake County).

Many areas that were previously designated nonattainment for the older $PM_{2.5}$ standard such as Plumas County, parts of the Los Angeles–South Coast Air Basin, San Joaquin Valley, and Imperial County (CA), are also under consideration for continued nonattainment, pending state submissions and final US EPA approval, slated for February 6, 2026.

Ambient PM_{10} and $PM_{2.5}$ vary seasonally, with the highest concentrations typically reported during summer months and during high-pressure, hazy days. Diurnal variations in $PM_{2.5}$ concentrations have also been reported, with maximum concentrations occurring during the morning and evening.

2.4.8 Long-Range Transport from Sources Outside the United States

It has been known for decades that both precursor gas-phase molecules and, to a lesser extent, particulate substances (e.g., acid sulfates) can be transported hundreds of kilometers or more. Because

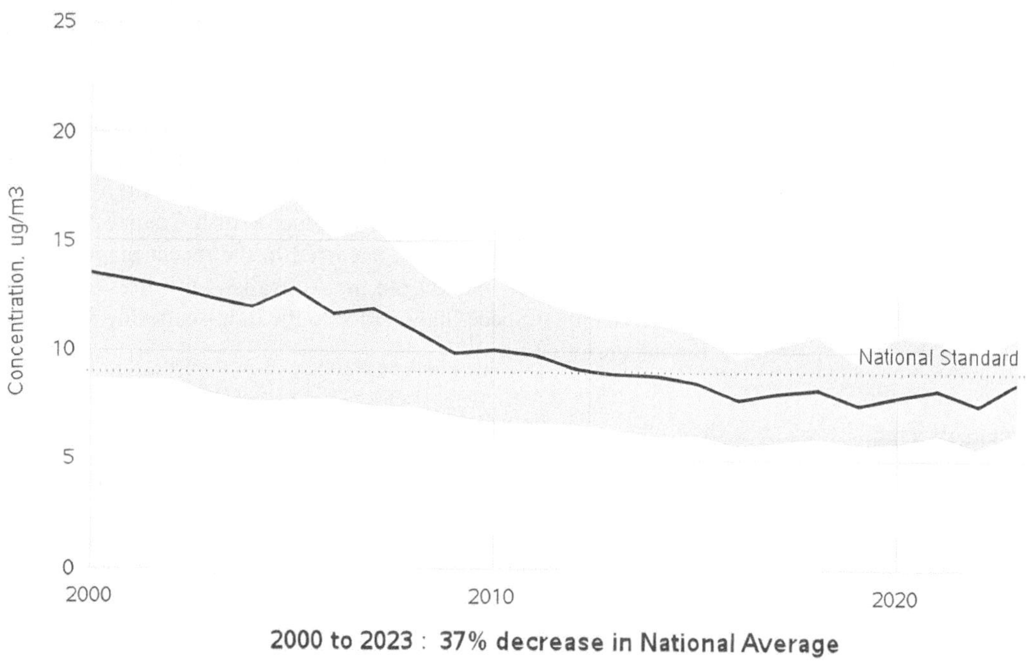

FIGURE 2.16 $PM_{2.5}$ air quality (annual mean, averaged over three years in the United States from 2000 to 2023). (From US EPA, https://www.epa.gov/air-trends/particulate-matter-pm25-trends)

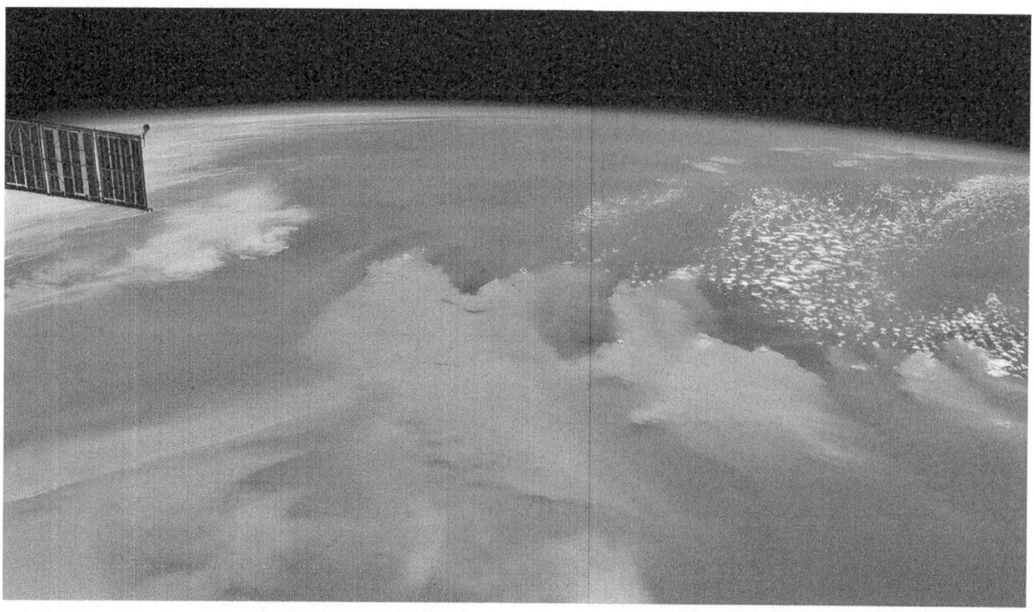

FIGURE 2.17 Photograph taken from the International Space Station on January 4, 2020, showing transport of particulate matter from Australian bushfires from a distance of 433 km above the Earth. (From NASA, *Earth Observatory Explosive Fire Activity in Australia*, EOSDIS/LANCE, Jan. 6, 2020.)

of the availability of data from intensive monitoring studies of particulate aerosols and analysis of their chemical constituents, as well as satellite imaging, there is substantial evidence that particles can be transported from very distant sources (thousands of kilometers). This includes transport from biomass burning in southern Mexico and Central America, which occasionally causes anomalously high PM levels in southern Texas and generally elevated levels in central and southeastern states; dust from Africa crossing the Atlantic and affecting air quality in Florida (African dust is the dominant constituent of ambient aerosol during the summer in Florida); and dust from the Gobi desert being transported to Washington, Oregon, California, Idaho, and British Columbia, resulting in elevated ambient concentrations. Significant transport occurred in the recent massive fires in both California and Australia. The transport of smoke from the Australian bush fires is readily seen in Figure 2.17 taken from the International Space Station (due to the light-scattering ability of submicron particles present within uncontrolled fires).

REFERENCES

Canadell, J.G., Monteiro, P.M.S., Costa, M.H., Cotrim da Cunha, L., Cox, P.M., Eliseev, A.V., Henson, S., Ishii, M., Jaccard, S., Koven, C., Lohila, A., Patra, P.K., Piao, S., Rogelj, J., Syampungani, S., Zaehle, S., and Zickfeld, K., Global carbon and other biogeochemical cycles and feedbacks, in *Climate Change 2021: The Physical Science Basis. Contribution of Working Group I to the Sixth Assessment Report of the Intergovernmental Panel on Climate Change. Chapter 5, Climate Change 2021: The Physical Science Basis. Contribution of Working Group I to the Sixth Assessment Report of the Intergovernmental Panel on Climate Change*, Masson-Delmotte, V., Zhai, P., Pirani, A., Connors, S.L., Péan, C., Berger, S., Caud, N., Chen, Y., Goldfarb, L., Gomis, M.I., Huang, M., Leitzell, K., Lonnoy, E., Matthews, J.B.R., Maycock, T.K., Waterfield, T., Yelekçi, O., Yu, R., and Zhou, B., Eds., Cambridge University Press, Cambridge, United Kingdom and New York, NY, 2021, pp. 673–816. https://doi.org/10.1017/9781009157896.007

Huang, G., Brook, R., Crippa, M., Janssens-Maenhout, G., Schieberle, C., Dore, C., Guizzardi, D., Muntean, M., Schaaf, E., and Friedrich, R., Speciation of anthropogenic emissions of non-methane volatile organic compounds: A global gridded data set for 1970–2012, *Atmos. Chem. Phys.*, 17, 7683–7701, 2017. https://doi.org/10.5194/acp-17-7683-2017.

Li, F., Tong, S., Zhang, W., Jia C., Wang, L., Ge, M., Characteristics of atmospheric non-methane hydrocarbon compounds (NMHCs) and their sources in urban typical secondary transformation in Beijing, China, *Appl. Geochem.*, 155, 105732, 2023. https://doi.org/10.1016/j.apgeochem.2023.105732

McDuffie, E. E., Smith, S. J., O'Rourke, P., Tibrewal, K., Venkataraman, C., Marais, E. A., Zheng, B., Crippa, M., Brauer, M., and Martin, R. V.: A global anthropogenic emission inventory of atmospheric pollutants from sector- and fuel-specific sources (1970–2017): an application of the Community Emissions Data System (CEDS), *Earth Syst. Sci. Data*, 12, 3413–3442, 2020. https://doi.org/10.5194/essd-12-3413-2020.

READINGS

Atkinson, R., Gas-phase tropospheric chemistry of organic compounds: A review, *Atmos. Environ.*, 24A, 1, 1990.

Brasseur, G.P., Orlando, J.J., and Tyndall, G.S., *Atmospheric Chemistry and Global Change*, Oxford University Press, Oxford, 1999.

Bridgman, H.A., *Global Air Pollution*, John Wiley & Sons, New York, 1994.

Brimblecombe, P., *Air Composition and Chemistry*, 2nd ed., Press Syndicate, Cambridge University Press, Cambridge, UK, 1996.

Elsom, D.M., *Atmospheric Pollution: A Global Problem*, 2nd ed., Blackwell Publishers, Oxford, 1992.

Environmental and Energy Study Institute, *Fact Sheet—A Brief History of Octane in Gasoline: From Lead to Ethanol*, 2016. https://www.eesi.org/papers/view/fact-sheet-a-brief-history-of-octane.

Finlayson-Pitts, B.J., and Pitts, J.N., Jr., *Chemistry of the Upper and Lower Atmosphere: Theory, Experiments and Applications*, Academic Press, Orlando, FL, 2000.

German Environment Agency, *Background March 2023: Air Quality 2022 (Preliminary Evaluation)*, 2023. https://www.umweltbundesamt.de/sites/default/files/medien/479/publikationen/2023_uba_hgp_luftqualitaet_engl_bf.pdf

Ghio, A.J., and Samet, J.M., Metals and air pollution particles, in *Air Pollution and Health*, Holgate, S.T., Samet, J.M., Koren, H.S., and Maynard, R.L., Eds., Academic Press, San Diego, CA, 1999, p. 635.

Giant, L.D., Shoaf, C.R., and Davis, J.M., United States and international approaches to establishing air standards and guidelines, in *Air Pollution and Health*, Holgate, S.T., Samet, J.M., Koren, H.S., and Maynard, R.L., Eds., Academic Press, San Diego, CA, 1999, p. 947.

Harrison, R.M., Ed., *Pollution: Causes, Effects and Control*, Royal Society of Chemistry, Thomas Graham House, Cambridge, UK, 1990.

Hidy, G.M., *Aerosols: An Industrial and Environmental Science*, Academic Press, Orlando, FL, 1984.

Houghton, J.T., Meira-Filho, L.G., Callender, B.A., Harris, N., Kallenberg, A., and Maskell, K., *The Science of Climate Change*, Cambridge University Press, Cambridge, UK, 1995.

IPCC, Climate change 2021: Working group I, *The Physical Science Basis*, 2021. https://www.ipcc.ch/report/ar6/wg1/?__cf_chl_jschl_tk__=pmd_82ab6d0e1d669a7ebc1b0e92472cc231efc29d49-1628502190-0-gqNtZGzNAeKjcnBszQp6#TS.

Isidorov, V.A., *Organic Chemistry of the Earth's Atmosphere*, Springer-Verlag, Berlin, 1990.

Jacob, D.J., *Introduction to Atmospheric Chemistry*, Princeton University Press, Princeton, NJ, 1999.

Jacobson, M.Z., *Air Pollution and Global Warming: History, Science, and Solutions*, 2nd ed., Cambridge University Press, New York, NY, 2012.

Lippmann, M., and Maynard, R.L., Air quality guidelines and standards, in *Air Pollution and Health*, Holgate, S.T., Samet, J.M., Koren, H.S., and Maynard, R.L., Eds., Academic Press, San Diego, CA, 1999, p. 983.

National Oceanic and Atmospheric Administration Climate Monitoring and Diagnostics Laboratory, 2020. www.esrl.noaa.gov/gmd/ccgg/trends/.

National Research Council, *Global Tropospheric Chemistry: A Plan for Action*, National Academy Press, Washington, DC, 1984.

National Research Council, *Ozone Depletion, Greenhouse Gases and Climate Change*, National Academy Press, Washington, DC, 1989.

National Research Council, *Rethinking the Ozone Problem in Urban and Regional Air Pollution*, National Academy Press, Washington, DC, 1991.

Pooley, F.D., and Milne, M., Composition of air pollution particles, in *Air Pollution and Health*, Holgate, S.T., Samet, J.M., Koren, H.S., and Maynard, R.L., Eds., Academic Press, San Diego, CA, 1999, p. 619.

Rowland, F.S., Chlorofluorocarbons and the depletion of stratospheric ozone, *Am. Sci.*, 77, 36, 1989.

Rowland, F.S., and Isaken, I.S.A., *The Changing Atmosphere*, John Wiley & Sons, Chichester, UK, 1987.

Ryan, R. G., Marais, E. A., Gershenson-Smith, E., Ramsay, R., Muller, J.-P., Tirpitz, J.-L., and Frieß, U., Measurement report: MAX-DOAS measurements characterise Central London ozone pollution episodes during 2022 heatwaves, *Atmos. Chem. Phys.*, 23, 7121–7139, 2023. https://doi.org/10.5194/acp-23-7121-2023.

Seinfeld, J.H., and Pandis, S.N., *Atmospheric Chemistry and Physics: From Air Pollution to Climate Change*, 3rd ed., John Wiley & Sons, New York, 2016.

Singh, H.B., Ed., *Composition, Chemistry and Climate of the Atmosphere*, Van Nostrand Reinhold, New York, 1995.

US EIA, *Annual Energy Outlook 2020 (With Projections to 2050)*, January 2020. www.eia.gov/aeo.

US EPA, *2008 National Emission Inventory: Review, Analysis and Highlights*, EPA-454/R-13-005, EPA, Research Triangle Park, NC, May 2013.

US EPA, *Air Quality Criteria for Particulate Matter*, Vol. 1, EPA 1600/AP-95/001, EPA, Washington, DC, 1996.

Van Loon, G.W., and Duffy, S.J., *Environmental Chemistry: A Global Perspective*, Oxford University Press, Oxford, 2000.

Velasco, E., Retama, A., and Molina, L., *Ozone pollution: A persistent challenge in large cities like Mexico City, Los Angeles and Beijing*, 2025. https://doi.org/10.14324/ucloepreprints.287.v1.

Warneck, P., *Chemistry of the Natural Atmosphere*, 2nd ed., Academic Press, Orlando, FL, 1999.

Watson, A.Y., Bates, R.R., and Kennedy, D., Eds., *Air Pollution: The Automobile and Public Health*, Health Effects Institute, National Academy Press, Washington, DC, 1988.

QUESTIONS

1. Describe sources of natural air pollution and their air quality significance.
2. How much of the total energy consumption (in quadrillions of Btu and percent) in the United States was used by the electric power sector in 2018 and what were the primary sources of the electricity in descending order?
3. How does ambient air pollution differ from other forms of air pollution?

4. What are the unique characteristics of Los Angeles–type smog?
5. How are smog and haze related?
6. What are the sources of primary and secondary pollutants?
7. Describe the significance of primary and secondary pollutants in the context of air pollution control.
8. Transportation is currently the major source of which of the primary pollutants?
9. Describe the relative success of regulatory programs in reducing emissions of CO, NO_x, and SO_2 between 1970 and 2018 in the United States.
10. Why are CO and CO_2 air quality concerns?
11. How are CO and CO_2 removed from the atmosphere?
12. Describe the relative significance of emissions of SO_x and reduced sulfur compounds to the atmosphere.
13. Three major nitrogen oxides are naturally present in the atmosphere and are released to it by natural and anthropogenic sources. These are N_2O, NO, and NO_2. Describe the environmental significance of each.
14. What roles do paraffinic, olefinic, and aromatic hydrocarbons play in atmospheric chemistry?
15. How do peroxy radicals affect tropospheric O_3 levels?
16. What role does ammonia play in atmospheric chemistry?
17. How do aldehydes and organic acids affect air quality?
18. Why are biogenically produced hydrocarbons important in the atmospheric chemistry of tropospheric O_3?
19. What are photochemical oxidants?
20. Describe the role that OH plays in the atmospheric sink processes involving CO, CH_4, and SO_2.
21. Describe types, uses, emissions, and environmental concerns associated with halogenated hydrocarbons.
22. What factor or factors determine the lifetime of a substance in the atmosphere?
23. Sink processes involve the conversion of substances to more chemically stable substances. What is the final chemical form to which the following pollutants are converted: SO_2, H_2S, O_3, and NO_2?
24. Particles differ in size and shape. Despite these differences, they are often described in the context of their aerodynamic diameter. Explain.
25. How do coarse and fine particles differ relative to their size, mechanisms of formation, and chemical composition?
26. Distinguish between ultrafine, Aitken nuclei, and accumulation mode particles.
27. Atmospheric particles vary in their chemical composition. Why is this the case?
28. Describe how the physical and chemical properties of particles affect their atmospheric behavior.
29. Describe the relative significance of metals found in particulate matter.
30. Describe wildfires and dust trends to increase $PM_{2.5}$ and ozone in the most recent years.

3 Atmospheric Dispersion, Transport, and Deposition

The atmosphere has served as a sink for emissions of volcanoes and a variety of geological processes, forest and grassland fires, decomposition and other biological processes, and ocean processes for hundreds of millions (if not billions) of years. It has also served as a sink for pollutants generated by human activities, proceeding from humans' first use of fire to the smelting of metal ores and use of fossil fuels such as coal, oil, and natural gas to motor vehicle, power generation, residential consumption, and other emissions from our very industrialized and technologically advanced modern times.

Despite its vastness, the atmosphere (at least in the short term) is not a perfect sink. Its ability to carry away (transport), dilute (disperse), and ultimately remove (deposition) waste products released to it is limited by various atmospheric motion phenomena. Pollutant concentrations may reach unacceptable levels as a result of local or regional overloading of the near-surface atmosphere, topographical barriers, and microscale, mesoscale, and macroscale air motion phenomena.

The atmosphere serves as a medium for atmospheric chemical reactions that ultimately serve to remove contaminants. These reactions may produce pollutants that may themselves pose significant environmental concerns. Levels of long-lived pollutants such as methane (CH_4), nitrous oxide (N_2O), and carbon dioxide (CO_2) may increase, causing global warming and, in the case of halogenated hydrocarbons, stratospheric ozone (O_3) depletion.

Because of technological and economic limitations, we have little choice but to use the atmosphere for the disposal of airborne wastes. Like other natural resources, our use of the atmosphere must be a wise one, recognizing its limitations and using it in a sustainable way.

3.1 DISPERSION AND TRANSPORT

Pollutants released from ground-level and elevated sources (smokestacks) are immediately subject to atmospheric processes, with dispersion in ever-increasing volumes of air by both vertical and horizontal transport. Transport is the process by which air motions carry gas-phase and particulate-phase species from one region of the atmosphere to another. Transport enhances dispersion and provides an opportunity for pollutants from different sources to interact.

Pollutant transport and dispersion are affected by atmospheric dynamics, fluid physical phenomena that occur in the atmosphere, and the physical laws that govern them. These may facilitate or constrain transport and dispersal.

Pollutants are initially released into the planetary boundary layer (PBL), that portion of the atmosphere most directly affected by the Earth's surface. The PBL is subject to fluxes of heat and water (H_2O) vapor from the surface, and other physical forces. Its depth, on average, ranges from a few hundred meters to 1 to 2 km. Above the PBL is a relatively stable layer of air that separates it from the free troposphere above.

Pollutant transport and dispersion are affected by different scales of motion. These are the microscale, mesoscale, synoptic scale, and macroscale or global scale (Table 3.1). Microscale refers to air motion in the near vicinity of a source and includes phenomena that affect plume behavior; mesoscale refers to atmospheric motions on the order of tens to hundreds of kilometers and such phenomena as fronts, airflows in river valleys, and coastal airflows; synoptic scale refers to systems on the order of 10^6 km^2 or more such as high-pressure and low-pressure systems responsible for day-to-day weather variations; and planetary scale refers to atmospheric motions on the order of

DOI: 10.1201/9781003664109-3

TABLE 3.1
Meteorological Scales of Air Motion

	Geographical		
Scale	**Area (km²)**	**Period**	**Phenomena**
Microscale	2–15	Minutes	Plume behavior Downwash
Mesoscale	15–160+	Hour to days	Sea, lake, and land breezes Mountain valley winds
Synoptic scale	> 10^6	Days	Migratory high-pressure and low-pressure systems Cold and warm fronts Semipermanent high-pressure systems
Global scale	Whole world	Weeks to months	Hadley cell flows Tropical storms Jet stream meanders Cold and warm fronts

continents or larger. These scales of air motion are useful in describing atmospheric phenomena. It is important to note that the atmosphere is one continuous flowing fluid and all motions are a part of this larger flow.

In the context of a few months, the PBL is relatively well mixed. However, on shorter timescales and near the Earth's surface (where pollutants are emitted), transport and dispersion are often limited by atmospheric conditions. Some atmospheric conditions result in increased ground-level pollutant concentrations that may potentially harm humans and our environment. Consequently, they are discussed in detail later. Particular attention is given to horizontal wind (speed and direction), turbulence, topography, atmospheric stability, and inversions.

3.1.1 Wind

Horizontal winds are characterized by both velocity (wind speed) and direction. As seen in Chapter 1, wind speed is affected by horizontal pressure and temperature gradients (the higher the pressure gradient, the higher the wind speed) and friction, which is proportional to the roughness of the Earth's surfaces (surface roughness). Relationships between surface roughness and wind speed for urban, suburban, and rural areas can be seen in Figure 3.1. The maximum height of each wind profile indicates where surface effects end and the gradient wind (wind affected by pressure differentials and the Coriolis effect) begins. For the urban area depicted, this occurs at approximately 500 m (1,650 ft); for the suburban area, at approximately 300 m (990 ft); and for the rural area, at approximately 250 m (825 ft).

For continuously emitting stack sources, dilution begins at the point of release. This plume dilution is inversely proportional to wind speed, that is, by doubling wind speed, pollutant concentration is decreased by 50% of its initial value. The effect of wind speed is to increase the volume of air available for pollutant dispersal. As seen in Figure 3.1, urban areas are characterized by relatively high surface roughness and, as a consequence, diminished wind speeds. This is ironic in the sense that urban areas, because of their relatively high pollutant emissions, are in greater, not lesser, need of being ventilated by the wind.

Winds have directional aspects. These include the prevailing northeasterly flows in the subtropics, southwesterly flows in the middle latitudes, and easterly flows at high latitudes in the northern hemisphere. They also include the cyclonic (clockwise) and anticyclonic flows associated with migrating low-pressure and high-pressure systems. Because flows are somewhat circular, wind direction will depend on one's position in the circulating pressure cell. It also depends on local

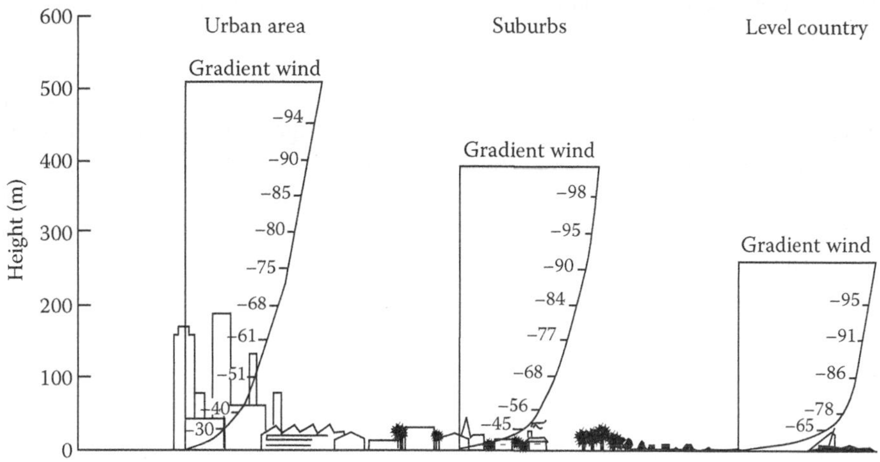

FIGURE 3.1 Effect of surface roughness on wind speed as a function of height over urban, suburban, and rural areas. (Adapted from Turner, D.B., *Workbook for Atmospheric Dispersion Emissions*, EPA Publication AP-26, EPA, Washington, DC, 1969.)

topography. At night in river valleys, airflows are downslope and downriver; they are upslope during daylight hours. Along sea and lake coasts, winds during clear weather flow inland during the day and waterward at night.

Wind direction is quite variable, with large changes often occurring over relatively short periods of time. A change in wind direction of 30° or more in 1 hour is not uncommon. Over a period of 24 hours, it may shift by 180°. Seasonal factors may cause wind direction variations of 360°.

Wind direction and variability can have significant effects on air quality. Areas downwind of point sources where winds are relatively persistent may experience relatively high ground-level concentrations compared with other areas at similar distances. If the wind is more variable, pollutants will be dispersed in a larger volume of air and be more equally distributed around the source; ground-level concentrations are therefore likely to be lower.

Wind direction is particularly important in the transport and dispersion of pollutants over large geographical areas. It is the southwesterly airflows that carry acid precursors from the US Midwest to the northeastern states and southeastern Canada. Similar flows have transported pollutants from countries in eastern Asia to the West Coast of the United States.

3.1.2 Turbulence

Airflows within the PBL are influenced by prevailing high-altitude air motion, frictional drag of the Earth's surface, and vertical airflows. Turbulence is characterized by circular eddies that may be vertical or horizontal or have various other orientations. These eddies represent air movements over shorter timescales than those that determine mean wind speeds. Turbulent eddies are produced by both mechanical and thermal forces.

Mechanical turbulence is induced by wind moving over and around structures and vegetation. It increases with wind speed and surface roughness. It is also produced by the shearing effect of fast-moving air aloft as it flows over air slowed by friction.

Thermal turbulence results from the heating or cooling of air near the Earth's surface. On clear days, solar heating of ground surfaces transfers heat to the air above it. Convection cells of rising warm air and descending cooler air develop. Under intensive surface heating, convective eddies are generated that extend vertically on the order of 1,000 to 1,500 m (~3,600 to 5,000 ft).

For the most part, mechanical and thermal turbulence are daytime phenomena. Both are dampened by nighttime radiative cooling of the ground and the air adjacent to it.

The effect of both mechanical and thermal turbulence is to enhance atmospheric mixing and pollutant dispersion. As a consequence, pollutant concentrations are significantly decreased. An exception is the downwash phenomena that causes plumes to be brought to the ground near smokestacks (by mechanical turbulence). In most cases, turbulence has a positive effect on air quality.

Downwash results in high pollutant concentrations in the turbulent wake downwind of a source. Downwash can also occur as a consequence of the shearing effect of high-velocity winds (> 70 kph, ~40 mph).

3.1.3 Atmospheric Stability

The atmosphere, particularly in the PBL, is characterized by highly variable horizontal and vertical air movements. In turbulent flows (described previously), the atmosphere is unstable and pollutants are rapidly dispersed. Turbulent flows associated with heating of the Earth's surface are dampened on cloudy days. Under such conditions, the atmosphere is more stable and pollutants are less rapidly dispersed.

When an entity is undergoing rapid change or has the immediate potential to do so, it is said to be unstable. When it is undergoing little or no change, and is even resistant to change, it is said to be stable. Stability and instability represent opposite ends in a continuum of possibilities. This continuum is implicit in the phrase *atmospheric stability.*

Vertical air motion is significantly affected by temperature gradients. The rate of temperature change with height is described as the lapse rate. Tropospheric temperatures, on average, decrease with height (Figure 1.7). This decrease, or normal lapse rate, is –6.5°C/km (18.9°F) or –0.65°C/100 m.

The normal lapse rate differs from what would be expected if a parcel of warm, dry air were released into a dry atmosphere. In this theoretical case, the buoyant parcel would rise and expand adiabatically (i.e., no energy is transferred to surrounding air). As it rises, its temperature decreases at a constant rate. This theoretical change of temperature with height is called the adiabatic lapse rate. It has values of –10°C/km (–29°F) or –1°C/100 m.

Because parcels of air released into the lower atmosphere contain H_2O vapor, the adiabatic lapse rate is used to describe how air cools when it rises in a dry atmosphere. Under real-world conditions, air contains a significant amount of H_2O vapor that cools as the air parcel rises. When air reaches its saturation vapor pressure, heat is released as H_2O vapor condenses (heat of vaporization is released). Air is warmed, resulting in a somewhat smaller decrease of temperature with height than that predicted for adiabatic conditions.

The normal lapse rate represents a summation and averaging of many different lapse rate conditions that vary from more negative (than the adiabatic lapse rate) to positive values. Individual lapse rates (environmental lapse rates) are determined from vertical temperature profile measurements. Although environmental lapse rates are reported as a single value, they represent a summation and averaging of temperature variations with height. Because they represent temperature changes with height, environmental lapse rates are used as indicators of atmospheric stability and the dispersion potential of pollutants.

The relationship between environmental lapse rates and stability can be seen in Figure 3.2. Line A indicates a slight decrease of temperature with height. Because in this case sources or sinks of thermal energy are present (there is little or no heating or cooling of the ground and adjacent air), air cools as it expands and its pressure decreases. This temperature change is close to the adiabatic rate. As a consequence, atmospheric conditions are described as neutral. A neutral lapse rate occurs in response to (1) cloudy conditions that inhibit incoming solar radiation and outgoing thermal radiation, (2) windy conditions that rapidly mix heated or cooled air near the Earth's surface, and (3) transitional circumstances near sunrise and sunset when changes in stability occur. Under such neutral conditions, dispersion is relatively good.

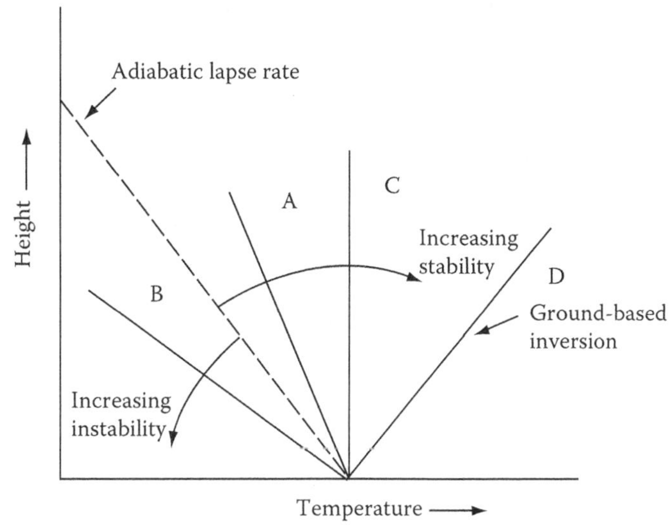

FIGURE 3.2 Vertical near-surface temperature profiles illustrating variations in lapse rate conditions and increasing–decreasing atmospheric stability.

The lapse rate characterized by line B shows a temperature decrease that is greater (–2°C/10 m) than those under neutral conditions (near the adiabatic lapse rate). A parcel of polluted air in such an environment will rise rapidly. The lapse rate is described as superadiabatic. Atmospheric conditions are very unstable, with strong vertical air motion. Such instability occurs on clear days with light winds at midday. Pollutant dispersion, as expected, is excellent.

Line C represents an isothermal lapse rate (i.e., temperature does not change with height). If a parcel of warm polluted air were released into this somewhat stable environment, it would rise slowly and soon cool to the temperature of its surroundings. Because the atmosphere is more stable, dispersion is more limited. Dispersion potential under these lapse rate conditions can be characterized as moderate.

Line D indicates lapse rate conditions in which temperature increases with height. As such, the temperature change is inverted from "normal." Under such conditions, the atmosphere is very stable. Because of warmer temperatures above the ground, a warm parcel of polluted air will quickly come into equilibrium with the temperature of its surroundings. Vertical air motion is suppressed and the dispersion potential of emitted pollutants is poor. Such stable ground-level inverted lapse rate conditions occur at night under clear skies with calm to light winds.

Atmospheric stability, described earlier, represents changes in near-surface air temperatures that occur over the course of a single day. They do not include larger-scale meteorological conditions associated with high-pressure systems.

3.1.4 Inversions

As indicated, when lapse rate temperatures increase with height, they are inverted from normal. Such atmospheric conditions are described as inversions. They can be surface based (occur near the ground) or elevated (occur aloft). Atmospheric processes may produce frontal, advective, radiational, and subsidence inversions.

In a frontal inversion, air from a warm front flows over cold air in an adjoining air mass. Inverted temperatures occur at the interface of the two fronts. Because of the movement of these air masses and the interaction between them, frontal inversions have only limited effects on air quality.

An advective inversion forms when warm air flows over a cold surface or cold air. They are commonly associated with land and sea breezes and may be surface based or elevated.

Radiational and subsidence inversions pose significant air quality concerns because they suppress vertical mixing over industrialized river valleys and urban areas, resulting in increased pollutant levels.

3.1.4.1 Radiational Inversions

Radiational inversions are produced as a result of the radiational cooling of the ground. Because they form at night, they are also called nocturnal inversions. Radiational inversions are, in most cases, ground based.

Radiational inversions only occur on clear nights. Surface-based inversions begin to form as the sun sets and intensify throughout the night until sunrise. Because the Earth is a net radiator of heat at night, it begins a cooling process that subsequently cools the air immediately above it. Relatively warm air overlays an increasingly deepening layer of cool air beneath. This inversion layer may be only 10 to 20 m (33–66 ft) deep over flat terrain. With the exception of ground-based sources, such inversions have only a limited effect on air quality.

Radiational inversions in river valleys are of major environmental importance because of their historically heavy industrialization and pollutant emissions and because such inversions are intensified as a result of the effects of topography.

River valleys were formed by the erosive force of water as it moved from higher elevations to the sea. Valleys serve as conduits for water as well as cool, dense air flowing downslope from the radiative cooling of ridges bordering the valley. Upon reaching the valley floor, this cool, dense air runs under warmer air and forces it aloft. The flow is downslope and downriver. Because of its density and volume, cool air floods the valley floor. This river of cool air deepens over the nighttime hours and reaches its maximum depth just before sunrise. In some mountain valleys, the top of the inversion layer may be 100 to 200 m (330–660 ft) above the valley floor. A radiational inversion in a mountain valley and its vertical temperature profile are illustrated in Figure 3.3.

The height of the inversion layer has a significant effect on how well or how poorly pollutants are dispersed. In river valleys, emissions from sources having stack heights of up to 100 m may be trapped in the inversion layer. As such, dispersion is very poor and pollutant concentrations near the top of the inversion layer are high. In this condition, higher stacks will need to be built to avoid trapping air pollutants in river valleys.

Polluted air emitted from a source will rise to an altitude where its temperature is the same as its surroundings. This occurs near the top of the inversion layer. As a consequence, a layer of intensely smoky or hazy air forms at this height. This is most evident in the early morning hours in a river valley before inversion breakup.

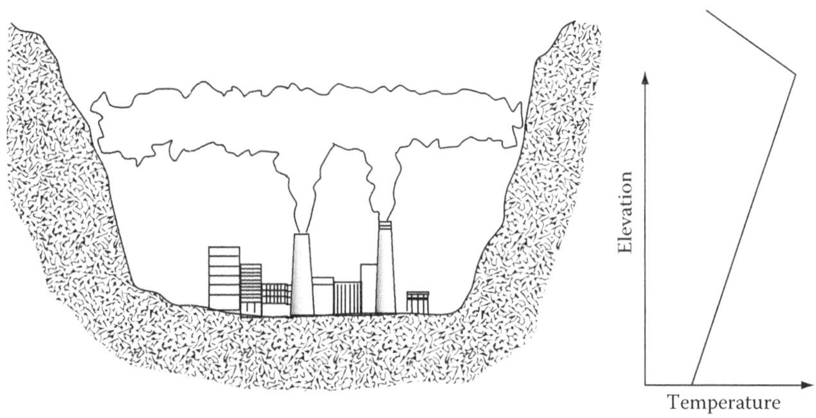

FIGURE 3.3 Nocturnal, radiational inversion in an industrialized river valley and associated temperature profile.

During the winter, when days are short, radiational inversions may persist up to 16 to 18 hours in northern latitudes; they are typically considerably shorter during the summer. Pollutant levels, as can be expected, are higher under more persistent inversion conditions.

Radiational inversions begin to break up as the sun starts to warm the ground and the air above it. Increasingly large convection cells of turbulent air are formed, which causes complete inversion breakup several hours after sunrise. At that time, the heavily polluted air mass near the top of the inversion is brought to the ground. This phenomenon is described as fumigation because of the high ground-level concentrations produced.

Despite radiative cooling of the ground, surface-based inversions generally do not form in urban areas located on flat terrain. This is because urban surfaces emit considerable quantities of heat that produce a well-mixed layer of air above them. Emitted heat, however, can be absorbed by the polluted air mass (often described as a dust dome) that forms above many cities. As heat is absorbed by pollutants, a layer or two of warm air forms aloft (Figure 3.4). Although these are elevated inversions, they are, nevertheless, produced by radiational cooling.

3.1.4.2 Subsidence Inversions

Subsidence inversions are formed over large geographical areas as the result of the subsidence of air in high-pressure systems. As air subsides (sinks) to lower altitudes, it compresses the air beneath it, causing temperatures to rise. Because turbulence almost always occurs near the ground, air in this part of the atmosphere is relatively unaffected by subsidence occurring above. As a consequence, an inversion layer (which may be 50 m [165 ft] thick) forms between the subsiding air and the relatively turbulent air below it (Figure 3.5).

The height of the inversion layer varies. It is highest near the center of the cell and lowest near the cell's periphery. As a result of turbulence, elevated inversion layers do not reach the ground. Although commonly associated with high-pressure systems, subsidence inversions only have significant effects on air quality when the inversion layer is relatively close to the ground (e.g., 300–400 m, ~990 to 1,300 ft) and persistent (3–5 days).

Most high-pressure systems are migratory, that is, they move over large expanses of the Earth's surface after they form. These migrating systems contribute to the hazy summer conditions over the American Midwest, Southeast, and Northeast. Occasionally, they have inversion layers that

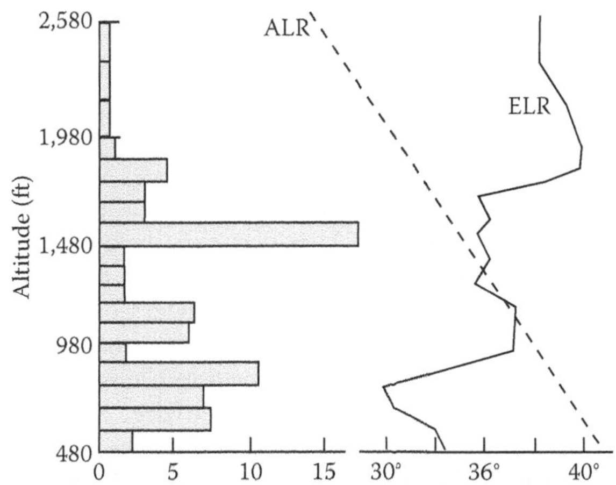

FIGURE 3.4 Effects of dust levels on environmental lapse rate over the Cincinnati, OH, metropolitan area. ALR, adiabatic lapse rate; ELR, environmental lapse rate. (From Bach, W., *Geogr. Rev.*, 61, 573, 1971. With permission.)

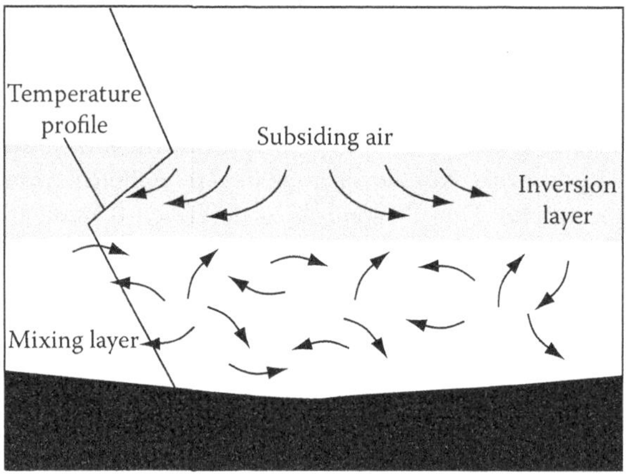

FIGURE 3.5 Formation of a subsidence inversion and its associated temperature profile.

are relatively low and more persistent than normal. Such stagnating systems may result in the increased production of pollutant levels near the ground. These occurrences are called episodes. Fortunately, severe pollution episodes associated with migrating high-pressure systems are relatively rare in North America and northern Europe. The most severe and persistent inversions occur in middle latitudes in autumn. Subsidence inversions of particular note are those associated with semipermanent marine high-pressure systems (see Chapter 1). The inversion layer comes closest to the ground on the easterly or continental side of such systems. As a consequence, west coasts of continents have relatively low and persistent inversion conditions. Inversions below 800 m (2,600 ft) occur over the southern coast of California approximately 90% of the time during the summer. Such inversions are primarily responsible for the smoggy and hazy atmospheric conditions over the Los Angeles Basin.

3.1.5 Mixing Height

Mixing height (MH) is the height of the vertical volume of air above the Earth's surface where relatively vigorous mixing and pollutant dispersion occurs. A definable MH is assumed to occur under unstable and neutral conditions. It cannot be defined when the air mass above the surface is stable. Elevated inversions frequently place a cap on the MH. Average MHs for selected US cities are indicated on the map in Figure 3.6.

The MH varies both diurnally and seasonally. It is markedly affected by topography and high-pressure systems. During the day, minimum MHs occur just before sunrise. The MH increases progressively as the sun warms the Earth and the Earth warms the air above it. Increasingly larger convective cells are formed so that the MH reaches its maximum value in the early afternoon (commonly several thousand meters). Maximum values occur during summer, with minimum values in late autumn and winter in the middle latitudes of the northern hemisphere.

Topographical features such as water surfaces, with their large heat-absorbing capacities, have lower maximum and average MHs than land surfaces with little vegetative cover (e.g., deserts). Not surprisingly, Phoenix, AZ, has a relatively high summertime MH, with coastal cities among the lowest (Figure 3.6).

MHs are affected by semipermanent marine high-pressure systems (note Los Angeles in Figure 3.6) and migrating high-pressure systems. Subsidence of air and formation of inversions cause MHs to decrease. MHs are an important variable used in air quality modeling (see Chapter 7).

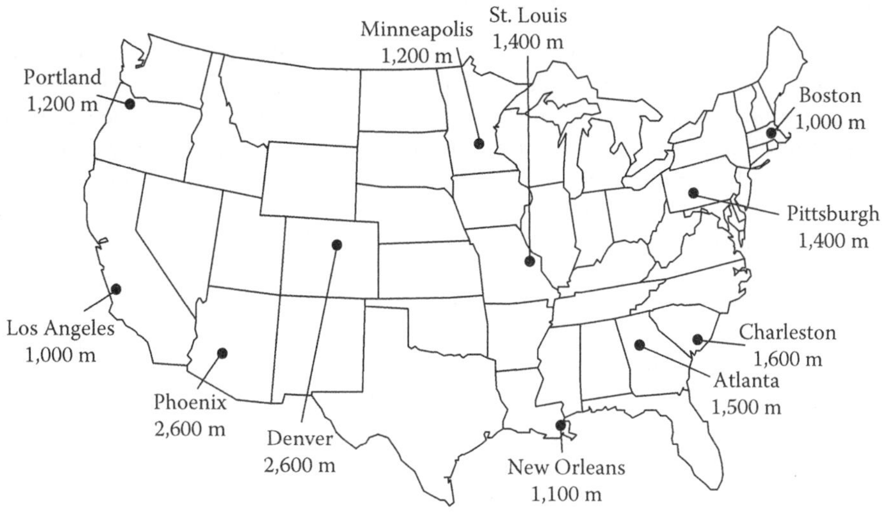

FIGURE 3.6 Average summertime MHs for selected US cities.

3.1.6 Topography

Both microscale and mesoscale air motions are affected by nearby topographical features. Topography can have significant effects on both air movement and pollutant levels. These include differential vertical airflows associated with forests, plowed agricultural fields, parking lots, and the like. Such flows affect the behavior of smokestack plumes.

In river valleys, downslope airflows at night intensify (deepen) surface-based inversions, and valley winds during the day help move pollutants upslope and out of the valley. Mountains also serve as barriers to air movement. In the Los Angeles Basin, the San Bernardino Mountains retard airflow in northerly and easterly directions, further intensifying smog and haze conditions. Mountains also increase surface roughness, thereby decreasing wind speeds.

The smog problem over the Los Angeles Basin is also affected by the adjacent cool waters of the Pacific. When sea breezes bring cool air in from the ocean, warmer air is pushed aloft, further intensifying the elevated inversion.

Mesoscale airflow patterns occur on relatively calm days in coastal areas as a result of differential heating and cooling of land and water surfaces. During summer, when skies are clear and prevailing winds are light, land surfaces warm more rapidly than sea and lake water. The subsequently warmed air flows up and waterward. As a consequence of temperature and pressure differences, air flows landward at the surface from the water, forming a sea or lake breeze. Air moving from the land cools and descends to form a weak circulation cell. At night, rapid radiational cooling of the land results in surface airflows toward water, forming a land breeze. These land breezes are generally lighter than lake and sea breezes.

Land, sea, and lake breezes, and the circulation patterns that form with them, occur only when prevailing winds are light. They are overridden when winds are strong. In the case of the south coast of California, sea breezes intensify subsidence inversions. They may also cause advective inversions, which commonly occur in late spring when large bodies of water are still cold relative to adjacent land areas. As water-cooled air moves inland, it warms; the inversion is broken up and replaced by superadiabatic lapse rate conditions. The weak circulation cells associated with land, lake, and sea breezes may allow pollutants to be recirculated to some degree and carried over from one day to the next.

3.1.7 Pollutant Dispersion from Point Sources

Point sources may occur at ground level, or as is often the case, pollutants are emitted from smokestacks that vary in height. The subsequent history of plumes formed depends on (1) the physical and chemical nature of pollutants, (2) meteorological factors such as wind speed and direction, humidity, precipitation, pressure, temperature, and atmospheric stability, (3) location of the source relative to physical obstructions, and (4) topographical factors that affect air movement. As these affect plume rise, its spread horizontally and vertically, and its transport, they also affect maximum ground-level concentrations (MGLCs) and the distance of MGLCs from the source.

3.1.7.1 Pollutants and Diffusion

In many cases, point source plumes are a mixture of gas-phase and particulate-phase substances. Particles with aerodynamic diameters of greater than or equal to 20 μm have appreciable settling velocities, and as a consequence, deposition occurs relatively close to their sources. Smaller particles, particularly those with aerodynamic diameters of less than or equal to 1 μm, have very low settling velocities and dispersion behavior similar to that of gases and vapors. The gaseous nature of plumes, given sufficient time, may allow for dispersion by simple diffusion, whereby the random motion of molecules results in pollutants migrating from areas of high concentration (the center of the plume) to areas of low concentration (the plume's periphery). Diffusion causes plumes to spread both horizontally and vertically. As a consequence, the effect of diffusion can be seen to increase with downwind distance.

3.1.7.2 Plume Rise and Transport

Dispersion from a smokestack source is significantly affected by its physical height as well as plume rise. In Figure 3.7, the plume is seen to rise to a maximum height and then level off. The distance from the top of the stack to the center of the plume is described as plume rise. The distance from the ground to the center of the plume (including the stack) is the effective stack height.

Dispersion is enhanced with increasing stack height and plume rise. Ground-level concentrations will be lower, and at constant wind speed, the distance at which MGLCs occur will be increased as effective stack height increases.

The height of the plume at the point it levels off depends on the exit temperature of stack gases, cross-sectional diameter of the stack, emission velocity, horizontal wind speed, and atmospheric stability (as indicated by the vertical temperature gradient).

The effective stack height for a source can be increased by building taller stacks. Tall-stack technology has been used since the 1960s by electrical utilities operating large coal-fired power

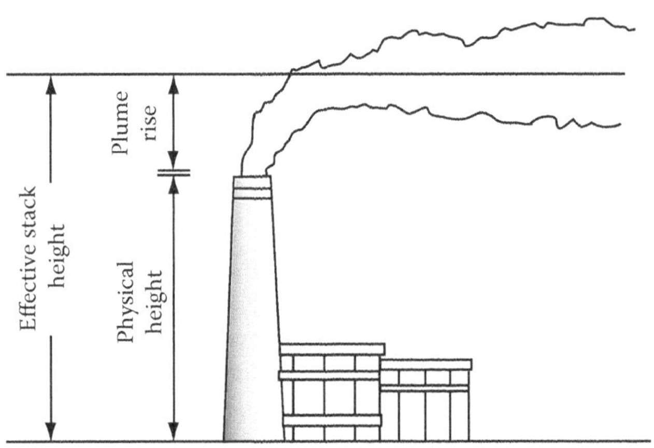

FIGURE 3.7 Plume rise and effective stack height.

plants. Such stacks are commonly 250 to 300 m (850–1,000 ft) high, with some stacks approximately 400 m (1,300 ft). They were designed and operated on the principle that pollutants could be dispersed from such facilities without causing unacceptable downwind ground-level concentrations.

Wind speed in the horizontal dimension significantly affects both plume rise and ground-level concentrations. Higher wind speeds decrease effective stack height. However, due to the increased volume of air associated with increasing wind speeds, ground-level concentrations are usually reduced. Higher wind speeds decrease the distance at which MGLCs occur. At very high speeds (~80 kph, 50 mph), plume rise may be negligible; the plume may be brought to the ground immediately downwind of the source.

Tall stacks are designed to take advantage of the higher wind speeds that occur aloft as the frictional drag of the Earth's surface is diminished. Pollutant dispersion is enhanced as a result of these higher wind speeds.

Plume rise, as indicated, is subject to atmospheric stability. Under unstable conditions, significant plume rise occurs; under stable conditions, plume rise is markedly reduced. In the latter case, dispersion is decreased and higher ground-level concentrations can be expected.

3.1.7.3 Plume Characteristics

As a plume moves downwind of a source, it expands by diffusion in both horizontal and vertical dimensions. Plumes take form and behavior patterns that reflect stability conditions in the atmosphere. Major plume types are illustrated in Figure 3.8. In the first case (Figure 3.8a), the lapse rate is superadiabatic with relatively calm winds. There is significant initial plume rise and a subsequent "looping" motion. This motion results from portions of the plume being buoyed up by convective rising, with subsequent descending of air. Upward and downward air motion is considerable. Because eddies that produce this motion are oriented in all directions, significant horizontal dispersion takes place as well.

A coning plume is illustrated in Figure 3.8b. Coning plumes form when lapse rates are neutral to isothermal. As such, the atmosphere is slightly unstable to slightly stable. Such lapse rates occur on cloudy or windy days or at night. Atmospheric turbulence is primarily mechanical; turbulent eddies may have different orientations that result in dispersion in a relatively symmetrical pattern.

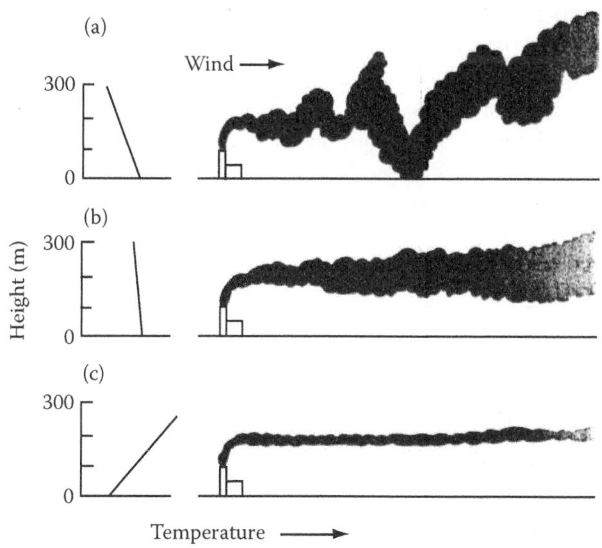

FIGURE 3.8 Plume form and behavior associated with different atmospheric stability conditions. (a) Looping, (b) coning, and (c) fanning.

A fanning plume is illustrated in Figure 3.8c. It is produced under stable conditions in which the top of a ground-based inversion is well above the stack. The plume rises slightly, there being little vertical air movement and dispersion. Horizontal motions, however, are not inhibited. As a consequence, the plume may be characterized by varying degrees of horizontal spreading.

Other plume forms and behaviors (not illustrated here) occur. Lofting, fumigating, and trapping plumes are associated with inversions. A lofting plume may be produced when the atmosphere above a surface inversion is unstable, with the stack above the surface-based inversion. The plume rises upward as its downward movement is restricted by the inversion beneath it. Lofting plumes are produced at sunset on clear nights, over open terrain. A fumigating plume is produced when a surface-based inversion breaks up after sunrise. Pollutants are brought to the ground by the downward movement of convective cells. When inversions occur both above and below a smokestack, the plume is trapped; in appearance, it is somewhat similar to, but deeper than, a fanning plume.

As plumes move downwind, they generally mix with air that is less polluted. Such mixing occurs as a result of diffusion, advection, displacement, convection, and mechanical turbulence. Consequently, concentrations decrease with downwind distance. Depending on their height and atmospheric stability, plumes may reach the ground within a few kilometers of a source or may remain airborne for extended distances. Plumes can be seen because of the presence of light-scattering particles. Plumes become less and less visible as particle concentrations decrease as a result of dilution.

3.1.8 Large-Scale Transport and Dispersion

3.1.8.1 Long-Range Transport

Historically, pollutant concerns have focused on the dispersion of pollutants in urban areas and those downwind of large point sources. It was once assumed that when pollutants were diluted to acceptable levels or plumes were no longer visible, they were not a problem. In the early 1970s, the phenomenon of long-range transport was identified as the cause of increased nocturnal rural O_3 levels. Both O_3 and its precursors were transported hundreds of kilometers from urban areas (urban plume). The phenomenon of acidic deposition and its attendant ecological effects and associations with the long-range transport of acid precursors in North America became known in the mid- to late 1970s. This was followed by the recognition that Arctic haze over Barrow, AK (in the northern hemispheric spring), was associated with long-range transport of pollutants over the North Pole from northern Europe and Asia. There was, in the late 1980s, increasing evidence that tropospheric pollutants were being transported into the stratosphere. In more recent times, we have come to better understand processes by which long-lived substances such as carbon dioxide (CO_2), CH_4, halogenated hydrocarbons, and particles move around the planet. Since the late 1990s, Asian dust and pollutants were observed and transported to North America. Long-range air pollution transport has no country boundary and can transport pollutants from one continent to another.

3.1.8.2 Urban Plume

Large urban centers include numerous individual point and mobile sources. These contribute collectively to large polluted air masses that affect air quality for tens to hundreds of kilometers downwind. This air mass is described as an urban plume. The transport and dispersion of pollutants in an urban plume occurs over larger geographical areas and timescales than those typically associated with individual sources. Of particular importance are airflows associated with migrating high-pressure and low-pressure systems.

3.1.8.3 Planetary Transport

A stable layer of air at the top of the PBL retards vertical mixing and isolates it from the free troposphere above. Because upward air movement is somewhat impeded, timescales on the order of a few hours to a few days are needed for transport of pollutants out of the PBL. As a consequence of convective energy flows, baroclinic (pressure-related) instability, and heat release from the condensation of water vapor, pollutants are transported to the top of the troposphere, with uniform mixing occurring in about a week.

Air in the troposphere is continuously stirred by convective air movement and other atmospheric phenomena. As a consequence, substances with lifetimes of several months or more are well mixed within the troposphere. However, significant concentration differences exist between northern and southern hemispheres.

Atmospheric phenomena at the equator retards airflow from one hemisphere to another. As a consequence, cross-equatorial mixing time is approximately 1 year.

3.1.9 Stratosphere–Troposphere Exchange

The troposphere and stratosphere are characterized, respectively, by decreasing and increasing temperatures with height. They are separated by an isothermal layer of air (the tropopause). The increasing temperatures with height in the stratosphere serve to limit the upward and downward movement of atmospheric gases between the troposphere and stratosphere. Trace gas measurements have shown, however, that such exchange takes place, albeit relatively slowly.

As a consequence of stratosphere–troposphere exchange processes, long-lived chemical species such as chlorofluorocarbons (CFCs), CH_4, and nitrous oxide (N_2O) are transported into the stratosphere, and chemical species originating in the stratosphere are transported into the troposphere. These include O_3, nitric oxides (NO_x), and substances such as chlorine nitrate ($ClONO_2$) and hydrogen chloride (HCl) from the photodestruction of CFCs.

The time required to exchange the mass of the entire troposphere with the stratosphere is estimated to be 18 years. Because of differences in mass, it takes about 2 years for the entire stratosphere to mix with the troposphere.

A number of potential pathways and mechanisms have been proposed to explain stratosphere–troposphere exchange. The simplest of these consists of a single Hadley-type cell in each hemisphere with a uniform rising motion across the tropical tropopause, poleward movement in the stratosphere, and return flow into the troposphere outside of the tropics. This is consistent with low H_2O vapor levels in the tropical stratosphere and high O_3 levels in the lower polar stratosphere. This circulation cell is described as a wave-driven "extratropical pump."

This, and a second pathway for stratosphere–troposphere exchange, are illustrated in Figure 3.9. In the second pathway, transport occurs from the lower stratosphere to the troposphere in midlatitudes and from the troposphere to the stratosphere along surfaces of constant potential temperature that cross into the tropopause. Exchange of air tends to occur in association with events known as tropopause folds. In this phenomenon, the tropopause on the poleward side of a jet stream (see Chapter 1) is distorted during the development of large weather systems. Large intrusions of stratospheric air occur, which become trapped and eventually mix with the troposphere. Air from the troposphere can also be trapped in the stratosphere during tropopause folds. In a third proposed pathway, air is transported convectively from the troposphere to the stratosphere.

3.1.10 Stratospheric Circulation

Due to strong inversion conditions, the stratosphere is very stable, with little vertical air motion. Because of differences in stratospheric temperature between the equator and poles, as well as those caused by diabatic heating (associated with O_3 absorption of ultraviolet light), zonal (east × west) and meridianal (north × south) flows characterize the stratosphere. Thermal gradients result in strong zonal winds, which reach peak speeds near solstices and reverse after equinoxes.

Despite diabatic heating, circulation in the stratosphere is wave driven. Atmospheric waves transport air poleward in the winter hemisphere. It subsides at the poles, where it warms by compression. A circulation cell develops throughout the middle and upper stratosphere. There is a slow rising of

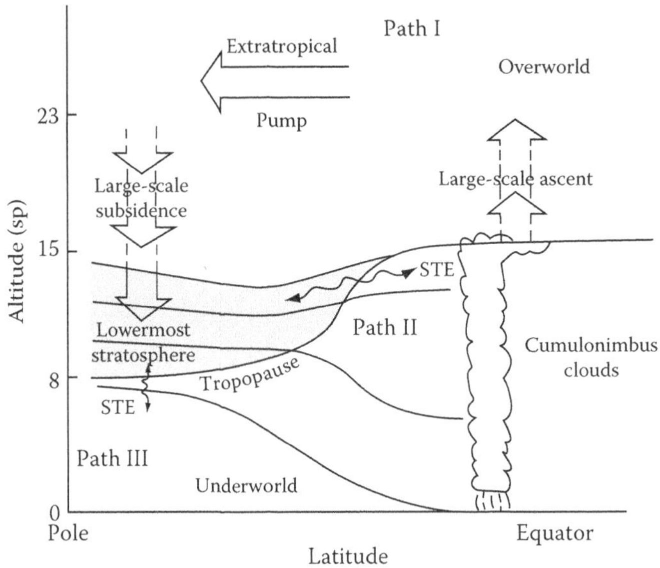

FIGURE 3.9 Stratosphere–troposphere exchange pathways and processes. (From Holton, J.R. et al., *Rev. Geophys.*, 33, 403, 1995. With permission.)

air in the summer hemisphere and tropics and more rapid sinking over a smaller area in the winter hemisphere.

This stratospheric circulation significantly affects the movement of chemical species. It results in the movement of O_3 (most O_3 is produced in the tropical stratosphere) and gases transported from the troposphere poleward; movement is strong in the winter hemisphere and less so in the summer hemisphere.

Such stratospheric airflows transport trace gases such as CFCs to the Antarctic and O_3-depleted air northward from the Antarctic. They also have significant effects on substances that contribute to the formation of polar stratospheric clouds (see Chapter 4).

3.2 ATMOSPHERIC REMOVAL AND DEPOSITION PROCESSES

3.2.1 Atmospheric Lifetimes

Reference was made to the average lifetimes or residence times of various atmospheric pollutants in Chapter 2. All gas-phase and particulate-phase pollutants have a life history in the atmosphere before they are ultimately removed. By averaging the life histories of all molecules of a substance, or all particles of a particular type or size, one can determine their average residence or lifetime. Residence time can also be described in the context of a pollutant's half-life, that is, the time required to reduce its concentration by 50% of its initial value.

Atmospheric lifetimes can be calculated using mass balance equations. Under steady-state conditions (emission rate = removal rate), the residence time or lifetime of a substance can be calculated from the following equation:

$$\tau = Q / P \tag{3.1}$$

where τ is the residence time, Q is the total mass of substance in the atmosphere (g), and P is the emission or removal rate (g/year).

If the total mass (Q) of a given atmospheric substance is 2×10^{12} g and $P = 100 \times 10^{12}$ g/year,

$$\tau = 2\times10^{12}\, g / \left(100\times10^{12}\, g / \text{year}\right) =\sim 1\,\text{week}$$

Lifetimes are also calculated based on the substance's reactivity with sink chemicals such as hydroxyl radical (OH·) and nitrate $\left(NO_3^-\right)$. In such cases, lifetimes are calculated by dividing the product of the OH concentration and the rate constant (k) into 1. For the reaction between CH_4 and OH,

$$CH_4 + OH\cdot \rightarrow CH_3\cdot + H_2O \tag{3.2}$$

an atmospheric lifetime of 5 years is calculated:

$$\begin{aligned}\tau &= 1/k\left[OH\cdot\right].\\ &= 1/\left[6.3\times10^{-15}\,\text{cm}^3\,\text{molecule}^{-1}\text{s}^{-1}\times\left(1\times10^6\,\text{OH}\cdot\text{molecules}\,\text{cm}^{-3}\right)\right]\\ &= 1.5^9\times10^8\,\text{s} =\sim 5\,\text{year}\end{aligned} \tag{3.3}$$

These atmospheric lifetime calculations are based on chemical kinetics. They assume that there are no competitors for OH· or, if there are, that OH· is in a steady-state concentration. If there are competing loss processes (such as photolysis and reaction with O_3), lifetimes may be shorter.

Although the atmospheric lifetimes of gas-phase substances are determined by chemical reactions with substances such as OH·, their by-products, unreacted molecules, and particulate-phase species will ultimately be removed by deposition processes.

3.2.2 Deposition Processes

Dry deposition is characterized by direct transfer of gas-phase and particulate-phase substances to vegetation, water, and other Earth surfaces. This transfer may take place by impaction, diffusion to surfaces, and, in the case of plants, physiological uptake of atmospheric contaminants. Particulate-phase substances may also be removed by sedimentation.

Dry deposition is characterized by a deposition velocity (Vg). It is a proportionality constant that relates the flux (F) of a chemical species or particle to a surface and its concentration (C) at some reference height:

$$Vg = -F / C \tag{3.4}$$

The deposition velocity is a positive number despite the fact that the flux is negative; it is given as centimeters per second (cm/s). Dry deposition velocities are given for coarse and fine particles and for gas-phase substances in Table 3.2. The highest deposition velocities have been observed for coarse particles and gas-phase nitric acid (HNO_3). Because of its high solubility, HNO_3 is readily absorbed into dew and other aqueous surfaces and rapidly taken up by plants.

Wet deposition includes all processes by which airborne gases and particles are transferred to the Earth's surface in aqueous form (rain, snow, fog, clouds, dew). These processes include (1) absorption of gas-phase substances in cloud droplets, raindrops, and the like; (2) in-cloud processes wherein particles serve as nuclei for the condensation of H_2O vapor to form cloud or fog droplets; and (3) collision of rain droplets and particles both within and below clouds. In the last case, collisions with, and subsequent incorporation of, gas-phase substances can also occur. This process is called washout. The removal of particles in the rain-making process is called rainout. Ozone deposition has gained emphasis in recent years because it plays a key role in understanding ecosystem

TABLE 3.2
Dry Deposition Velocities (cm/s) for Particles and Selected Gases

Pollutant	Surface	Range (cm/s)
Particles > 2 μm	Exterior	0.5–2
Particles ≤ 2 μm	Exterior	< 0.5
SO_2	Exterior and interior leaf surfaces	0.2–1 dry foliage, open stomates; > 1 for wet surface
HNO_3	Primarily exterior leaf surfaces	1–5 or greater
NO_2	Primarily leaf interiors; also exterior leaf surfaces	0.1–0.5 when stomates open
O_3	Primarily leaf interiors; also exterior leaf surfaces	0.1–0.8

damage, improving air quality and climate models, and assessing the impacts of land-use changes. It helps quantify how ozone is removed from the atmosphere, informs regulations, and is crucial for predicting both ecological and human health outcomes.

3.3 METEOROLOGICAL APPLICATIONS: AIR POLLUTION CONTROL

As seen in previous discussions in this chapter, a variety of meteorological factors affect the dispersion of pollutants from individual sources as well as urban areas. Thus, meteorology has significant applications in air pollution control programs. These include episode prediction, planning, and community responses. These also include the determination of whether, under worst-case atmospheric conditions, proposed sources will be in compliance with air quality standards and visibility protection requirements for the prevention of significant deterioration provisions of clean air legislation. Compliance is determined from air quality modeling.

3.3.1 Episode Planning

In the late 1950s and early 1960s, the US Weather Service (now the National Weather Service) observed that increased pollutant levels in urban areas were associated with slow-moving migratory high-pressure systems. As a consequence, it began a program of issuing air pollution potential forecasts when such systems covered an area of at least 90,000 km^2 (~35,000 mi^2) and were expected to persist for at least 36 hours.

Despite the fact that regulatory efforts to control air pollution have significantly reduced health threats associated with ambient air pollution, such stagnant high-pressure systems occur periodically, as do the semipermanent high-pressure systems over southern California. As such, they can cause significant increases in ground-level concentrations that may adversely affect the health of individuals at special risk (e.g., asthmatics, those ill with respiratory or cardiovascular disease, and the elderly). Therefore, there is a continuing need to forecast air pollution episodes. When the National Weather Service forecasts a developing episode, regulatory authorities at the local, state, and national levels begin to implement episode control plans, which may include requiring phased reductions in emissions of one or more primary pollutants and issuing community health warnings (see Chapter 8).

3.3.2 Air Quality Modeling

New and existing stationary sources regulated under New Source Performance Standards, and National Emission Standards for Hazardous Air Pollutants provisions of clean air legislation are required to demonstrate that they will comply with National Ambient Air Quality Standards even under atmospheric conditions that are unusually favorable for increased ground-level concentrations. Such compliance (in new source reviews) can only be demonstrated by using dispersion models to evaluate the source's effect. Primary inputs to dispersion models are emission and source information, meteorological data, and receptor information. Meteorological data and information required for these models include stability class, wind speed, ambient temperature, and MH. Point source and other sources such as from mobile dispersion, as well as other models, are described in detail in Chapter 7.

READINGS

Bourne, N.E., Atmospheric dispersion, in *Handbook of Air Pollution Technology*, Calvert, S., and Englund, H., Eds., John Wiley & Sons, New York, 1984.

Brasseur, G.P., Orlando, J.J., and Tyndall, G.S., Eds., *Atmospheric Chemistry and Global Change*, Oxford University Press, Oxford, 1999.

Briggs, G.A., *Plume Rise*, AEC Critical Review Series, Oak Ridge National Laboratory, Oak Ridge, TN, 1969.

Clifton, O. E., Fiore, A. M., Massman, W. J., Baublitz, C. B., Coyle, M., Emberson, L. et al. Dry deposition of ozone over land: Processes, measurement, and modeling. *Rev. Geophys.*, 58, e2019RG000670, 2020. https://doi.org/10.1029/2019RG000670.

Dobbins, R.A., *Atmospheric Motion and Air Pollution*, John Wiley & Sons, New York, 1979.

Finlayson-Pitts, B.G., and Pitts, J.N., Jr., *Chemistry of the Upper and Lower Atmosphere: Theory, Experiments, and Applications*, Academic Press, Orlando, FL, 2000.

Holton, J.R., *Introduction to Dynamic Meteorology*, 3rd ed., Academic Press, Orlando, FL, 1992.

Lutgens, F.K., and Tarbuck, E.G., *The Atmosphere*, 7th ed., Prentice Hall, Saddlebrook, NJ, 1998.

Scorer, R.S., *Meteorology of Air Pollution: Implications for the Environment and Its Future*, Ellis Horwood, New York, 1990.

Turner, D.B., *Workbook of Atmospheric Dispersion Estimates: An Introduction to Dispersion Modeling*, Lewis Publishers/CRC Press, Boca Raton, FL, 1994.

QUESTIONS

1. Is the atmosphere a sink?
2. What is the planetary boundary layer? What role does it play in tropospheric air motion?
3. What is turbulence and how is it formed?
4. How does turbulence affect the dispersion of pollutants from a source?
5. What is the effect of wind direction on pollutant concentrations downwind of a source?
6. The velocity of wind moving past a constant emission source changes from 1 to 4 m/s. What is the relative quantitative effect of this change of wind velocity on the pollutant concentration?
7. How is atmospheric stability related to lapse rate conditions?
8. Describe dispersion characteristics of the atmosphere under the following lapse rate conditions: ~2°C, 0°C, and 1°C/100 m.
9. Describe radiational inversion formation.
10. Indicate differences in the forms of radiational inversions in river valleys and over cities on flat, open terrain.
11. How do sea, lake, and land breezes affect air quality over a city?
12. Why can't polluted air, in most cases, penetrate an inversion layer? Indicate the physical principles involved.
13. How are subsidence inversions produced?
14. Characterize subsidence inversions relative to their vertical temperature profile, geographical scale, and persistence.
15. What meteorological factors affect plume rise?
16. Under what lapse rate conditions are looping, coning, and fanning plumes produced?
17. When do maximum ground-level concentrations of pollutants occur in mountain valleys? Why?
18. What is an urban plume? How is it formed?
19. What air quality problems are associated with long-range transport?

20. Generally, how long does it take for a long-lived pollutant to be uniformly mixed vertically in the troposphere? Horizontally throughout the troposphere?
21. By what mechanisms are pollutants transported into the stratosphere?
22. Characterize air circulation in the stratosphere.
23. What factors affect the lifetime of pollutants in the atmosphere?
24. What is dry deposition? What factors contribute to increased deposition rates?
25. Describe processes that result in rainout and washout of pollutants from the atmosphere.
26. What meteorological conditions produce pollution episodes?
27. How is meteorological information used in the implementation of episode plans?
28. Describe why ozone deposition has received increased attention in recent years.

4 Atmospheric Effects

Although the atmosphere affects the fate of pollutants via various sink processes, it experiences significant short-term and long-term pollution-induced changes. These may be local, regional, or global in scale, depending on factors such as sources, long-range transport, movement into the stratosphere, and accumulation of long-lived pollutants. Atmospheric effects may include changes in (1) visibility, (2) emissions, (3) air quality including surface ozone (O_3) and particulate matter (PM), (4) urban climate, (5) quantity and frequency of rainfall and associated meteorological phenomena, (6) atmospheric deposition, (7) stratospheric ozone (O_3) depletion, and (8) global climate change. The significance of these effects may be slight (visibility impairment) to potentially very serious (stratospheric O_3 depletion and changes in climate).

4.1 VISIBILITY

The single most recognizable effect of air pollution on the environment is visibility reduction, which results from the scattering and absorption of light by suspended particles and, to a lesser extent, absorption of light by atmospheric gases.

Changes in visibility may result from natural processes and phenomena as well as anthropogenic activities such as human-made air pollution from surface ozone and particulate matter formation.

We humans see objects and distinguish their shape and color characteristics because we have evolved the ability to receive reflected energy in the visible (400–700 μm) portion of the electromagnetic spectrum. Our ability to "see" is determined by factors associated with our individual vision system, our unique psychology, and a variety of physical factors in the atmosphere. These include (1) illumination of the observed object or area by the sun as mediated by clouds, ground reflection, and atmosphere; (2) reflection, absorption, and scattering of incoming light by objects and the sky; (3) scattering and absorption of light from objects and the source of illumination by the atmosphere and contaminants; (4) physiological response of the eye and brain to the resulting light energy received; and (5) subjective judgment of images received by the brain.

The human eye perceives the brightest object in its sight path as white, and the brain determines the color of other objects by comparison. The eyes' and brain's ability to perceive contrasts (the color difference between lightest and darkest objects) changes in response to illumination and the setting. Without contrast, or where contrast is limited, an object or scene may not be visible. The threshold contrast is that atmospheric condition at which an object or scene begins to be perceptible against its surroundings. Contrast determines the maximum distance at which an object or the various aspects of a scene can be discerned. Visibility is usually described in the context of light intensity; it can also include color. The human visual system detects color differences in response to the spectral distribution of incoming light.

4.1.1 Natural Visibility Reduction

Our ability to see objects clearly in an "unpolluted" atmosphere is limited by blue sky scattering, curvature of the Earth's surface, and suspended natural aerosols. At sea level, a particle-free atmosphere scatters light and limits visual range to 330 km (205 mi). The scattering of light by air molecules is called Rayleigh scattering and is responsible for the blue color of the sky on "clear" days. Because Rayleigh scattering decreases with decreasing atmospheric density, visual range increases with altitude. At approximately 4,000 m (12,000 ft), the potential visual range is approximately 500 km (310 mi).

DOI: 10.1201/9781003664109-4

When dark objects such as distant mountains are viewed through a relatively particle-free atmosphere, they appear blue because blue light is scattered preferentially into the line of sight. Clouds on the horizon may appear yellow-pink as the atmosphere scatters more blue light from bright targets out of the line of sight, resulting in longer wavelengths being seen.

In addition to Rayleigh scattering, which in conjunction with the Earth's curvature serves to place a theoretical limit on how far one can see, other natural factors, for example, fog, rain, snow, windblown dust, and natural hazes, limit visibility. Natural hazes are formed from aerosols produced from volcanic emissions of sulfur dioxide (SO_2) and from chemical reactions involving hydrocarbons (HCs) and various sulfur (S) species produced biogenically and geogenically. Reactions involving terpenes produce particles with diameters of less than 0.1 μm. Such particles preferentially scatter blue light and are the likely cause of bluish hazes over heavily forested areas such as the Blue Ridge and Great Smoky Mountains.

4.1.2 Visibility Concerns Associated with Human Activities

Pollution-induced changes in visibility that have received both scientific and regulatory interest include intense smog conditions such as those in Los Angeles and Houston and, to a lesser extent, other cities; seasonal regional hazes that affect large areas in the midwestern, eastern, and southeastern United States; plume blight and hazes produced by large point sources in the pristine air quality regions of the American Southwest and Rocky Mountains; springtime haze over the Arctic Circle; and, more recently, the Asian brown cloud.

4.1.2.1 Smog

Smog problems in Los Angeles and other cities are long-standing. Despite three decades of pollution control efforts, smog problems persist and, in cases such as Houston, TX, have become worse with time.

4.1.2.2 Regional Haze

The problem of regional haze in the midwestern, eastern, and southern United States grew progressively worse from the mid-1950s to the 1990s (visibility has not worsened in the past decade). This was particularly evident during the summer months. Regional haziness in the US during the summer is associated with slow-moving high-pressure systems. These often track up the Ohio Valley, where major point sources emit large quantities of SO_2 that, in combination with warm temperatures, abundant sunlight, high relative humidity (RH), and reduced vertical mixing, produce conditions favorable for regional haze production. Such haze is visible from orbiting Earth satellites. The decline in visibility observed during the summer months in the Northeast, Midwest, and Southeast is strongly correlated with increased use of coal to produce electrical power to serve air conditioning needs in residential and nonresidential buildings.

4.1.2.3 Pristine Air

Visibility impairment has historically been a problem in US urban centers; it is increasingly a problem on a broad regional scale in many areas east of the Mississippi River. In the regulatory sense, however, it has been the pristine areas of the western United States that have received the most attention. The scenic beauty associated with our national parks, monuments, and wilderness areas is due in great measure to the high air quality, high visibility (visual ranges of 100–164 km [62–102 mi] are not uncommon), and contrast characteristics of these regions. Major concerns include visibility-reducing and aesthetics-degrading power plant plumes and regional hazes produced by the photochemical conversion of SO_2. The former is commonly described as plume blight. In pristine areas, power plant plumes may maintain their distinct character for tens of kilometers downwind. The plume, and the haze it will eventually produce, causes blighted vistas, destroying the unique character of some of America's most beautiful landforms and environments.

The problems of visibility degradation of pristine air and the protection of areas of high air quality have been addressed in clean air legislation and subsequent regulatory actions by the US Environmental Protection Agency (US EPA; see Chapter 8).

4.1.2.4 Arctic Haze

The phenomenon of Arctic haze was first reported by weather reconnaissance crews flying over the Arctic in the 1950s. Significant springtime visibility reduction was observed over areas such as Barrow, AK, where the environment was considered to be relatively pristine. The Arctic haze phenomenon has intensified over the past half century. It now covers a linear expanse of 800 to 1,300 km (496–806 mi). It occurs at altitudes below 9 km (5.8 mi), with a maximum intensity at approximately 4 to 5 km (2.5–3.1 mi).

The primary cause of Arctic haze is industrial emissions of SO_2, elemental carbon (EC), HCs, metals, and other gases and particles from northern European and Eurasian sources. These pollutants are transported northward into an environment where removal processes are less efficient. A steep inversion layer (up to 30°C–40°C [86°F–104°F] differential between cooler surface and warmer air layers several hundred meters above) further contributes to the phenomenon. Sulfates, at concentrations of 2 μg/m^3, are on the order of 10 to 20 times greater than would be expected from sources in the affected region.

4.1.2.5 South Asian Haze

Although the Arctic haze phenomenon has been known and studied for decades, the "discovery" of the South Asian haze or Asian brown cloud was first reported in 1999. The haze, a brownish layer 3 km thick, covers most of the tropical Indian Ocean and South, Southeast, and East Asia (Figure 4.1). This is the most densely populated region in the world. Its discovery and recent

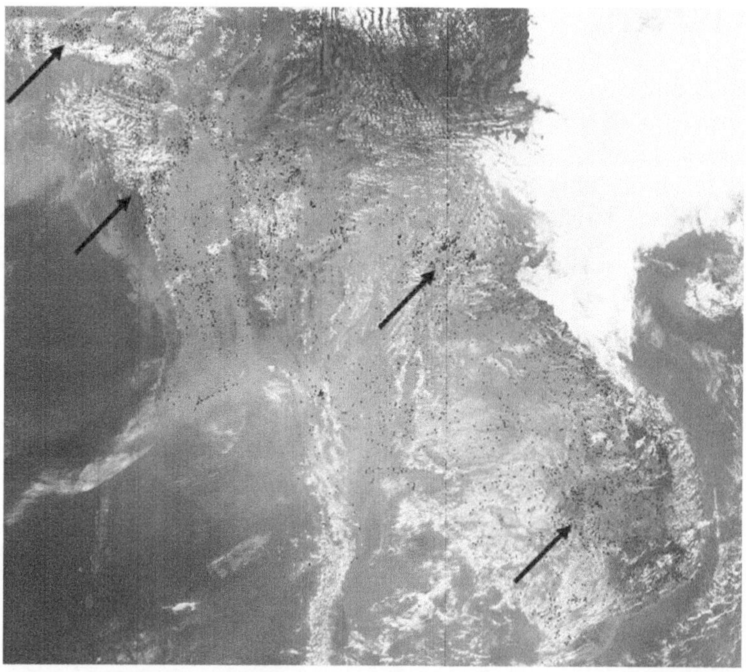

FIGURE 4.1 Satellite image of haze over Southeast Asia. (From National Aeronautics and Space Administration, https://www.nasa.gov/image-feature/goddard/2019/southeastern-asia-ignited-by-agricultural-fires.)

understanding of the phenomenon were outcomes of the Indian Ocean Experiment (INDOEX) studies conducted by an international team of scientists.

Haze particles and precursors are transported far beyond their source region, particularly during the dry season (December to April). South Asian haze includes elevated concentrations of sulfates, nitrates, organic compounds, carbon (C), fly ash, and other pollutants. Major sources include biomass burning and industrial emissions.

Direct effects include a significant reduction (10%–20%) in solar radiation reaching the surface, a 50% to 100% increase in solar heating of the lower atmosphere, rainfall suppression, reduced agricultural productivity, and adverse health effects.

4.1.3 Factors Affecting Visibility Reduction

Visibility reduction associated with anthropogenic activities is due primarily to light scattering by particles and, to a lesser extent, absorption of light by atmospheric gases such as nitrogen dioxide (NO_2). Nitrogen dioxide absorbs short-wavelength blue light, with longer wavelengths reaching the eye. Its color, therefore, appears yellow to reddish brown. Light can also be absorbed by particles. Such absorption is important when soot (C) or large amounts of windblown dust are present. In general, most atmospheric particles are relatively inefficient in absorbing visible light.

Fine fraction solid or liquid particle aerosols are responsible for most atmospheric light scattering. Particles with diameters similar to the wavelengths of visible light (0.4–0.7 μm) are particularly effective in light scattering (Figure 4.2). The scattering of all visible wavelengths equally is called Mie scattering. This wavelength-dependent interaction of light with the atmosphere can be described mathematically by radiative transfer equations:

$$-\mathrm{d}I = BeId_x \tag{4.1}$$

where $-\mathrm{d}I$ is the decrease in intensity (extinction), Be is the extinction coefficient, I is the initial beam intensity, and d_x is the length of light path.

Equation 4.1 can be used to describe the effect of the atmosphere on a light beam as it passes from a source to an observer. The value of Be, the extinction coefficient, is determined by the scattering and absorption of light by particles and gases and varies as a function of wavelength and contaminant concentrations.

The intensity of a beam in the direction of an observer decreases with distance as light is scattered or absorbed. Over a short distance, this decrease is proportional to the path length and intensity

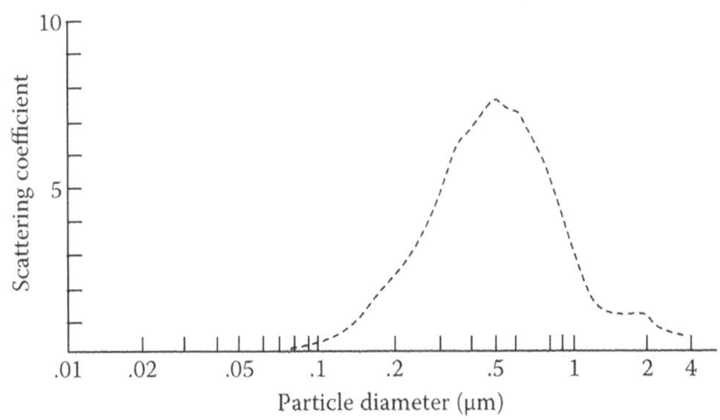

FIGURE 4.2 Relationship between light scattering and particle size. (From US EPA, EPA/600/8-82-029, EPA, Washington, DC, 1982.)

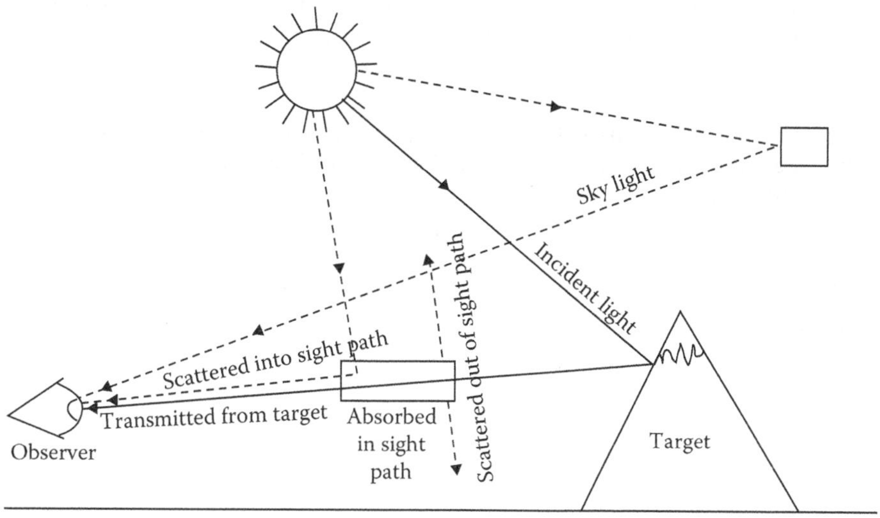

FIGURE 4.3 Relationships between light scattering and visibility of a target object. (From US EPA, EPA/600/P-99/002aB, EPA, Washington, DC, 2001.)

of a beam at that point. Similarly, when an observer is looking at a distant object such as a mountain, light from the mountain is diminished by absorption and scattering caused by the intervening atmosphere. In addition, light is also scattered into the sight path from the surrounding atmosphere. It is this "sky light" that forms the optical phenomenon we know as haze. These relationships can be seen in Figure 4.3.

With increasing distance, extinction and added sky light cause both dark and bright objects to be "washed out" and approach the brightness of the horizon. Consequently, the contrast of an object relative to the horizon and other objects decreases.

The apparent contrast and visual range of a large, dark target object can be calculated from the following series of exponential equations:

$$C = C_o e^{-BeX} \tag{4.2}$$

where C is the apparent contrast at observer distance, C_o is the initial contrast at the object, e is the natural log base (2.718), Be is the extinction coefficient, and X is the observer-to-object distance.

For a black object, the initial contrast is −1, and Equation 4.2 becomes:

$$C = (-1)e^{-BeX} \tag{4.3}$$

By assuming a standard threshold of perception of 0.02, we can calculate the visual range (V_r), which is equal to the observer-to-object distance:

$$0.02 = -e^{-BeV_r} \tag{4.4}$$

$$V_r = 3.92 / Be \tag{4.5}$$

Equation 4.2 is called the Koschmieder equation. It is used to determine the visual range from measurements of the extinction coefficient. The visual range is an inverse function of the extinction coefficient, which is directly related to atmospheric aerosol concentrations.

The extinction coefficient (*Be*) is the sum of both absorption and scattering components expressed as inverse lengths of the atmosphere (megameters). Extinction coefficients range from 10 Mm^{-1} in clean air to 1,000 Mm^{-1} in very polluted environments. These correspond to visual ranges of 400 km to 4 km (see Section 4.1.5).

Light scattering by aerosol particles includes several physical phenomena: reflection, which causes backscatter; and diffraction and refraction, which cause scattering in the forward direction (Figure 4.4). In diffractive scattering, a light ray is bent by the edge effect of an aerosol particle into its shadow. In refraction, two effects are produced. For particles with a refractive index of $n > 1$, the speed of the light wave is reduced, causing a reduction in the wavelength inside the particle. This results in a phase shift, which produces positive and negative interferences. Refraction also bends light waves, causing them to pass through and out of a particle at a different angle from that at which they entered. This refractive bending is responsible for the separation of the color components of white light by a prism. Refractive bending of light by aerosol particles can also have chromatic effects. Coarse particles (> 2 μm) scatter light in the forward direction primarily by diffraction and refraction. For particles in the accumulation mode (0.1–1.0 μm), physical interactions with light are complex and enhanced. Although coarse particles are efficient in scattering light, their contribution to visibility reduction is limited due to their relatively small numbers.

Light scattering may be affected by the chemical composition of aerosol particles. As a rule, particles of differing chemical composition will have different light-scattering efficiencies. In addition, hygroscopic substances sorb water (H_2O), which allows them to grow into the critical diameters that scatter light most effectively. Sulfate aerosols are very hygroscopic and therefore absorb light efficiently.

4.1.4 Meteorological Effects

Meteorological factors affect visibility in several ways. Sunlight affects visibility by promoting secondary aerosol formation. Atmospheric photochemistry produces the major visibility-reducing aerosols, that is, sulfates, nitrates, and oxyhydrocarbons.

Wind speed and atmospheric stability affect visibility because they determine atmospheric dispersion and therefore the concentration of aerosol particles. In general, as wind speed increases, visibility improves. This wind-induced atmospheric mixing results in lower aerosol concentrations.

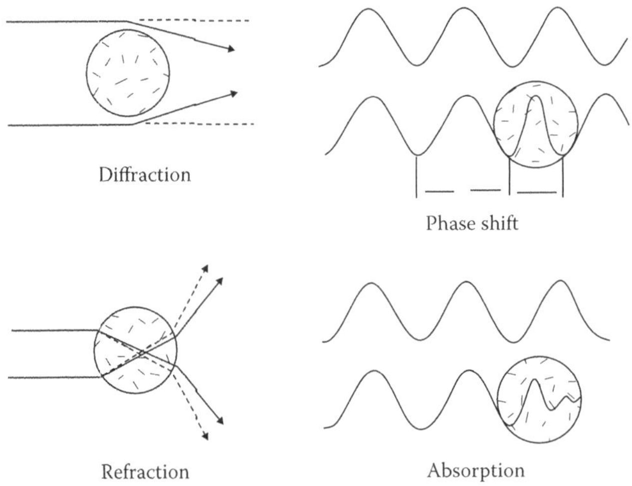

FIGURE 4.4 Light-scattering phenomena. (From US EPA, EPA/600/8-82-029, EPA, Washington, DC, 1982.)

During periods of atmospheric stagnation (associated with slow-moving high-pressure systems), vertical mixing is suppressed, aerosol concentrations increase, and visibility is significantly reduced. The resultant haze may cover hundreds of thousands of square kilometers. It is not uncommon to have a major portion of the Midwest, Southeast, or East Coast covered by a "blanket" of haze.

Visibility may be significantly affected by RH. Hygroscopic particles such as sulfates (described earlier) take up moisture and increase in size under humid conditions. Marked visibility reduction has been reported when RH exceeds 70%, although the importance of this effect extends over a range of RHs. In general, regions and periods of low RH are associated with good visibility.

4.1.5 Visibility Measurements

Qualitative visibility measurements (reported as prevailing visibility) have been routinely made (since 1939) by human observers at commercial and military airfields in conjunction with, or in addition to, the National Weather Service. Observers viewed the horizon noting whether dark objects (daytime) or lights (nighttime) could be seen at known distances. These were reported as the greatest visibility attained around at least one-half of a horizon circle. Such readings are dependent on the individual observer and the availability of "targets" at varying distances. In most cases, such visibility readings are poorly related to air quality.

In the past three decades, quantitative techniques (using instrumental methods) have been developed that more accurately measure visibility reduction and its relationship to atmospheric pollutants. These methods have been incorporated into visibility monitoring networks.

The largest instrumental visibility monitoring network has been installed at airports in the United States to provide real-time data for runway visibilities. The network includes automated surface observing systems (ASOS) and automated weather observing systems (AWOS). These use a forward-scattering visibility meter that measures air clarity. Clarity readings are converted to sensor equivalent visibility values, which are a measure of what would be perceived by the human eye. The instrument samples 1.9 cm (0.75 in.) of the atmosphere, providing average 1-minute values. By using an algorithm, relatively accurate visibility readings for within 3 to 5 km (2–3 mi) of the measurement location can be determined.

Visibility measurements are also routinely made by the National Park Service, US Forest Service, Bureau of Land Management, and the Fish and Wildlife Service in and around Class 1 visibility protection areas subject to regulation under the prevention of significant deterioration provisions of clean air legislation (see Chapter 8). Visibility measurements in Class 1 regions (see Chapter 8) have been made using instrumental methods for more than a decade. The US EPA has developed a collaborative program with these federal agencies in a large monitoring network that includes visibility and air quality measurements. This network is called the Interagency Monitoring of Protected Visual Environments (IMPROVE).

The most frequently used approach to characterize the relationship between visibility and air quality is the measurement of the extinction coefficient using a transmissometer. Visual range is then calculated from the Koschmieder equation (Equation 4.2) by assuming that the atmosphere and daytime illumination is uniform over the sight path and the threshold is 2%. Recall that visual range is inversely related to the extinction coefficient, which is directly related to atmospheric aerosol concentrations.

The extinction coefficient, *Be*, can be used to calculate a deciview index. The deciview index expresses uniform changes in haziness in increments from pristine conditions to very impaired visibility. The relationship between extinction coefficients, deciviews, and visual range can be seen in Figure 4.5. A 10% to 20% change in the extinction coefficient represents a 1- to 2-deciview change; a doubling of the extinction coefficient increases the deciview value by 7 units and decreases the visual range by 50%.

4.1.6 Visibility Patterns and Trends

Visibility patterns have been evaluated at approximately 280 airports over the past 50 years using visual range data. These data indicate that summertime visibility impairment has been particularly

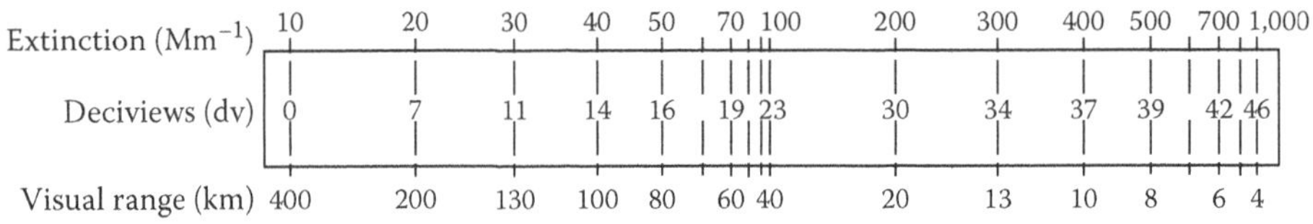

FIGURE 4.5 Relationships between extinction coefficients, deciviews, and visual range. (From US EPA, EPA/600/P-99/002aB, EPA, Washington, DC, 2001.)

TABLE 4.1
Pollutant Contributions (%) to Visibility Impairment at IMPROVE Sites in the Eastern and Western United States

Pollutant	Eastern US	Western US
Sulfates	60–86	25–50
Organic carbon	10–18	25–40
Nitrates	7–16	5–45
Elemental carbon	5–5	5–15
Crustal material	5–15	5–25

Source: US EPA, EPA 454/K-03-001, EPA Office of Air Quality and Standards, Research Triangle Park, NC, 2003.

significant in areas east of the Mississippi River in the lower Midwest, in middle Atlantic and southeastern states, and on the southern coast of California, which, for the most part, includes the Los Angeles Basin. Summertime haze in the eastern United States increased significantly from 1970 to 1980 and decreased slightly between 1980 and 1990. In 1970, the highest summertime haze conditions were reported for Indiana, Ohio, Pennsylvania, New Jersey, southeast New York, Delaware, Maryland, Kentucky, West Virginia, Virginia, and North Carolina. By 1980, high haze conditions expanded through most of the southeastern states, including Florida and westward into Missouri. Haziness in the Southeast has increased 80% since the late 1950s.

In addition, visibility trends have been evaluated from instrumental monitoring of light extinction at national parks and wilderness areas over the past 25 years. These monitoring efforts were expanded under the IMPROVE program, which began in 1987. Monitoring data include visibility and pollutant measurements.

IMPROVE data show that visibility impairment is significantly worse in the rural East than in the rural West. The average visual range in eastern Class 1 regions is 38 to 146 km (24–91 mi) compared with an estimated natural visibility of 145 km (90 mi). The average annual visibility is 102 to 272 km (64–170 mi) in most Class 1 areas (which are mostly in the western states) compared with an estimated natural visibility of 226 km (140 mi).

No significant declines in visibility and air quality were observed at IMPROVE sites during the period 1992 to 2008. Seventy percent of these sites showed improvements for annual best and median visibility days. Little or no visibility improvement was observed in the number of worst visibility days at sites such as the Grand Canyon, Great Smoky Mountains, Mount Rainier, Shenandoah, Yellowstone, and Yosemite National Parks.

Air quality data from the IMPROVE program indicate that visibility reduction is associated with particulate-phase sulfates, nitrates, elemental C, and crustal material. The relative contribution of each of these particle components can be seen for eastern and western rural sites in Table 4.1. Note the dominance of sulfates in causing visibility impairment in the eastern United States and their significant contribution even in western states (65% on the worst days). Both organic C and nitrates are seen to be significant causes of visibility reduction at western sites. Organic C contributes up to 40% to 45% of light extinction in the Pacific Northwest, with nitrates being the primary cause of light extinction (~40%) in southern California.

4.2 TURBIDITY

Visibility is a measure of horizontal light scattering. Turbidity, on the other hand, is a measure of vertical extinction. Vertical extinction is dependent on the same atmospheric light-scattering characteristics of aerosols associated with visibility reduction. The principal effect of atmospheric turbidity is the diminution of sunlight received on the ground. Although most light scattering is in a forward direction, intensity is reduced.

Turbidity, or optical thickness of the atmosphere, is directly related to aerosol concentrations. Aerosols in both the tropospheric and stratospheric layers of the atmosphere contribute to turbidity. In general, tropospheric extinction can be directly attributed to aerosols from human activity. Stratospheric turbidity is largely attributable to aerosols produced from volcanoes and, increasingly, aerosols associated with human activities.

The turbid air over major cities is often described as a dust dome. In studies in the Cincinnati, OH, metropolitan area, a complex stratification of the dust dome has been observed, with alternating polluted and cleaner air layers that correlate with different lapse rate conditions (Figure 3.4). High turbidity occurs between 11 a.m. and 12 p.m. on both relatively clean and polluted days. On relatively clean days, the lowest 1,000 m (~3,000 ft) contributed 30% of light extinction; on polluted days, 65%. Other turbidity-related phenomena in the troposphere include (1) a commonly observed turbid air mass centered over the eastern part of the United States; (2) maximum turbidity during the summer and minimum turbidity in winter; (3) losses of direct radiation of up to 50% in the rural midwestern United States during summer, resulting in up to a 20% reduction of total radiation received at the ground; and (4) an increasing upward trend in turbidity over Washington, DC, and Davos, Switzerland, over the past four decades.

In the stratosphere, an aerosol layer described as the Jange layer forms from volcanic eruptions and is maintained by particle formation and transport in the upper tropical troposphere. This stratospheric aerosol layer varies in height from 9 to 28 km (5.5–17 mi), with maximum concentrations at approximately 18 km (11 mi). This aerosol consists of highly concentrated liquid sulfuric acid (H_2SO_4) droplets, typically 0.1 to 0.3 μm in diameter. Most of the mass of this aerosol layer is formed from SO_2 and reduced S gases such as carbon disulfide (CS_2) and carbonyl sulfide (COS). They are oxidized in the stratosphere to H_2SO_4, which condenses on preexisting nuclei.

The stratospheric aerosol layer can have a variety of atmospheric effects. In backscattering sunlight, it reduces the amount of solar radiation reaching the ground, thus cooling the atmosphere. Cooling of portions of the troposphere by such turbidity has been observed after major volcanic eruptions, particularly those that have high S gas emissions. In addition, aerosol particles provide sites for heterogeneous chemical reactions that repartition nitrogen (N) species and decrease O_3 concentrations. In polar regions, reactions on these sulfate particles can result in catalytic destruction of O_3 by chlorine (Cl) atoms.

4.3 THERMAL AIR POLLUTION

Historically, only gases and aerosol particles were considered to be atmospheric pollutants. However, because of the effect of surfaces and waste heat on urban heat balance and the resultant alteration of the atmospheric environment, heat or thermal energy is certainly a form of air pollution.

In making temperature measurements on a clear night (with light winds) in an urban area as well as in the surrounding countryside, it is possible to detect and characterize a phenomenon known as the urban "heat island." In this heat island, areas of elevated temperatures (Figure 4.6) can be identified, with the entire urban area being warmer than the surrounding countryside. The urban areas appear to

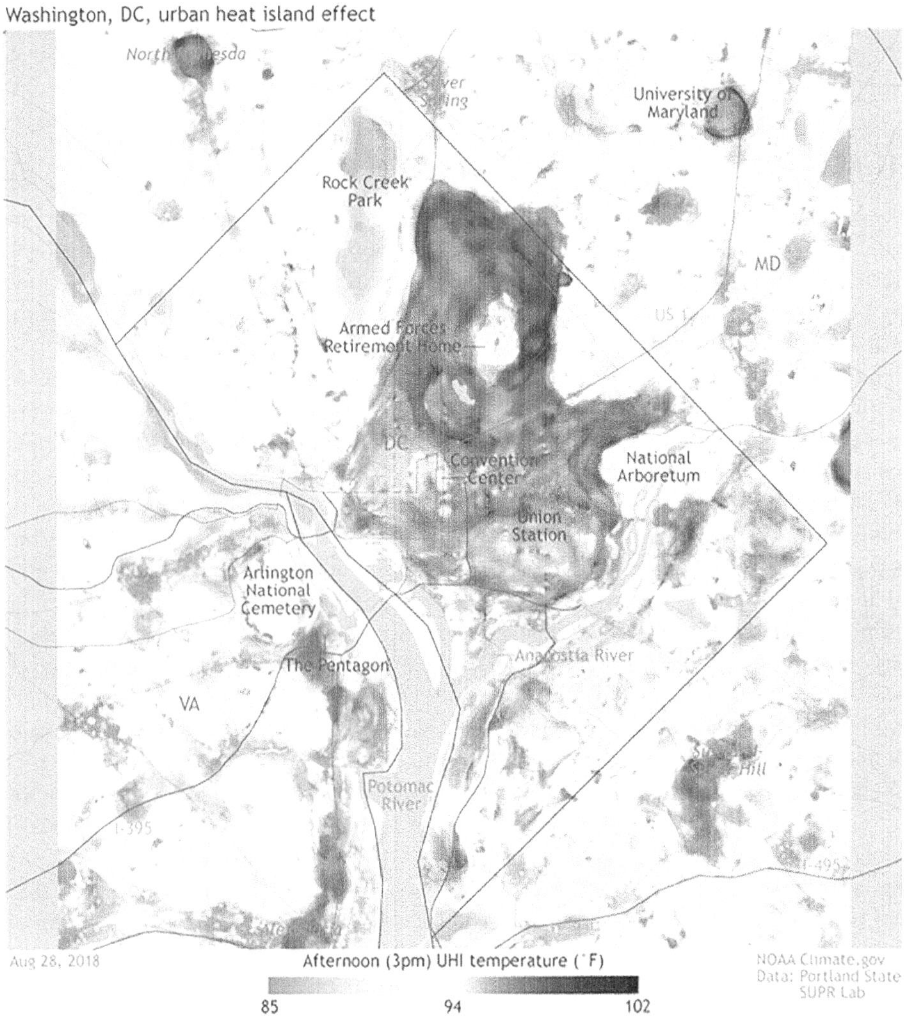

FIGURE 4.6 Urban heat island associated with Washington, DC. (From NOAA, 2018. NOAA Education Accomplishments Report. Available online at https://www.noaa.gov/sites/default/files/legacy/document/2019/Jun/FY_2018_NOAA_Education_Accomplishments_Report.pdf)

be approximately 6°C (10°F) warmer on average than rural/park areas, and urban areas appear to be approximately 2°C (4°F) warmer on average than suburban. The principal contributors to heat island formation include (1) waste heat dumped into the environment from energy-utilizing processes, (2) solar energy absorption and heat storage properties of urban surfaces, and (3) decreased urban ventilation resulting from the increased surface roughness of buildings. Urban surfaces such as buildings, sidewalks, and streets more readily absorb and store heat from sunlight during daylight hours than do vegetated areas external to the city. In addition, urban surfaces cool more gradually at night.

In cities such as New York, anthropogenic heat may be 2.5 times greater than that received from the sun during winter; during summer, it represents a significant portion of the total city energy budget (on the order of 17%). On an annual basis, anthropogenic heat in many cities may exceed 10% of that received from the sun.

Urban heat islands may have significant effects on weather and local climate within urban centers and, in some cases, downwind. These may include (1) longer frost-free seasons, (2) less frequent fog, (3) decreased snow accumulation, (4) lower RH, (5) decreased likelihood of surface-based nocturnal inversions, (6) increased rainfall downwind, and (7) a distinctive nocturnal metropolitan air circulation pattern.

This nocturnal circulation pattern is much like a sea breeze on an island. As warm air rises from the urban center it flows outward toward rural areas. Cooler air outside the urban center sinks and flows inward toward the city. This thermally induced airflow creates a turbid mass over the city that tends to be highest in the middle of the heat island. The resultant "dust dome" forms best under conditions of light wind and compact city form.

Because of thermal turbulence and vertical mixing, surface-based inversions generally do not occur over urban heat islands. However, elevated nocturnal inversions of longer duration may frequently occur. These elevated inversions may result, in part, from the dust dome absorption of thermal energy. They may also result from the advection of stable rural air over warmer urban surfaces. As this air moves over the city, it is heated from below, resulting in an inversion layer above the lower, relatively warm air.

Increased rainfall downwind of cities (discussed in a subsequent section) may be due in part to the heat island effect. Enhanced vertical convective mixing tends to make precipitation more probable.

One of the more notable potential effects of the heat island phenomenon is the virtual disappearance of the fogs that had been a common part of the environment of London, England.

4.4 EFFECTS ON PRECIPITATION AND PRECIPITATION PROCESSES

Pollutants emitted in urban areas may increase precipitation and affect other related meteorological phenomena. In some cases, the opposite effect has been observed; that is, pollution sources have decreased downwind precipitation.

Atmospheric particles can serve as sites for condensation of H_2O vapor. Condensation around nucleating particles is an important factor in cloud development and precipitation. Only a small fraction of the total aerosol load of the atmosphere serves as weather-active or cloud condensation nuclei (CCN).

The concentration of CCN determines the initial size and number of H_2O droplets, which determine the frequency and amount of precipitation. An inverse relationship exists between CCN concentrations and initial droplet size; that is, as CCN concentration increases, droplet size decreases. Water droplets increase in size by coalescence or agglomeration. Because larger droplets coalesce more effectively, they have a higher probability of falling as rain. The more numerous, and smaller, droplets produce less rainfall. This phenomenon may explain the decreased rainfall reported downwind of sources such as sugarcane fires in Australia and forest fires in the northwestern United States These sources apparently contribute large numbers of CCN, which decrease droplet size.

Where pollution sources contribute significant numbers of giant condensation nuclei, small CCN may be less important in affecting precipitation. The increase in rainfall observed downwind of some cities may be due to the contribution of giant nuclei (> 1 μm) that promote coalescence and increase precipitation probability.

Pollutants may also affect precipitation in cold cloud processes by influencing the concentration of freezing nuclei. Because atmospheric levels of freezing nuclei are frequently low, the addition of freezing nuclei in the form of pollution particles may enhance precipitation under cold cloud conditions. An application of this principle is cloud seeding with silver iodide crystals.

In previous paragraphs, it was noted that increased precipitation has been observed downwind of some US cities. Studies downwind of St. Louis, MO, indicate that summer rainfall in the 1970s increased 6% to 15% as a consequence of urban effects. Precipitation increases seem to be associated with elevated particulate pollutant levels.

4.5 ATMOSPHERIC DEPOSITION

Pollutants affect not only the frequency and quantity of precipitation, but also its chemical composition. Historically, the primary focus of environmental and regulatory concern related to changes in precipitation chemistry has been acidification. The effects of acidic deposition have been extensively studied. As a result of its known adverse environmental effects, a significant regulatory effort to reduce emissions of acid precursors has been implemented under provisions of the Clean Air Act Amendments of 1990 (see Chapter 8). Although some improvements in precipitation acidity are already evident in North America, acidic deposition remains a significant environmental problem in northern Europe and is an increasing problem in Asia.

Environmental concerns associated with the effects of acidic deposition have, to some degree, overshadowed other major atmospheric deposition phenomena. These include the deposition of nitrate N, mercury (Hg), and long-lived organic compounds such as pesticides, polychlorinated biphenyls (PCBs), and dioxins. These emerging atmospheric deposition concerns are treated here, with potential effects on ecosystems discussed in Chapter 6.

4.5.1 Acidic Deposition

4.5.1.1 Characteristics of the Phenomenon

Changes in the hydrogen ion (H^+) concentration or pH of precipitation associated with atmospheric pollution have been reported for more than a century. However, significant scientific attention to the phenomenon only began in Scandinavian countries in the mid-1960s and in North America in the early to late 1970s.

When pollutant-caused acidification of precipitation was first described, it was called acid rain. As additional studies were conducted, it became apparent that snow was also being affected. Thus, it became more appropriate to describe the phenomenon as acid precipitation. Because acidic materials are deposited on land and water surfaces in the absence of precipitation (dry deposition), the concept was further broadened. It also includes acid fog (with a pH as low as 1.8), acid clouds, acid dew, and probably acid frost as well. To take all these into account, the term *acidic deposition* is used in the scientific community to describe the phenomenon. The term *acid rain*, however, is commonly used by political leaders and the general public. Thus, in public discourse, the terms *acid rain* and *acidic deposition* are synonymous.

When H_2O condenses after evaporation or distillation, it has a pH of 7 and is neither acidic nor alkaline. As it comes into equilibrium with carbon dioxide (CO_2) in the atmosphere, it rapidly changes to a pH of 5.65. Cloud H_2O, rainwater, and snow, in the absence of pollution and in equilibrium with CO_2, would naturally be acidic (pH = 5.65).

The term *pH* refers to the negative log of the hydrogen ion (H^+) concentration. A solution with a pH of 6 has a H^+ concentration of 0.000001 M, and one with a pH of 5 has a H^+ concentration of 0.00001 M. Values of pH are geometric; that is, a pH of 5 is 10 times more acidic than a pH of 6, and a pH of 4 is 100 times more acidic than a pH of 6.

A pH of 5.65 has been used as a reference to assess the significance of pollution-related precipitation acidification. Studies in remote areas of the world where anthropogenic effects on precipitation are likely to be minimal indicate that the average background pH value is closer to 5.0 than 5.65.

Average precipitation deposition values (pH) for the United States in 2023 are illustrated in Figure 4.7. Areas with high acidic deposition are clustered around the Ohio Valley, where high-S coal is used to generate electricity. In this area, pH values of rainfall average approximately 5.3 to 6.0. Using a reference value of pH = 5.0, sulfuric acid's contribution to precipitation acidity has decreased substantially (≈70%–90%). Nitric acid's contribution has decreased moderately (≈30%–50%), but its relative share increased as sulfur fell away since the 1990s.

In the northeastern United States, decreased precipitation acidity seems to be due to sulfuric (65%) and nitric (30%) acids. The apparent major sources of precursors (S and N oxides) for these strong acids include fossil fuel–fired power plants, industrial boilers, metal smelters, and automobiles. In the western states, where sulfur oxide (SO_x) sources are few, precipitation acidity is predominantly due to nitric acid (HNO_3), with motor vehicles being the primary source of acid precursors.

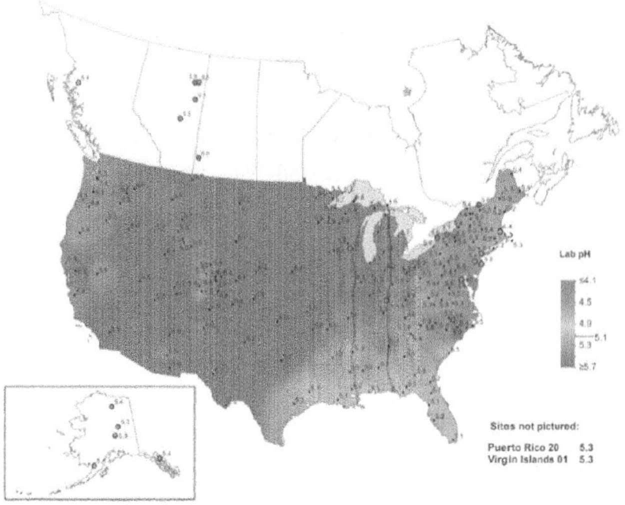

FIGURE 4.7 Geographic distribution of precipitation H^+ (expressed as pH) concentration over the continental United States in 2023. (From National Atmospheric Deposition Program/National Trends Network, https://nadp.slh.wisc.edu/filelib/annual-summary/2023as.pdf.)

The major source region for acid precursors in the eastern United States is the middle Ohio Valley. Approximately 60% to 70% of SO_x emissions have been associated with large electric power plants that burn high-S coal. These plants, with their tall stacks (~200 to 400 m, 660–1,260 ft), disperse emissions at relatively high altitudes (> 300 m, ~1,000 ft), providing longer residence times for the conversion of acid precursors to strong acids. Consequently, emission of SO_x from the Midwest and northern Appalachian states has affected acidic deposition over large geographical areas downwind, including transboundary transport into Canada. In the 1980s and early 1990s, approximately 50% of the acid-forming pollution burden falling on eastern Canada originated in the United States. Canadian sources, to a lesser extent, contribute to acidic deposition in the United States.

Long-range transport results in acidic deposition on land and water surfaces at considerable distances downwind of the sources of acid precursors. However, acidic deposition at a particular receptor site is likely to be more strongly affected by local sources. The relative contribution to acidic deposition of distant, compared with local, sources has been a matter of intense debate. Nevertheless, it is apparent that long-range transport of acid precursors has a profound effect on acidic deposition hundreds of kilometers downwind of precursor sources.

Although HNO_3 is responsible for only approximately 35% of the H^+ concentration in acidic precipitation and dry deposition in the eastern United States, the importance of nitrogen oxides (NO_x), and their contribution to acidic deposition, may be far more significant than this number indicates. In the northeastern United States, S emissions have not increased much in the past quarter century, although significant increases have occurred in the Southeast. What has changed substantially since 1960 has been the emission of NO_x from automobiles and electrical power generation. It is the absorption of sunlight by NO_2 that initiates photochemical reactions that promote atmospheric production of H_2SO_4 and HNO_3. The problem of acidic deposition is not simply a matter of emissions of acid precursors, but also of atmospheric processes that result in their conversion to strong acids.

4.5.1.2 Environmental Impacts

Acidic deposition poses a threat to ecosystems in which, because of local or regional geology (crystalline or metamorphic rock), soils and surface waters cannot adequately neutralize acidified rain, snow, and dry deposited materials. Such soils and waters contain little or no calcium or magnesium carbonates and therefore have low acid-neutralizing capacity (ANC). Acid-sensitive ecosystems in the USA are illustrated in Figure 4.8. These include the mountainous regions of New York and New England; the Appalachian Mountain chain, including portions of Pennsylvania, West Virginia, Virginia, Kentucky, and Tennessee; upper Michigan, Wisconsin, and Minnesota; the pre-Cambrian Shield of Canada, including Ontario, Quebec, and New Brunswick; and various mountainous regions of the North American West.

4.5.1.2.1 Chemical Changes in Aquatic Systems

Acidic deposition has been linked to the chronic acidification of thousands of lakes and episodic acidification of thousands of streams in the Scandinavian peninsula and southeastern Canada. In the United States, surface H_2O acidification by acidic deposition has been more limited. The National Surface Water Survey (NSWS), conducted in acid-sensitive regions in the 1980s, revealed that the chemical composition of approximately 75% (880) of acidic lakes and 47% (2,200) of acidic streams was dominated by sulfate from atmospheric deposition, indicating that acidic deposition was the primary cause of acidification. Most surface waters acidified by acidic deposition were found in six regions: New England, the southwest Adirondacks, small forested watersheds in the mid-Atlantic mountains, the mid-Atlantic coastal plain, the northcentral portion of Florida, and the upper peninsula of Michigan and northeastern Wisconsin. Virtually no acidic surface waters were observed in the western United States or the southeastern mountains.

Although a lake or stream may have low ANC and therefore be acid-sensitive, acidification, whether it is chronic or episodic, will depend on source factors as well as processes occurring within a watershed. As rain or snowmelt waters move through a watershed, their chemistry is altered by processes that neutralize acids. The likelihood of neutralization increases as the distance to the receiving waters increases.

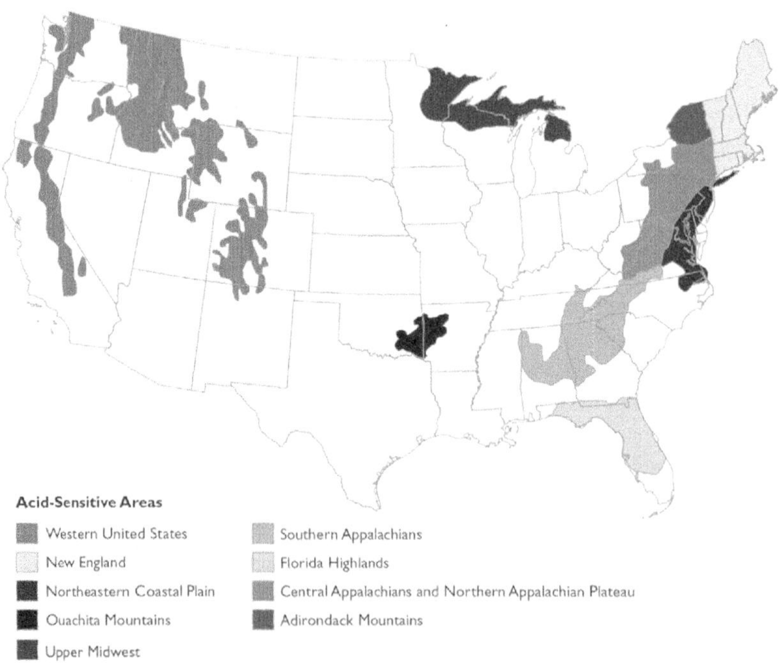

FIGURE 4.8 Acid-sensitive ecosystems in the United States of America. (National Acid Precipitation Assessment Program Report to Congress 2011: An Integrated Assessment, December 11, 2011. https://obamawhitehouse.archives.gov/sites/default/files/microsites/ostp/2011_napap_508.pdf)

Surface H_2O chemistry changes associated with acidic deposition may include (1) increases in sulfate concentration, (2) little change in nitrate levels (because watersheds effectively retain nitrate), (3) increases in base cations such as calcium (Ca^{2+}) and magnesium (Mg^{2+}) associated with acid neutralization processes, (4) decreases in pH and ANC, (5) increases in aluminum Al^{3+}, and (6) decreases in dissolved organic carbon (DOC).

Although chronic acidification of lakes has received the most attention, it is important to note that episodic acidification is a common phenomenon in regions that receive high levels of acidic deposition. In lakes and streams where chronic acidification has lowered ANC levels, small decreases in ANC caused by pulses of nitrate, sulfate, or organic anions can lower pH to critical levels. Pulses of nitrate commonly contribute to acidic episodes in portions of New England, the mid-Atlantic mountains, and the Adirondacks. Pulses of sulfate associated with acidic deposition have been shown to be significant contributors to episodic acidification in some streams in the mid-Atlantic mountains.

4.5.1.2.2 Changes in Soil Chemistry

Acidic deposition can affect soil chemistry, particularly in regions with low ANC. Changes include leaching of base cations such as Ca^{2+}, Mg^{2+}, and potassium (K^+). These changes result in reduced base cation exchange capacity of the soil. Acidity also results in the leaching of toxic metals such as Al, lead (Pb), Hg, and others.

4.5.1.3 Trends in Sulfate and Acidic Deposition

In 2023, the ninth year of operation for the Cross-State Air Pollution Rule (CSAPR) SO_2 programs, sources regulated under both the CSAPR annual SO_2 programs and the Acid Rain Program (ARP) collectively reduced SO_2 emissions by 15.1 million tons (96%) compared to 1990 levels (prior to ARP implementation), 10.6 million tons (94%) compared to 2000 levels (ARP Phase II), and 9.6 million tons (94%) compared to 2005 levels (before CAIR and CSAPR implementation). In total, ARP and CSAPR sources emitted approximately 650,000 tons of SO_2 in 2023. These decreases were on the order of 94% to 96%. The largest reductions were observed along the Ohio River and its tributaries. Reductions in H^+ concentrations were similar to those of sulfate concentrations in both magnitude and location.

4.5.2 Nitrogen Deposition

Although NO_x, ammonia (NH_3), and organic N compounds occur naturally in the atmosphere, their concentrations significantly increase as a consequence of human activities. Fossil fuel combustion and combustion of fuels by mobile sources are the major sources of NO_x. The largest sources of anthropogenic-related emissions of NH_3 are agricultural fertilizers and domesticated animals.

Nitrogen oxides are removed from the atmosphere by sink processes wherein NO_x is converted to nitrate compounds such as HNO_3 and ammonium nitrate (NH_3NO_3). Ammonia neutralizes strong acids such as HNO_3 and H_2SO_4.

Atmospheric deposition of N compounds is significant. In 2023, sources regulated under both the CSAPR NO_x annual program and the Acid Rain Program (ARP) emitted a combined total of 639,000 tons. This represents a reduction of 5.8 million tons (90%) from 1990 levels, 4.5 million tons (88%) from 2000 levels, and 3.0 million tons (82%) from 2005 levels. The highest nitrate deposition rates have been reported in Pennsylvania, Ohio, Indiana, Illinois, and portions of New York, Delaware, Michigan, West Virginia, Missouri, and Kentucky (Figure 4.9). Elevated nitrate deposition is common in most states east of the Rocky Mountains. Maximum NH_3 deposition is relatively low. The highest NH_3 levels occur over farming states.

Anthropogenic emissions of NO_x are approximately three times those that occur naturally. These emissions are expected to increase by approximately 60% in the next 50 years. Presumably, most NO_x is returned to the Earth's surface as nitrates.

Outside agricultural areas, atmospheric deposition is the major source of N in surface soils and H_2O. It is a significant source of nitrate in many watersheds and estuaries. The following

FIGURE 4.9 Geographical distribution of nitrate (NO_3^-) nitrogen over the continental United States in 2023. (From National Atmospheric Deposition Program/National Trends Network, https://nadp.slh.wisc.edu/filelib/annual-summary/2023as.pdf.)

are representative atmospheric N inputs for Albemarle-Pamlico Sound in North Carolina (38%), Chesapeake Bay (21%), Delaware Bay (20%), Massachusetts Bay (5%–27%), Tampa Bay, FL (28%), and the Potomac River (5%).

Because N is a major plant nutrient, atmospherically deposited N has the potential to cause significant ecological changes in forest and freshwater ecosystems and brackish water estuaries. Potential ecosystem effects are discussed in Chapter 6.

4.5.3 Mercury Deposition

Atmospheric deposition of Hg is increasingly becoming a major environmental and public health concern because of its acute and chronic toxicity and movement through food chains.

Mercury is found in small quantities in many rocks and minerals. Although natural sources are responsible for some atmospheric contamination, anthropogenic sources account for 75% of all emissions to the atmosphere. These sources include utility and industrial combustion of coal (44%); municipal, medical, and hazardous waste incineration (33%); manufacturing processes (10%); and a variety of miscellaneous uses (13%).

Mercury's unique chemistry affects its movement between the Earth's surface and the atmosphere. Elemental mercury (Hg^0) is environmentally mobile, traveling significant distances. It is not readily absorbed by tissues and is relatively nontoxic. Divalent mercury (Hg^{2+}) is typically found bound to particles. The atmospheric lifetime of particle-bound Hg varies as a function of particle size. The smaller the particle of which it is a constituent, the longer it remains airborne and the farther it travels.

Mercury emissions from most sources are likely to be in the form of Hg^0, Hg^{2+}, and particle-bound Hg. Ionic mercury dissolves readily in water. As such, it is the most common form found in surface waters. Microorganisms transform Hg^{2+} to methylmercury (CH_3Hg), a much more toxic form. Methylmercury tends to remain dissolved in water. On conversion back to Hg^0, it is readily emitted to the atmosphere. By being transformed from one form to another over and over again (Equation 4.6), Hg can be transported over great distances.

$$Hg^0 \leftrightarrow Hg^{2+} \leftrightarrow CH_3Hg \tag{4.6}$$

Mercury moves in the environment until it is deposited at either high altitudes or high latitudes, where its mobility is markedly decreased.

In 2023, units regulated under Mercury and Air Toxics Standards (MATS) emitted 27 tons less mercury compared to 2010, representing a 91% reduction.

Atmospheric Hg is the major source of Hg loading to aquatic ecosystems. Approximately 85% of the Hg entering the Great Lakes and half of the load to the Chesapeake Bay is atmospheric in origin. In aquatic ecosystems, Hg is bioaccumulated, that is, concentration increases as it moves up through food chains. Mercury concentrations in fish and fish-eating birds may be tens of thousands of times greater than in waters supporting their prey species.

Nearly all US states now issue fish consumption advisories due to elevated mercury levels, particularly methylmercury found in recreationally caught fish. As of the most recent data, 29 states have statewide advisories covering lakes and rivers, while 16 states including many along the Gulf Coast and in the Midwest, Northeast, and Mid-Atlantic also have advisories for coastal waters. Although Hg emissions per unit of coal burned are very small, such emissions are a concern because of the enormous quantities of coal that are consumed and the tendency of Hg to both accumulate and move in the environment. Consequently, in 2001, the US EPA began initial steps to develop and promulgate Hg emission reduction requirements from coal-fired power plants under hazardous pollutant provisions of clean air legislation. Emission standards were implemented in 2012 and are discussed in Chapter 8.

4.5.4 Deposition of Pesticides, PCBs, Dioxins, and Furans

A variety of semivolatile organic compounds are found in atmospheric deposition samples at relatively low concentrations. The more important of these are organochlorine pesticides, PCBs, and families of related compounds known as dioxins and furans.

These compounds share important characteristics. They are highly persistent, which means they have long environmental lifetimes (half-lives on the order of 10 to 20 years). Because they are semivolatile, they go through many cycles of volatilization and subsequent condensation. Like Hg, this process facilitates the movement of these substances around the planet.

Organochlorine pesticides, PCBs, dioxins, and furans can move up through food chains. As such, the highest concentrations are found in species at the top of food chains, most notably fish and fish-eating birds. Several states have fish advisories for PCBs.

These organochlorine substances are neurotoxins, potential human carcinogens, and endocrine disruptors. Dioxins are reported to be very potent teratogens; that is, they can cause congenital birth defects. The most toxic dioxin compound (2, 3, 7, 8-tetrachlorodibenzo-*p*-dioxin [TCDD]) has an LD_{50} (dose required to kill 50% of animals under test) of 0.022 to 0.045 mg/kg on rats and 1 mg/kg on hamsters. These warnings are especially critical for sensitive populations such as pregnant women and children. In addition to mercury, other contaminants like polychlorinated biphenyls (PCBs) with advisories in approximately 84% of states and per- and polyfluoroalkyl substances (PFAS), found in at least 17 states, have prompted further restrictions, with some states recommending no fish consumption from certain waterbodies. This expanding advisory landscape reflects growing concerns over both legacy pollutants and emerging contaminants in US waters.

Although most organochlorine pesticides have either been banned or are subject to use restrictions in the United States, they continue to be significant contaminants due to their long lifetimes and movement from countries that continue their use. Although PCBs have not been used in the United States since the late 1970s, they continue to be a major environmental contamination problem. Dioxins and furans differ from organochlorine pesticides and PCBs in that their production is unintentional. They are produced as by-products in the (1) manufacture of certain pesticides, (2) bleaching of wood pulp in the making of paper, and (3) combustion incineration of wastes that contain chlorinated plastics and paper. As such, medical and municipal incinerators have been major sources of dioxin and furan emissions. Although emission rates are low, such substances accumulate over time.

4.6 STRATOSPHERIC OZONE DEPLETION

4.6.1 Ozone Depletion Concerns

The specter of anthropogenically induced stratospheric O_3 depletion was first raised by atmospheric scientists in the late 1960s and early to mid-1970s. Initial concern was expressed about the building, and subsequent extensive use, of high-altitude supersonic transport (SST) planes in commercial aviation. Such aircraft would fly in the lower stratosphere (to reduce friction), where they would release substantial quantities of nitric oxide (NO) that can catalytically destroy O_3. Scientific and public concern associated with potential O_3 depletion resulted in the cancellation of US government support for the development of an SST commercial aviation capability in the United States.

At approximately the same time, atmospheric scientists proposed that nuclear weapons testing in the atmosphere could cause stratospheric O_3 depletion. High temperatures associated with nuclear detonations can produce large quantities of O_3-destroying NO that will flow into the stratosphere.

After the cancellation of the US SST program, stratospheric O_3 concerns temporarily abated. In 1974, the O_3 depletion specter arose anew when two scientists, Sherwood Rowland and Mario Molina, proposed that chlorofluorocarbons (CFCs) could be broken down in the stratosphere, releasing Cl atoms that catalytically destroy O_3 molecules.

The Rowland–Molina theory of O_3 depletion received considerable scientific and public policy attention and was, in a major way, responsible for the ban on CFC use as aerosol propellants by the United States, Canada, and Sweden in 1978. Other developed countries that used CFCs industrially and commercially viewed the Rowland–Molina hypothesis as an unverified theory and thus continued CFC use without limit.

NO_x (NO and NO_2) is another natural source that could destroy the ozone in the upper stratosphere. The major source of NO in the stratosphere is transport from the troposphere, which comes from surface anthropogenic and natural emissions. In the late 1970s, Rowland proposed that the O_3 layer may be threatened by the increasing release of nitrous oxide (N_2O), a colorless gas emitted to the atmosphere from denitrification processes of ammonium-based and nitrate-based fertilizers, anaerobic bacteria in soils, sewage, and oceans, biomass burning, automobile and aircraft combustion, nylon manufacturing, and aerosol spray cans. In this new theory, N_2O would be transported to the stratosphere because its loss rate from the troposphere is slow and could be well mixed there, with an average mixing ratio of 0.33 ppmv in 2011. Then, in the stratosphere, it would be photolytically destroyed to produce NO, an O_3-destroying chemical associated with previous concerns such as SSTs and atmospheric nuclear testing.

Besides NO_x, hydrogen-containing compounds, particularly the hydroxyl radical (OH) and the hydroperoxy radical (HO_2), will also help shape the ozone profile in the lower stratosphere. The hydroxyl radical destroys the ozone in the lower stratosphere by an HO_x catalytic ozone destruction cycle. However, methane (CH_4) and carbon monoxide (CO) can react with hydroxyl radicals and contribute to ozone production in the stratosphere in turn, although it is relatively small.

Recent assessments show that the ozone layer continues to recover steadily due to global compliance with the Montreal Protocol and its amendments. Stratospheric levels of chlorine and bromine key drivers of ozone depletion have declined by over 11% and 14%, respectively, since their 1990s peaks. While the 2023 Antarctic ozone hole was unusually large and long-lasting, influenced in part by the 2022 Tonga volcanic eruption, the 2024 ozone hole was among the smallest on record, reflecting a return to normal recovery trends. Scientists project that, if current regulations remain in place, the ozone layer will return to pre-1980 levels by around 2040 globally, by 2045 in the Arctic, and by 2066 over Antarctica. However, ongoing climate variability, volcanic activity, and emissions of other greenhouse gases (e.g., N_2O, CH_4, CO_2) continue to affect the pace and pattern of ozone recovery.

4.6.2 Ozone Layer Dynamics

The O_3 layer represents an atmospheric environment in which O_3 is continually produced and destroyed. The natural formation and destruction of O_3 in the stratosphere is initiated by high-intensity, shortwave solar ultraviolet (UV) radiation (< 242 nm), as shown in the following equations:

$$O_2 + hv \rightarrow O + O \tag{4.7}$$

$$O + O_2 + M \rightarrow O_3 + M \tag{4.8}$$

$$O + O_3 \rightarrow 2O_2 \tag{4.9}$$

$$O_3 + hv \rightarrow O_2 + \left({}^1D\right) \tag{4.10}$$

where hv is a photon of UV light and M is an energy-absorbing molecule. This series of reactions is called the Chapman cycle. Ozone formation and destruction in the stratosphere include reactions involving molecular oxygen (O_2), atomic oxygen (O), and O_3. Ozone is initially formed by the UV photolysis of O_2 into two O atoms that react with O_2 to produce O_3 (Equations 4.7 and 4.8). Reactions between O_3 and O result in the breakdown of O_3 and the regeneration of O_2 (Equation 4.9). Ozone absorption of UV light also results in the destruction of O_3 and the regeneration of O_2 (Equation 4.10). Heat produced by the absorption of UV radiation is responsible for the inverted temperature profile that characterizes the stratosphere (Figure 1.7). Chapman cycle reactions (see earlier) produce more O_3 than is present in the stratosphere. As a consequence, stratospheric O_3 levels must be maintained at observed levels by loss mechanisms. The O_3 loss mechanisms responsible for maintaining stable stratospheric O_3 levels are reactions with NO_x (described later).

Considerable latitudinal and seasonal differences in stratospheric O_3 levels occur (Figure 1.2). Although more O_3 is produced over the tropics than at the poles, total column O_3 levels (because of transport mechanisms) have historically been higher in polar regions than over the tropics. Variability in hemispheric O_3 levels results from what is described as the quasi-biennial oscillation, a cycle of varying wind direction in the equatorial stratosphere that affects northward or southward movement of O_3 in different phases of the cycle.

4.6.3 Role of Nitrogen Oxides

Nitrogen oxides (primarily NO) associated with human activities were perceived to be a threat to the O_3 layer because of emissions from SSTs and from nuclear weapons testing. They are primarily responsible for maintaining stratospheric O_3 at equilibrium levels. Nitric oxide can react with O_3 to produce NO_2 and O_2. Nitrogen dioxide can subsequently react with $O(^3P)$ to regenerate NO. Thus, NO catalytically destroys O_3 because NO is repeatedly regenerated:

$$NO + O_3 \rightarrow NO_2 + O_2 \tag{4.11}$$

$$NO_2 + O(^3P) \rightarrow NO_2 + O_2 \quad (4.12)$$

$$NO_2 + h\nu \rightarrow NO + O(^3P) \quad (4.13)$$

Nitrogen oxides such as NO_2 and NO are emitted into the lower atmosphere in large quantities. Because of a variety of chemical reactions and sink processes, these emissions pose little threat to the O_3 layer. Ozone depletion associated with NO_x would require significant direct emissions to the stratosphere from extensive high-altitude aircraft traffic, nuclear explosions, or the destruction of stable N compounds such as N_2O. Nitrous oxide is transported into the stratosphere, where it is oxidized by singlet oxygen, $O(^1D)$, yielding two molecules of NO:

$$O(^1D) + N_2O \rightarrow 2NO \quad (4.14)$$

It can also be formed by reactions initiated by the absorption of cosmic rays and solar protons. Nitrous oxide is the major source of NO_x in the stratosphere.

A major sink mechanism for NO is reaction with hydroperoxy radical (HO_2) to produce hydroxyl radical (OH) and NO_2 (which is further oxidized to HNO_3):

$$HO_2\cdot + NO \rightarrow OH\cdot + NO_2 \quad (4.15)$$

$$NO_2 + OH\cdot + M \rightarrow HNO_3 + M \quad (4.16)$$

Nitric acid serves as a means by which NO_x is removed from the stratosphere and also serves as a reservoir for subsequent NO_x regeneration.

The reaction of NO_x with chlorine monoxide (ClO) scavenges Cl in the stratosphere. The resultant product can serve as a reservoir for Cl atoms (see next section).

4.6.4 Role of Chlorofluorocarbons and Other Halogenated Hydrocarbons

Chlorofluorocarbons were first identified as a potential threat to the O_3 layer in the early 1970s when it was observed that free Cl catalytically destroys O_3 and that CFCs (which have no tropospheric sink) can be photolytically destroyed to yield free Cl and fluorine (F) by UV light in the spectral region of less than 230 nm.

The wavelength distribution of solar radiation and UV light shifts to shorter wavelengths with altitude. Consequently, CFC destruction occurs primarily in the upper stratosphere where high irradiance of CFCs by UV wavelengths of less than 230 nm occurs. Because these UV wavelengths do not pass into the lower stratosphere, CFCs must be transported to the upper stratosphere by molecular diffusion. Because Cl production from CFC destruction occurs above the altitude where most O_3 and O_2 are present, Cl atoms must diffuse downward into the middle of the O_3 layer where they participate in catalytic chain reactions that destroy O_3:

$$Cl\cdot + O_3 \rightarrow ClO\cdot + O_2 \quad (4.17)$$

$$ClO\cdot + O(^1D) \rightarrow Cl\cdot + O_2 \quad (4.18)$$

When O_3 reacts with Cl, it is converted to O_2 and ClO. Chlorine monoxide reacts with $O(^1D)$ to regenerate Cl. Continual regeneration of Cl by this and other reaction pathways can, in theory, result in the destruction of 100,000 O_3 molecules before Cl is removed from the stratosphere as hydrochloric acid (HCl) or chlorine nitrate ($ClONO_2$).

The reaction cycle in Equations 4.17 and 4.18 is important primarily in the middle and upper stratospheres where O(^{1}D) levels are high. At 15 km (9.3 mi), it accounts for only 5% of halogen-associated loss; at 21 km (13 mi), it accounts for approximately 25%.

Most O_3 depletion in the lower stratosphere associated with halogens occurs as a result of the following series of chemical reactions:

$$Cl\cdot + O_3 \rightarrow ClO\cdot + O_2 \quad (4.19)$$

$$ClO\cdot + HO_2\cdot \rightarrow HOCl + O_2 \quad (4.20)$$

$$HOCl + h\nu \rightarrow Cl\cdot + OH\cdot \quad (4.21)$$

$$\frac{OH\cdot + O_3 \rightarrow HO_2\cdot + O_2}{\text{Net}: 2O_3 \rightarrow 3O_2} \quad (4.22)$$

In this cycle, ClO reacts with HO_2, with O_3 destruction also resulting from reactions with OH. It accounts for approximately 30% of the halogen-associated loss in the lower stratosphere.

Several other halogens have the potential to destroy stratospheric O_3. Major O_3-destroying chemicals include carbon tetrachloride (CCl_4), methyl chloroform (CH_3CCl_3), methyl chloride (CH_3Cl), methyl bromide (CH_3Br), and halon fire-extinguishing agents such as $CBrClF_2$ and $CBrF_3$.

Brominated compounds are photolytically destroyed in the stratosphere yielding Br atoms that react catalytically with O_3 in a manner similar to that in Equations 4.19 and 4.20:

$$Br\cdot + O_3 \rightarrow BrO\cdot + O_2 \quad (4.23)$$

$$\frac{BrO\cdot + O\left[{}^1D\right] \rightarrow Br\cdot + O_2}{\text{Net}: O_3 + O \rightarrow 2O_2} \quad (4.24)$$

This Br cycle results in an approximately 20% to 30% halogen-related O_3 loss in the lower stratosphere.

The ClO and BrO cycles are interconnected by the following reactions:

$$ClO\cdot + BrO\cdot \rightarrow Br + ClOO \quad (4.25)$$

$$ClO\cdot + BrO\cdot \rightarrow BrCl + O_2 \quad (4.26)$$

The ClO–BrO cycles seem to be responsible for approximately 20% to 25% of O_3 loss due to halogen chemistry at altitudes of 16 to 20 km (9.9–12.4 mi) in the middle latitudes and are significant in the lower polar stratosphere.

Sink processes for Cl and Br involve reactions that yield HCl, bromic acid (HBr), $ClONO_2$, and bromine nitrate ($BrONO_2$). Hydrochloric acid forms as a result of the reaction of Cl with methane (CH_4) transported from the troposphere:

$$Cl\cdot + CH_4 \rightarrow HCl + CH_3\cdot \quad (4.27)$$

Bromic acid is formed on reaction with HO_2:

$$Br\cdot + HO_2\cdot \rightarrow HBr + O_2\cdot \tag{4.28}$$

ClO forms $CLONO_2$ by reaction with NO_2:

$$ClO\cdot + NO_2 + M \rightarrow ClONO_2 + M \tag{4.29}$$

BrO forms $BrONO_2$ in the same fashion:

$$BrO\cdot + NO_2 + M \rightarrow BrONO_2 + M\cdot \tag{4.30}$$

Both HCl and $ClONO_2$ are temporary reservoirs for Cl because it can be easily regenerated via the following reactions:

$$HCL + OH\cdot \rightarrow Cl\cdot + H_2O \tag{4.31}$$

$$ClONO_2 + hv \rightarrow Cl\cdot + NO_3 \tag{4.32}$$

Bromine atoms are similarly regenerated from HBr and $BrONO_2$.

On an atom-to-atom basis, Br is more efficient than Cl in destroying stratospheric O_3. Unlike Cl, it reacts relatively slowly with organic compounds such as CH_4. Consequently, Br sink processes are slower. In addition, photolysis of brominated compounds to produce Br is much faster than the corresponding photolysis of chlorinated compounds.

4.6.5 Antarctic Ozone Hole

In 1985, British scientists conducting ground-based measurements of total column O_3 above Halley Bay (76°S latitude) observed unusually large declines in O_3 levels associated with the beginning of the southern hemispheric spring. Because such declines had not been previously reported (measurements were also being made by satellites), observed O_3 declines were thought to be anomalous. After careful reevaluation of the accuracy and precision of their instruments, British scientists concluded that the observed O_3 declines were real. Examination of raw NASA satellite data (which excluded extreme results in data summaries) revealed that the phenomenon began to occur in the mid-1970s. Satellite data also indicated that the area affected by the Antarctic O_3 hole extended over several million square kilometers (~1.5 million mi^2). Since its discovery, the geographical area affected by the Antarctic O_3 hole has increased from a few million square kilometers in 1979 to more than 25 million km^2 (15 million mi^2) in 2000, with lowest column values of less than 90 Dobson units (DU) and O_3 depletions of nearly 100% (~0 DU) at altitudes of 15 to 20 km (~9 to 12 mi). Its persistence has increased from weeks to months.

Currently, the Antarctic ozone hole is defined as the area where the ozone column abundance decreases to less than 220 DUs. Such decreases historically have occurred only over the Antarctic during the southern hemisphere (SH) spring (September–November); this reduction mechanism has been explored by scientists. In the SH winter (June–September), there is no sunlight over Antarctica and it becomes very cold there. In addition, the polar front jet stream wind system (the polar vortex) encircles the Antarctic continent, confining cold air within the Antarctic polar region and preventing the warm, ozone-rich air outside from penetrating into the Antarctic stratosphere, which further cools down the air. When the temperature decreases to less than 195 K in the stratosphere, polar stratospheric clouds (PSCs) start to form and, on the surface of these clouds, "inactive" chlorine reservoirs, HCl and $ClONO_2$, can react with water ice and produce photochemically active forms such as Cl_2, HOCl, and $ClNO_2$ through heterogeneous reactions. Later, when the sun rises in spring, sunlight will break down new molecules into "active" chlorine, which strongly destroys ozone and then the ozone hole appears.

Changes in O_3 levels reported as an ozone hole area from August to December can be seen in Figure 4.10. Recent observations indicate that the O_3 hole persisted into early December. Sulfate aerosol from the Mt. Pinatubo volcanic eruption in the Philippines in 1991 is believed to have contributed to its expansion and to historically low values in 1992 and 1993.

The Antarctic O_3 hole surprised atmospheric scientists. Their O_3 depletion models were based on gas-phase chemistry and focused on the stratosphere over and near source regions where temperatures were, in theory, more favorable for O_3-destroying chemistry. Several theories were initially proposed to explain its formation. Some scientists suggested that the unique polar circulation patterns (i.e., the polar vortex) were responsible, whereas others hypothesized that it was being produced by chemical reactions involving anthropogenic pollutants. Scientists project that, if current regulations remain in place, the ozone layer will return to pre-1980 levels by around 2040 globally, by 2066 over Antarctica. According to the 2022 WMO/UNEP Scientific Assessment, recovery in the Antarctic ozone layer is underway but slower at lower altitudes. Although upper-stratospheric ozone is rebounding, total column and lower-altitude depletion persists in springtime, delaying full recovery in those layers until around mid-century or later.

Numerous studies since 1986 have demonstrated the presence of elevated concentrations of Cl and ClO in the region of the O_3 hole, providing strong evidence that it is caused by chemical reactions involving Cl in conjunction with unique meteorological conditions in the Antarctic. Ozone destruction occurs because of complex heterogeneous chemical reactions involving PSCs, HCl, $ClONO_2$, molecular chlorine (Cl_2), and hypochlorous acid (HOCl). Both HCl and $ClONO_2$ are produced in sink processes that remove Cl from O_3-destroying processes.

Polar stratospheric clouds form in the very cold darkness of the long Antarctic night. The PSC particles, which consist of frozen solutions of HNO_3 and H_2O (type 1 PSC) or frozen water (type 2

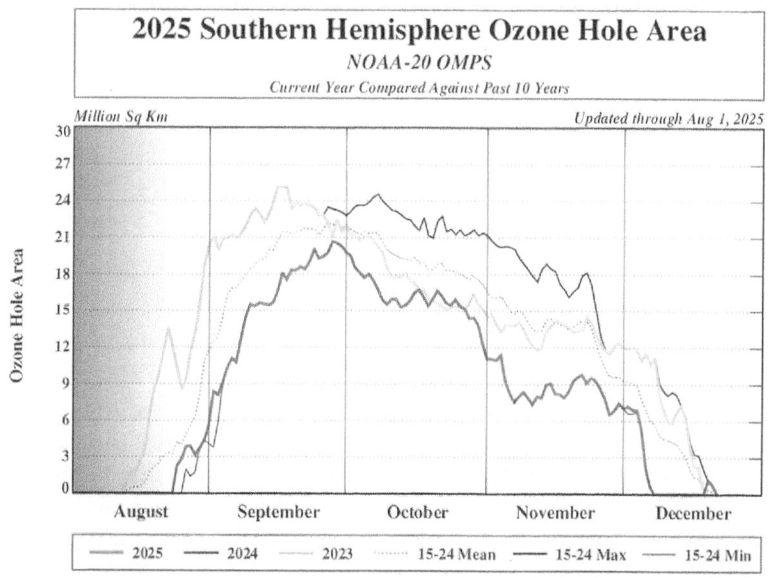

FIGURE 4.10 Changes in altitudinal O_3 concentrations over the South Pole during August to August 1 of 2025. (From https://www.cpc.ncep.noaa.gov/products/stratosphere/polar/gif_files/ozone_hole_plot_N20.png.)

PSC), absorb sink substances such as $ClONO_2$, HCl, HOCl, and nitrogen pentoxide (N_2O_5), which are then activated to release Cl_2 and HOCl. These heterogeneous-phase processes and subsequent chemical reactions are illustrated in Figure 4.11 and the following equations:

$$ClONO_2 + HCl_{(s)} \rightarrow HNO_{3(s)} + Cl_2 \quad (4.33)$$

$$ClONO_2 + H_2O_{(s)} \rightarrow HNO_3 + HOCl \quad (4.34)$$

$$HOCl + HCl_{(s)} \rightarrow Cl_2 + H_2O_{(s)} \quad (4.35)$$

In addition, NO_x (in the form of N_2O_5) is removed from the system by the production of HNO_3.

$$N_2O_5 + H_2O_{(s)} \rightarrow HNO_{3(s)} \quad (4.36)$$

$$N_2O_5 + HCl_{(s)} \rightarrow ClNO_2 + HNO_{3(s)} \quad (4.37)$$

The subscript (s) indicates that the substance is in the solid phase.

When the sun comes up over the Antarctic at the beginning of the austral spring, O_3 destruction occurs in two different pathways. In the first of these, Cl_2 and HOCl absorb sunlight that results in the production of Cl atoms:

$$Cl_2 + hv \rightarrow 2Cl\cdot \quad (4.38)$$

$$HOCl + hv \rightarrow Cl\cdot + OH\cdot \quad (4.39)$$

$$Cl\cdot + O_3 \rightarrow +ClO\cdot + O_2 \quad (4.40)$$

Much greater O_3 destruction or loss results from reactions involving ClO dimers (ClOClO):

$$ClO\cdot + ClO\cdot + M \rightarrow +ClOClO + M \quad (4.41)$$

$$ClOClO + hv \rightarrow Cl\cdot + ClOO \quad (4.42)$$

$$\frac{ClOClO + M \rightarrow Cl\cdot + O_2 + M}{Net: 2\left(Cl\cdot + O_3 \rightarrow ClO\cdot + O_2\right)} \quad (4.43)$$

As indicated, PSCs play a critical role in O_3 depletion chemistry in the Antarctic. Type 1 PSCs form at temperatures of 2 to 5 K above the ice frost point. These type 1 particles are believed to contain large quantities of HNO_3 and H_2O. There is evidence that they may be ternary solutions of HNO_3, H_2O, and H_2SO_4. They range in size from 0.1 to approximately 5 μm in diameter. Type 2 PSCs are believed to consist mostly of H_2O. They are formed at temperatures corresponding to the frost point of H_2O (−188 K) or below. Particle sizes are large, ranging from 5 to 50 μm.

In addition to PSC particles, stratospheric sulfate aerosol (SSA) produced by volcanoes such as El Chinchon (1982) and Mt. Pinatubo (1991) plays a significant role in stratospheric O_3 depletion

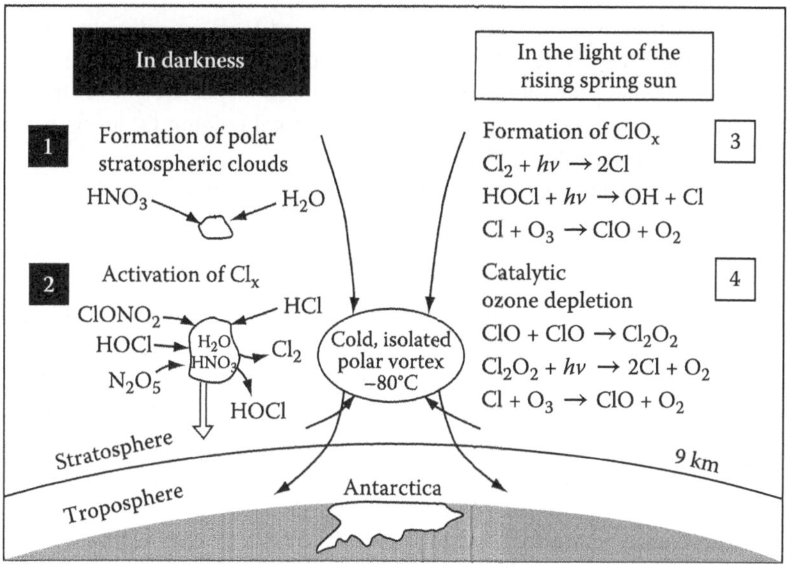

FIGURE 4.11 Ozone depletion processes over Antarctica. (From Brasseur, G.P. et al., *Atmospheric Chemistry and Global Change*, Oxford University Press, Oxford, 1999. With permission.)

chemistry. Stratospheric sulfate aerosol particles provide surfaces for heterogeneous-phase chemistry and serve as nuclei for PSC formation. It could potentially lead to lower temperatures in the atmosphere.

4.6.6 Ozone Depletion in the Arctic

Although O_3 depletion occurs in the Arctic because of chemical reactions that are similar to those in the Antarctic, an O_3 hole does not form. This is because Arctic stratospheric temperatures are warmer by 10 K and the polar vortex is less stable and less isolated from mixing with external air masses than in the Antarctic. In addition, O_3 concentrations usually increase in December and early spring due to long-range transport from the tropics. Any decreases associated with O_3 destruction would be superimposed on normal increases and may be masked.

Nevertheless, significant reductions in stratospheric O_3 were observed in the Arctic in six of the last nine years of the twentieth century during late winter and early spring (January–March). Reductions on the order of 20% to 25% were considerably smaller than those observed each spring in the Antarctic. These Arctic O_3 losses have been associated with unusually cold Arctic winters. Significant Arctic O_3 depletion occurs primarily in the lower stratosphere. Scientists project that, if current regulations remain in place, the ozone layer will return to pre-1980 levels by around 2040 globally, and by 2045 in the Arctic. However, ongoing climate variability, volcanic activity, and emissions of other greenhouse gases (e.g., N_2O, CH_4, CO_2) continue to affect the pace and pattern of ozone recovery.

4.6.7 Trends

Ozone depletion occurs at high and midlatitudes. It has not been observed in the latitudinal zone 30°N to 30°S. Although the Antarctic O_3 hole continues unabated, the extent of depletion has remained relatively unchanged since the early 1990s.

The rate of O_3 depletion over the midlatitudes has slowed from an estimated linear downward trend of 4.0%, 1.8%, and 3.8% per decade for northern midlatitudes in winter–spring, northern midlatitudes in summer–fall, and year-round in southern latitudes, respectively, over the period 1979 to 1991. Since the recovery from the Mt. Pinatubo eruption, total column O_3 levels have been nearly constant in midlatitudes, as reported in the late 1990s.

Total levels of O_3-depleting compounds in the troposphere have been declining since 1994. Although total Cl was declining, total Br was still increasing as of 1998. The peak total Cl abundance was 3.7 ppbv between 1992 and 1994. The combined abundance of stratospheric Cl and Br peaked at the beginning of the millennium.

Ozone depletion has resulted in cooling of the lower stratosphere since 1980. The trend was 0.6°C (~1°F) per decade during the period 1979 to 1994. Model simulations indicate that this cooling trend was strongly correlated with O_3 depletion at these altitudes.

In the Intergovernmental Panel on Climate Change (IPCC) Fourth Assessment Report (AR4), IPCC Special Report, on Emissions Scenarios (SRES) have been developed to simulate the time evolution of global ozone trend in the future. New projections of ozone precursors that account for the stricter emission reductions from mobile sources by developing countries include the current legislation (CLE) scenario, maximum feasible reduction (MFR) scenario, and the business as usual called A2 scenarios. The changes in NO_x emissions for these three scenarios are +12%, −27%, and +55%, respectively, relative to year 2000. In consequence, the ensemble-mean burdens in the tropospheric ozone change by +6%, −5%, and +18% for the MFR, CLE, and A2 scenarios, respectively. The ozone decreases throughout the troposphere in the MFR scenario, but the zonal annual mean concentrations increase by up to 6 ppb in the CLE scenario and by typically 6 to 10 ppb in the A2 scenario.

In the IPCC AR5 report, the time evolution of the globally averaged stratospheric ozone column is calculated under the new Representative Concentration Pathways (RCP) scenarios. The stratospheric ozone column is estimated to be approximately −12 DUs compared with the 1850 levels. Since then, it gradually starts to recover under different RCP scenarios, with the lowest speed under RCP 2.6 and fastest under RCP 8.5. At the end of the twenty-first century, the RCP 2.6/4.5/6.0 scenarios have all recovered to at least the 1850 level. For RCP 8.5, the globally averaged stratospheric ozone column is simulated to be approximately 7.5 DUs more than the 1850 value.

In the IPCC AR6 report, Arctic ozone depletion is episodic and highly variable, largely influenced by winter polar stratospheric dynamics and temperatures. During particularly cold Arctic winters, total column ozone loss in spring can reach up to ~30% due to halogen-induced reactions on polar stratospheric clouds. In warmer winters, depletion is much smaller or negligible. The decline and gradual recovery of ozone-depleting substances (ODSs), especially chlorine and bromine compounds, is progressing. Stratospheric chlorine is near peak levels, and bromine may still be incrementally increasing. These trends are slowing Arctic ozone losses as regulations take effect. Overall, Arctic ozone remains below pre-1980 baseline levels in midlatitudes, while total midlatitude ozone started slowly increasing after the late 1990s as the Montreal Protocol restrictions came into effect.

4.6.8 Changes in Surface UV Light Radiation

UV light in the UV-A and UV-B spectral regions received at the surface of the Earth varies geographically. As would be expected, higher irradiance of the ground occurs in the tropics, with progressive decreases toward the poles. The amount of UV light received at a particular location depends on the angle of incidence of the sun, column O_3 levels, cloud cover, and atmospheric pollution.

Increases in UV radiation associated with the Antarctic O_3 hole have been well documented. In the late southern hemisphere spring, UV-B levels over the Antarctic exceed those that occur in San Diego, CA, where the sun's angle of incidence is considerably less.

In regions of the planet where smaller decreases in stratospheric O_3 depletion have occurred, it has been more difficult to measure increases in surface UV-B irradiance. Surprisingly, between 1974 and

1985, a decrease in UV-B irradiance was observed in eight geographical areas in the United States. This anomaly has only recently been determined to have been due to instrument error.

A subsequent study conducted near Toronto, Canada, demonstrated that UV-B irradiance increased significantly in the period 1989 to 1993. The intensity of UV at 300 nm increased by 35% per year in winter and 7% per year in summer.

The results of the Canadian study have been confirmed and firmly established by numerous ground-based measurements. In other midlatitude studies (~40°N), increases in UV-B of 1.5% per year at 300 nm and 0.8% per year at 305 nm have been reported for the period 1989 to 1997.

Ground-based measurements of UV irradiation increases were consistent with estimates from satellite data and radiative forcing models that include cloud, surface reflectivity, and aerosol effects. Increases in UV-B irradiance per decade from 1979 to 1992 were estimated to be 3.7% ± 3% at 60°N and 3% ± 2.8% at 40°N latitude, and 9% ± 6% at 60°S and 3.6% ± 2% at 40°S latitude.

The relationship between decreased column O_3 levels and increases in UV-B irradiance for clear-sky measurements can be seen from the model line (Figure 4.12) fitted to data from Canada, Germany, Greece, Hawaii, New Zealand, and the South Pole. The portion of the model line in the left third of the graph was based on data collected at the South Pole.

In 2022, WMO Scientific Assessment of Ozone Depletion and IPCC AR6 report, although ozone-depleting substances (ODS) remain present, their atmospheric concentrations are declining under sustained global controls. As a result, future surface UV levels are expected to decrease gradually, particularly outside polar regions. Direct links between ozone depletion and surface UV-B radiation are well established: a 1% reduction in column ozone corresponds to roughly a 0.7%–1% increase in tropospheric hydroxyl radicals (OH) and notable increases in UV-B exposure. During polar spring seasons, UV-B levels increased historically by as much as ~130% over Antarctica and ~22% over the Arctic. In midlatitude regions, reconstructed UV radiation trends from 1979 through the early 2010s indicate modest increases of 4%–7% in UV-B exposure, matching periods when ozone was still declining. Post-2010, surface UV trends have flattened or slightly reversed in many regions due to ozone recovery and decreasing aerosol pollution.

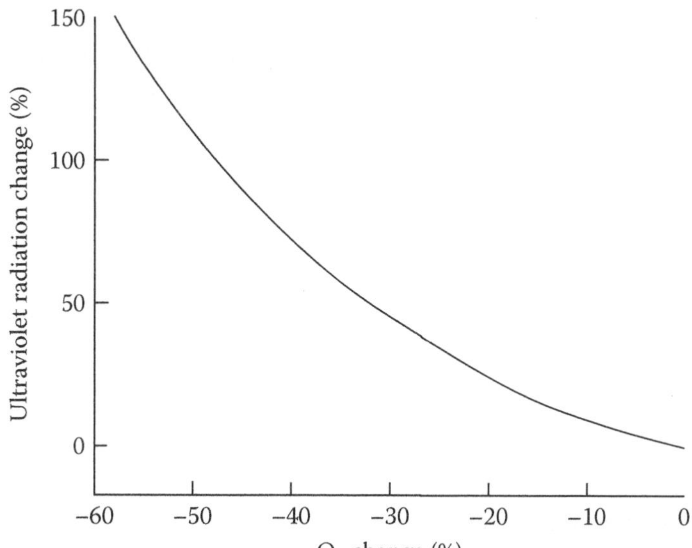

FIGURE 4.12 Relationship between change in stratospheric O_3 concentrations and change in surface-received UV light. (From World Meteorological Organization, *Scientific Assessment of Ozone Depletion*, WHO Global Ozone Research Monitoring Project, Report 44, WMO, Geneva, Switzerland, 1998.)

4.6.9 Ozone Depletion Effects on Human Health

The O_3 layer absorbs 99% of UV wavelengths at less than 320 nm. Ozone absorption in the UV-B region (290–320 nm) is, however, less efficient than at shorter wavelengths, and relatively significant fluxes of both UV-B and UV-A (320–400 nm) reach the Earth's surface. UV-B, upon absorption by organisms, can cause significant adverse biological effects. Therefore, it is often described as damaging UV radiation (DUV).

Significant short-term and long-term environmental exposures to DUV may cause sunburn, cataracts, or skin cancer. There are three basic forms of skin cancer: basal cell carcinoma, squamous cell carcinoma, and melanoma. Carcinoma-type skin cancers have clearly been demonstrated to be associated with exposure to sunlight, with the greatest prevalence observed among light-complected individuals living in so-called sunbelt latitudes in North America. Skin carcinomas, fortunately, are the least life-threatening of all forms of cancer, with a mortality rate of less than 5%. Because tumors must be surgically removed, skin cancers are often disfiguring.

Melanoma tumors develop from the darkly pigmented melanin structures or moles that many Caucasians develop on their skin, particularly during midlife. Melanoma is a much more deadly form of skin cancer because the tumors metastasize rapidly, spreading malignant cells to other parts of the body through the bloodstream and lymphatic system. The mortality rate is approximately 30%. The incidence of melanoma in the United States has increased eightfold since 1960.

No dose–response relationship between melanoma and exposure to sunlight and UV radiation has been demonstrated. Rather, melanoma appears to be associated with severe sunburn experienced early in life. Melanoma has also been linked to building environments with fluorescent lighting. Paradoxically, melanoma tumors apparently associated with UV radiation exposures from fluorescent lighting occur on nonexposed parts of the body.

Increases in UV-B radiation fluxes at the Earth's surface because of stratospheric O_3 depletion by even a few percent are expected to cause a significant increase in all forms of skin cancer among Caucasians and other light-pigmented individuals. This could mean tens of thousands of new cases of skin cancer in the United States each year.

4.6.10 Recovery of the Ozone Layer

Stratospheric O_3 depletion caused by anthropogenic emissions of O_3-destroying chemicals is expected to gradually abate over the next half century as they are slowly removed by natural processes. This anticipated recovery will be due in great measure to the reduction in emissions of Cl-containing and Br-containing gases that has resulted from a series of international agreements that began with the Montreal Protocol in 1987. In the early austral spring of 2002, the Antarctic O_3 hole was smaller (15 vs. 25 million km^2) than any other year in over a decade. Whether this was the beginning of a trend or due to normal meteorological variations (the polar vortex was also much smaller) was not known at the time of this writing.

The stratospheric O_3 layer is likely to return to what it was prior to 1980. Full recovery to 1980 ozone levels is expected: by ~2040 globally (excluding Antarctica), by ~2045 in the Arctic, and by ~2066 over Antarctica.

4.7 GLOBAL WARMING

4.7.1 Climate and Global Climate

The atmosphere, driven by the energy of the sun, is in continual motion and change. This change or flux viewed in the context of hours, days, or weeks is called weather. Weather includes precipitation events; fluctuations in temperature, humidity, and wind speed; the degree of cloud cover; and incoming solar radiation. These phenomena, viewed within the context of geographical areas and extended periods of time (years, decades, centuries, millennia), describe climate. Climate is characterized by average temperatures, temperature extremes, precipitation, the quantity and periodicity of solar radiation, and other atmospheric phenomena. Although climate is often described as being average weather, it is, in fact, a much broader concept.

Weather, because it is based on short-term events, can be quite variable. Climate is less variable because it is based on averaged values. Climate is subject to change, but the time frame of such change may be relatively long. Evidence of small to large climatic changes may be found in historical accounts, various aspects of the geological record, and analytical results of measurements made on deep-sea sediments, polar ice, and tree rings.

Climate is usually viewed in the context of different geographical regions. It can be viewed in the global context as well. Global climate is described by the average global surface temperature. The primary determinants are the (1) absolute amount of energy received from the sun, (2) amount of solar energy reflected back to space (albedo), and (3) absorption of incoming solar radiation and outgoing thermal radiation by atmospheric constituents.

Global surface temperatures, and thereby climate, have changed significantly over time. Studies of oxygen isotope ratios in carbonate ocean sediments indicate that the Earth's temperature decreased by 15°C (59°F) in the past 65 million years. This decrease has been attributed to the changes in positions of continents and oceans (continental drift), formation of high mountains, intensification of volcanism, and decrease in atmospheric CO_2 levels. Changes in global climate may also occur through a variety of other physical mechanisms, for example, (1) quasi-periodic ocean temperature changes, (2) increased sea salt formation associated with decreased Arctic Ocean salinity, (3) variation in solar radiation, (4) changes in surface albedo, and (5) changes in the Earth's orbital geometry.

The effects of orbital variation and variations in solar irradiance are discussed next to illustrate the effects of natural phenomena on the Earth's climate system.

4.7.1.1 Orbital Changes and Climatic Variation

The Earth's orbit relative to the sun varies in three distinctively different ways. First, the Earth's axis is tilted 23.5° from the perpendicular to the plane described by the Earth's orbit around the sun. In summer, the northern hemisphere tilts toward the sun; in winter, it tilts away from the sun.

The obliquity or tilting of the Earth's axis varies from 22.1° to 24.5°, with a complete cycle every 41,000 years. The obliquity of the Earth's axis is responsible for seasonal climatic changes over the Earth's surface. As it changes from 22.1° to 24.5°, contrasts between seasons are likely to become greater.

Second, the Earth's axis changes direction over time. This precession or wobbling of the Earth describes a circle among the stars, with the center at due north. Currently, the axis points toward Polaris, the North Star. In about 12,000 years, it will point at Vega, which will become the North Star. When the axis is tilted toward Vega, the orbital positions at which summer and winter solstices occur will be reversed. The northern hemisphere will experience summer near perihelion (when the Earth is closest to the sun) and winter near aphelion (when the Earth is farthest from the sun). Winters can be expected to be colder and summers warmer than at present. The precession cycle is about 20,000 years.

Lastly, the nature of the Earth's orbit changes from a somewhat circular to a more oblong ellipse over a period of 100,000 years. This orbital change is called eccentricity. Because the orbit is not exactly symmetrical, the Earth is currently farthest away from the sun (aphelion) on July 4 and closest to the sun on January 3 (perihelion). This difference in distance is approximately 3%. As such, the Earth receives 6% more solar energy in January than in July. When the orbit is more elliptical, the amount of solar energy received at perihelion would be on the order of 20% to 30% more than at aphelion. Orbital variations, obliquity (a), precession (b), and eccentricity (c), are illustrated in Figure 4.13. These variations do not increase or decrease the amount of solar energy received by the Earth; they do determine how it is distributed over the Earth's surface and change the degree of contrast between the seasons.

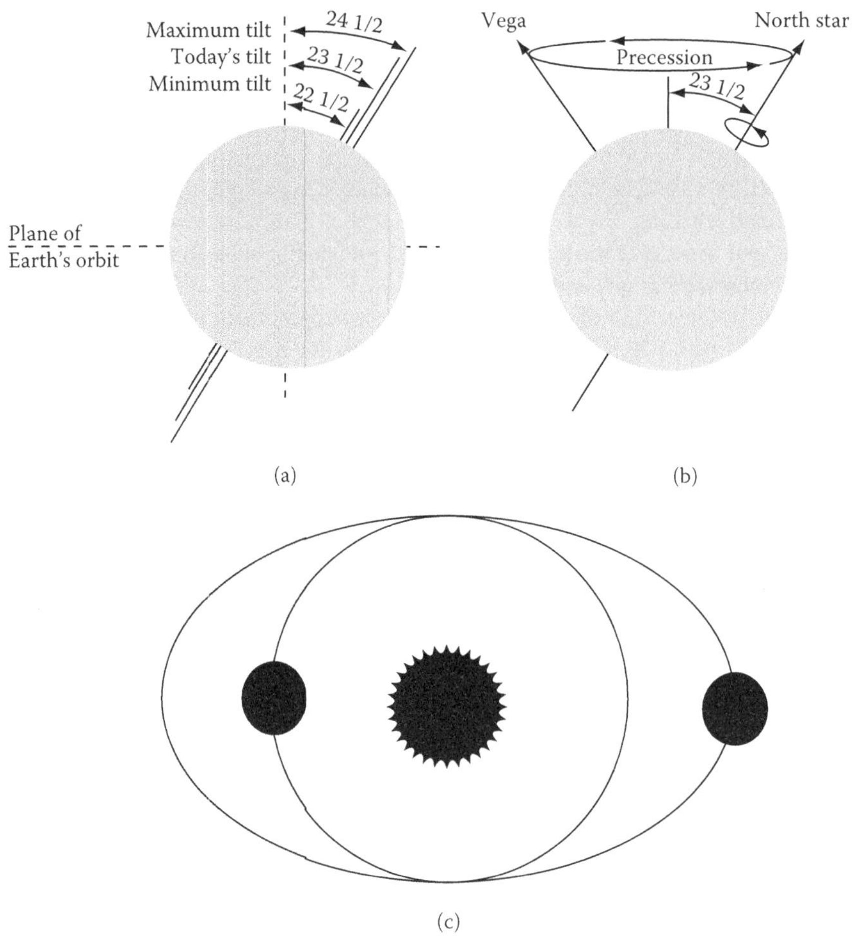

FIGURE 4.13 Variations in the Earth's orbit.

Orbital variations have been linked to cycles of glaciation experienced by the Earth over the past 2.5 million years. Serbian scientist Milutin Milankovitch proposed such a linkage in the early twentieth century. Therefore, the 20,000, 40,000, and 100,000-year periodicities associated with precession, obliquity, and eccentricity are called Milankovitch cycles. Evidence of periods of significant global surface cooling, with periodicities of 20,000, 40,000, and 100,000 years, have been reported from O isotope studies of marine carbonate sediments. As a consequence, a number of scientists have concluded that these orbital changes were the fundamental cause of glaciations in the past 2.5 million years and that they are "pacemakers of the ice ages."

Glacial periods lasting thousands of years were characterized by the development and movement of massive ice sheets (several kilometers thick) over parts of the northern hemisphere. They were followed by climatic warming and glacial retreat. The last large glacier over North America receded approximately 12,000 years ago. The Earth is presently in an interglacial warming period.

4.7.1.2 Variations in Solar Radiation

Average global surface temperatures are determined in part by the amount of solar radiation reaching the Earth's atmosphere and surface. Astronomers have theorized that at least for long periods, solar output and hence the quantity of solar radiation received by the Earth is essentially constant. The solar constant (1,370 $W/m^2/s$) is the total solar radiation received on a flat surface oriented perpendicular to the sun external to the Earth's atmosphere. Its value is 1.94 cal/cm^2/s. Satellite measurements have recently shown that solar radiation does vary and that the solar constant is not constant when viewed in the short term. Short-term changes in solar output are directly related to sunspot cycles. The peak value of solar irradiances occurs near the maximum of solar activity. The peak-to-peak variation in the sun's total irradiance during the sunspot cycle is approximately 0.08%.

Sunspots are dark, relatively cool regions on the sun's surface with limited heat flow. Interestingly, the sun's irradiance is greater rather than less during peak periods of sunspot activity. Apparently, there is a concurrently active source of solar emission that more than compensates for the decreased output from sunspot regions. Increased solar emissions are associated with faculae and plages, that is, bright, active regions where emissions of solar radiation are enhanced relative to those for the surrounding quieter regions of the sun. Plages make up an order of magnitude or more of the sun's surface and are longer-lived than sunspots.

Changes in the sun's luminosity associated with sunspots and other sun surface activity have been proposed as a cause of significant climatic change. The Little Ice Age, a period of global cooling observed worldwide during the fourteenth to nineteenth centuries, was associated with a significant decline in sunspot activity. The Little Ice Age was accompanied by (1) the advance of European mountain glaciers, (2) lower average winter temperatures in central Britain, (3) the abandonment of cereal cultivation in Iceland and vineyards in England, (4) the abandonment of settlements in Greenland, and (5) winter freezing of the North Sea and canals in Venice. It is probable that global surface temperatures were lower by 1°C (1.8°F) or more during that period, and that precipitation patterns were also changed.

Changes in the sun's luminosity associated with sunspot cycles have, from time to time, captured the interest of climatologists in their attempts to explain climatic change. Ice core and ocean temperature analyses show significant correlations between temperature changes and the 11-year and 22-year sunspot cycles. These correlations suggest that changes in solar luminosity may, in part, have contributed to the global warming observed in the last century. However, the potential relationship between changes in sunspot activity and climate changes has been subject to considerable controversy.

4.7.2 Greenhouse Effect and Global Warming

As indicated in Chapter 1, solar energy not absorbed by the atmosphere or reflected by clouds passes through the atmosphere; most of it is then absorbed by the Earth's surface. On absorption, it is converted into longer-wavelength infrared radiation. Although the atmosphere is relatively transparent to visible light, it is considerably less so for infrared energy.

The lower atmosphere is warmed by sorption of this infrared energy and reradiation from atmospheric gases and clouds. The atmosphere, acting much like the panes of a greenhouse, allows shortwave visible light to pass through while retarding the flow of longer-wave infrared energy back to space. This is called the greenhouse effect (Figure 4.14) and is responsible for the relatively moderate temperatures that occur near the Earth's surface (~15°C, 59°F).

The greenhouse effect is primarily due to the sorption and reradiation of thermal energy by CO_2 and H_2O vapor. Other trace gases also have a greenhouse role. These include CH_4, N_2O, tropospheric O_3, and human-made substances such as CFCs, perfluorocarbons, and sulfur hexafluoride.

The ability of the atmosphere to retard thermal flows back to space is described as emissivity. Atmospheric emissivity varies across the infrared spectrum. As can be seen in Figure 1.5, most infrared absorption is due to H_2O vapor and CO_2. These gases, as well as O_3, absorb relatively little infrared energy in the spectral region of 7 to 13 μm, the so-called atmospheric window.

4.7.3 Radiative Forcing

Although the atmosphere retains thermal energy as a result of the greenhouse effect, incoming radiation is balanced by outgoing terrestrial radiation. Factors that alter electromagnetic radiation

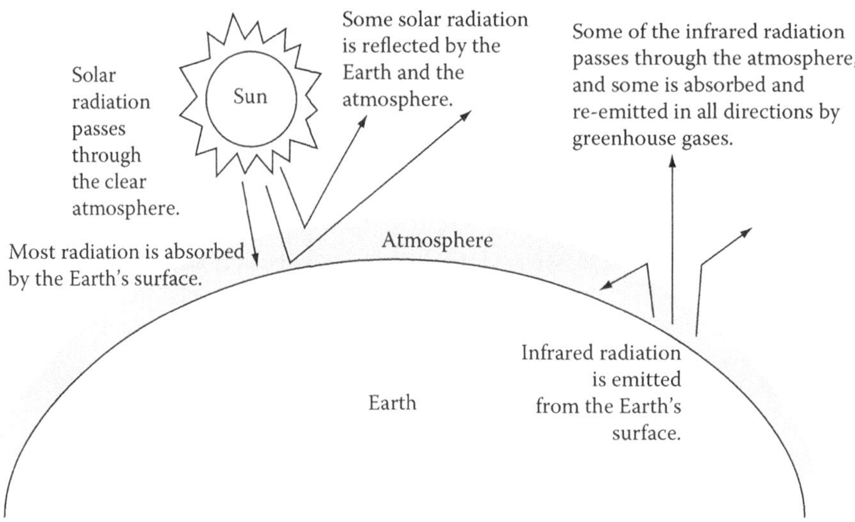

FIGURE 4.14 Greenhouse effect.

received from the sun and lost to space or alter its redistribution within the atmosphere and among the atmosphere, land, and oceans have the potential to affect global and regional climates. As defined by the IPCC, any change in the net radiative energy available to the global Earth atmosphere is termed *radiative forcing*. Positive forcings warm the Earth; negative forcings cool it.

An increase in greenhouse gases reduces the Earth's ability to radiate heat back to space. The atmosphere absorbs more outgoing thermal energy and re-emits it at higher altitudes and lower temperatures. Consequently, this positive radiative forcing warms the lower atmosphere and surface. Thus, the greenhouse effect is enhanced. The degree to which radiative forcing occurs depends on the magnitude of increases in greenhouse gases and the concentrations and radiative properties of individual gases.

Some aerosols, either naturally or anthropogenically produced, reflect sunlight. As such, they can cause negative forcing and climatic cooling. Black carbon aerosol, on the other hand, absorbs solar radiation and warms the climate system. Changes in aerosol concentrations can affect the amount of clouds and their reflectivity. In most cases, tropospheric aerosols contribute to negative radiative forcing. However, aerosols have much shorter atmospheric lifetimes than greenhouse gases.

Climate forcing is any change imposed on the Earth's energy balance. It can be measured and expressed in watts per square meter. Climate responses are the changes that result from changes in climate forcing.

4.7.4 Greenhouse Gases

Greenhouse gas concentrations affected by human activities (for 2019 and the preindustrial period prior to 1800) are summarized in Table 4.2. Also presented are annual percentage increases (based on year 2019), radiative effectiveness (ability to absorb thermal energy) of greenhouse gases relative to CO_2, and estimated contributions to global warming (radiative forcing).

Carbon dioxide concentrations have increased by more than 31% since 1750. The rate of increase has averaged 1.5 ppmv over the past two decades (0.4%/year). Much of this increase is due to the combustion of fossil fuels (70%–90%). Land use changes (where high–C density ecosystems such as forests are replaced by lower–C density agricultural or grazing land) have also had significant effects on CO_2 emissions to the atmosphere (10%–30%). Current CO_2 levels are the highest they have been in almost a half million years.

TABLE 4.2
Greenhouse Gases, Radiative Effectiveness, and Radiative Forcing

	CO_2	CH_4	Halocarbons (as CFCs)	N_2O	O_3
Concentration					
Preindustrial	~280 ppmv	~0.665 ppmv		~270 ppbv	13 ppbv
2025	430 ppmv	1.94 ppmv	~800 pptv	339 ppbv	40 ppbv
Annual increase					
Percent/year, 2023	10–11	~1.5	~0	0.20	—
Concentration/year	3.37 ppmv	13 ppbv	—	1.0 ppbv	—
Radiative effectiveness	1*	21*	13,000–16,000*	206*	2,000*
Relative contribution to global forcing (%)	79	20	15	8	13

Source: *Adapted from Elsom, D.M., *Atmospheric Pollution: A Global Problem*, 2nd ed., Blackwell Publishers, Oxford, 1992. With permission. And NOAA https://www.esrl.noaa.gov/gmd/dv/ftpdata.html.

Most CO_2 removed from the atmosphere is absorbed by the upper 70 to 100 m (230–330 ft) of the Earth's oceans. Forests seem to be an important sink as well. As a consequence, only approximately 40% to 50% of CO_2 emitted to the atmosphere remains there. The atmospheric lifetime for CO_2 is estimated to be 50 to 100 years.

Although CH_4 and N_2O are strong infrared absorbers, their role as greenhouse gases compared with CO_2 and H_2O vapor has historically been relatively limited. However, their importance has increased as their atmospheric concentrations have increased. Methane and N_2O absorb strongly in the atmospheric window with relative radiative effectiveness of 21 and 206, respectively; that is, on a molecule-to-molecule basis, they absorb and retain heat 21 and 206 times, respectively, more effectively than CO_2.

Methane concentrations in the atmosphere have increased 292% from preindustrial times. The average annual increase was on the order of 12 ppbv/year from 1984 to 2007 and about 15–18 ppbv/year in the last five years, and the concentrations continue to increase. Major emission sources of CH_4 seem to include anaerobic bacterial activity in swamps and rice paddies, digestive systems of ruminant animals (particularly livestock) and termites, and landfill sites. Other sources include coal mines and leaking natural gas pipelines.

Nitrous oxide is produced by biological processes in soil. It is also emitted to the atmosphere as a result of fossil fuel and biomass burning, soil disturbance, and animal and human wastes. Use of nitrogen fertilizers may also be a significant source of atmospheric N_2O. Atmospheric N_2O levels have increased by about 63 ppbv, or 24%, since 1750, with a current annual average increase of 0.85 ppbv, or 0.25%. Present-day N_2O concentrations are the highest they have been in 1,000 years.

Chlorofluorocarbons are human-made and therefore new to the atmosphere. Although most of the environmental concern associated with increasing atmospheric CFC levels has focused on stratospheric O_3 depletion, CFCs are strong thermal absorbers (not surprising, given their widespread use as refrigerants). Although concentrations of the two most widely used CFCs (CFC-11 and CFC-12) are in the pptv range, the potential contribution of CFCs to global warming is significant. Chlorofluorocarbons absorb strongly in the atmospheric window. On a molecule-to-molecule basis, they absorb infrared energy 12,000 to 16,000 times more effectively than CO_2. With the phaseout of CFC production and use, CFC levels in the troposphere have been declining.

As seen in Table 4.2, tropospheric O_3 has significant potential to absorb thermal radiation, with a radiative effectiveness of 2,000. Tropospheric O_3 levels have increased by approximately 68% since 1750.

Radiative forcing associated with increases in well-mixed greenhouse gases since 1750 is estimated to be 3.34 W/m^2. Of this, 2.16 W/m^2 has been apportioned to CO_2, 0.54 W/m^2 to CH_4, 0.41 W/m^2 to halocarbons, and 0.21 W/m^2 to N_2O. A positive radiative forcing of 0.40 W/m^2 has been estimated for the tropospheric increase in O_3, and −0.05 W/m^2 for stratospheric O_3 depletion. Including O_3, the increase in radiative forcing would be 3.7 W/m^2. This would correspond to the percentage contributions to global climatic warming indicated in Table 4.2. Although CO_2 has the lowest radiative effectiveness, its relatively high emissions and increasing atmospheric levels are apparently responsible for more than 50% of the radiative forcing that has taken place over the last two and a half centuries. As CFC emissions continue to be restricted and CFCs destroyed by stratospheric processes, their contribution to radiative forcing is expected to diminish significantly.

Radiative forcing for aerosols is estimated to range from −0.4 to −1.1 for sulfate and 0.2 to 0.6 for elemental C.

4.7.5 Evidence of Climate Change

The IPCC, in its 2021 assessment report, unequivocally concluded that the global climate had undergone significant changes in the twentieth century. In the northern hemisphere from 1994 to 2023, it is likely the warmest 30-year period of the past 1,400 years. Figure 4.15 shows the estimated changes in the observed globally and annually averaged surface temperature anomalies relative to 1850 to 1900 compared with the range of projections from the previous IPCC assessments since 1950. The range of global annual mean surface air temperature changes from 1950 to 2100 for models used in the IPCC AR5 2021.

The year 2023 was the 40th consecutive year that the average global surface temperature was above the climate mean temperature for the period 1961 to 1990. The seven warmest years in the past 140 years occurred in the 1990s and 2000. According to IPCC AR6, these warming trends have only intensified: by 2011–2020, global temperatures had exceeded the 1961–1990 baseline for more than 30 consecutive years, with the decade warming by about +0.9°C, and now exceeding +1.1°C above the 1850–1900 preindustrial baseline. Recent years, specifically 2016, 2020, 2023, and 2024, rank among the warmest ever observed globally.

Between 1950 and 2020, average nighttime minimum air temperatures over global land surfaces have continued to increase at a rate of approximately 0.2°C per decade, a trend that remains consistently higher than the rate of increase in daytime maximum temperatures.

The largest increases in global surface temperatures have occurred over continents in the midlatitudes and high latitudes of the northern hemisphere. These increases have been greater than at any time in the past 1,000 years.

Satellite temperature data (collected since 1979) for the troposphere up to an altitude of 7.7 km (4.8 mi) show little evidence of the warming trend observed at the ground. Although balloon measurements have shown a tropospheric warming trend since 1958 (Figure 4.16), balloon data in the period 1979 to 2000 resemble satellite data (with balloon data somewhat higher). This anomaly is cited by opponents of proposed global warming control programs and contrarian scientists who contend that human-induced global warming is not occurring. The causes of these apparent discrepancies in atmospheric and surface temperature trends are not known. Suggestions have included short-term temperature variations due to stratospheric O_3 depletion, El Niño, and the short period of measurement involved.

Changes in atmospheric temperature are supported by new analyses showing that the heat content of oceans has significantly increased since the late 1950s. More than half of this increase has occurred in the upper 300 m (~990 ft). The equivalent rate of temperature increase in this layer is approximately 0.09°C (0.16°F)/decade. The oceans represent the largest reservoir of heat in the world's climate system.

Changes in other climate indicators have been reported. These include an increase in annual land precipitation in the northern hemisphere in the midlatitudes and high latitudes. Atmospheric

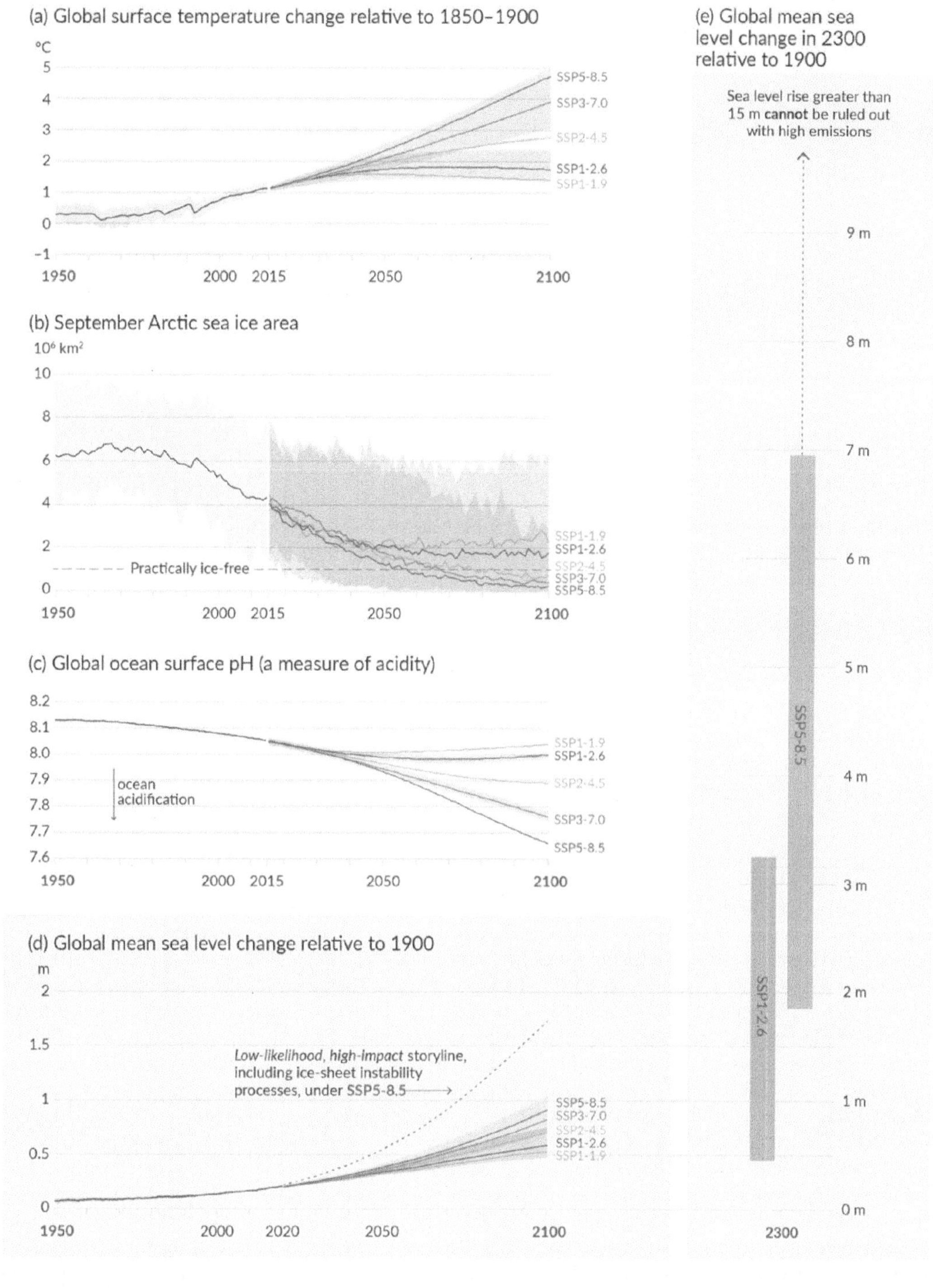

FIGURE 4.15 Panel (a) Global surface temperature changes in °C relative to 1850–1900. These changes were obtained by combining Coupled Model Intercomparison Project Phase 6 (CMIP6) model simulations with observational constraints based on past simulated warming, as well as an updated assessment of equilibrium climate sensitivity (see Figure SPM.1 in the Summary for Policymaker of the Working Group I of the AR6 in the link later). Changes relative to 1850–1900 based on 20-year averaging periods are calculated by adding 0.85°C (the observed global surface temperature increase from 1850–1900 to 1995–2014) to simulated changes relative

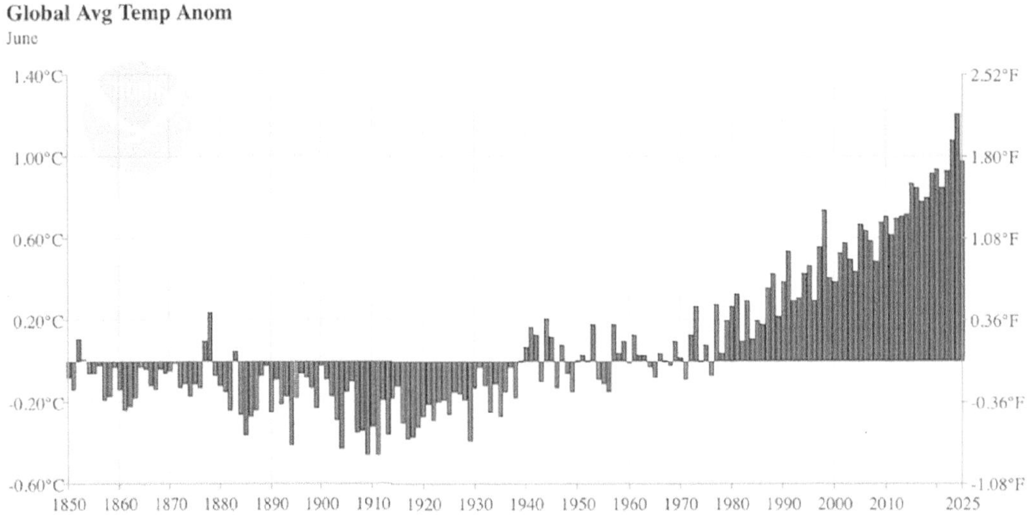

FIGURE 4.16 Surface and tropospheric temperature (satellite data) trends for the period 1850 to 2025. (From NOAA NCEI Climate at a Glance, Global Time Series, published July 2025, https://www.ncei.noaa.gov/access/monitoring/climate-at-a-glance/global/time-series/globe/land_ocean/tavg/1/6/1850-2025.)

H_2O vapor has reportedly increased several percent per decade over many regions of the northern hemisphere over the past 25 years.

Increases in cloud cover on the order of 2% for midlatitude to high-latitude continental regions over the northern hemisphere have also been reported for the twentieth century. This change has been positively correlated with decreases in the range of diurnal temperatures.

---

to 1995–2014. Very likely ranges are shown for SSP1–2.6 and SSP3–7.0. Panel (b) September Arctic sea ice area is 106 km^2 based on CMIP6 model simulations. Very likely ranges are shown for SSP1–2.6 and SSP3–7.0. The Arctic is projected to be practically ice-free near mid-century under intermediate and high GHG emissions scenarios. Panel (c) Global ocean surface pH (a measure of acidity) based on CMIP6 model simulations. Very likely ranges are shown for SSP1–2.6 and SSP3–7.0. Panel (d) Global mean sea level change in meters, relative to 1900. The historical changes are observed (from tide gauges before 1992 and altimeters afterwards), and the future changes are assessed consistently with observational constraints based on emulation of CMIP, ice-sheet, and glacier models. Likely ranges are shown for SSP1–2.6 and SSP3–7.0. Only likely ranges are assessed for sea level changes due to difficulties in estimating the distribution of deeply uncertain processes. The dashed curve indicates the potential impact of these deeply uncertain processes. It shows the 83rd percentile of SSP5–8.5 projections that include low-likelihood, high-impact ice-sheet processes that cannot be ruled out; because of low confidence in projections of these processes, this curve does not constitute part of a likely range. Changes relative to 1900 are calculated by adding 0.158 m (observed global mean sea level rise from 1900 to 1995–2014) to simulated and observed changes relative to 1995–2014. Panel (e) Global mean sea level change at 2300 in meters relative to 1900. Only SSP1–2.6 and SSP5–8.5 are projected at 2300, as simulations that extend beyond 2100 for the other scenarios are too few for robust results. The 17th–83rd percentile ranges are shaded. The dashed arrow illustrates the 83rd percentile of SSP5–8.5 projections that include low-likelihood, high-impact ice-sheet processes that cannot be ruled out. Panels (b) and (c) are based on single simulations from each model and so include a component of internal variability. Panels (a), (d), and (e) are based on long-term averages, and hence the contributions from internal variability are small. (From Climate 2021 Synthesis Report, https://www.ipcc.ch/report/ar6/wg1/downloads/report/IPCC_AR6_WGI_SPM.pdf). Permission from IPCC.)

Increases in land surface temperatures have been positively correlated with decreasing snow cover and land ice. There is increasing evidence that the retreat of alpine and continental glaciers is occurring in response to twentieth-century global warming. Additionally, ground-based measurements have shown an approximate 2-week reduction in the annual duration in lake and river ice in the midlatitudes to high latitudes in the northern hemisphere.

Based on the latest NOAA and NSIDC data, Arctic sea ice has continued to decline significantly. Since 1979, the September minimum extent has decreased by approximately 12.2% per decade, while the March maximum has declined more modestly at about 2.6% per decade. The overall annual extent is shrinking at a rate of 4%–5% per decade. In 2024, the September minimum reached 4.28 million km^2—one of the lowest on record. Additionally, sea ice thickness and volume have declined, particularly with a sharp loss of multiyear ice since 2007. These trends reflect ongoing warming in the region and contribute to broader climate feedbacks such as Arctic amplification.

Global mean sea level has risen about 20 cm (±5 cm) between 1901 and 2018, with an accelerated increase from 1.3 mm/year (1901–1971) to 1.9 mm/year (1971–2006) and further up to 3.7 mm/year (2006–2018). Satellite altimetry data (1993–2023) show that sea level rise has intensified—from approximately 2.1 mm/year at the start to about 4.5 mm/year by 2023—totaling ~11.1 cm over that period. Human-driven thermal expansion and melting ice have been the primary contributors, with ocean warming accounting for roughly 50% of the rise from 1971–2018 and glacier and ice sheet loss contributing the remainder.

The strongest evidence to date to support the occurrence of global warming is the reported changes in plants and animals in terms of (1) shifts in species' ranges and (2) the timing of events such as migration, egg laying, and flowering. Recent analyses of hundreds of studies (and more than 1,700 species) have revealed consistent temperature-related shifts in species distribution poleward and up in elevation (6.1 km/decade toward the poles) and advancement of spring events by 2.3 days/decade. Such biological changes are further described in Chapter 6.

4.7.6 Linking Climate Change to Anthropogenic Influences

Changes in global climate are becoming well documented. The question before us is, "Why is it changing?" Is it changing because of natural oscillations in the Earth's climatic system, increased solar radiation associated with sunspot activity, increases in greenhouse gases, or some combination of these?

Significant greenhouse gas emissions have occurred and will continue to occur. Because the principle of greenhouse warming of the Earth's atmosphere is well established, it is logical to hypothesize that climate changes that have taken place in the past century are at least in part due to radiative forcing associated with greenhouse gas emissions. Such a hypothesis can only be "proven" by (1) evaluating past changes in the Earth's climate in relationship to geological indicators of past atmospheric temperatures and greenhouse gas concentrations determined from air trapped in Arctic and Antarctic ice, and (2) simulating the effects of greenhouse gas increases on global and regional climate by using sophisticated models.

Climate models are based on physical laws that are represented by mathematical equations. Models are designed to predict changes in both global and regional climate in response to perturbations such as increases in one or more greenhouse gases, increased solar radiation, increased albedo, and the like.

Climate models have been undergoing development, use, and continued refinement for almost a decade and a half. To simulate climate, major components of the Earth's climate system are represented in submodels—for example, atmosphere, land surfaces, oceans, cryosphere (frozen ground), and biosphere—as well as the processes that go on within them. Models in which the atmosphere and ocean components have been coupled are known as atmosphere–ocean general circulation models (AOGCMs). Comprehensive AOGCMs are very complex and require supercomputers to run.

Coupled models provide credible simulations of climate down to subcontinental scales and over temporal scales from seasons to decades. They cannot simulate all aspects of climate and do not fully account for surface–air temperature differences observed since 1979. Clouds and humidity are sources of significant uncertainty. Nevertheless, AOGCMs have had a relatively good track record in reproducing warming trends in the twentieth century when driven by greenhouse gases and sulfate aerosols.

Various lines of evidence indicate that temperature changes in the past century are not due to natural variability. These include (1) reconstructions of temperatures in the past 1,000 years, (2) new model estimates of internal variability and responses to natural forcing, (3) qualitative consistencies between observed climate changes and model responses to anthropogenic forcing, (4) evidence of human influence using a wide range of detection techniques, and (5) the fact that most attribution (cause-and-effect relationships with defined levels of statistical confidence) studies have found that the estimated rate and magnitude of global warming due to increasing greenhouse gases alone are comparable to or greater than observed global warming over the past 50 years.

4.7.7 Projections of the Earth's Future Climate

Using estimates of greenhouse gas emissions and climate models, the IPCC has projected a suite of future climate changes to the year 2100. These scenarios lead to very different estimates of CO_2 concentrations in 2100 (from 500 to 1500 ppmv). Estimated increases in concentrations of other greenhouse gases vary considerably. In most scenarios, radiative forcing due to CO_2, CH_4, N_2O, and tropospheric O_3 continues to increase. However, the fraction of total forcing due to CO_2 is projected to increase from approximately one-half to three-quarters.

Climate sensitivity (equilibrium response of global surface temperature to an equivalent doubling of the CO_2 concentration) is predicted to be in the range of 1.5°C to 4.5°C (2.7°F–8.1°F). This range of potential global surface temperatures reflects uncertainties, particularly those related to clouds and other processes.

Global climate models predict significant changes in other broadscale climate variables. These include

1. Greater relative warming at high latitudes due to faster warming of land surfaces compared with oceans
2. Decrease in daily temperature differences, with higher nighttime low temperatures
3. General decrease in daily surface temperature variability in winter and increases in summer in the northern hemisphere
4. Global increase in H_2O vapor, evaporation, and precipitation, with precipitation increases over midlatitudes to high latitudes and Antarctica, and changes in the variability of Asian monsoon precipitation
5. Decrease in snow and ice cover over the northern hemisphere, recession of glaciers and ice caps, decreased Greenland ice sheet mass, and an increase in the mass of Antarctic ice sheets due to increased precipitation
6. Rise in sea level due to thermal expansion and melting of glaciers and polar ice caps, with projected increases of 0.09 to 0.88 m (0.30–2.8 ft)
7. Increase in extreme weather and climate events, for example, higher maximum temperatures, more hot days, more intense precipitation, increased risk of drought, and increases in tropical cyclones and peak and mean precipitation

A comprehensive National Climate Assessment study was released by the US Global Change Research Program in November 2023 (Crimmins et al. 2023) that is recommended for the most current and ongoing discussion of the complexity of climate change. It can be downloaded at https://repository.library.noaa.gov/view/noaa/61592.

4.7.8 Global Warming Uncertainties

The recently confirmed role of sulfate aerosol (associated with regional haze over developed areas of the world) in producing a cooling effect points out how difficult it is to predict what the effects of anthropogenic activities may be on something as complex as the Earth's climatic system. As a consequence, a fair amount of uncertainty exists in predicting climatic changes on global and regional scales. There are, nevertheless, a number of remaining uncertainties. These include (1) anomalous tropospheric vertical temperature profiles in balloon and satellite observations, (2) estimates of internal climate variability from models and observations, (3) reconstruction of solar and volcanic forcing, (4) effects of aerosols and cloud cover, and (5) differences in the response of different models to the same radiative forcing potential.

One major uncertainty is cloud behavior. Global warming is expected to increase cloud formation as well as change cloud distribution and characteristics. Differences in the nature of cloud cover can affect the predicted increase in global warming. By increasing cloud cover, particularly low-lying clouds over the oceans, more sunlight would be reflected back to space. An increase of the Earth's albedo of 0.5% would be expected to halve the greenhouse warming associated with a doubling of CO_2. If cirrus clouds were to increase in abundance, they would absorb heat, thus warming the atmosphere.

Other uncertainties are associated with the various feedback mechanisms among land, oceans, atmosphere, and the Earth's climatic system. Global warming could, for example, increase evaporation, resulting in increased H_2O vapor concentrations. This is expected to enhance the greenhouse effect. Global warming is expected to melt sea ice and snow cover, allowing underlying surfaces to absorb more solar radiation and thus decreasing the Earth's albedo and increasing evaporation. By increasing the temperature of the world's oceans, the solubility of CO_2 would be decreased. As upper layers of the world's oceans become saturated with CO_2, less CO_2 would be removed from the atmosphere. In addition, as global temperatures rise, there is a potential for large amounts of CH_4 to be released from frozen tundra in the northern hemisphere.

A variety of both positive and negative feedback mechanisms exist that can either enhance or mitigate global warming. Most proposed feedback mechanisms would enhance warming.

4.7.9 Environmental Impacts

A variety of systems may be sensitive to climate change. These include H_2O resources, agriculture and forestry, coastal zones and marine systems, human settlements, and biological communities.

Changes in rainfall, stream flow, and groundwater recharge are predicted for many regions as the planet warms. Consequently, human populations living in H_2O-stressed countries could increase from 1.7 to 5 billion by 2025 (depending on the rate of population growth). Flood frequency and magnitude could also increase in many regions as a result of the greater frequency of heavy rainfall events.

Crop yield reductions are predicted to occur as a result of increased drought conditions and expansion of the world's arid and semiarid zones. These reductions in crop yields could be counterbalanced in part by CO_2 stimulation of plant growth.

Coastal zones are predicted to be adversely affected as a result of sea level rises as well as the increase in storm frequency and severity. As a consequence, many coastal areas are expected to experience increased flooding, accelerated erosion, loss of wetlands, and saltwater intrusion into freshwater sources.

Human settlements may be adversely affected, particularly in low-lying coastal areas. Of particular concern is flooding from a rise in sea level, coastal storm surges, and runoff from more intense storms. Such changes pose risks to homes, businesses, infrastructure, and humans themselves.

Significant changes in biological communities are already occurring, and the number of species affected is expected to increase.

4.7.10 Stratospheric Gases and Atmospheric Temperature Changes

While the atmosphere has been warming near the Earth's surface, a significant cooling trend (~0.6°C [1.08°F]/decade) has been occurring in the lower stratosphere (~16–21 km [~9.6–12.6 mi]) since 1980. At middle latitudes, it has been even larger (~0.75°C [1.35°F]/decade), with an even more substantial cooling trend (~3°C [5.4°F]/decade) in the polar regions during late winter and spring. This polar trend has been observed in the Antarctic since the early 1980s and in the Arctic since the early 1990s. Model simulations indicate that O_3 depletion is the primary contributing factor. Greenhouse gases (e.g., CO_2) may also be contributing to stratospheric cooling (~25% of that observed).

Methane, upon photolytic decomposition, is a major source of stratospheric H_2O. Increased stratospheric H_2O is expected to significantly enhance global warming. A doubling of stratospheric H_2O vapor has the same global warming potential as a 70 ppmv increase in CO_2 in the troposphere.

Because coupling between climatic phenomena in the stratosphere and troposphere is known to occur, observed stratospheric cooling and potential future stratospheric warming are likely to have some effect on global surface temperatures. The nature and magnitude of these effects are not known.

REFERENCE

USGCRP, *Fifth National Climate Assessment*, Crimmins, A.R., Avery, C.W., Easterling, D.R., Kunkel, K.E., Stewart, B.C., and Maycock, T.K., Eds., US Global Change Research Program, Washington, DC, USA, 2023. https://doi.org/10.7930/NCA5.2023.

READINGS

Adams, D.D., and Page, W.P., Eds., *Acidic Deposition: Environmental, Economic and Policy Issues*, Plenum Publishers, New York, 1985.

Barrie, L.A., Arctic air pollution: An overview of current knowledge, *Atmos. Environ.*, 20, 643, 1986.

Beaty, C.B., The causes of glaciation, *Am. Sci.*, 66, 452, 1978.

Brasseur, G.P., Orlando, J.J., and Tyndall, G.S., *Atmospheric Chemistry and Global Change*, Oxford University Press, Oxford, 1999.

Bridgman, H.A., *Global Air Pollution: Problems in the 1990s*, John Wiley & Sons, New York, 1994.

Brown, C.A., Hamill, P., and Wilson, J.C., Particle formation in the upper tropical troposphere: A source of nuclei for the stratosphere, *Science*, 270, 1650, 1995.

Charlson, R.J., Schwartz, S.E., and Hales, S.E., Climate forcing by anthropogenic aerosols, *Science*, 255, 423, 1992.

Climate 2014 Synthesis Report. https://www.ipcc.ch/report/ar6/syr/downloads/report/IPCC_AR6_SYR_FullVolume.pdf.

Deming, D., Climate warming in North America: Analysis of borehole temperatures, *Science*, 268, 1576, 1995.

Douglas, J., Mercury and the global environment, *EPRI J.*, January/February, 14, 1994.

Elsom, D.M., *Atmospheric Pollution: A Global Problem*, 2nd ed., Blackwell Publishers, Oxford, 1992.

Finlayson-Pitts, B.J., and Pitts, J.N., Jr., *Chemistry of the Upper and Lower Atmosphere: Theory, Experiments and Applications*, Academic Press, San Diego, CA, 2000.

Gates, D.M., *Climate Change and Its Biological Consequences*, Sinauer Associates, Sunderland, MA, 1993.

Graedel, T.E., and Crutzen, P.J., *Atmospheric Change: An Earth System Perspective*, Freeman, New York, 1993.

Hidy, G.M., Hansen, D.A., Henry, R.C., Ganesan, K., and Collins, J., Trends in historical acid precursor emissions and their airborne and precipitation products, *JAPCA*, 34, 333, 1984.

Hidy, G.M., Source-receptor relationships for acid deposition: Pure and simple? A critical review, *JAPCA*, 34, 518, 1984.

Houghton, J.T., Ding, Y., Griggs, D.J., Noguer, M., van der Linden, P.J., and Xiaosu, D., Eds., *Climate Change 2001: The Scientific Basis: Third Assessment Report of the Intergovernmental Panel on Climate Change*, Cambridge University Press, Cambridge, UK, 2001.

Husar, R.B., and Patterson, D.E., *Haze Climate of the United States*, National Technical Information Service PB 87–141057/AS, Springfield, VA, 1987.

IPCC, The sixth report: Climate change 2021, *The Physical Science Basis*, 2021. https://www.ipcc.ch/report/ar6/wg1/.

Jennings, S.G., Ed., *Aerosol Effects on Climate*, University of Arizona Press, Tucson, AZ, 1993.

Kahn, J.R., The surface temperature of the sun and changes in the solar constant, *Science*, 242, 908, 1998.

Kerr, R.A., Global pollution: Is the Arctic haze actually industrial smog?, *Science*, 205, 290, 1979.

Kiehl, J.T., and Briegleb, B.B., The relative roles of sulfate aerosols and greenhouse gases in climate forcings, *Science*, 260, 311, 1993.

Kraljic, M.A., *The Greenhouse Effect*, H.W. Wilson Co., New York, 1992.

Legge, A.H., and Srupa, S.V., Eds., *Acidic Deposition: Sulphur and Nitrogen Oxides*, Lewis Publishers, Chelsea, MI, 1990.

Lucier, A.A., and Haines, S.G., Eds., *Mechanisms of Forest Response to Acidic Deposition*, Springer-Verlag, New York, 1990.

McCarthy, J.J., Canziani, O.F., Leary, N.A., Dokken, D.J., and White, K.S., Eds., *Climate Change 2001: Impacts, Adaptation and Vulnerability: Third Assessment Report of the Intergovernmental Panel on Climate Change*, Cambridge University Press, Cambridge, UK, 2001.

McElroy, M.B., and Salawitch, R.J., Changing composition of the global stratosphere, *Science*, 243, 763, 1989.

Metz, B., Davidson, O., Swort, R., and Pan, J., Eds., *Climate Change 2001: Mitigation: Third Assessment Report of the Intergovernmental Panel on Climate Change*, Cambridge University Press, Cambridge, UK, 2001.

National Acid Precipitation Assessment Program, *Interim Assessment: The Causes and Effects of Acidic Deposition*, US Government Printing Office, Washington, DC, 1987.

National Oceanic and Atmosperic Administration Education Accomplishments Report. noaa.gov/sites/default/files/atoms/files/FY_2018_NOAA_Education_Accomplishments_Report.pdf.

National Research Council, *Acidic Deposition: Long-Term Trends*, National Academy Press, Washington, DC, 1988.

National Research Council, Ozone depletion, greenhouse gases and climate change, in *Proceedings of the Joint Symposium Board of Atmospheric Science & Climate & Committee on Global Change*, National Academy Press, Washington, DC, 1989.

National Research Council, *Protecting Visibility in National Parks and Wilderness Areas*, National Academy Press, Washington, DC, 1993.

National Research Council, *Rethinking the Ozone Problem in Urban and Regional Air Pollution*, National Academy Press, Washington, DC, 1991.

Peters, R.L., and Lovejoy, T.E., *Global Warming and Biological Diversity*, Yale University Press, New Haven, CT, 1992.

Rodriguez, J.M., Probing stratospheric ozone, *Science*, 261, 1128, 1993.

Rowland, F.S., Chlorofluorocarbons and the depletion of stratospheric ozone, *Am. Sci.*, 77, 36, 1989.

Saitoh, T.S., Shimade, T., and Hoshi, H., Modeling and simulations of the Tokyo urban heat island, *Atmos. Environ.*, 30, 3431, 1996.

Schneider, S.H., Detecting climate change signals: Are there any "fingerprints"? *Science*, 263, 341, 1994.

Schneider, S.H., The greenhouse effect: Science and policy, *Science*, 243, 771, 1989.

Schwartz, S.E., Acidic deposition: Unraveling a regional phenomenon, *Science*, 243, 753, 1989.

Seinfeld, J.H., and Pandes, S.N., *Atmospheric Chemistry: From Air Pollution to Climate Change*, John Wiley & Sons, New York, 1998.

Stolarski, R., Bojkov, R., Bishop, L., Zerefos, C., Staehlin, J., and Zawodny, J., Measured trends in stratospheric ozone, *Science*, 256, 342, 1992.

Thompson, D.J., The seasons, global temperature, and precession, *Science*, 268, 59, 1995.

Tolbert, M.A., Sulfate aerosols and polar stratospheric cloud formation, *Science*, 264, 527, 1994.

Turekian, K.K., *Global Environmental Change: Past, Present and Future*, Prentice Hall, Saddlebrook, NJ, 1996.

US EPA, *Air Quality Criteria for Particulate Matter*, EPA/600/P-99/002bB, EPA, Washington, DC, 2001.

US EPA, *Air Quality Criteria for Particulate Matter*, Vol. 1, EPA/600/AP-95/001a, EPA, Washington, DC, 1995.

US EPA, *Air Quality Criteria for Particulate Matter and Sulfur Oxides*, EPA/600/8-82-029a-c, EPA, Washington, DC, 1982.

US EPA, *Latest Findings on National Air Quality: 1999 Status and Trends*, EPA/454/F-00-002, EPA, Washington, DC, 2000.

US EPA, *Protecting Visibility: An EPA Report to Congress*, EPA/450/5-79-008, EPA, Washington, DC, 1979.

US EPA, *Research Summary: Acid Rain*, EPA/600/8-79-028, EPA, Washington, DC, 1979.

US EPA, *Third United States Climate Action Report*, 2002. http://Yosemite.epa.gov/globalwarming.nsf/contract/ResourceCenterPublicationsUSClimateAction Report.html.

Van de Kamp, J., *Health Effects of Global Warming*, Department of Health and Human Services, Bethesda, MD, 1992.

Van Loon, G.W., and Duffey, S.J., *Environmental Chemistry: A Global Perspective*, Oxford University Press, Oxford, 2000.
Warnek, P., *Chemistry of the Natural Atmosphere*, Academic Press, New York, 1988.
World Meteorological Organization, *Scientific Assessment of Ozone Depletion*, WMO Global Ozone Research and Monitoring Project, Report 44, WMO, Geneva, Switzerland, 1998.
Wuebbles, D.J., Grant, K.E., Cornell, P.S., and Penner, J.E., The role of atmospheric chemistry in climate change, *JAPCA*, 39, 22, 1989.

QUESTIONS

1. What will be changed by atmospheric effects?
2. Significant changes in visibility have occurred in the eastern part of the United States in the past half century. Describe their cause.
3. Even if western states had particle aerosol levels similar to those in the east, visibility reduction would be less. Why?
4. Describe methods by which visibility reduction can be quantified.
5. How do atmospheric particles reduce our ability to see objects clearly?
6. What is atmospheric turbidity? How is turbidity related to visibility?
7. What is Arctic haze and how is it formed?
8. What is a heat island? What is its environmental significance?
9. How do particle size and number affect the probability of precipitation?
10. Why is the term *acid rain* not appropriate to describe acid phenomena in the atmosphere?
11. Describe the relationship between pH and H^+ concentrations.
12. Why are some ecosystems acid-sensitive whereas others are not?
13. Emissions of mercury to the atmosphere are relatively small. Why are these emissions increasingly becoming an environmental concern?
14. Why is atmospheric deposition of pesticides, PCBs, and dioxin an environmental concern?
15. Why is the atmospheric deposition of nitrogen an environmental concern?
16. What are the effects of stratospheric O_3 on the atmosphere?
17. Chlorine atoms in the stratosphere are present at an average concentration of approximately 4 ppbv. Relatively speaking, this is a very low concentration. Why is this chlorine such an environmental concern?
18. What proof is there that chlorine and bromine atoms are causing stratospheric O_3 depletion?
19. Why does the Antarctic O_3 hole occur in the southern hemispheric spring?
20. What evidence is there that stratospheric O_3 depletion is occurring outside the Antarctic?
21. What are polar stratospheric clouds? What role do they have in O_3 depletion?
22. Describe two direct effects of stratospheric O_3 depletion.
23. How do variations in the Earth's orbit cause changes in climate?
24. What is the greenhouse effect? Describe the role of major greenhouse gases.
25. What is radiative forcing?
26. Describe the radiative effectiveness of CFCs, CH_4, N_2O, and tropospheric O_3 relative to CO_2.
27. Describe evidence that global warming associated with greenhouse gas emissions is already occurring.
28. What pollutant has the greatest impact on radiative forcing? Why?
29. What are some of the potential impacts of global warming on coastal and agricultural areas?
30. Although global surface temperatures have increased, the lower stratosphere has cooled. Why?
31. Why are global warming effects greater in the Arctic than at middle latitudes?

5 Health Effects

5.1 AIR POLLUTION EPISODES

Air pollution episodes are characterized by significant short-term increases in atmospheric pollutant concentrations above normal daily levels. Episodes vary from those that pose relatively limited public health concern to those more extreme. Extreme air pollution episodes were reported for the Meuse Valley, Belgium, in 1930; Donora, PA, and the Monongehela River Valley in 1948; and London in 1952. In each of these cases, a persistent (3–6 days) thermal inversion combined with significant industrial and, in the case of London, domestic pollutant emissions resulted in high ground-level concentrations that caused acute illness and, in some cases, death in exposed populations.

Sixty deaths were reported in the Meuse Valley disaster, 20 in Donora, and approximately 4,000 in London. In each case, death claimed individuals with existing respiratory or cardiovascular disease. Pneumonia was the primary cause of death in the 1952 London disaster. Based on retrospective studies, "excess mortality" was associated with "fogs," going back to as early as 1873. In each of these extreme episodes, many individuals reported becoming ill. Illness characterized by cough, shortness of breath, chest pain, and eye and nose irritation was reported by people of all ages. In the Donora area, more than half of the 14,000 residents reported experiencing acute illness symptoms.

Other somewhat less extreme episodes have been reported for London in 1956, 1957, 1959, 1962, 1975, and 1991; Dublin in 1982; the Ruhr Valley of Germany in 1962, 1979, 1982, and 1985; the Netherlands in 1985, 1987, and 1991; and New York City in 1953, 1962, and 1966.

These episodes are significant in that they provided solid scientific documentation that exposure to elevated ambient pollutant levels can cause acute illness and even death. The more disastrous episodes contributed to significant efforts to control ambient air pollution. This was particularly the case in London, where clean air legislation was enacted as early as 1956.

Despite significant control efforts, air pollution episodes continue to occur in developed countries. Although pollutant levels are now considerably lower than in the disastrous episodes of the past, application of increasingly more sophisticated epidemiological techniques to what could be described as historically minor episodes shows significant associations between elevated pollutant levels, acute illness such as asthmatic attacks, and even increased death rates.

Air pollution episodes are not limited to developed countries. Increased industrialization, limited pollution control efforts, and population growth pose significant exposure risks on a day-to-day basis as well as during episode periods in many cities in China, India, Mexico, and other countries.

5.2 POLLUTANT EXPOSURES AND CAUSE–EFFECT RELATIONSHIPS

Ambient air pollution exposures that result in severe community-wide illness and death are, for the most part, relatively rare in the United States and other developed countries. Pollutant exposures in our cities, towns, and even countrysides vary considerably, both in types of pollutants present and prevailing concentrations. Exposures reflect the nature and extent of industrial activities, dispersion conditions of the surrounding environment, motor vehicle emissions, atmospheric chemistry, and, in some mountain valleys, wood space heating.

Pollutant exposures to the "normal" range of concentrations may result in a variety of health effects. These include acute but transient symptoms such as eye, nose, and throat irritation and, in some sensitive individuals, asthmatic attacks. Acute asthmatic responses to ambient pollutants are a major regulatory concern. The primary focus of most regulatory activity in North America and

DOI: 10.1201/9781003664109-5

other developed countries has been to protect exposed or potentially exposed populations from the chronic effects of relatively low-level exposures. These may include respiratory and cardiovascular disease, neurotoxic effects, and cancer.

A good understanding of the observed or potential health effects of pollutants at levels typical of urban–suburban communities is requisite to all health-based regulatory programs. Ideally, such regulatory activities are supported by scientific data that adequately establish a cause-and-effect relationship between pollutant exposures and illness symptoms among those exposed so that an adequate margin of safety can be incorporated into air quality standards. Because humans experience a variety of exposures and other stresses, establishing definitive cause–effect relationships between adverse health effects and one or more atmospheric pollutants is a difficult undertaking.

A strong relationship between individual or combined pollutant exposures and health effects is indicated when there is a convergence of evidence from epidemiological, toxicological, and occupational exposure studies. Information from such studies is of particular importance in evaluating the potential health effects of chronic exposures to ambient pollutants.

5.2.1 Epidemiological Studies

Epidemiological studies are conducted to determine potential relationships between a variety of environmental factors and human disease. They are characterized by the application of statistical analysis to data collected on the health status of populations of individuals, pollutant exposures, and potential confounding factors. Such studies often provide evidence of possible causal relationships between pollutant exposures and observed or reported health effects. In general, epidemiological studies become more important as the risk attributable to atmospheric pollutants becomes smaller and the duration of exposure required to produce effects becomes longer. They have been particularly useful in identifying the acute effects of elevated short-term pollutant exposures. These effects may include pulmonary function changes, asthmatic attacks, and increased mortality.

Epidemiological studies vary in design. They may be cross-sectional (assess the relationship between pollutant exposures and health effects over a cross section of a population), longitudinal (assess exposures and health effects over time), or case-control (exposed and non- or less-exposed populations are compared). They may be prospective (data to be collected) or retrospective (evaluation of existing data). Study designs differ in their power to identify potential causal relationships and statistical confidence in results obtained. Over the past decade, increasingly sophisticated statistical techniques have been applied to the study of short-term pollutant exposures. These have included time-series analyses of daily health data collected over a period of days or weeks. Such studies are longitudinal, and as such, each participant serves as his or her own control. Consequently, the effect of confounding variables (described next) is significantly reduced when data are statistically analyzed. Such studies are characterized by a strong study design, good exposure and health effects data, and large study populations.

5.2.1.1 Confounding Factors

Epidemiological assessment of potential relationships between pollutant exposures and observed parameters of human health has, in many cases, been confounded by a variety of coexisting factors. These have included individual sensitivity, age, existing disease, gender, race, socioeconomic status, tobacco smoking, lifestyle, occupation, and meteorological conditions. Assessment of potential effects may also be confounded by interaction between (1) two or more pollutants, (2) pollutants and meteorological variables such as temperature and relative humidity, (3) pollutants and other exposures (occupational, smoking, and indoor air pollution), and (4) pollutants and infectious disease. Such interactions may, in part, explain the differences observed between epidemiological and toxicological studies.

5.2.1.2 Interaction Effects

5.2.1.2.1 Interactions between Pollutants

Because community air is commonly contaminated by complex mixtures of gases and particulate matter (PM), it is likely that some of these may interact to modify the physiological effects of others. This may occur in several different ways. For example, one pollutant may affect the site of deposition of another. This is apparently the case for gases such as sulfur dioxide (SO_2). Because of its solubility in watery fluids, SO_2 is normally removed in the upper respiratory system. Absorption and adsorption of SO_2 on particles provides a transport mechanism for SO_2 into the pulmonary system, where SO_2 toxicity occurs. Consequently, particles can potentiate the effect of SO_2. Harmful aerosols such as sulfuric acid (H_2SO_4) can be produced as a result of interaction of gaseous pollutants with the moist environment of the lungs.

Pollutant interactions may result in health and physiological effects that are additive, antagonistic, or synergistic. Exposure to pollutants with similar toxic effects may result in an additive response. The total toxic response is the sum of individual responses to each pollutant. Antagonism occurs when one pollutant interferes with the toxic effects of another, thus mitigating the effects of the more toxic pollutant. Reactions of ammonia (NH_3) and SO_2 in human breath are an example of such an effect. Synergism describes health responses that are significantly greater than the sum of the individual effects of toxic pollutants. This is particularly true for pollutants that inhibit pulmonary defense mechanisms (e.g., tobacco smoke). Inhibition of respiratory clearance increases the risk of disease (e.g., cancer) from pollutants such as asbestos and radon (Rn). In many cases, synergistic effects are multiplicative; that is, risk is increased by the product of each individual exposure risk.

Epidemiological and toxicological information on additive, antagonistic, and synergistic effects is limited. Because of the large variety of pollutants present in metropolitan atmospheres, it is probable that each of these interaction effects occurs. In day-to-day pollutant exposures, they cannot, for the most part, be evaluated using the scientific tools currently available. It is standard regulatory practice to set exposure limits on individual pollutants as if they were acting alone. Such exposure limits are subject to intense legal, political, and scientific challenges from those whose activities would be subject to regulation. Use of exposure limits based on multiple pollutant exposures would pose an enormous regulatory challenge.

5.2.1.2.2 Interactions between Pollutants and Meteorological Factors

Most major air pollution episodes were associated with meteorological extremes that confounded assessment of the causal role of ambient air pollution in the genesis of mortality and morbidity. During the many "London fogs" and disasters in the Meuse Valley and Donora, weather was cold and damp. The independent effect of temperature extremes on mortality has been well established. Indeed, studies in the United Kingdom have indicated that mortality was more closely related to cold and humidity than to fog frequency. Studies in Los Angeles have noted a marked increase in mortality during periods of elevated photochemical oxidant levels and high temperature. In such cases, investigators commonly concluded that observed excess mortality was due to high temperatures rather than pollutant exposures. Heat waves, in general, are major stressors of the elderly and infirm and usually result in excess mortality. With the application of more powerful research designs and statistical techniques, epidemiologists are increasingly able to separate the independent effects of temperature extremes and pollutant exposures in evaluations of excess mortality.

5.2.1.2.3 Interaction between Ambient Air Pollutants and Other Pollutant Exposures

In addition to community air pollution exposures, many individuals are exposed to elevated pollutant levels from cigarette smoking and in their occupational and domestic environments. Such exposures may mask health risks associated with ambient pollutants. They may also increase health risks associated with ambient pollution.

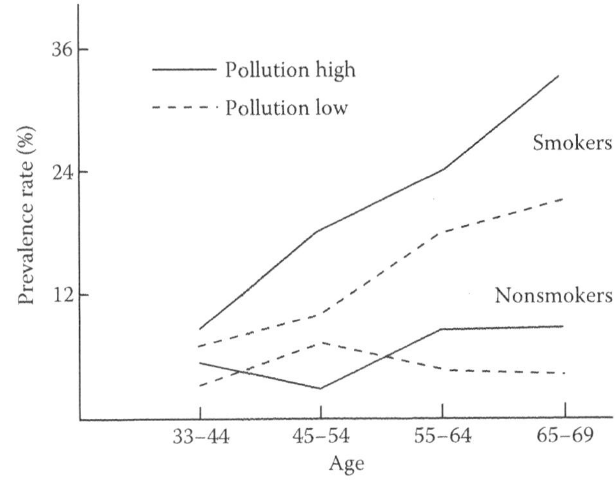

FIGURE 5.1 Prevalence of chronic bronchitis associated with tobacco smoking and exposure to ambient air pollution. (From Lambert, P.M. and Reid, D.D., *Lancet*, 5, 853, 1970. With permission.)

Cigarette smokers may be at higher risk for adverse effects associated with exposures to ambient pollution. This increased risk may be due to one or more factors. First, cigarette smoking may initiate respiratory and cardiovascular disease. Individuals with such disease, no matter what the cause, are at increased risk of illness or death when exposed to elevated pollutant levels. Smoking may increase the effect of community air pollution exposures (Figure 5.1) by impairing respiratory clearance mechanisms.

Many individuals are exposed to relatively high levels of gaseous and particulate-phase pollutants in their place of employment. These exposures are usually much greater than those associated with breathing polluted community air. Because they are occupationally exposed, some individuals may be at higher risk from ambient pollutant exposures. This increased risk may be due to occupationally induced respiratory or cardiovascular disease or impaired defense mechanisms.

Assessment of health effects of community air pollutants is confounded, to some degree, by the fact that individuals spend more time indoors than in the ambient environment. On average, Americans only spend 2 hours/day outdoors. Therefore, their ambient pollution exposure may be quite different than a police officer's or construction worker's. An expanded discussion of indoor air pollution and its health implications can be found in Chapter 11.

5.2.1.3 Exposure Assessment

Most early epidemiological studies of air pollution and related health effects were limited by inadequate information on pollutants present and data on population exposures. As a result, many of these studies used the place of residence as an index of exposure. The place of residence often introduces bias into statistical analyses because it may be associated with unique ethnic or cultural traits, living standards, occupations, and exposure to infectious agents. Exposure assessment in many recent longitudinal studies is carried out using personal exposure monitors, time activity diaries, and exposure models based on fixed-site monitors and dispersion modeling of various emission sources. Exposure assessment, like other aspects of epidemiological studies, has become more sophisticated and better serves the needs of pollutant effects studies.

5.2.1.4 Population Susceptibility

Epidemiological assessment of air pollution–related health effects has historically been complicated by variations in population susceptibility. Such assessments are now simplified by identifying the populations at risk, for example, the aged, the very young, those with existing respiratory or cardiovascular disease, those occupationally exposed, and cigarette smokers. These populations/individuals may be more sensitive to normally encountered pollutant levels than the average healthy individual. Epidemiological studies of sensitive populations are more powerful in identifying air pollution–health effects associations when they indeed exist. Present air quality standards (Chapter 8) promulgated in the United States are intended to protect those at special risk as well as the general population.

Despite the confounding factors and historical limitations described, epidemiological studies have been able to provide valuable insight into defining the probability that air pollution exposures contribute to the incidence or prevalence of specific community health problems. A strong association or high probability of disease causation can be inferred when (1) there are a number of different populations in which a similar association is observed, including different population groups, locations, climate, and times of year; (2) the incidence or severity of the health effect increases with increasing exposure and, conversely, decreases with decreasing exposure; and (3) a plausible biological mechanism can be hypothesized for the observed association.

The value of epidemiological studies is increased when they are used in conjunction with controlled biological (toxicological) studies on animals and humans. They are important in identifying possible associations that can be tested under controlled laboratory conditions. In addition, they can be used to evaluate potential human health risks identified in laboratory exposures.

5.2.2 Toxicological Studies

Toxicological, or controlled biological, studies are conducted on animals and, less commonly, humans to determine the functional, structural, and biochemical effects of toxic substances. Historically, toxicological studies have been conducted to determine the toxicity of a substance administered in varying doses, with dose being a function of the concentration of the substance and duration of exposure. Significant differences in toxic response occur when the same dose of toxic material is administered over different exposure periods. Acute exposures result when an organism is subjected to a high dose rate, that is, a high concentration of a substance over a relatively short period. Response to the exposure is sudden and severe and usually lasts for a brief period. If the dose rate is sufficiently high, death may result. Indeed, the term *toxic* implies that a substance has the potential to cause death. Lower doses, that is, lower concentrations over longer periods, generally do not directly cause mortality. As dose rate decreases, the response is less severe, may take longer to develop, and may be more persistent. In chronic exposures, adverse effects may take years, if not decades, to develop.

In traditional toxicology, animals are exposed to varied doses of toxic substances to determine their ability to cause physiological and pathological changes and to characterize dose–response relationships. In addition to evaluating such physiological and pathological changes, the scope of toxicology in recent years has been expanded to include studies of carcinogenesis (cancer induction), teratogenesis (induction of birth defects), mutagenesis (production of mutations), gametotoxicity (damage to sex cells), and endocrine disruption.

One of the major advantages of toxicological studies is that the investigator can control many variables that may confound epidemiological studies. Controllable variables in animal studies include age, sex, environmental conditions such as temperature and relative humidity, genetic constitution, nutrition, and pollutant dose. By controlling dose as a function of exposure duration, dose–response relationships can be developed. Human exposures are also controllable, but to a more limited degree. In addition to those variables described for animals, factors peculiar to humans such as smoking, occupation, and health status are, to some degree, under the experimenter's control.

5.2.2.1 Human Studies

Ideally, toxicological studies of humans should provide the strongest scientific evidence available for establishing a cause–effect relationship between air pollutant exposures and adverse health effects. In such studies, it is possible to expose humans under controlled conditions and monitor them for signs indicating the onset of disease. Most human studies are, however, limited to short-term exposures. Because of the long exposure periods required to induce chronic responses and the irreversibility of some of these responses, most human studies are not conducted at realistic concentrations of pollutants that occur in ambient atmospheres. In addition, ethical considerations, which are imperative when designing human exposure studies, limit the kind of information that can be acquired.

In theory, a true cause–effect relationship can only be developed by human experimentation. Ethical and other limitations make it virtually impossible to conduct such studies.

5.2.2.2 Animal Studies

Animal exposures can provide valuable information on the effects of pollutants when no other acceptable means of obtaining such information are available. Because animals can be sacrificed for pathological and biochemical analyses, they are ideally suited for toxicological studies. Additionally, chronic exposures can easily be conducted.

Use of animals as models for humans does have limitations; animals are not humans. Although they may have the same general organ systems, they may be structurally and physiologically different. In addition, species differ in their sensitivity to pollutants. Dogs, for example, are 10 times more resistant to nitrogen dioxide (NO_2) than rodents are. Also, most animals are not as long-lived as humans. Consequently, results obtained from animal studies can only be extrapolated to humans with some degree of uncertainty. Uncertainty is reduced when similar results can be demonstrated in more than one mammalian species, especially primates, and the organ systems under study are closely analogous to those of humans.

To provide observable toxicological responses, exposures used in controlled studies are usually much higher than those occurring in ambient environments. In general, controlled exposures are limited to one toxic pollutant.

5.2.3 Occupational Exposure Studies

Although ethical and legal constraints limit the use of human volunteers for experimental evaluation of the effects of toxic substances, such constraints have not limited the voluntary or involuntary exposure of industrial workers. For many air pollutants of interest, exposures in the work environment occur at much higher levels than in ambient air. Because of potential health risks and legal concerns, some industries maintain detailed exposure and health data on employees. As a consequence, occupational exposures may be relatively well defined and often provide an important source of information on illness development and expression, job-related disease, and dose–response relationships. In the absence of human toxicological studies, occupational exposures often provide the only available human dose–response data.

Studies of occupationally exposed workers have limitations. Although one or more pollutants may be common to both occupational and ambient environments, the overall pollutant mix varies considerably. In addition, occupational exposures are limited to workday and workweek cycles. Community exposures are not only more heterogeneous and variable; they are further complicated by differences in indoor and ambient air quality and the amount of time individuals spend in such environments.

A major limitation of occupational studies is that the population of workers does not reflect the general population. In many industries, workers are nominally healthy males between the ages of 18 and 65. Studies of such workers may provide important information on disease syndromes induced by specific pollutants or pollutant combinations. They cannot, however, provide adequate information on those who are not industrially exposed and who may be at special risk. In such instances, information on health effects on healthy individuals may not be as important as responses of individuals who are more sensitive. Such individuals may be excluded from the workforce or, because of existing health problems, be excluded from specific occupations.

5.3 EFFECT OF POLLUTANTS ON THE HUMAN BODY

Pollutant effects are normally manifested in specific target organs. These may be direct; that is, pollutants come in intimate contact with the organ affected. Such is the case for eye and respiratory irritation. Effects may be indirect. For example, pollutants may enter the bloodstream from the lungs or gastrointestinal system by respiratory clearance. Effects may then be distant from the immediate organ of contact. Indeed, a target organ may not have immediate intimate contact with air contaminants. The principal target organs or organ systems are eyes and the respiratory and cardiovascular systems (Figure 5.2).

5.3.1 Eye Irritation

Eye irritation is one of the more prevalent manifestations of pollutant effects on the human body. It is most often associated with exposure to aldehydes and photochemical oxidants. The threshold for eye irritation by oxidants is approximately 0.10 to 0.15 ppmv (reported as ozone [O_3]). Eye irritation increases with increasing oxidant concentration, although O_3 and NO_2, the two principal oxidants, do not cause eye irritation. Apparently, oxidant levels are indicators of the eye irritation potential of photochemical pollutants such as peroxyacyl nitrate (PAN), acrolein, formaldehyde (HCHO), and other photochemically produced compounds. There is some question whether eye irritation should be categorized as a significant health effect because no other physiological changes are detectable and eye irritation resolves quickly after exposure ceases.

5.3.2 Effects on the Cardiovascular System

Pollutants such as carbon monoxide (CO) and lead (Pb) are absorbed into the bloodstream and may have both direct and indirect effects on the cardiovascular system. These effects will be discussed

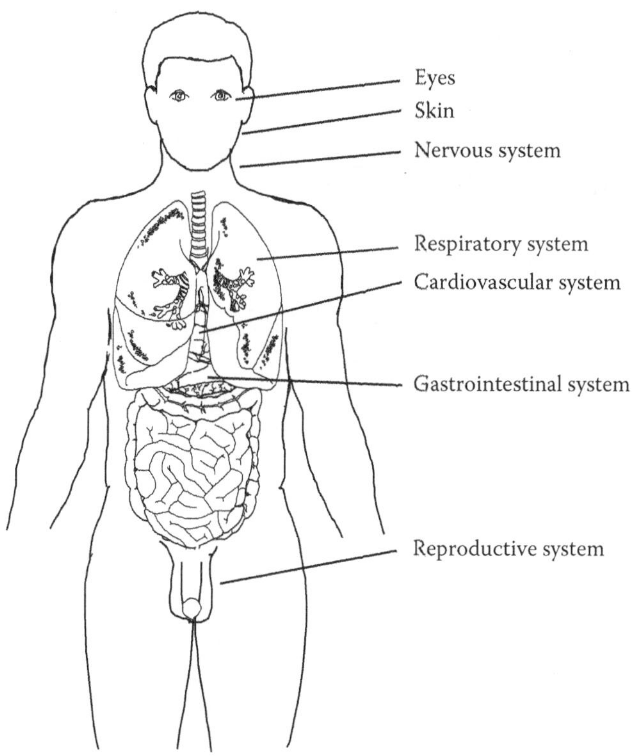

FIGURE 5.2 Target organs for air pollutant exposures.

in subsequent sections. Cardiovascular disease may also result from the indirect effects of other air pollution–incited disease. For example, some individuals die of cor pulmonale, heart failure resulting from the stress of severe chronic respiratory disease. Recent epidemiological studies have shown that premature cardiovascular system–related mortality is strongly associated with exposures to small (≤ 2.5 μm) particles.

5.3.3 Effects on the Respiratory System

The respiratory system is responsible for gas exchange and, therefore, receives direct exposure to airborne contaminants. Effects on the respiratory system are determined in great measure by its structural and functional anatomy as well as respiratory defense mechanisms.

The principal function of the respiratory system is to supply the body with oxygen (O_2) for metabolism. In addition, it removes waste carbon dioxide (CO_2) from the bloodstream and tissues. These functions are facilitated by the three major units of the respiratory tract: the nasopharyngeal, tracheobronchial, and pulmonary regions (Figure 5.3).

5.3.3.1 Units of the Respiratory System

5.3.3.1.1 The Nasopharyngeal Region

The nasopharyngeal region, or upper airway, consists of nasal passages, nasopharynx, oropharynx, and glottis. It begins at the terminus of the bony turbinates and extends to the region of the vocal

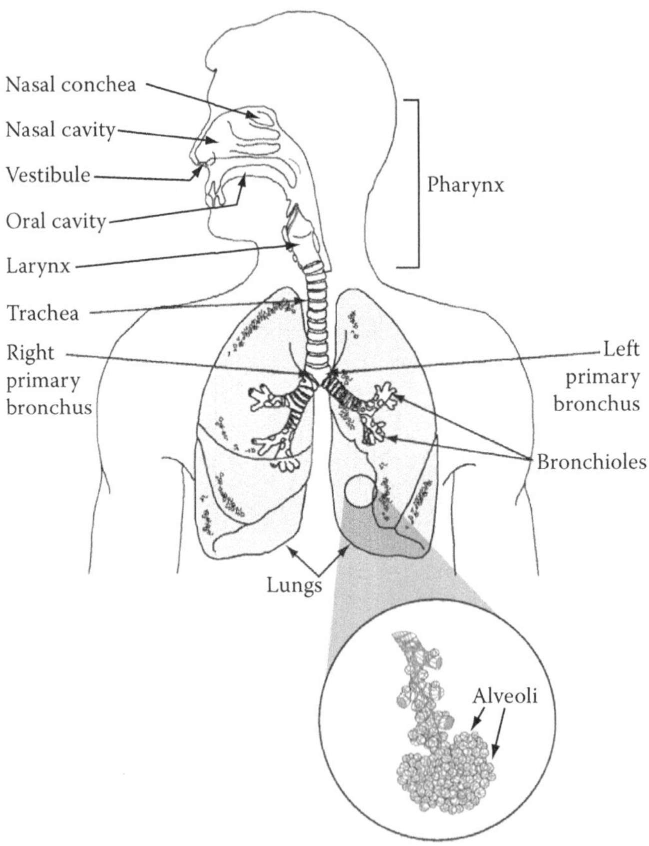

FIGURE 5.3 General anatomy of the human respiratory system.

cords. The nasal area is subdivided into two cavities by a septum. These cavities, just beyond the nostrils, are ringed with coarse hairs. Projecting from the wall of the cavities are bony turbinates that force inspired air to flow along a winding course where, on contact with the large surface area of the nasal passages, it is warmed and moistened. Nasal surfaces are covered by a mucus layer and hair-like projections called cilia that serve to remove the larger particles in the thoracic range (≤ 10 μm).

The oropharynx extends from the soft palate to the glottis, located at the junction of the trachea and esophagus, where eating and breathing functions are separated. The glottis is the space between the vocal cords that are found in the larynx, or voice box.

5.3.3.1.2 The Tracheobronchial Region

The tracheobronchial region of the respiratory system is a series of tubes or ducts that transport inspired air to lung tissue, where O_2 and CO_2 are exchanged. This region consists of a large tube, the trachea, which subdivides to form smaller-diameter tubes, bronchi. Bronchi in turn branch to form smaller and smaller airways. From the trachea to the site of gas exchange, there may be as many as 23 to 32 generations of branching, with each generation being smaller in diameter and length. Bronchi comprise the first 12 to 22 branching generations. Their walls have cartilaginous plates and a muscular layer that constricts when foreign substances enter bronchi. Air flows from bronchi into the smaller bronchioles (which do not have cartilaginous plates and are usually < 1 mm in diameter). The interior surfaces of the trachea, bronchi, and bronchioles are lined by ciliated cells interspersed with mucus-secreting cells. The mucus layer produced by these cells is moved toward the oropharynx by the rhythmically beating cilia. Terminal bronchioles are the smallest bronchioles to have cilia.

5.3.3.1.3 The Pulmonary Region

The pulmonary region of the respiratory system consists of respiratory bronchioles, alveolar ducts, and alveoli (Figure 5.3). Alveoli are membranous air sacs at the terminus of the alveolar ducts. A fully developed lung is estimated to have approximately 300 million alveoli. The large alveolar surface area facilitates efficient exchange of O_2 and CO_2 between the lungs and the blood capillaries that cover the alveoli.

Movement of air into (inspiration) and out of (expiration) the lungs is accomplished by muscles of the chest and diaphragm, which alternately compress and expand the lungs. During inspiration, air pressure in the alveoli decreases, becoming slightly negative. Because of this negative pressure, air flows into the lungs. During expiration, alveolar air pressure rises, causing air to flow out of the lungs.

5.3.3.2 Defense Mechanisms of the Respiratory System

The respiratory system is protected from airborne contaminants by a variety of defense mechanisms. In the nasal region, the stiff nasal hairs or impaction on the mucus layer of the winding passages of the turbinates may remove large particles. Cilia sweep the mucus layer and entrapped particles toward the back of the throat where they are swallowed or expectorated. Contaminants may also be removed from upper airways by the sneeze reflex.

Other defense mechanisms serve to remove or prevent contaminants (mainly particles) from entering the tracheobronchial and pulmonary regions. When, for example, particles enter the upper portion of the bronchial tree, muscle layers constrict the bronchi, thus narrowing bronchial diameter and reducing the number of particles entering the lungs. The cough reflex, similar to the sneeze reflex, assists this process by eliminating particles trapped in mucus in the tracheobronchial region. Cilia that line the tracheobronchial region clean the airways by propelling mucus and particles upward where they are removed by coughing and expectoration or swallowing. In the latter way, some particulate-phase pollutants may enter the gastrointestinal system.

Despite these defense mechanisms, many fine particles and some relatively insoluble gases enter pulmonary tissue. Within the alveoli and bronchioles of the pulmonary region, specialized cells called phagocytes and macrophages ingest deposited matter (phagocytes and macrophages, derived from white blood cells, consume bacteria, viruses, and other particles and, as a result, serve as a pulmonary defense mechanism). These phagocytes and the matter they contain are normally transported out of the lungs in mucus by the cilia. However, they may also move through the alveolar membrane and enter either the lymphatic or circulatory system. Additionally, soluble components deposited in the alveoli may be absorbed by adjacent tissue and enter the lymphatic or circulatory system without being phagocytized. Although these mechanisms affect clearance from the respiratory system, they may expose other body systems to toxic materials. They may also result in tissue damage when phagocytes are destroyed and spill their contents when attempting to ingest toxic particles.

5.3.3.3 Air Pollution and Respiratory Disease

Community air pollution has been implicated as a causal or aggravating agent in diseases of the respiratory system such as chronic bronchitis, pulmonary emphysema, lung cancer, bronchial asthma, and infections.

5.3.3.3.1 Chronic Bronchitis

Bronchitis is a respiratory disease characterized by inflammation of the membrane lining the bronchial airways. Bronchitis may be caused by pathogenic infections or respiratory irritants, for example, those that occur in cigarette smoke, industrial exposures, and ambient and indoor air pollution.

When bronchial inflammation persists for 3 months or longer, it is classified as chronic bronchitis. Extended irritation of the bronchial membrane is the primary cause of chronic bronchitis. It is characterized by a persistent cough and excessive mucus or sputum production, often accompanied by destruction of cilia and thickening of bronchial epithelium. Chronic bronchitis significantly affects respiratory function, as airway resistance is increased by occlusion of the bronchi, swelling of the inflamed membrane, and concomitant excess mucus production. Consequently, persons afflicted with the disease have difficulty breathing.

Severe cases of chronic bronchitis are often followed by the development of pulmonary emphysema. Combined chronic bronchitis and pulmonary emphysema are usually diagnosed and described as chronic obstructive lung disease (COLD) or chronic obstructive pulmonary disease (COPD).

Some epidemiological studies suggest that ambient pollutants may initiate or aggravate chronic bronchitis. The problem of identifying ambient air pollutants as causal agents in the etiology of chronic bronchitis is confounded by the significant causal role of cigarette smoking in initiating this disease. It is further obscured by the effects of occupational exposures. However, available epidemiological and toxicological evidence suggest that long-term community exposures to a combination of ambient pollutants such as PM and SO_2 may contribute to the initiation of chronic bronchitis.

5.3.3.3.2 Pulmonary Emphysema

Whereas chronic bronchitis is a disease that affects the upper portion of the respiratory airway system, pulmonary emphysema is a disease of lung tissue that normally facilitates gas exchange with blood.

Emphysema is often a disease of older individuals. It is characterized by the destruction or degeneration of walls of the alveolar sacs. As a consequence, the total surface area of pulmonary tissue is reduced, diminishing aeration of blood. Individuals afflicted with emphysema usually develop pulmonary hypertension. In addition to elevation of arterial pressure, pulmonary resistance is increased. Pulmonary emphysema is consequently accompanied by shortness of breath and other breathing difficulties.

Patients with emphysema usually have problems exhaling because air remains trapped in the lungs and overinflates damaged lung tissue. This overinflation contributes to the "barrel chest" characteristic of most patients with emphysema. Exhalation difficulties are due to the compression and collapse of some of the smaller airways and the overall decrease in lung elasticity common to those afflicted by emphysema.

Little epidemiological evidence exists to implicate ambient air pollutants as a contributing factor to the initiation and development of pulmonary emphysema. Animal studies suggest, however, that chronic exposures to NO_2 can initiate preemphysematous lung changes.

5.3.3.3.3 *Lung Cancer*

Cancer is one of the leading causes of death in the United States, accounting for approximately 370,000, or one in five, deaths each year. Of these, approximately 150,000 are due to lung cancer, which is the leading cause of cancer deaths. Common to all cancers is the characteristic unrestrained cell growth that produces malignant tumors, which have a higher rate of growth than normal tissue that surrounds them. Most cancers metastasize, that is, spread to other parts of the body. The tendency for lung cancer cells to metastasize is particularly great; as a consequence, the prognosis for patients with lung cancer is poor, with an approximate 90% mortality.

Cancer is usually a disease of the elderly and individuals in late middle age. The pattern is due in part to the fact that cancers are commonly latent diseases; that is, there is a delay of several decades or more between the initiation of exposures to a carcinogenic agent and development of the disease syndrome. For lung cancer, the latency period may be as long as 30 years or more, with peak incidence at age 50. The long interval between initial exposure to the carcinogenic (i.e., cancer producing) agent and disease onset makes it especially difficult to conclusively identify a specific causal agent. During this latency period, an individual may undergo changes in occupation, socioeconomic status, place of residency, nutrition, and personal habits such as smoking. These changes tend to confound the epidemiological evaluation of urban air pollution in the etiology of lung cancer. The latency period not only confounds the identification of causal agents in diseases such as lung cancer but also tends to give individuals a false sense of security relative to the safety of a variety of chemical or physical exposures.

The most common form of lung cancer is bronchogenic; that is, it originates in the bronchial lining and invades tissues of the bronchial tree. The malignancy may spread to the rest of the lung and eventually to other parts of the body. Available epidemiological evidence indicates that nearly 90% or more of the approximately 150,000 lung cancer deaths in the United States each year are caused by chronic exposure to tobacco smoke. Lung cancer has also been linked to passive tobacco smoke exposures and occupational exposures to asbestos, arsenic, and radioactive gases (Rn) and dusts.

A causal role for ambient air pollution in the development of some lung cancers is suggested from several types of epidemiological investigations, including (1) comparisons of urban versus rural populations and migrants from high to low air pollution countries, and (2) statistical regression analyses of the relationships between lung cancer deaths and various indices of air pollution.

Comparisons of rural and urban nonsmoking populations indicate that urban lung cancer death rates are approximately twice those found in rural areas. These studies suggest that there is an apparent urban factor that contributes to the overall incidence of lung cancer. However, they do not identify the characteristics of the urban environment that are primarily responsible.

Studies of individuals migrating from countries that had high ambient air pollution levels to those where air pollution levels were much lower indicate that lung cancer rates in such migrants tend to be intermediate between those of the country of origin and their new homeland. Even when differences in smoking habits are considered, lung cancer rates of migrants from historically high air pollution countries such as the United Kingdom were greater than those from countries such as New Zealand. These data suggest that ambient air pollution exposures may produce effects that persist long after exposures have ceased.

The strongest evidence for a link between ambient air pollution and lung cancer comes from an ongoing prospective mortality study of 1.2 million adults started in 1982. Based on an epidemiological risk factor analysis of 500,000 of these adults, lung cancer mortality was shown to be significantly linked with exposures to fine particles ($PM_{2.5}$), with an 8% increase in lung cancer mortality for each 10 $\mu g/m^3$ increase in $PM_{2.5}$.

5.3.3.3.4 *Bronchial Asthma*

Asthma is an acute respiratory ailment characterized by the constriction of muscles and swelling of the lining of respiratory airways, excessive mucus production, and increased resistance to airflow. Such reactions are episodic, being described as asthmatic attacks. They may occur suddenly and without warning. Overt symptoms include severe shortness of breath, wheezing, chest tightness, and coughing. Attacks may occur for a few minutes to several hours.

Asthma is a relatively common respiratory ailment with a prevalence rate of approximately 4.3% in the United States, or approximately 10 to 12 million individuals. Asthma prevalence in children is relatively high, with a rate of 8.3% in midwestern children and 11.8% in southern children ages 3 to 11 years. Highest asthma rates are reported for African-American children (13.4%). The overall prevalence rate of asthma has increased by 50% since 1980, whereas the annual death rate has doubled to 5,000 per year.

Asthma has been strongly linked with exposure and sensitization to inhalant allergens. There is limited evidence that exposure to air pollutants may potentiate allergenic responses that cause asthma.

Asthmatic attacks may be caused by exposure to inhalant allergens and nonspecific irritant gas-phase and particulate-phase substances. In controlled exposures, SO_2 has been observed to cause asthmatic attacks, and a number of epidemiological studies have linked admissions to hospital emergency rooms for treatment of asthmatic symptoms with high-O_3 pollution days. Increased asthmatic symptoms are reported for children in "asthma camps" on such days. Recent studies indicate that O_3 exposures not only initiate asthmatic attacks but also contribute to asthma development.

5.3.3.3.5 *Respiratory System Infections*

Several animal studies over the past three decades have shown that exposure to pollutants such as O_3, NO_2, SO_2, and PM increases the risk of respiratory system infections. Such infections include pneumonia and bronchitis and are primarily caused by bacteria. The effect that pollutants have on viral infections such as influenza is less clear.

The primary factors involved in increased infection risks may include modulation of pulmonary immune cells (e.g., macrophages), lung surfactants, and physical factors such as ciliary beating and particle clearance. Because of differences in exposure conditions and sensitivity, it is difficult to extrapolate infection risks from animals to humans. Nevertheless, results from human exposure studies and retrospective epidemiological studies of air pollution episodes strongly suggest that young and aged members of an exposed population are at greatest risk of infection, particularly of the upper respiratory tract.

5.4 HEALTH EFFECTS OF REGULATED AIR POLLUTANTS

Although polluted atmospheres may contain hundreds of different substances, only a relatively small number have been identified as having the potential to cause adverse health effects under community-wide exposure conditions encountered in ambient environments. Most notable are those for which National Ambient Air Quality Standards (NAAQSs) have been promulgated. These are CO, sulfur oxides (SO_x), O_3, PM with aerodynamic diameters of $\leq$ 10 and $\leq$ 2.5 μm (PM_{10} and $PM_{2.5}$, respectively), NO_2, and Pb. Although nonmethane hydrocarbons (NMHCs) are regulated under NAAQSs, their regulation is not based on any human health risks. Rather, they are regulated to achieve the O_3 standard. A large number of other substances, commonly referred to as air toxics, are regulated under National Emissions Standards for Hazardous Air Pollutants (NESHAP) and Air Toxics provisions of clean air legislation. Most pollutants regulated under NESHAP and Air Toxics provisions are associated with sources that result in localized, rather than community-wide, exposures. Because of their ubiquitous presence, population exposure potential, and potential to cause adverse health effects, pollutants regulated under NAAQS provisions are discussed here in detail.

5.4.1 Carbon Monoxide

Potentially harmful exposures to CO occur from a variety of sources and environments. At very high concentrations (> 1,000 ppmv), CO exposures may be lethal, with death resulting from asphyxia. At lower concentrations (several hundred ppmv), CO may cause neurological-type symptoms, including headache, fatigue, nausea, and, in some cases, vomiting. Asphyxiation and sublethal symptoms are usually caused by exposures associated with poorly vented combustion appliances, idling motor vehicles in closed environments, excessive CO production and inadequate ventilation associated with occupational industrial activities, and smoke inhalation from structural fires. Carbon monoxide poisoning, although not uncommon, represents exposures that are considerably higher than those that occur in the ambient air of our cities and towns, the kind of exposure conditions addressed by the NAAQS for CO. Such community exposures rarely exceed the 35 ppmv, 1-hour standard.

The principal mechanism of CO toxicity is tissue hypoxia associated with preferential bonding of CO with hemoglobin (Hb) to form carboxyhemoglobin (COHb). Carbon monoxide has an Hb affinity 240 times that of O_2. In addition, CO binds more tightly with Hb than does O_2. As a consequence, the blood's capacity to carry O_2 is decreased. Carbon monoxide also binds to several intracellular proteins, which may result in extravascular effects.

The quantity of COHb formed depends on a variety of factors, for example, the concentration and duration of CO exposure, exercise, temperature, health status, and metabolism of the individual exposed. At low exposure concentrations, the quantity (%) of Hb tied up by CO is a linear function of exposure concentration and duration (Figure 5.4). Note the effect of exercise on lung ventilation rate and COHb formation in blood.

The formation of COHb is reversible. Depending on initial COHb levels, the elimination half-time is approximately 2 to 6.5 hours. Because elimination is slow, COHb levels in blood may increase on continuous exposure. Approximately 10% to 50% of the body store of CO is external to the vascular system.

The primary health concerns associated with low-level ambient (< 50 ppmv) exposures to CO are effects on the cardiovascular system (particularly in sensitive individuals) and neurobehavior. There are more than 500,000 deaths in the United States each year from heart attacks, and 5 to

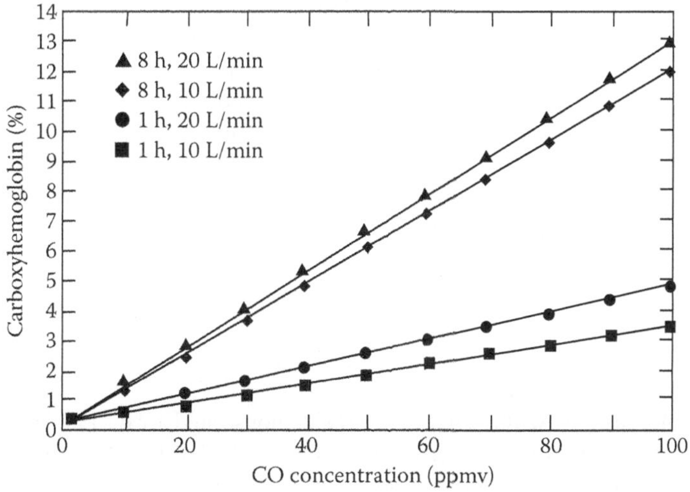

FIGURE 5.4 COHb levels in the blood as a function of exposure concentration, duration, and ventilation rate. (From US EPA, EPA/600/8-90/045f, EPA, Washington, DC, 1991.)

7 million people are estimated to have a history of heart attack, angina (heart-related chest pain), or both. Scientific support for the current CO NAAQS comes from studies of patients with stable angina associated with cardiovascular disease. The lowest observed physiological effect level in patients with exercise-induced ischemia (tissue oxygen deficiency) is somewhere between 3% and 4% COHb. Baseline values in nonexposed nonsmokers are approximately 0.5% to 0.7%. Because nonsmokers may be exposed from a variety of sources, average COHb levels are approximately 1%; COHb levels in smokers vary from 3% to 8%, with an average of approximately 4%.

The literature on the effects of CO on the cardiovascular system is extensive. Many studies have been confounded by the effects of cigarette smoking. Cigarette smoking significantly increases the risk of atherosclerosis and myocardial infarction (heart attack). It is not certain whether chronically elevated COHb levels in smokers are major causal factors for these effects. It has been suggested that chronic exposure to CO may cause atherosclerosis, whereas acute exposures may trigger myocardial arrythmias and infarction. Evidence that CO exposure from sources other than smoking can produce chronic effects is limited.

Several clinical studies of volunteers have been conducted over the past several decades. Of note are results of the Health Effects Institute's (HEI) multicenter study. Carbon monoxide exposures sufficient to elevate COHb to 2% produced significant angina-related effects in exercising patients with coronary artery disease. A dose–response relationship was observed, with no obvious threshold. The significance of the HEI study is that exposure at the 8-hour CO NAAQS of 9 ppmv is sufficient to produce a COHb level of 2.1%. If one assumes that 10% of the US population may reach a COHb concentration of 2.1% (independent of tobacco smoking) and couples this with the widespread prevalence of coronary disease, it is apparent that a significant number of people may be at risk even when CO concentrations do not exceed the 8-hour NAAQS. The effects of environmental CO exposures on smokers are likely to be additive to those associated with smoking. Smokers, who are, as a group, at greater risk of coronary disease, would therefore be subject to an increased risk of developing coronary symptoms.

Several epidemiological studies in the 1970s and better-designed studies conducted in the 1990s have shown significant associations between CO levels and daily mortality, frequency of admission to hospital emergency departments for cardiorespiratory complaints, and hospital admissions for congestive heart failure. Other potential health effects such as chronic flu-like illness symptoms have been associated with short-term repeated exposure to CO. Complaints of headache, irritability, and malaise have been described for children and adults. Symptoms have been shown at measured COHb concentrations of only 5%. It has also been suggested that 3% to 5% of individuals who seek medical attention for headache or dizziness in urban hospital emergency rooms may be suffering from effects of CO exposure. Such effects have been described as "chronic occult CO poisoning syndrome." The nature of exposures responsible for this syndrome is unknown. It is probable that they may be associated with malfunction or improper operation of gas furnaces, stoves, or ovens, augmented by ambient exposures.

The brain is the body organ most sensitive to reduced tissue O_2 levels associated with CO exposures. Although it can, under normal circumstances, initially compensate for CO-induced hypoxia by increased blood flow and tissue O_2 extraction, some neurological or neurobehavioral effects can be anticipated. Behaviors that require sustained attention or performance seem to be very sensitive to COHb-induced effects. Human studies of hand–eye coordination, vigilance, and continuous performance have reported consistent relationships with COHb as low as 5%. The effects at lower COHb values seem to be minimal.

Neurological and cardiovascular effects clearly demonstrated to be associated with different COHb blood concentrations are summarized in Table 5.1. The levels of COHb required to produce these changes are likely to occur only in worst-case atmospheric conditions in the United States, that is, in downtown areas with heavy traffic and very poor atmospheric mixing conditions. They are also likely to occur in underregulated environments of large cities in developing countries. High-population, high-altitude Mexico City would be particularly at risk.

TABLE 5.1
Neurological and Cardiovascular Responses of Humans at Various COHb Saturation Levels

Blood COHb (%)	Effect
0–1	None known
2.0	Quicker onset of angina pain in exercising patients; pain duration lengthened
2.5	Impairment of time interval discrimination
3.0	Change in relative brightness thresholds
4.5	Increased reaction time to visual stimuli
10	Performance changes in driving simulation
10–20	Headache, fatigue, dizziness, coordination loss

Source: NACPA, USDHEW, Publication AP-62, Washington, DC, 1970.

The NAAQS for CO, like other NAAQSs, is designed to protect the most sensitive populations; in this case, individuals with existing cardiovascular disease. Based on both theoretical and experimental research, other populations may be at risk. These would include the fetus, infants, the elderly, individuals with preexisting disease that decreases the availability of O_2, and individuals using certain medications and recreational drugs. Exposure to other pollutants and residence at high altitude may also increase CO exposure health risks.

5.4.2 Sulfur Oxides

The sulfur oxides include both gas-phase and particulate-phase chemical species. Of these, only SO_2 is present in sufficient ambient concentrations to be a public health concern. Particulate-phase SO_x includes strongly to weakly acidic substances such as H_2SO_4 and its neutralization products, such as acid ammonium sulfate (NH_4HSO_4) and ammonium sulfate ($[NH_4]_2SO_4$); most acid sulfate studies have focused on H_2SO_4.

5.4.2.1 Sulfur Dioxide

Much of the available information on acute and chronic effects of SO_2 is based on laboratory animal studies conducted at exposure concentrations considerably higher than those found in polluted atmospheres. Because of its solubility in watery fluids, SO_2 is efficiently removed in the upper respiratory tract. It forms sulfurous acid (H_2SO_3), which readily dissociates into bisulfite and sulfite ions that move into the systemic circulation. Less than 1% of inspired SO_2 reaches the alveoli. Exposure of the lower airways and alveoli increases during exercise, mouth breathing, and under low ambient SO_2 levels. The major physiological effects of SO_2 are changes in mechanical function of the upper airways, including an increase in nasal flow resistance and a decrease in nasal mucus flow rate. Animal studies indicate that SO_2 exposures can adversely affect pulmonary defense mechanisms such as mucociliary transport, alveolar clearance of particles, and macrophage function. There is some evidence that exposures may cause chronic bronchitis.

When the primary air quality standard for SO_2 (0.03 ppmv [80 μg/m³] annual mean, 0.14 ppmv [365 μg/m³] maximum 24-hour concentration) was promulgated in 1971, no health effects had been reported for short-term exposures (< 1 hour) at levels observed in the ambient environment (≤ 1 ppmv, 2.67 mg/m³). In 2010, the US Environmental Protection Agency (US EPA) revised the primary SO_2 standard by establishing a new 1-hour standard at a level of 75 ppb. Since 1980, numerous challenge studies conducted on asthmatic individuals have shown that when asthmatics and others with hyperactive airways are briefly exposed to SO_2 at concentrations of 0.25 to 0.50 ppmv (0.66–1.3 mg/m^3), they exhibit acute responses characterized by bronchoconstriction (airway narrowing) with increased airway resistance and decreased forced expiratory flow rate, and clinical symptoms of shortness of breath and wheezing. Bronchoconstriction occurs within 5 to 10 minutes of exposure and is brief in duration, with lung function returning to normal for most subjects within an hour of exposure. Such responses occur even at lower SO_2 levels during moderate exercise. This results from increases in breathing rate and oral breathing and thus exposure. Oral breathing increases SO_2 penetration to the lower respiratory tract and results in airway drying.

Children and adults with mild to moderate asthma are at greatest risk for short-term SO_2-induced respiratory effects. Individuals with more severe asthma are at lower risk because their low exercise tolerance deters them from engaging in sufficiently intense exercise that could cause any effects.

A substantial percentage (20%–25%) of individuals with mild to moderate asthma exposed for 5 to 10 minutes to concentrations of 0.6 to 1.0 ppmv (1.60–2.67 mg/m^3) of SO_2 during moderate exercise or activity would be expected to have changes in respiratory function and symptom induction that clearly would exceed those associated with daily variation in lung function or in response to other stimuli. At SO_2 exposures of less than 0.5 ppmv, the severity of effects may be sufficient to cause disruption of ongoing activities, increase use of bronchodilator medication, and possibly increase the need for medical attention. Less than 20% of individuals with mild to moderate asthma exposed to 0.2 to 0.5 ppmv of SO_2 during moderate exercise are likely to experience lung function changes greater than they experience daily. Such responses are less likely to be perceptible and of immediate public health concern.

Available epidemiological data do not show any evidence of excess asthma mortality associated with SO_2 exposures in urban areas. Viewed from a national perspective, it seems that the probability of asthmatic individuals being exposed to peak levels of SO_2 greater than 0.25 ppmv for 5 to 10 minutes is low. However, some monitoring data indicate that peak SO_2 concentrations higher than 0.25 ppmv do occasionally occur, suggesting that individuals with mild to moderate asthma residing in the vicinity of major sources may be at increased risk.

5.4.2.2 Acid Sulfate

Sulfuric acid is a strong irritant. However, consistent effects on pulmonary function or respiratory symptoms have not been shown in acute exposure studies of healthy adults. There is some evidence that asthmatic individuals are more sensitive to the effects of acid sulfate on lung mechanical function. They may experience moderate bronchoconstriction after exposure to H_2SO_4 at concentrations of less than 1 mg/m^3. These effects are not as consistent or dramatic as those reported for SO_2 exposures. Children with allergic asthma seem to be more sensitive to H_2SO_4 compared with older asthmatics.

Sulfuric acid, like SO_2, can affect pulmonary defense mechanisms. Acute exposures as low as 0.1 mg/m^3 have been shown to alter mucociliary transport in normal humans. Clearance of deposited particles seems to be stimulated at low concentrations and retarded at higher concentrations. Sulfate exposure can also adversely affect macrophage functioning. There is also some evidence that acid sulfates can enhance sensitization to potentially allergenic substances and modulate the activity of cells involved in allergic responses.

Biological responses to acid sulfates are likely due to the deposition of hydrogen ions (H^+) on the surface tissues of respiratory airways. The relative potency of acid sulfates has been shown in toxicological studies to be directly related to its degree of acidity (increased potency with increased H^+ concentration). Because H^+ may be responsible for acid sulfate toxicity, the biological response is likely to be affected by how it is inhaled. Acid sulfates are more likely to be neutralized by ammonia (NH_3) as a result of mouth, as compared with nasal, breathing.

There is insufficient evidence to definitively establish a causal relationship between acid aerosol exposure and adverse health effects. This may reflect, in part, the fact that few epidemiological studies have included measurements of H^+ and that acid aerosols vary colinearly with PM_{10} and $PM_{2.5}$

concentrations. Long-term and population-based annual assessments of mortality (comparing different communities) indicate only a weak effect of acid aerosol. Recent studies with direct measurements of acid aerosols and other major pollutants have shown consistent relationships between acid aerosol levels and hospital admissions for respiratory disease. Acid aerosol has also been related to a significantly higher prevalence of bronchitis in children and reduced lung growth (determined from lung function measurements) as a result of chronic exposures.

5.4.3 Particulate Matter

Before atmospheric particles can cause adverse health effects, they must enter and be deposited in the human respiratory system. Respiratory penetration and deposition depend on the aerodynamic size of particles, respiratory defense mechanisms, and breathing patterns. Particles can be described as inhalable (IPM), thoracic (TPM), or respirable (RPM), based on their penetration and potential for deposition. Size ranges and fractional penetration of IPM, TPM, and RPM are illustrated in Figure 5.5. Inhalable PM can enter the upper airways of the head; TPM, the airways and gas exchange regions of the lung. Respirable particles are a subset of TPM with a high probability of being deposited in the pulmonary region.

Particles larger than 10 μm generally do not pass through the nasal hairs and defense mechanisms of the upper respiratory system. Save for allergens, they are not a public health concern. Thoracic and respirable particles are significant public health concerns because they enter respiratory airways and are deposited in lung tissue.

The aerodynamic diameter of particles is the primary determinant of respiratory deposition. Deposition occurs by inertial impaction, sedimentation, and diffusion. Impaction is particularly effective in the particle size range of 5 to 10 μm; sedimentation at 3 to 5 μm; and diffusion at less than 0.1 μm.

Inertial impaction is the process by which larger particles traveling in an airstream collide with a surface when the airstream changes direction; sedimentation occurs when large particles fall to a surface under the force of gravity. In diffusion, very small particles come into contact with surfaces as a result of their random motion. These deposition mechanisms are affected by breathing patterns.

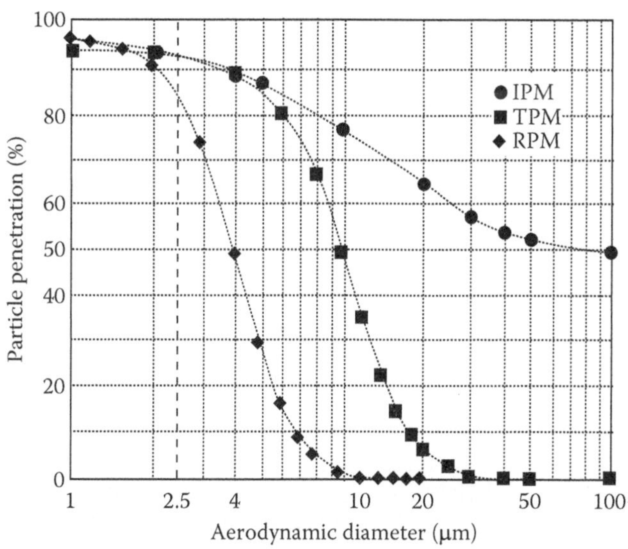

FIGURE 5.5 Particle penetration (%) for inhalable, thoracic, and respirable particles through an ideal inlet. (From US EPA, Third External Review Draft of Air Quality Criteria for Particulate Matter, April 2002, Vol. 1, nepis.epa.gov/Exe/ZyPDF.cgi/30005CR7.PDF?Dockey=30005CR7.PDF)

Deep breathing increases the number of particles that come into contact with respiratory epithelia and, as a consequence, increases deposition by impaction, sedimentation, and diffusion. Breathing depth and rate influence the (1) size of particles deposited, (2) regional location of deposition, and (3) deposition mechanisms.

Regional deposition patterns are particularly important because the sites of particle deposition determine the mechanisms and rates of clearance, the potential for retention and particle dissolution, and the location and nature of tissue injury that may occur.

Deposition is described in the context of regions of the respiratory system (nasopharyngeal, tracheobronchial, and pulmonary) and, based on particle size, the fraction deposited. Fractional deposition efficiency for three TPM particle sizes can be seen in Figure 5.6. Particle deposition is greatest in the pulmonary region for the smallest particle size (2.5 μm), with high nasopharyngeal deposition for the largest particle size (7.5 μm).

Because of changes in flow patterns in the tracheobronchial zone, many particles tend to be deposited at or near airway bifurcations (branchings). Because nerve endings are concentrated at these sites, deposited particles mechanically stimulate reflex coughing and bronchoconstriction. The sensitivity of nerve endings to chemical stimuli causes an increase in breathing rate and reduced pulmonary compliance (ability of the lungs to yield to increases in pressure without disruption).

The deposition of particles is also influenced by mass concentrations, molecular composition, pH, and solubility. Deposition varies among nonsmokers, smokers, and individuals with lung disease. Tracheobronchial deposition is slightly higher in smokers and greatly increased in individuals with lung disease. It is also increased in individuals with varying levels of airway obstruction, such as smokers, asthmatics, and individuals with COPD.

Particle retention depends on clearance rates, which vary among different regions of the respiratory tract. In the ciliated airways of the nose and upper tracheobronchial zone, clearance in healthy individuals is achieved in less than 1 day. In general, clearance time increases as the site of deposition becomes deeper. Clearance from the alveoli may take weeks to months. Slow particle clearance is considered to be detrimental because harmful substances are in contact with sensitive tissue for longer periods.

Until the late 1970s and early 1980s, it was widely held that exposures to high concentrations of PM that were prevalent in the United States and Europe before 1970 were a significant health hazard

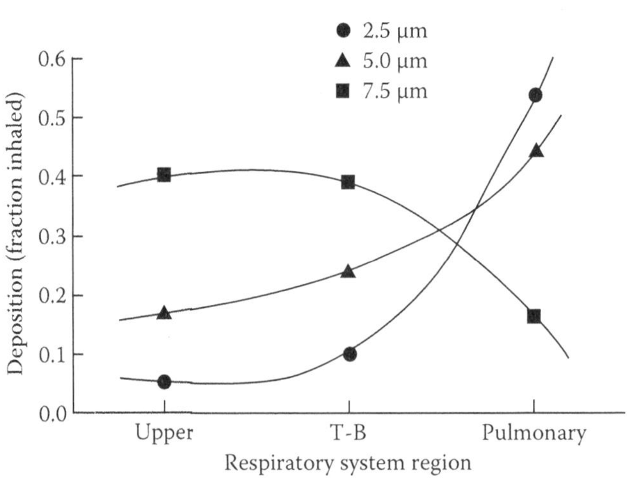

FIGURE 5.6 Deposition of particles of varying size in nasopharyngeal, tracheobronchial, and pulmonary regions of the lung. (From Foster, W.M. in, *Air Pollution and Health*, Holgate, S.T., Samet, J.M., Koren, H.S., and Maynard, R.L., Eds., Academic Press, San Diego, 1991, p. 304. With permission.)

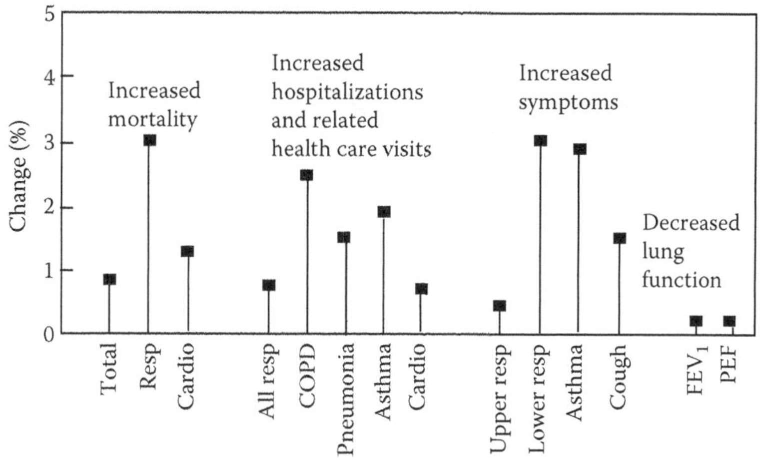

FIGURE 5.7 Estimated percent change in health endpoints associated with a 10 μg/m³ increase in PM_{10} levels in acute exposure epidemiological studies. (From Pope, C.A., in *Air Pollution and Health*, Holgate, S.T., Samet, J.M., Koren, H.S., and Maynard, R.L., Eds., Academic Press, San Diego, 1999, p. 688. With permission.)

and that evidence for adverse effects at the much lower concentrations present in the atmosphere after 1970 was relatively weak.

By the mid-1980s and 1990s, investigators began to report significant associations between ambient PM levels and various adverse health effects. These reports included (1) the Harvard Six Cities Study, which showed that long-term PM exposure was associated with increased risk of respiratory illness in children and cardiopulmonary mortality in adults; (2) the Utah Valley studies, which showed that PM levels were significantly associated with various illness indices, including respiratory hospitalizations, lung function changes, respiratory symptoms, school absenteeism, and premature mortality; and (3) a series of studies that showed a relationship between daily changes in ambient PM levels and mortality.

Most recent epidemiological investigations have focused on short-term, acute exposures. They show consistent effects across a range of related health outcomes. A summary of acute effect estimates as a percentage change with a 10 μg/m³ increase in PM_{10} is presented in Figure 5.7. These indicate that a 10 μg/m³ increase in average 24-hour PM_{10} concentration is associated with an increase of approximately 0.8% in daily mortality, a somewhat larger association with cardiovascular mortality, and a much larger association with respiratory mortality. Increased hospitalizations and related health care visits are significant for various respiratory diseases and, to a lesser extent, cardiovascular disease. Increased symptom prevalence is associated with PM_{10} levels; these include lower respiratory system symptoms, asthma, and cough.

Chronic studies indicate that long-term exposure to PM, particularly in the small particle size range ($PM_{2.5}$), is associated with various cardiac and pulmonary health effects (Figure 5.8). $PM_{2.5}$ includes ammonium sulfate, ammonium nitrate, organic matter, elemental carbon, crustal material, and sea salts. Total mortality seems to increase by 2% to 4% for every 5 μg/m³ increase in $PM_{2.5}$, with somewhat larger associations for cardiopulmonary mortality. Chronic exposure studies have also reported significant increases in respiratory disease (bronchitis) and small declines in lung function.

Although epidemiological evidence linking both acute and chronic PM exposures to significant health effects is strong, toxicological mechanisms for these effects are less clear. Considerable research has focused on the biological mechanisms that may be responsible for the cardiopulmonary and respiratory effects reported in epidemiological studies.

Associations of mortality and morbidity (illness) with ambient PM levels suggest that a variety of tissues may be affected by particles in the PM_{10} range in ways that cause lung disease and systemic

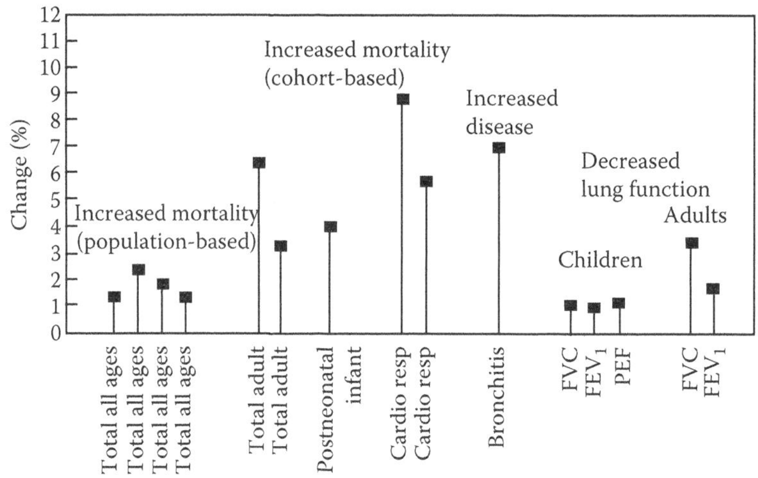

FIGURE 5.8 Estimated percent change in health endpoints associated with a 5 μg/m³ increase in $PM_{2.5}$ levels in chronic exposure studies. (From Pope, C.A., in *Air Pollution and Health*, Holgate, S.T., Samet, J.M., Koren, H.S., and Maynard, R.L., Eds., Academic Press, San Diego, 1999, p. 694. With permission.)

effects. Several biologically plausible mechanisms have been proposed to explain these results. They consistently include some form of inflammatory response. Because the production of clots in coronary vessels (coronary thrombosis) is the cause of myocardial infarction (heart attack), the apparent causal relationship between coronary thrombosis and particle deposition in the airways is somewhat problematic. It has been proposed that inflammatory responses in the lungs produce procoagulant factors locally or as a result of mediators that, when released, act on the liver to increase the levels of procoagulant factors. There is limited epidemiological evidence to support this, as increased blood viscosity has been observed at elevated pollutant levels.

Ultrafine particle (UFP; ≤ 0.01 μm in diameter) research provides a biologically plausible explanation for the toxicity of ambient particles. Such particles induce significant toxic effects in lung tissue even when they are formed from materials that are not toxic as larger respirable particles. Factors that affect or are associated with UFP toxicity include surface area and chemistry, particle number, oxidative stress, and particle interstitialization (i.e., movement of particles from lung spaces into surrounding tissues).

The large surface area of UFPs seems to facilitate the adsorption and absorption of substances from the environment or from the fluid of lung epithelia onto particle surfaces, which then increases particle reactivity. Iron (Fe), one such substance, produces reactive oxygen species. Macrophages attempting to phagocytize UFPs seem to be stimulated to release inflammatory mediators such as tumor necrosis factor, and on sustained stimulation, epithelial cells release chemokines that contribute to inflammation. UFPs present in large numbers commonly undergo interstitialization and therefore cannot be removed by normal pulmonary clearance mechanisms.

Particles may cause lung tissue injury because of oxidative stress. The production of free radicals (including hydroxyl radical, OH) may occur in response to exposures to small particles. Oxidative stress seems to result from the localized release of high concentrations of transition metals that stimulate the release of reactive oxygen species.

5.4.4 Hydrocarbons

Hydrocarbons (HCs) in the atmosphere pose two major exposure and health concerns. The more important of these is the precursor role that reactive (and less reactive) HCs have on tropospheric O_3

production. The HC or, more appropriately, the NMHC standard has been promulgated to achieve the NAAQS for O_3; it is not designed to protect public health from exposure risks that may be associated with specific NMHCs.

As a general rule, exposures to individual NMHC species are relatively low and health risks on a community-wide basis are considered to be small to negligible. NMHCs such as the aldehydes (e.g., HCHO) and acrolein are potent irritants. Exposure to these substances, individually or collectively, may cause eye, nose, throat, and sinus irritation. Such symptoms are transitory, resulting in no apparent long-term adverse health effects.

Polycyclic aromatic hydrocarbons (PAHs) are of some health concern. The PAHs include a group of substances characterized by multiple benzene rings. They are commonly produced as a result of incomplete combustion of coal and wood and, to a lesser extent, other fuels. The PAHs are carcinogenic, with benzo-α-pyrene being one of the most potent carcinogens and common PAHs found in aerosol samples.

As a result of fuel use changes, PAH levels in urban areas in the United States have declined to about one-third of the levels present in the mid-1950s. Similar declines have been reported for northern Europe. Although PAHs have been suspected of contributing to increased lung cancer rates in urban areas, epidemiological studies have not shown such a definitive relationship.

Many NMHCs are regulated as hazardous or toxic air pollutants. One of these is benzene, one of the seven pollutants initially regulated under the hazardous pollutant provision of the 1970 Clean Air Act (CAA) Amendments. They also include most of the 182 other substances regulated under the Air Toxics provision of the 1990 CAA Amendments.

Pollutants regulated under Hazardous/Toxics provisions include substances that vary in their toxicological properties and potential risks to humans and the environment. Many are suspected human carcinogens, mutagens, teratogens, and the like; some are irritants, and some neurotoxic.

5.4.5 Nitrogen Oxides

Nitric oxide is a relatively nonirritating gas believed to pose little or no health risk at ambient exposure levels. Its importance lies in the fact that it is rapidly oxidized to NO_2, which has a much higher toxicity.

Unlike SO_2, which is rapidly absorbed in fluids of the upper tracheobronchial region, NO_2 is less soluble and thus passes into the pulmonary region where tissue damage may occur. In individuals occupationally exposed to high NO_2 levels, adverse effects such as pulmonary edema only manifest themselves hours after exposure has ended. Elevated NO_2 exposures have been known to occur in the manufacture of nitric acid (HNO_3), in farm silos, from electric arc welding, and from use of explosives in mining.

In toxicological studies of animals, abnormal pathological and physiological changes have only been observed at concentrations higher than those found in ambient environments. At the lowest NO_2 exposure levels at which adverse effects have been detected (0.5 ppmv, 1,000 μg/m^3), pathological changes have included destruction of cilia, alveolar tissue disruption, and obstruction of respiratory bronchioles. The exposure of rats and rabbits at higher levels has caused severe tissue damage resembling emphysema.

Results of animal toxicological studies support the hypothesis that NO_2 may be a causal or aggravating agent in respiratory infections. There is evidence that NO_2 can damage respiratory defense mechanisms, allowing bacteria to multiply and invade lung tissues. In mice, hamsters, and monkeys infected with pneumonia-causing bacteria, exposure to elevated NO_2 levels resulted in reduced ability to clear bacteria from the lungs, reduced life span, and increased mortality.

Epidemiological studies have been conducted to determine whether NO_2 exposures at ambient levels are associated with respiratory symptoms and disease or other adverse health effects. Most have focused on the potential effects of either outdoor or indoor NO_2 exposures on children's health. A few have attempted to evaluate the potential adverse effects of combined indoor and outdoor exposures.

Results from a variety of outdoor exposure studies on children's health have been reported. The earliest of these (known as the Chattanooga School Children Study) investigated potential health effects associated with emissions of NO_2 from a trinitrotoluene (TNT)-manufacturing facility. Lung function (ventilation rate) of second-grade children in the high-NO_2 area was significantly lower than that of children in the control area. Respiratory illness rates of families in the exposed area were also higher. Later studies indicated that respiratory symptom duration increased with increasing annual mean NO_2 concentrations, more frequent hospitalizations with increasing 2-year means, and more treatments in hospitals for lower respiratory illness with higher lifelong NO_2 exposures in female children. Similar results have also been reported for Swedish school children in 10 communities where chronic cough and respiratory infections (bronchitis or pneumonia) were associated with increased annual mean NO_2 exposures.

Studies of adults have yielded mixed results. Positive associations between NO_2 and respiratory illness have been reported in some investigations but not in others. In the European Study on Air Pollution and Health (APHEA), significant positive associations were observed between daily death rates and NO_2 concentrations. An increase of 25 ppbv (50 μg/m^3) 1-hour maximum concentration was associated with a 1.3% increase in the daily number of deaths. Other less definitive studies indicate that increased death rates are associated with increased NO_2 exposures.

Significant public health concern was expressed relative to potential health effects of NO_2 exposures on children's health after the publication of a 1970s epidemiological study in the United Kingdom that observed increased respiratory symptoms among children living in homes using gas cooking stoves (compared with those using electric stoves).

In 1992, the US EPA conducted a review and meta-analysis of 11 epidemiological studies that evaluated potential effects of indoor NO_2 exposures on respiratory illness among children younger than 12 years of age. Relatively good agreement was observed among these studies, with an approximate 20% increase in respiratory illness associated with the presence of gas cooking stoves in households. This increase in respiratory symptoms was described as being equivalent to a 15 ppbv (30 μg/m^3) increase in NO_2.

In a recent residential Australian study, the presence of a gas stove and exposure to NO_2 were observed to be significant risk factors for asthma and respiratory symptoms. The presence of a gas cooking stove in a household increased the likelihood of asthma threefold and respiratory illness twofold.

Several studies indicate that NO_2 exposures associated with kerosene heater usage increase the risk of respiratory illness in children younger than 7 years of age. Nitrogen dioxide exposures of more than 16 ppbv (32 μg/m^3) were reported to more than double the risk of lower respiratory symptoms and illness (fever, chest pain, productive cough, wheeze, chest cold, bronchitis, pneumonia, or asthma) compared with children at lower levels of exposure in nonkerosene stove–using households. Increased risk of upper respiratory symptoms (sore throat, nasal congestion, dry cough, croup, or head cold) was also reported.

The annual average NAAQS for NO_2 in the United States is 53 ppbv (106 μg/m^3). The World Health Organization (WHO), however, recommends a much lower annual average guideline (21 ppbv, 42 μg/m^3), as well as a 1-hour guideline of 100 ppbv (200 μg/m^3). These recommendations indicated that the US NAAQS for NO_2 may not have been sufficiently protective of public health and that a short-term standard was needed to protect individuals (particularly children) from peak exposures. In 2010, the US EPA adopted a 1-hour NAAQS of 100 ppbv.

5.4.6 Ozone

Ozone is one of the most ubiquitous and toxic pollutants found in ambient atmospheres. Depending on the year, approximately 20% or more of the US population lived in areas where the 1-hour NAAQS (regulatory requirement prior to 2004) of 120 ppbv (235 μg/m^3) was exceeded.

Pathological and physiological effects of O_3 exposures (at the high end, or higher, of those reported in urban areas) have been evaluated in animal (usually rodent) exposure studies. Ciliated cells in respiratory airways as well as squamous epithelial cells in alveoli were injured or killed. This was followed by cell proliferation and tissue repair, with a loss of damaged epithelial cells and formation of exudates (fluid) and inflammatory cells. Longer-term exposures are characterized by continued hyperplasia (excess epithelial cell proliferation), low-grade inflammation with exudate, and synthesis of collagen (scar-forming substance).

Ozone may injure tissue membranes by oxidizing amino acids, sulfhydryl (SH) groups on enzymes and other proteins, and polyunsaturated fatty acids (lipids). This lipid peroxidation, along with the resultant production of free radicals, increases membrane permeability, with a concomitant leakage of essential electrolytes and enzymes. It also results in swelling and disintegration of cell organelles such as lysosomes and mitochondria, and inhibition of metabolic pathways. Significant increases in activity of various antioxidant enzymes and enzyme systems have been reported at continuous exposure concentrations in the range of 0.4 to 0.8 ppmv (783–1,567 $\mu g/m^3$).

At the physiological level, O_3 exposures result in significant lung function changes. These include increased respiratory rates and pulmonary resistance, decreased tidal volume, and changes in respiratory mechanics. Changes seem to be transient, returning to normal after exposure ceases or after several days of adaptation to continuous exposures. Such changes are dose-dependent and increase with exercise-associated lung ventilation.

Ozone exposures have a variety of effects that impair the body's ability to defend itself against infection. These include (1) interference with mucociliary transport as a result of injury to the ciliated epithelium lining the upper respiratory system, (2) impaired killing of bacteria, and (3) impaired macrophage function. On the other hand, O_3 can inactivate certain viruses, including those that cause influenza.

There is limited information available on O_3's potential to cause cancer. Studies on microorganisms, plant root tips, and tissue culture indicate that O_3 is mutagenic and therefore genotoxic. Mutagenic and genotoxic substances, as a rule, have a high potential for causing cancer. There is, however, no conclusive evidence that O_3 causes cancer in laboratory animals or humans.

There seems to be an adaptation to, or attenuation of, the effects of O_3 with intermittent, repeated exposures to health-affecting O_3 levels. In humans, the effect of O_3 is greatest on the second day of exposure, with normal function returning by the fifth day. In California, studies of human pulmonary function changes associated with relatively high exposures (~0.50 ppmv, 979 $\mu g/m^3$) showed a more severe response when these levels occurred in a high-pollution season rather than during a low-O_3 season.

Recent epidemiological studies have shown significant associations between daily O_3 levels and a range of adverse health effects. These include decreases in pulmonary function, aggravation of preexisting disease such as asthma, increases in daily hospital admissions and hospital admission for respiratory ailments, and premature mortality.

Most recent O_3 epidemiological investigations are longitudinal and follow a single population over time. Such studies, with time-series statistical analyses, have been able to discriminate between ambient temperature and O_3 exposure effects. This is especially true for mortality.

Significant associations between ambient O_3 levels and pulmonary function changes and other respiratory responses have been reported from studies of children attending summer camps. These have included significant decreases in FEV_1 (forced expiratory volume in the first second after a maximal inhalation), asthma exacerbation, chest symptoms, and decreased lung function.

Studies conducted in the United States, Canada, and Europe designed to evaluate potential relationships between hospital admissions for respiratory disease and atmospheric pollutants reported significant associations between respiratory (e.g., pneumonia, COPD) and asthma admissions and daily 1-hour maximum O_3 levels.

Emergency room visits for treatment of respiratory symptoms (which occur an order of magnitude more frequently than hospital admissions) significantly increase on high-O_3 days.

Acute morbidity effects experienced by the general public include minor restricted activity days (due to illness), acute respiratory symptom days, and impaired athletic performance. Although less severe than the health effects described previously, they are likely to occur more frequently and affect a larger percentage of the population. The mortality rates caused by O_3 health effects were calculated for various cities in the United States (Table 5.2; Figure 5.9).

Several recent epidemiological studies indicate that chronic exposure to ambient O_3 is associated with an increased prevalence of asthma in humans as well as declines in mid-expiratory lung function. Thus, O_3 not only exacerbates asthma, but also seems to cause it.

As is apparent, recent epidemiological studies have shown that exposure to O_3 at current ambient levels has significant adverse health effects on human respiratory health and includes a small risk of premature mortality. There seems to be no threshold dose for these effects.

5.4.7 Lead

Lead is a heavy metal that occurs naturally in soil, water, and air. It has been used for thousands of years because of its properties and relative ease of smelting from mineral ores. As a consequence, humans have been, and continue to be, exposed to Pb levels significantly above background. Notable historical exposures have included the ingestion of (1) liquids stored in Pb containers (particularly during Roman times), (2) food from Pb-soldered cans, and (3) Pb paint and dusts by children; as well as the inhalation of (1) airborne Pb dusts from motor vehicles, and (2) primary and secondary Pb as well as other metal smelter emissions. Significant exposures have also occurred (and continue to

TABLE 5.2
Baseline Mortality Rates (per 100,000 Population) Used in O_3 Risk Assessment

		Type of Mortality (ICD-9 Codes)		
City	**Counties**	**Nonaccidental (< 800)**	**Cardiorespiratory (390–448; 490–496; 487; 480–486; 507)**	**Respiratory (460–519)**
Atlanta	Fulton, Dekalb	623	131	—
Boston	Suffolk	736	—	—
Chicago	Cook	781	189	—
Cleveland	Cuyahoga	1,058	268	—
Detroit	Wayne	913	234	76
Houston	Harris	533	123	—
Los Angeles	Los Angeles	569	155	—
New York	Bronx, Kings, Queens, New York, Richmond, Westchester	704	199	—
Philadelphia	Philadelphia	1,057	242	—
Sacramento	Sacramento	686	—	—
St. Louis	St. Louis	1,147	—	—
Washington, DC	Washington, DC	942	—	—
National	—	790	196	80

Source: Review of the National Ambient Air Quality Standards for Ozone: Policy Assessment of Scientific and Technical Information, EPA-452/R-07-003, January 2007, US Environmental Protection Agency, Office of Air Quality Planning and Standards, Research Triangle Park, North Carolina.

Note: Data for the year 2002 from the United States Department of Health and Human Services (US DHHS), Centers for Disease Control and Prevention (CDC), National Center for Health Statistics (NCHS), Compressed Mortality File (CMF) Compiled from CMF 1968–1988, Series 20, No. 2A, 2000; CMF 1989–1998, Series 20, No. 2E, 2003; and CMF 1999–2002, Series 20, No. 2H, 2004 on CDC WONDER On-line Database. See http://wonder.cdc.gov/.

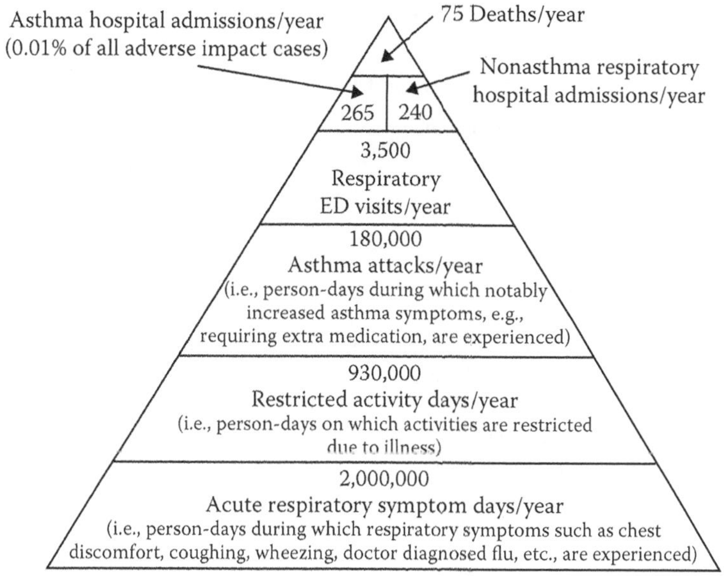

FIGURE 5.9 Estimates of adverse health effects avoided in New York City as a consequence of compliance with the 80 ppbv, 8-hour O_3 NAAQS. (From Thurston, G.D. and Ito, K., in *Air Pollution and Health*, Holgate, S.T., Samet, J.M., Koren, H.S., and Maynard, R.L., Eds, Academic Press, San Diego, 1999, p. 504. With permission.)

occur) among Pb workers. Much of our understanding of its adverse effects on humans is based on workers' exposures and on the exposures of children who ingested Pb paint dust.

Unlike other trace elements, Pb is not required by the body to carry out any metabolic or physiological functions. Because of its widespread use, total human body burdens today are approximately 300 to 500 times greater than those of our preindustrial ancestors.

Upon exposure, Pb is absorbed and transported by the bloodstream to all parts of the body. Although commonly found in blood and soft tissues, because of its similarities to calcium (Ca), it tends to be deposited in bone; bone thus serves as a storage depot.

The amount of Pb extracted from food, water, ingested dust, and airborne particles depends on its form and on an individual's nutritional status (poorly nourished individuals absorb more Pb), metabolic activity, and prior exposure history. Upon inhalation exposure, approximately 20% to 40% of inhaled Pb particles are deposited in lung tissue, where 50% or more is absorbed and enters the bloodstream. Some inhaled particles are removed by pulmonary clearance mechanisms and, upon swallowing, enter the bloodstream through the gastrointestinal tract. Lead absorption in the gastrointestinal system averages approximately 40% in children and 10% in adults.

Acute or chronic Pb poisoning may occur as a result of significant exposures. Principal target organs are blood, the brain and nervous system, and the renal and reproductive systems. Acute exposures (blood levels of > 60 μg/dL) may produce colic, shock, severe anemia, nervousness, kidney damage, irreversible brain damage, and death. Chronic poisoning may result in a variety of symptoms depending on the exposure level. Acute and chronic symptom responses based on blood lead levels (BLLs) are summarized in Table 5.3. Exposures on the more extreme end of BLLs may cause severe brain damage and damage to kidneys and blood-forming tissue. At the lower end, health effects may include neurodevelopmental changes in children, increased blood pressure and related cardiovascular effects in adults, and possibly cancer.

Because Pb accumulates and is only slowly removed from the body, repeated exposures (over months to years) commonly result in elevated BLLs. Because it is stored in bone, BLLs reflect

TABLE 5.3
Blood Lead Levels and Associated Health and Physiological Effects in Children and Adults

BLLs (μg/dL)	Children	Adults
< 10		Early signs of hypertension ALA-D inhibition
10–15	Crosses placenta Neurodevelopmental effects ALA-D inhibition Impairment of IQ Increased erythrocyte protoporphyrin Reduced gestational age and birth weight	
15–20		Increased erythrocyte protoporphyrin
20–30	Altered CNS electrophysical response Interference with vitamin D metabolism	
30–40	Reduced Hb synthesis Peripheral nerve dysfunction	Systolic hypertension
40–50		Reduced Hb synthesis
60	Peripheral neuropathies	Reproductive effects in females
70	Anemia	
80		Anemia Encephalopathy symptoms
80–100	Encephalopathy symptoms Chronic nephropathy Colic and other gastrointestinal symptoms	Chronic nephropathy

Source: Centers for Disease Control, *Preventing Lead Poisoning in Children*, DDHS, Washington, DC, October 1991.
Note: ALA-D, aminolevulinic acid dehydratase; CNS, central nervous system.

relatively recent exposures (past 1–3 months) and mobilization from bone and other depots. Although not indicative of the total body burden, blood Pb is relatively closely correlated with exposure levels.

Hematological changes, that is, effects on blood chemistry and associated physiological changes, are the earliest manifestations of chronic exposure. Lead interferes with the synthesis of heme (the O_2-carrying component of Hb in red blood cells), which may result in anemia, a symptom commonly associated with chronic Pb poisoning. Inhibition of enzymes involved in heme synthesis has been observed at BLLs as low as 10 μg/dL. Lead also inhibits heme-dependent liver enzymes, increasing an individual's vulnerability to the harmful effects of other toxic substances. It can also interfere with vitamin D production and cause neurological effects. Inhibition of vitamin D production has been observed at BLLs of 30 μg/dL in adults and as low as 12 μg/dL in children.

Lead exposures may cause adverse sexual and reproductive effects. Sperm abnormalities, reduced fertility, and altered testicular function have been observed in male industrial workers at BLLs of 40 to 50 μg/dL. Because it readily crosses the placenta, Pb may also pose significant risks to the fetus in exposed pregnant females. Lead is mobilized from bone during pregnancy; BLLs in pregnant females are reportedly higher than population averages. Associations between maternal BLLs and preterm delivery and low birth weight have been reported.

The nervous system is adversely affected by Pb. At high blood levels (> 80 μg/dL), it causes encephalopathy (brain damage). Children with BLLs in this range may experience permanent neurological damage such as severe mental retardation and recurrent convulsions. There is evidence to suggest that Pb may impair peripheral nerve conduction in children at BLLs as low as 20 to 30 μg/dL. Brain wave changes have been observed at levels as low as 15 μg/dL, with no apparent threshold.

Prospective epidemiological studies have shown an association between general measures of intelligence (intelligence quotient [IQ]) in children and prenatal and postnatal blood levels as low as 10 to 15 μg/dL. These studies suggest that up to at least age 7, exposure to relatively low Pb levels (10–40 μg/dL) may result in neurodevelopmental effects: decreased intelligence, short-term memory loss, reading and spelling underachievement, impairment of visual motor function, poor perception integration, disruptive classroom behavior, and impaired reaction time. As a consequence, the US Centers for Disease Control (CDC) has issued a guideline value of 10 μg/dL as the lowest BLL of public health concern in children.

An association between Pb exposure (at low BLLs) and elevated blood pressure in adults has been reported. A doubling of blood Pb appears to increase systolic pressure by 1 mmHg and diastolic pressure by 0.6 mmHg. In other studies of systolic blood pressure in men, a doubling of blood Pb from 5 to 10 μg/dL was associated with an increase of 1.25 mmHg systolic pressure.

Lead has been shown to be genotoxic, that is, it can cause gene mutations. It can also cause cell transformation and interfere with DNA synthesis in mammalian tissue culture. Lead compounds can induce kidney tumors in rodents. Although not conclusive, studies are suggestive of a causal relationship between Pb exposures and cancer. Based on available evidence, the US EPA has designated Pb as a Group B2 human carcinogen; that is, there is sufficient evidence of carcinogenicity from animal studies, but epidemiologic studies are not conclusive.

Young children (< 7 years old) are subject to the greatest health risk from Pb exposures because they have (1) a less developed blood–brain barrier and therefore greater neurologic sensitivity, (2) a faster resting inhalation rate, (3) a tendency to breathe through their mouth during play, (4) a greater intake dose rate due to their smaller weight, and (5) hand-to-mouth behaviors that result in ingestion of Pb from soil and dust. They also absorb more Pb through the intestines, particularly at less than 2 years of age. Children from low-income neighborhoods are frequently at very high risk because of diets low in calcium (Ca) and iron (Fe), elements that suppress Pb absorption.

Lead inhalation exposures have declined considerably since the mid-1970s. In survey studies conducted by the CDC from 1976 to 1980 and 1989 to 1991, BLLs decreased by approximately 78% in all age groups. Decreases in the period 1976 to 1980 are illustrated in Figure 5.10. During that time, 88% of children ages 1 to 5 were estimated to have had BLLs of less than or equal to 10 μg/dL; in the period 1988 to 1991, less than 10% of children were determined to have BLLs of less than or equal to 10 μg/dL. Decreases in BLLs have been observed across subgroups stratified by race/ethnicity, gender, urban status, and income levels. Declines in BLLs similar to those in the United States have been reported in Sweden, Belgium, and the United Kingdom.

Observed declines in BLLs coincided in time with the phasedown of Pb in gasoline, reduction in use of Pb-soldered cans for food and soft drinks (beginning in 1980), and limits on Pb in paint (1978). The CDC has concluded that the major contributor to the observed decline in BLLs has been reduced Pb use in gasoline. As of 2025, the US EPA has implemented significant updates to strengthen lead exposure protections. New dust-lead clearance levels now require "no detectable lead" post-abatement, with action levels tightened to 5 μg/ft^2 for floors and 40 μg/ft^2 for window-sills. Revised residential soil lead guidance lowers screening thresholds to better protect children near contaminated sites. EPA also released the All Ages Lead Model (AALM) v3.0, enabling more accurate life-course exposure assessments across multiple lead sources. These efforts are part of EPA's broader Lead Strategy, which includes reviewing the Lead National Ambient Air Quality Standards (NAAQS), drafting new smelter rules, and expanding community-focused lead mitigation.

A strong and consistent association exists between ambient concentrations of Pb and BLLs in children and adults. The blood Pb/air Pb exposure relationship has been estimated to be 1.8 μg/dL per μg/m^3 for adults and 4.2 μg/dL per μg/m^3 for children. Assuming no threshold exists for neurodevelopmental effects in children and hypertension in adults, even airborne Pb exposures of 1 μg/m^3 may be expected to have some adverse effects on children and adults. Fortunately, ambient concentrations in urban areas unaffected by emissions from stationary

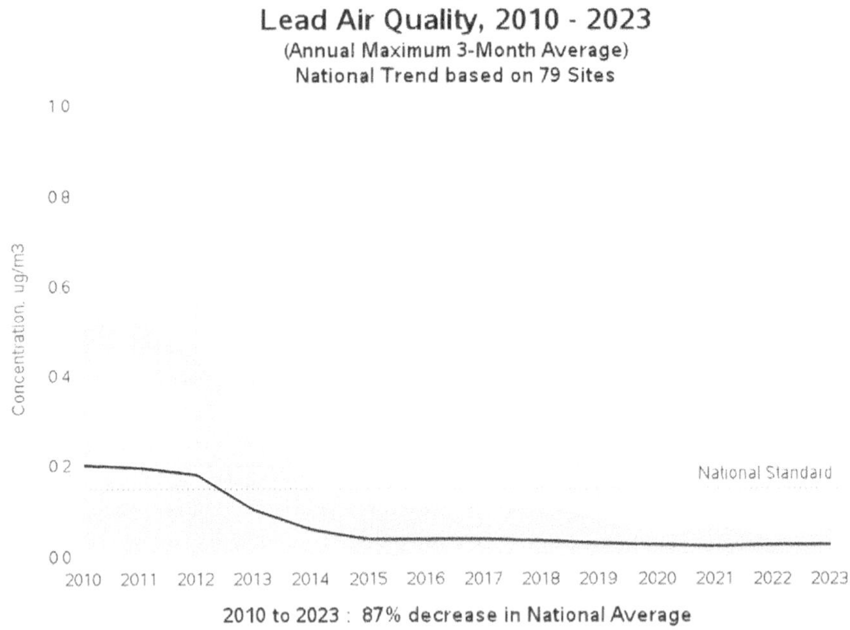

FIGURE 5.10 National trends in lead levels. (From US EPA, https://www.epa.gov/air-trends/lead-trends)

sources are now relatively low in the United States. In California, a state with a significant motor vehicle population, average ambient Pb levels of 0.06 μg/m^3 are at their lowest values in a half century.

Figure 5.10 shows an 87% reduction in Pb concentrations since 2010. The three curves on the graph represent the 10th, average, and 90th percentile concentrations for the sites. Although both ambient concentrations and BLLs have decreased significantly in the United States over the past three decades, Pb exposure from ambient air pollution is likely to be a significant public health threat in both developed and developing countries where it is still widely used as a gasoline additive and major stationary sources are unregulated.

5.4.8 Hazardous Air Pollutants

Under Section 112 of the 1970 CAA Amendments, the US EPA was authorized to promulgate national emission standards for substances designated as hazardous air pollutants. Such substances were given special regulatory status because they were deemed to pose health risks of a more serious nature than pollutants regulated under NAAQSs. These substances were thought to cause unique exposure risks because of their innate toxicity and no apparent threshold for adverse effects. Additionally, exposures from such substances were usually limited to the near vicinity of a source and were, in general, not considered to be community-wide concerns.

By the time the CAA Amendments were passed in 1990, only seven hazardous pollutants had been designated and emission standards promulgated. These were mercury (Hg), beryllium (Be), asbestos, vinyl chloride, benzene, inorganic arsenic (As), and radionuclides. Of these, only Hg and Be are uniquely toxic in the classical sense. The other five are presumed human carcinogens. Asbestos exposures, for example, have been definitively linked to increased risk of lung cancer and mesothelioma (a cancer of the chest cavity) in asbestos workers and their families; vinyl chloride to angiosarcoma (a form of liver cancer) and other cancers among rubber and plastics workers;

benzene to leukemia among a variety of workers exposed to it, as well as to lung cancer among smelter workers; and radionuclides to a number of different cancers.

Congress, in the 1990 CAA Amendments, listed another 182 substances and groups of substances for regulation as hazardous and toxic pollutants using technology-based rather than health standards. Major health concerns associated with these newly listed hazardous and toxic air pollutants are acute and chronic toxicity, neurotoxicity, reproductive toxicity, carcinogenicity, mutagenicity, and teratogenicity.

Although the toxic effects of many of these substances are well known from animal studies and workplace human exposures, little is known about ambient concentrations and health risks associated with ambient exposures.

Later, the US EPA identified and regulates 187 HAPs which include common chemicals used in various industries, such as benzene and perchloroethylene, and heavy metals, such as mercury and chromium mentioned previously. The EPA regularly monitors and evaluates HAPs through its periodic National Air Toxics Assessment (NATA), the sixth and most recent of which was conducted in 2014. HAP emission data is compiled as part of the National Emissions Inventory (NEI) and used to model and estimate ambient HAP concentrations; these estimates are then used to assess exposure and risk factors for populations, identify pollutants and sources of concern, and inform monitoring efforts and programs.

5.5 PERSONAL AIR POLLUTION

It is paradoxical that, as a nation, the United States has expended more than a trillion dollars over the past 30+ years in an effort to control ambient air pollution and thereby safeguard public health, whereas nearly 40 million Americans voluntarily expose themselves to a form of air pollution whose health consequences are orders of magnitude greater than those from ambient exposures. This is more of a paradox when one considers that scientific evidence for a causal role of cigarette smoking in several health problems is overwhelming.

Approximately 500,000 deaths each year can be attributed to tobacco smoking. Of these, 130,000 can be attributed to lung cancer, 25,000 to other cancers, 300,000 to cardiovascular disease, and approximately 20,000 to chronic pulmonary disease. According to reports issued by the Surgeon General, tobacco smoking is the single most important environmental factor contributing to premature mortality in the United States.

The excess mortality of smokers (compared with nonsmokers) is 70%. It is a function of quantity smoked and smoking duration. Life expectancy decreases with increased tobacco consumption and years of smoking; it is decreased by 8 to 9 years for a male smoking two packs of cigarettes per day.

In addition to the increased risk of premature death, smokers have higher morbidity rates. Before clinical disease is manifested, smokers usually suffer the discomfort of excessive coughing and phlegm production, and hoarseness.

One of the major effects of tobacco smoking is the induction of chronic respiratory diseases such as chronic bronchitis and emphysema. It is probable that tobacco smoking has caused a high percentage of diagnosed cases of such diseases as COPD because smokers have a higher incidence of these diseases than nonsmokers.

Tobacco smoking is a major cause of human cancer. Strong epidemiological evidence exists to implicate smoking as the causal factor in lung cancer and cancers of the larynx, oral cavity, esophagus, urinary bladder, kidney, and pancreas. Overall, lung cancer risk for smokers is about 10 to 12 times greater than that for nonsmokers.

In the public mind, lung cancer is the most important health consequence of smoking. It is not generally realized that smoking-related mortality from cardiovascular disease is approximately twice that for all forms of smoking-related cancer. Smoking is a major independent risk factor for heart attacks in adults. It also seems to contribute to more severe and extensive atherosclerosis of the aorta and coronary arteries.

The voluntary risk accepted by smokers becomes less than voluntary to the fetus of a pregnant female. Significant scientific evidence exists that maternal smoking retards fetal growth and increases the risk of spontaneous abortion and fetal and neonatal death in otherwise normal infants. There is also increasing evidence that children of smoking mothers may have measurable deficiencies in physical growth and intellectual and emotional development.

Of the thousands of identified chemical constituents in tobacco smoke, CO, nicotine, and tar are likely to be the major contributors to health hazards associated with smoking. Carbon monoxide levels are especially high, with levels in mainstream smoke of 1.5% to 5.5% (15,000–55,000 ppmv). An increased risk of myocardial infarction and sudden death in patients with cardiovascular disease seems to be associated with tobacco smoke CO exposures. However, because tobacco smoke contains numerous harmful substances, it is difficult to specifically assess the harmful effects of CO and its role in cardiovascular disease. Indeed, the effects of tobacco smoking on the incidence of clinical cardiovascular disease may not be due to CO alone; other factors such as nicotine, cyanide, and trace elements may be important. Nicotine is considered to be a major contributor to coronary atherosclerotic disease. Tar and nicotine make up the particulate phase of cigarette smoke. Tar consists of a variety of PAHs considered to be the likely causes of smoking-related cancers.

5.6 RISK ASSESSMENT AND MANAGEMENT

Risk assessment is a tool used by the US EPA and several other federal agencies to quantify risk associated with exposure to substances in the environment. It is a systematic process whereby the hazardous properties of a substance, human exposure to it, and dose–response relationships are evaluated to characterize risks of human environmental exposures; it serves as a basis for risk management decisions (Figure 5.11). The term *risk assessment* implies that environmental exposures pose a probability of an unwanted event taking place, for example, disease or even death. The assessment process has the following components: hazard identification, dose–response assessment, and exposure assessment. These three lead to a final risk characterization.

Hazard identification involves a delineation of whether exposure to a substance can cause a particular health effect, as well as an increased incidence of that effect (e.g., cancer, birth defects, and liver toxicity), and characterization of the nature and strength of a causal relationship.

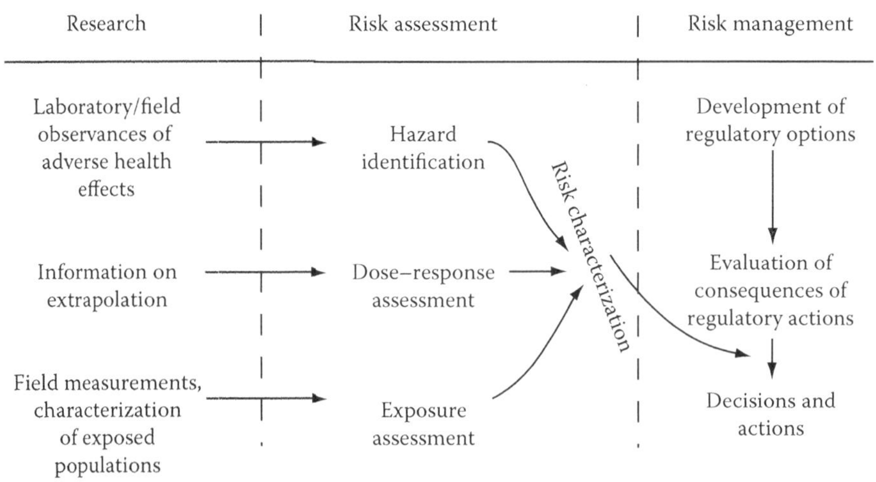

FIGURE 5.11 Relationships between different components of risk assessment and risk management. (Adapted from Samet, J., in *Air Pollution and Health*, Holgate, S.T., Samet, J.M., Koren, H.S., and Maynard, R.L., Eds., Academic Press, San Diego, 1999, p. 841. With permission.)

Dose–response assessments attempt to characterize the relationship between an exposure dose and the severity and incidence of observed health effects. Factors that influence dose–response relationships are considered, including intensity and pattern of exposure, age, and lifestyle variables that may affect susceptibility. They also include extrapolation from responses at high doses to those at lower ones, and from animals to humans.

Exposure assessments involve determinations of the level, frequency, and duration of exposures of human populations in question. They may include direct measurements of personal exposures, measurements from fixed-area monitoring sites, or modeling from source emissions. Modeling is used when monitoring data are not available.

Risk characterization combines the results of hazard identification and dose–response and exposure assessments to estimate the probability of a specific adverse health effect to an exposed individual or human population.

Historically, risk management has been seen as a process distinct from risk assessment. In risk management, results of risk assessments are integrated with political, social, economic, and technological factors to determine the need and methods for reducing risk. In the early history of risk assessment, the National Research Council (NRC) stressed the importance of maintaining a distinction between risk assessment and risk management to prevent biasing risk assessments based on the political feasibility of regulating substances in question.

Risk management involves decision making. This includes decisions on whether to regulate the substance and, if so, what specific regulatory options should be applied. At this stage, the process includes risk communication, the process by which results of risk assessments and any proposed risk management options are conveyed to those who are likely to be participants in the public policy-making process and may be directly or indirectly affected by regulatory decisions. Risk communication is a particularly difficult process due to the inherent limitations and uncertainties associated with scientific data, uncertainties associated with risk estimates, technical complexity, and strong views held by those to be regulated (and their supporters) and those who believe more stringent regulation is required. In most cases, risk communication (even when handled well) will subject risk assessment and subsequent risk management processes to considerable controversy.

In 1986, the US EPA developed and issued risk assessment guidelines consistent with earlier NRC recommendations. These, for the most part, focus on assessing risks of carcinogenicity, mutagenicity, developmental toxicity, and effects of chemical mixtures. They include default options that are policy judgments of how to accommodate uncertainties associated with risk characterization. These options are applied when scientific evidence is not sufficiently convincing as to which of several competing toxicological models and theories are correct. Historically, the US EPA has conducted risk assessments that incorporate conservative default options that are more likely to overstate rather than understate human health risks. This is particularly true with suspected human carcinogens. In such cases, the US EPA assumes a no-threshold relationship between dose and response, much like that used for exposure to radioactivity. Because some cancer-causing substances may have thresholds, use of the most conservative toxicological model in the face of uncertainty, not surprisingly, is strongly criticized by the regulated community.

As a result of the controversies that risk assessments and subsequent regulatory decision making give rise to, as well as obvious limitations of the risk assessment process, an NRC panel in 1996 redefined risk characterization as "a synthesis and summary of information about a potentially hazardous situation that addresses the needs and interests of decision makers and interested and affected parties. Risk characterization is a prelude to decision making and depends on an iterative, analytical–deliberative process."

Risk assessment, until relatively recently, received limited application in ambient air pollution regulatory activities. Congress, in revising the hazardous pollutant provisions of the 1990 CAA Amendments, required the US EPA to conduct risk assessments secondarily to technology-based regulation of 182 designated hazardous air pollutants. After applying technology-based emission limits, the US EPA must set residual risk standards that protect public health (with an adequate margin of safety) if technology-based standards are determined not to do so. A residual risk standard is required if the technology-based standard leaves a lifetime risk greater than one in a million for the most exposed person.

As indicated, many stakeholders are involved in risk assessment processes that may result in regulation of substances for which a quantifiable and potentially significant public health risk has been identified. These span the spectrum of concerns about risk, from the regulated community, who may believe that risks are negligible and control requirements onerous, to environmental activists who believe that public health is being unduly compromised. All stakeholders publicly agree that policy decisions should be based on "good science." However, good science, like beauty, is in the eyes of the beholder. Risk assessments by their very nature contain uncertainties. Because of such uncertainties, risk characterization and resulting policy decisions are heavily criticized.

The risk assessment process, as it has evolved over the past two decades, attempts to quantify risk, that is, describe the probability of disease in a population associated with specific pollutants and exposed populations. Those who oppose environmental regulation specific to a given situation or more generally often propose that the risk of a specific pollutant be compared with other risks in society (e.g., death from operating a motor vehicle). In such comparative risk evaluations, relatively low risks are associated with exposures to environmental pollutants.

The implication of comparative risk evaluations is that environmental regulation of many substances is not justified. Regulators, however, view risk in a broader context. Risk involves activities that are, in many cases, voluntary. Driving a motor vehicle and smoking tobacco are voluntary risks. Breathing air is essential for life and is not a voluntary action. Breathing polluted ambient air is therefore an involuntary risk. It is an involuntary risk to healthy adults and children, asthmatics, individuals with respiratory or cardiovascular disease, the old and infirm, and infant children.

READINGS

Ackermann-Liebuch, U., and Rapp, R., Epidemiological effects of oxides of nitrogen, especially NO_2, in *Air Pollution and Health*, Holgate, S.T., Samet, J.M., Koren, H.S., and Maynard, R.L., Eds., Academic Press, San Diego, CA, 1999, p. 559.

Boushey, H.A., Ozone and asthma, in *Susceptibility to Inhaled Pollutants*, ASTM STP 1024, Utill, M.Z., and Frank, R., Eds., American Society for Testing Materials, Philadelphia, PA, 1989, p. 214.

Brunekreef, B., Dockery, D.W., and Krzyzanowski, M., Epidemiological studies on short-term effects of low levels of major ambient air pollution components, *Environ. Health Perspect.*, 103, Suppl., 3, 1995.

California Air Resources Board/California Environmental Protection Agency, *Proposed Identification of Inorganic Lead as a Toxic Air Contaminant, Part B Health Assessment*, California Air Resources Board/ California Environmental Protection Agency, Sacramento, CA, 1996.

Carnow, B.W., and Meir, P., Air pollution and pulmonary cancer, *Arch. Environ. Health*, 27, 207, 1973.

Centers for Disease Control, *Preventing Lead Poisoning in Children*, DHHS, Washington, DC, October 1991.

Cohen, A.J., and Nikula, K., The health effects of diesel exhaust: Laboratory and epidemiologic studies, in *Air Pollution and Health*, Holgate, S.T., Samet, J.M., Koren, H.S., and Maynard, R.L., Academic Press, San Diego, CA, 1999, p. 707.

Frampton, M.W., and Roberts, N.J., Respiratory infection and oxidants, in *Susceptibility to Inhaled Pollutants*, ASTM STP 1024, Utill, M.Z., and Frank, R., Eds., American Society for Testing Materials, Philadelphia, PA, 1989, p. 182.

Goldstein, B., Toxic substances in the atmospheric environment: A critical review, *JAPCA*, 33, 454, 1983.

Horstmann, D.H., and Folinsbee, L.J., Sulfur dioxide-induced bronchoconstriction in asthmatics exposed for short durations under controlled conditions: A selected review, in *Susceptibility to Inhaled Pollutants*, ASTM STP 1024, Utill, M.Z., and Frank, R., Eds., American Society for Testing Materials, Philadelphia, PA, 1989, p. 195.

Koenig, Z.Q., Covert, D.S., and Pierson, W.E., Acid aerosols and asthma: A review, in *Susceptibility to Inhaled Pollutants*, ASTM STP 1024, Utill, M.Z., and Frank, R., Eds., American Society for Testing Materials, Philadelphia, PA, 1989, p. 207.

Maynard, R.L., and Waller, R., Carbon monoxide, in *Air Pollution and Health*, Holgate, S.T., Samet, J.M., Koren, H.S., and Maynard, R.L., Eds., Academic Press, San Diego, CA, 1999, p. 749.

McClellan, R.O., and Jackson, T.E., Carcinogenic responses to air pollutants, in *Air Pollution and Health*, Holgate, S.T., Samet, J.M., Koren, H.S., and Maynard, R.L., Eds., Academic Press, San Diego, CA, 1999, p. 381.

McNee, W., and Donaldson, K., Particulate air pollution: Injurious and protective mechanisms in the lungs, in *Air Pollution and Health*, Holgate, S.T., Samet, J.M., Koren, H.S., and Maynard, R.L., Academic Press, San Diego, CA, 1999, p. 431.

National Research Council, Committee on Risk Assessment of Hazardous Air Pollutants, *Science and Judgment in Risk Assessment*, National Academy Press, Washington, DC, 1994.

National Research Council, Committee on Risk Characterization, *Understanding Risk: Informing Decisions in a Democratic Society*, Stern, P.C., and Fineberg, H.V., Eds., National Academy Press, Washington, DC, 1996.

Paige, R.C., and Plopper, C.G., Acute and chronic effects of ozone in animal models, in *Air Pollution and Health*, Holgate, S.T., Samet, J.M., Koren, H.S., and Maynard, R.L., Eds., Academic Press, San Diego, CA, 1999, pp. 531–557.

Pope, C.A., and Dockery, D.W., Epidemiology of particle effects, in *Air Pollution and Health*, Holgate, S.T., Samet, J.M., Koren, H.S., and Maynard, R.L., Eds., Academic Press, San Diego, CA, 1999, p. 673.

Samet, J.M., Risk assessment and air pollution, in *Air Pollution and Health*, Holgate, S.T., Samet, J.M., Koren, H.S., and Maynard, R.L., Eds., Academic Press, San Diego, CA, 1999, p. 881.

Samet, J.M., and Cohen, A.J., Air pollution and lung cancer, in *Air Pollution and Health*, Holgate, S.T., Samet, J.M., Koren, H.S., and Maynard, R.L., Eds., Academic Press, San Diego, CA, 1999, p. 841.

Samet, J.M., and Jaakkola, J.J.K., The epidemiological approach to investigating outdoor air pollution, in *Air Pollution and Health*, Holgate, S.T., Samet, J.M., Koren, H.S., and Maynard, R.L., Eds., Academic Press, San Diego, CA, 1999, p. 431.

Schlesinger, R.B., Toxicology of sulfur oxides, in *Air Pollution and Health*, Holgate, S.T., Samet, J.M., Koren, H.S., and Maynard, R.L., Eds., Academic Press, San Diego, CA, 1999, p. 585.

Speizer, F.E., Acid sulfate aerosols and human health, in *Air Pollution and Health*, Holgate, S.T., Samet, J.M., Koren, H.S., and Maynard, R.L., Eds., Academic Press, San Diego, CA, 1999, p. 603.

Thurston, G.D., and Ito, K., Epidemiological studies of ozone exposure effects, in *Air Pollution and Health*, Holgate, S.T., Samet, J.M., Koren, H.S., and Maynard, R.L., Eds., Academic Press, San Diego, CA, 1999, p. 485.

US EPA, *Air Quality Criteria for Carbon Monoxide*, EPA/600/8-90/045f, EPA, Washington, DC, 1991.

US EPA, *Air Quality Criteria for Lead*, EPA/600/8-83-028a-f, EPA, Washington, DC, 1986.

US EPA, *Air Quality Criteria for Lead*, Supplement to the 1986 Addendum, EPA/600/8-90/049f, EPA, Washington, DC, 1990.

US EPA, *Air Quality Criteria for Particulate Matter*, Vol. III, EPA/600/AP-95-001c, EPA, Washington, DC, 1995.

US EPA, *Air Quality Criteria for Particulate Matter and Sulfur Oxides*, EPA/600/8-82-029a-c, EPA, Washington, DC, 1982.

US EPA, *Review of the National Ambient Air Quality Standards for Ozone: Policy Assessment of Scientific and Technical Information*, EPA-452/R-07-003, US Environmental Protection Agency, Office of Air Quality Planning and Standards, Research Triangle Park, NC, January 2007.

US EPA, *Review of the National Ambient Air Quality Standards for Sulfur Oxides: Assessment of Scientific and Technical Information*, Supplement to the 1986 OAQPS Staff Paper Addendum, EPA/452/R-94-013, EPA, Washington, DC, 1994.

Wedge, A., Lead, in *Air Pollution and Health*, Holgate, S.T., Samet, J.M., Koren, H.S., and Maynard, R.L., Eds., Academic Press, San Diego, CA, 1999, p. 797.

QUESTIONS

1. What is the public health significance of air pollution episodes?
2. Describe differences between acute and chronic health effects.
3. How do environmental and public health authorities establish a cause-and-effect relationship between observed symptoms and disease in a community and exposure to atmospheric pollutants?
4. Describe additive, synergistic, and antagonistic interactions associated with human exposures to pollutants.

5. The human respiratory system has defense mechanisms that protect it from airborne particles. Describe each of these mechanisms and how they function.
6. Asthma is a very common and severe respiratory disease. How is asthma affected by exposure to ambient pollutants?
7. What, if any, is the relationship between exposure to atmospheric pollution and infectious respiratory disease?
8. Describe evidence to support a causal relationship between exposure to atmospheric pollutants and the development of lung cancer in humans.
9. Describe adverse health effects associated with exposure to CO.
10. Describe health effects associated with exposures to SO_2 and acid sulfates.
11. Describe relationships reported for particle size and human health effects.
12. Distinguish between inhalable, respirable, and thoracic particulate matter.
13. What is chronic bronchitis? What causes and exacerbates it?
14. Exposure to elevated levels of fine particles has been observed to be statistically associated with premature mortality. What biological mechanisms are likely responsible?
15. What are the health concerns associated with regulation of emissions of nonmethane hydrocarbons under air quality standard provisions of clean air legislation?
16. What are the primary health concerns associated with exposures to hazardous or toxic air pollutants?
17. What are the various components of $PM_{2.5}$?
18. Describe health concerns associated with exposures to ambient O_3 levels.
19. Describe how lead exposures have changed over the past three decades in North America and the health benefits of such changes.
20. Asbestos is regulated as a hazardous air pollutant. Why?
21. Compare health risks associated with exposures to ambient air pollution and tobacco smoking.
22. What are the differences between risk assessment and risk management?
23. Describe how exposure and disease risks are evaluated and quantified.
24. Describe the public health and regulatory significance of voluntary and involuntary risk.
25. What is a comparative risk assessment?

6 Welfare Effects

Under the Clean Air Act (CAA) Amendments of 1970, National Ambient Air Quality Standards (NAAQSs) were to be promulgated to protect public health and welfare. The former were to be primary standards, and the latter secondary. Public health protection has had the highest priority, whereas public welfare is less well defined. Welfare effects are air pollution effects not related to human health. Although it has always been clear what some welfare effects are (e.g., damage to crops, plants, and materials; malodors), others have evolved over time. In the latter case, these include visibility concerns, atmospheric deposition, possible ecosystem changes associated with stratospheric ozone (O_3) depletion, and global warming.

Welfare effects within the regulatory context are limited to pollutants for which air quality standards have been promulgated, the so-called criteria pollutants: carbon monoxide (CO), particulate matter (PM), sulfur dioxide (SO_2), nitrogen oxides (NO_x), O_3, nonmethane hydrocarbons (NMHCs), and lead (Pb). The concept of welfare effects, as it is discussed in this chapter, is used more broadly. It includes non-health effects associated with other pollutants as well.

Most of the issues discussed in Chapter 4 (Atmospheric Effects) are welfare effects. Ecological changes and concerns associated with most of those issues, save visibility, are discussed in this chapter.

6.1 EFFECTS ON AGRICULTURAL CROPS, ORNAMENTAL PLANTS, AND TREES

Air pollution can impact agricultural crops in several ways depending upon the type of pollutant, but damage typically occurs to foliage. Gaseous pollutants such as NO_x, SO_x, and ozone enter plant cells through stomata, which are openings on leaf surfaces that allow uptake of CO_2 during plant metabolism. Deposition of NO_x and SO_x particulate can also acidify soil and cause plant damage. Soil acidification impacts soil chemistry and can reduce or increase the availability of key plant nutrients; reduced nutrients can lead to slow growth or plant death while excessive nutrients can lead to toxic conditions that can damage plant roots.

Plants have long served as sentinels of the biological injury that air pollutants are capable of producing as a result of acute and chronic exposures. Phytotoxic responses to pollutants such as SO_2, hydrogen chloride, and hydrogen fluoride have been reported in Europe since the middle of the nineteenth century. Particularly severe injury to, and destruction of, vegetation was associated with emissions of SO_2 and heavy metals from primary nonferrous metal smelters. In the United States, this included the Copper Hill Smelter in Ducktown, TN, a copper (Cu) smelter in Kellogg, ID, a zinc (Zn) smelter in eastern Pennsylvania, and other smaller smelters that have been long closed and would have been forgotten save for the despoiling of the surrounding environment that persists even though those people and their machines are long gone.

Smelter emissions have been a problem in Canada. In the 1920s, emissions from a smelter in Trail, British Columbia, were carried down the Columbia River Valley into the United States, causing significant damage to agricultural crops. This was one of the earliest recorded instances of a transboundary air pollution problem. The once-forested and subsequently devastated landscape downwind of the giant nickel (Ni) smelter near Sudbury, Ontario, is, to this day, a reminder of the significant effect that atmospheric pollutants can have on the environment. (On a historical note, emissions from the Sudbury smelter have been significantly reduced over the past two decades.)

Injury to agricultural plants and forests has been reported near uncontrolled nonferrous metal smelters in many parts of the world. Significant plant injury also occurs in developing countries around industrial sources, including coal-fired and oil-fired power plants, phosphate fertilizer mills, and glass plants.

DOI: 10.1201/9781003664109-6

Until the 1940s and 1950s, damage to agricultural crops, ornamental plants, and forests was, for the most part, a problem associated with point sources. As a result of intensive scientific investigation, the widespread injury to agricultural crops observed (as early as the mid-1940s) in the Los Angeles Basin was determined to be due to phytotoxic air pollutants such as O_3 and peroxyacyl nitrate (PAN), produced in the atmosphere as a result of photochemical reactions. Plant biochemist Dr. Arie Hagen-Smit, often referred to as the father of air pollution, initiated studies to unravel the chemical complexities of smog formation.

Ozone injury on sensitive vegetation has since been observed throughout the United States and in many parts of the world. Ozone levels sufficient to cause injury on very sensitive vegetation are reported in most areas east of the Mississippi River. Because of its ubiquitous distribution and high phytotoxicity, O_3 is the most important phytotoxic air contaminant.

Control efforts and changes in operating practices have resulted in a significant reduction in the localized plant damage that had been associated with many point sources. Paradoxically, one of these changes, the use of tall stacks for more effective dilution of emissions from coal-fired power plants, has, as expected, resulted in decreased injury to vegetation in the vicinity of these sources but inadvertently contributed to the problem of long-range transport and atmospheric deposition of strong acids and other pollutants.

Pollutants that have a history of causing significant plant injury under ambient exposure conditions include O_3, SO_2, PAN, fluoride (F), and ethylene (C_2H_4). Although it has been suspected of being a major plant-injuring pollution problem for some time, acidic deposition has only recently been determined to be, at least in part, causally related to plant injury under ambient conditions (e.g., the decline of red spruce in high-elevation forests).

Other pollutants are also known to cause injury to plants. They include nitrogen dioxide (NO_2), chlorine (Cl_2), hydrochloric acid (HCl), ammonia (NH_3), and PM. They are classified as minor pollutants because their emissions from continuous point sources at levels sufficient to cause injury have, for the most part, been relatively limited. Plant injury associated with exposures to these pollutants is commonly associated with accidental releases in industrial operations or transportation. On the other hand, injury induced by particulate dusts is usually associated with continuous emissions from point sources.

6.1.1 Plant Injury

Injury to plants can be manifested as visible or subtle effects. The former are identifiable changes in leaf structure, which may include chlorophyll destruction (chlorosis), tissue death (necrosis), and pigment formation. Visible symptoms may result from acute or chronic exposures. Acute injury occurs from brief exposures (several hours) to elevated levels of a phytotoxic pollutant. Tissue necrosis is the dominant symptom pattern associated with acute exposures, with necrotic foliar patterns usually specific for a given pollutant.

Chronic injury results from intermittent or long-term exposures to relatively low pollutant concentrations, with chlorophyll destruction and chlorosis the major symptoms. Chlorosis, however, is a nonspecific symptom associated with a number of pollutants, natural senescence (aging), and injury caused by other factors such as nutritional deficiencies.

Some pollutants produce symptom patterns other than, or in addition to, the classic patterns of necrosis and chlorosis associated with acute and chronic injury. These are described as growth abnormalities. They may be subtle, such as growth reduction, or visible, such as accelerated senescence of flowers, bolting (breaking) of flower buds, abscission (breaking off) of plant parts, and a curvature of the leaf petiole called epinasty.

The severity of injury, that is, the amount of leaf surface affected, depends on pollutant concentration and duration of exposure. The greater the exposure, the more severe the injury. Entire leaves or, in extreme cases, the whole plant may be killed. The extent of injury also depends on the sensitivity of the species, variety, or strain to the pollutant. Considerable variability in the response of plants to different phytotoxic pollutants, as well as responses to a given pollutant, occurs. As a general rule, actively growing plants and the most actively growing tissue are most sensitive to pollutant exposures and are more likely to experience injury.

6.1.1.1 Sulfur Dioxide

Plants exposed to increased SO_2 levels develop characteristic acute symptoms within 48 hours of exposure and cause plant stomata to close to protect plants from damage, effectively halting photosynthesis and limiting plant growth, metabolic function, and transpiration, contributing to water stress. Tissue in the internal center of leaves of broad-leaved species is killed, resulting in a pattern of ivory-to-white, red, or dead brown tissue between the veins. This interveinal necrosis extends outward from the veins to the leaf margin and is observable on both surfaces. In narrow-leaved species, for example, cereal grains, lilies, irises, and gladiolas, injury appears as irregular bifacial necrotic streaks between longer veins, and in some species, injury occurs at the tip of the leaf. Injury on coniferous plants is a reddish brown to brown tip necrosis with a banded appearance from repeated exposure. Chlorosis may occur on older needles under moderate exposures; such chlorotic needles are often prematurely shed. Interveinal necrosis on SO_2-injured pinto bean leaves is illustrated in Figure 6.1. SO_2 enters plant cells and dissolves to form bisulfite ions, which are generally harmless, and sulfite ions which are toxic and must be metabolized to sulfate, a plant nutrient. However, at low concentrations, absorbed NO_x and SO_x are effectively metabolized by plants and can be an additional source of nutrients.

6.1.1.2 Ozone

Unlike SO_2, which initially injures tissues in the internal center of leaves, O_3 preferentially injures cells close to the upper surface (in broad-leaved species). The most common O_3-induced symptom is "flecking," which is visible on the upper surface (Figure 6.2). These flecks are produced when groups of cells below the epidermis (surface cell layer) are injured or killed. They may be white, tan, or yellow in color. The upper surface may look chlorotic or "bronzed" if flecking is extensive. A variation of this symptom is stippling, wherein injured cells become unevenly pigmented, producing a red–purple to black–brown appearance (Figure 6.3). This injury is confined to interveinal areas. Injury in narrow-leaved species appears as small necrotic or chlorotic flecks between veins.

Ozone is highly reactive and the most harmful air pollutant to plants, altering plant metabolism by binding with and damaging plasma membranes and disrupting stomatal regulation and inhibiting photosynthesis. Ozone injury on coniferous species such as pine is common. It is the causal factor of emergence tipburn of eastern white pine, a disease observed on this species since the early twentieth century. The disease syndrome is characterized by the death of the tips of young elongating needles in spring and early summer. Killed needle tips become reddish brown. Ozone not only causes tip necrosis, but also other symptoms, ranging from chlorotic flecks to chlorotic mottling

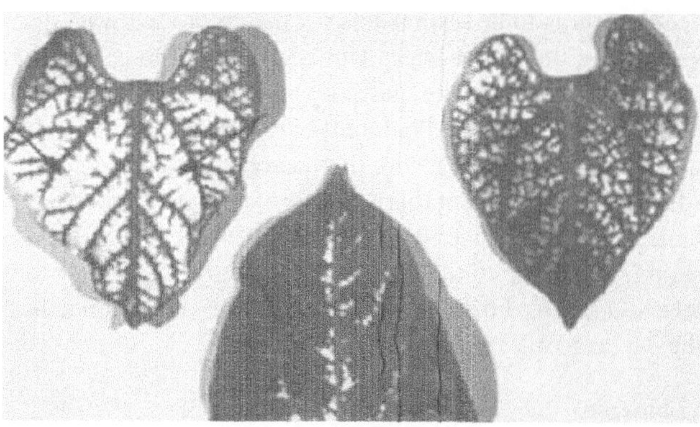

FIGURE 6.1 Interveinal necrosis on pinto bean associated with exposure to SO_2.

FIGURE 6.2 Ozone injury on tobacco.

FIGURE 6.3 Ozone injury on grape.

(small chlorotic patches). On O_3-affected trees, older needles prematurely age, discolor, and are shed. Consequently, affected trees may possess only the current year's needles.

Ozone, along with SO_2, seems to be responsible for the chlorotic dwarf disease of eastern white pine that has been observed in the northeastern United States for most of the twentieth century. In young, sensitive trees, current-year needles develop chlorotic flecks and mottling. Older needles become prematurely chlorotic and are shed before the new set of needles is fully developed. These trees are severely stunted and usually die before they reach 15 years of age.

In the San Bernardino Mountains of southern California, large numbers of ponderosa pines have been affected by what is called ozone needle mottle. The earliest symptoms are small chlorotic tissue patches that develop from the tip to the base in older needles. Death of the needle tip may also occur. As a consequence, O_3-injured trees may possess only 1-year-old needles during the summer months rather than the 3-year to 5-year normal complement.

6.1.1.3 Peroxyacyl Nitrate

Peroxyacyl nitrate is the causal agent of what was described as smog injury on vegetable crops in the Los Angeles Basin from the 1940s to the early 1960s, where it was a major problem. Injury due to PAN has been observed around other western cities, but to a lesser degree; it has also been found in New Jersey and southeastern Canada.

PAN destroys tissue near the lower surface of leaves, resulting in what has been described as glazing on the underside of the leaf. A distinct pattern of banding is observed after repeated intermittent exposures. Injury is not commonly reported on narrow-leaved or coniferous species.

6.1.1.4 Fluorides

Fluoride injury may result from the uptake of gaseous hydrogen fluoride (HF) through leaves or from soluble particulate fluorides absorbed through leaves or roots. Whatever the route of exposure, fluorides are transported through veins to leaf margins and leaf tips, where they accumulate. Fluoride injury results from exposures to elevated tissue concentrations that occur as a result of the accumulation of F over a period of weeks to months. The severity of injury is directly related to the level of F accumulated in marginal leaf tip tissue.

In broad-leaved species, F injury appears as marginal (Figure 6.4) or tip necrosis. Necrotic tissue is often separated from healthy tissue by a narrow, sharply defined reddish band. It may have a characteristic wavy color pattern due to successive F-induced tissue death. Necrotic tissue may break away from uninjured tissue near the reddish band, giving the leaf a ragged appearance.

Fluoride injury on narrow-leaved species is characterized by tip necrosis that extends in irregular streaks down the leaf. Necrotic tissue may vary in color from ivory to various shades of brown. A band of dark brown tissue sharply demarks injured from healthy tissues. Symptoms consist of chlorotic mottle at the margins and leaf tips. Small irregular chlorotic patches form between the veins and merge to form continuous bands as injury becomes extensive. Necrosis occurs when tissue injury is severe.

Tip necrosis is the characteristic symptom of F injury in conifers. Necrosis begins at the tip of current-year needles and progresses downward. The necrotic tip is usually reddish brown. This tip necrosis usually results in premature shedding of affected needles.

6.1.1.5 Ethylene

Plant injury caused by C_2H_4 was first reported when it was used in illuminating gas in greenhouses at the turn of the twentieth century. Because it is emitted in automobile exhaust, C_2H_4 is a ubiquitous

FIGURE 6.4 Fluoride injury on Oregon grape.

pollutant in urban and suburban areas. Its concentrations in ambient air, although relatively low, may be sufficient to cause significant effects on plants.

Ethylene is a natural maturation hormone in plants, controlling the ripening and aging of fruits. The effects of ambient levels of C_2H_4 on plants growing in the urban environment are potentially great because C_2H_4 has significant biological activity at very low concentrations (e.g., ppbv).

One of the best-documented effects of C_2H_4 exposure is dry sepal of orchids, associated with use of C_2H_4 illuminating gas in greenhouses. Dry sepal has also been reported in urban areas, presumably from elevated levels of C_2H_4 associated with automobile emissions.

Other plants in the flowering stage may be sensitive to C_2H_4 exposures. Common C_2H_4-induced responses include premature bud break, inhibition of flowering, and accelerated flower aging.

Because of its hormonal properties, C_2H_4 has the potential to cause widespread adverse effects on plants. However, field identification of injury resulting from such exposure may be nearly impossible because injury mimics natural senescence and changes induced by a variety of environmental factors. An interesting example of this problem is the widely observed premature pigmentation of tree and shrub leaves along many major highways.

6.1.1.6 Acidic Deposition

Acidic substances deposited on plants or the plant soil environment have the potential to cause direct injury or measurable changes in plant growth. Laboratory and greenhouse studies with simulated acidic rain events have shown that a wide variety of plants can be injured by exposure to acidic rainfall with a pH of approximately 3.0. Reported symptoms include (1) small (< 1 mm) necrotic leaf lesions or marginal tip necrosis; (2) pitting, curling, wrinkling, shortening, and death of leaves; (3) premature abscission; (4) erosion of the waxy upper leaf surface; (5) chlorosis; (6) excessive adventitious (side) budding; (7) premature senescence; and (8) galls, hypertrophy (tissue swelling), and hyperplasia (excessive cell production) in the specific species exposed.

The most commonly reported symptom of acidic deposition injury on plants in laboratory studies is small necrotic lesions. These seem to be a result of the deposition of acidified water on the leaf surface and its subsequent evaporation. Repeated exposure to simulated acid rain events results in an increased number of lesions on affected leaves. Foliar injury has been observed at pH levels associated with recorded severe ambient deposition events. However, to this author's knowledge, there have been no confirmed reports of atmospheric deposition causing similar symptoms in plants under field conditions. Other effects not manifested as visible injury include reduced yield and leaching of metabolites and inorganic nutrients from plant surfaces. Leaching of such substances seems to be a common response to acid deposition. Leachates include amino acids, free sugars, organic acids, vitamins, a variety of essential mineral nutrients—nitrogen (N), calcium (Ca), phosphorous (P), magnesium (Mg)—alkaloids, growth regulators, and pectic substances.

A variety of potential effects of acidic deposition on forest element cycles and tree nutrition have been proposed. These include

1. Accelerated leaching of base cations from foliage and soils that may result in nutrient deficiency, root damage, and reduced drought tolerance
2. Increased mobilization of aluminum (Al) and other metals, also resulting in nutrient deficiency, root damage, and reduced drought tolerance
3. Inhibited microbiological processing of soil, possibly leading to reduced decomposition of organic matter, damage to mycorrhizae, nutrient deficiency, and altered pathogen–host relationships
4. Increased bioavailability of N, leading to accelerated organic matter decomposition and increased susceptibility to natural stresses such as freezing temperatures, drought, and so on

The importance of these mechanisms in altering tree and forest growth is not known.

6.1.1.7 Interaction Effects

Pollutants do not occur in the atmosphere singly. Therefore, it is possible, and even probable, that simultaneous, sequential, or intermittent exposures to several phytotoxic pollutants may result in plant injury. Available evidence indicates that simultaneous exposures to gaseous mixtures can produce synergistic, additive, or antagonistic effects. The type of interaction response is related to the plant species exposed, pollutant combinations and concentrations, duration of exposure, and amount of injury caused by pollutants when exposed singly.

Plant exposures to mixtures of $SO_2 + O_3$ and $SO_2 + NO_2$ have reportedly decreased injury thresholds. Injury produced by exposure to gaseous mixtures may be more severe than the sum of the injury caused by exposure to individual pollutants (a synergistic response). Antagonistic responses are generally observed when injury caused by pollutants applied singly is severe; in these cases, the effect of mixtures is to reduce the severity of injury. Visible symptoms produced by gaseous mixtures are usually characteristic of a single pollutant.

6.1.2 Economic Losses

Numerous cases of significant pollutant-caused plant injury have been reported in the twentieth century. Excluding the photochemical oxidant problem in southern California and some areas of the northeast, most reports of air pollution–induced plant injury have been associated with point sources. With the exception of primary metal smelters, economic losses associated with point sources have often been insignificant. Many offending point sources in the United States have since been controlled. Nevertheless, because of photochemical oxidants such as O_3 and PAN, significant air pollution injury to vegetation is still widespread; associated economic losses, however, are often undetermined.

Scientists generally agree that O_3 causes 90% or more of the air pollution injury to crops in the United States. This recognition has led to the establishment of a National Crop Loss Assessment Network (NCLAN). From summaries of O_3 monitoring data; determination of the O_3 sensitivity of major crop plants such as corn, soybeans, wheat, cotton, grain, sorghum, and barley; and economic data, NCLAN-participating scientists have modeled the economic effect of O_3 on US agriculture. They have estimated that a 25% reduction in ambient tropospheric O_3 would result in a $1.71 billion annual increase in agricultural production; a 40% reduction would result in a $2.52 billion annual increase.

Quantification of economic loss is a difficult task. It requires surveys of suspected air pollution injury, confirmation of the causal factor, and estimation of the dollar costs. Loss estimates may be confounded by a number of factors. For example, the presence of visible injury on plants may not be translatable into economic loss. In many cases, plants with severe foliar injury may quickly recover, replacing injured leaves with younger healthy ones with no apparent lasting effects. In other instances, even a slight amount of injury can decrease the marketability of a crop. To deal with this reality, plant scientists use the concept of plant damage rather than plant injury when it relates to potential economic losses. It is implicit in the damage concept that economic losses have resulted. Plant injury does not have this connotation.

The impacts of air pollution on the economy can be classified as market and non-market costs. Market costs mainly result from reduced labor productivity, additional health expenditure, and crop and forest yield losses. The Organisation for Economic Co-operation and Development (OECD) projects that these costs will increase to reach about 0.5% of European gross domestic product (GDP) in 2060, leading to a reduction in capital accumulation and a slowdown in economic growth.

Non-market costs are those associated with increased mortality and morbidity, degradation of air and water quality, the health of ecosystems, and climate change. For the OECD as a whole, the annual welfare costs related to non-market health impacts of outdoor air pollution amount to almost USD 1.6 trillion by 2015, and rise to USD 3.8–3.9 trillion in 2060, of which more than 90% stem from the welfare loss of premature deaths (Table 6.1) (OECD 2016).

TABLE 6.1
Total Welfare Costs of Air Pollution, Central Projection in OECD

	OECD	
	2015	**2060**
Total market impacts (billions USD)	90	390
Share of income (percentage)	0.3%	0.5%
Per capita (USD per capita)	70	270
Total non-market impacts (billions USD)	1,550	3,750–3,850
Share of income (percentage)*	5%	5%
Per capita (USD per capita)	1,210	2,610–2,680

* Welfare costs from non-market impacts are not related to expenditures and therefore not an integral part of the calculation of income; the expression of these welfare costs as share of income is therefore only for illustrative purposes.

It is reasonable to expect that when a phytotoxic pollution problem continues for some time, economic losses will decline, as no grower can sustain such losses indefinitely. He or she may discontinue agricultural production altogether or utilize pollution-resistant varieties or other pollutant-tolerant crops.

6.1.3 Forest Declines

In many real-world cases, particularly in forest ecosystems, the relationship between observed injury and exposures to atmospheric pollutants has not been clearly established. This has been particularly true of many "forest declines."

The term *decline* is used to describe the process by which large numbers of trees die. In a decline, tree death occurs progressively; that is, trees are weakened, become less vigorous, and eventually die.

Declines may occur as a result of a variety of natural or anthropogenic stress factors. Natural phenomena initiating forest declines are drought, insects, and freezing temperatures. Weakened trees may succumb to other factors, including root rot, insects, and disease. In many cases, no single factor can explain the observed death of trees.

A number of forest declines have occurred in North America and central Europe in the past five decades. In some cases, forest declines have been definitively linked to atmospheric pollutants. In others, one or more atmospheric pollutants are suspected, at least in part, of contributing to the death of forest species.

6.1.3.1 California Pines

Declines in ponderosa and Jeffrey pines were first noted in the San Bernardino National Forest east of Los Angeles in the 1950s. Older needles on affected trees lost chlorophyll and died prematurely. Injured pines had reduced radial growth and tolerance to the western pine beetle and other stresses. Similar declines were later seen in the Laguna Mountains east of San Diego and the Sierra Nevada and San Gabriel Mountains. Both ponderosa and Jeffrey pines are sensitive to elevated O_3 levels; decline of these California pines has been definitively linked to O_3.

6.1.3.2 Eastern White Pine

Sensitive white pine genotypes have shown evidence of decline throughout their range in eastern North America. This decline has been a selective one, with only the most sensitive trees showing evidence of foliar injury, as well as reduced height and diameter growth. These effects are most pronounced in high-O_3 regions, and a causal connection between elevated O_3 levels and decline of eastern white pine appears evident.

6.1.3.3 Red Spruce and Fraser Fir

Declines in red spruce forests have been reported in the eastern mountains of the United States from New England to North Carolina. Radial growth has decreased sharply, with reductions in the Northeast observed as early as the 1960s. The most dramatic diebacks occurred at elevations above 800 m (2,400 ft), with declines of 50% or more occurring in the past three decades. Symptoms include dieback of branch tips from the top and inward from newly grown shoots and yellowing of the upper surfaces of older needles.

As a general rule, the percentage of severely damaged and dying trees increases with elevation. On three New England mountains, approximately 60% of red spruce were reported to be dead or dying at 1,000 m (~3,300 ft) compared with 20% at 700 m (~2,300 ft). At lower elevations, foliar symptoms, reduced tree growth, and tree death have been reported but are less prevalent than at higher elevations.

In the mountains of North Carolina, forest declines include both red spruce and Fraser fir. Growth of red spruce and, to a lesser extent, Fraser fir began to decrease in the 1960s. By 1987, most red spruce and Fraser firs on Mt. Mitchell's (located in the Great Smoky Mountains) western-facing slope had died. These coniferous species have also experienced high mortality on mountaintops in Tennessee and Virginia.

In the high-elevation mountains where declines are occurring, pollutant levels (particularly O_3) are relatively high compared with those at lower elevations. In addition, deposition rates of hydrogen ion (H^+) are often 10 times greater than at lower elevations. Where forest damage is greatest, trees are exposed to high-acidity, high-O_3 fogs and clouds for upwards of 100 days/year.

6.1.3.4 Pines in the Southeast

Surveys conducted by the US Forest Service in the southeastern states of Virginia, North and South Carolina, Georgia, and Florida have shown that average radial growth rates for loblolly and slash pines less than 63 cm (16 in.) in diameter have decreased by 30% to 50% over the past 40 years.

6.1.3.5 Hardwoods in the Northeast

Crown dieback and increased mortality rates in sugar maple were reported in southeastern Canada in the late 1970s. Similar dieback symptoms have been reported for sugar maples in Pennsylvania, New York, Massachusetts, and Vermont. Yellow birch, American beech, and white birch have also shown decline symptoms.

6.1.3.6 Declines in Central Europe

Multispecies forest declines were first reported in central Europe on low mountains in the 1970s. Silver fir in the Black Forest of Germany lost needles from the inside of branches outward and from the bottom of the tree upward toward the crown and died in large numbers. Initially, needles yellowed, followed by defoliation. Similar symptoms were subsequently observed on pines, beeches, and oaks. Discoloration and premature leaf fall have now been reported in all major forest species in western Germany. Decline effects on a coniferous forest can be seen in Figure 6.5.

Although damage to trees was first reported at 800 m (~2,600 ft) and above, it also occurs at lower elevations. Older trees on west-facing slopes that face prevailing winds are most affected.

A survey conducted in 1983 in western Germany indicated that approximately 34% of the country's trees were affected; by 1986, 54%. Damage to firs was extensive, with 83% showing some symptoms and more than 60% having moderate to severe damage. There was also major injury to spruce, beech, and oak.

Similar injury to forest trees was reported throughout central Europe at all elevations and on all soil types. In surveys conducted between 1983 and 1986, symptoms of decline were seen in 14% of

FIGURE 6.5 Decline of coniferous species in Harz Mountains, Germany, 1981. (Courtesy of Krupa, S.R., Department of Plant Pathology, University of Minnesota.)

Swiss forests, 22% of Austrian forests, 22% of coniferous and 4% of deciduous trees in France, and 24% of forest areas in Holland.

Research studies indicate that air pollutants, through mechanisms that vary by site, are a significant causal factor in forest declines. A link to air pollution is suggested by the large number of species affected, rapid onset of symptoms, large geographical areas affected, and wide range of associated climates and soil conditions involved. The scientific consensus on these declines is that they are caused by a combination of direct foliar damage and nutrient imbalance, both due to pollutant exposures. Ozone, along with acid fogs, seems to be the principal cause of tree damage and decline of European forests.

6.2 EFFECTS ON DOMESTICATED ANIMALS

Retrospective studies of air pollution disasters, such as that which occurred in Donora, PA, in 1948, indicate that in addition to human deaths and illnesses, domesticated animals such as dogs and cats were affected. From epidemiological studies, it is evident that humans chronically exposed to high community air pollution levels have a higher incidence of respiratory disease. Although similar studies have not been done on domesticated animals, dogs and cats, which share the same basic environment as humans, are likely to be affected by such exposures.

6.2.1 FLUORIDE

Although it is no longer a problem in North America, F once caused more air pollution injury to domesticated animals than any other pollutant. Cases of F toxicity (fluorosis in cattle, sheep, horses, and pigs) resulted primarily from the ingestion of forage contaminated by gas, particulate, and soil sources. Fluorosis in livestock once was prevalent in Florida, Tennessee, Utah, Washington, and Oregon.

Fluorosis can be acute or chronic. Acute fluorosis is rare because livestock do not voluntarily consume heavily contaminated forage. Chronic fluorosis is still observed in livestock that ingest F-contaminated forage over time.

Because F interferes with normal Ca metabolism, chronic toxicity is characterized by dental and skeletal changes. There may be incomplete formation of enamel, dentine, or teeth themselves; teeth soften so that excessive dental wear occurs, interfering with proper mastication of food. These dental changes take place during tooth development and are not reversible.

Skeletal changes are usually manifested as diffuse thickening of long bones and calcification of ligaments, resulting in stiffness and lameness. Lame animals eat less because they have difficulty standing for any length of time. Affected animals become emaciated as a result of reduced food intake and difficult mastication. In dairy cattle this results in decreased milk production.

Although F poisoning is no longer a problem in North America and other developed countries, it may be occurring in developing countries where there is little or no environmental regulation.

6.2.2 Lead

Lead emissions from uncontrolled sources were a cause of livestock poisoning in North America in the recent past. Such Pb poisoning was also common in the early part of the twentieth century when lead arsenate insecticide was used as an orchard spray.

Air pollution–related poisoning of livestock in the United States was reported in the vicinity of primary and secondary Pb smelting operations up until the late 1960s. Poisoning was, for the most part, due to ingestion of Pb dust–contaminated forage. Lead poisoning may be acute or chronic depending on exposure level and duration. In cattle, early stages of acute Pb poisoning are characterized by excessive thirst, salivation, loss of appetite, constipation, delirium, and reduced milk production. Affected animals die when exposures are very high. Chronic poisoning results in a spectrum of symptoms, including diarrhea, colic, nervous disorders, swollen joints, lethargy, incoordination, bellowing stupor, rough coat, and emaciation, as well as a variety of metabolic changes similar to those observed in chronically exposed humans.

6.2.3 Other Toxic Exposures

In most cases when air pollution injury to livestock is confirmed, such injury has been associated with stationary sources processing minerals. These minerals contain biologically active elements that, when ingested in excessive amounts, produce toxic effects. Other elements in mineral dusts such as arsenic (As), selenium (Se), and molybdenum (Mb) have reportedly poisoned livestock and other animals.

6.3 ECOLOGICAL EFFECTS

Phytotoxic pollutants can cause injury to individual plants of a species, widespread injury to individual species (ponderosa pine, red spruce, etc.), and, by inference, whole ecosystems. In an ecosystem, the existence of all species is, to some extent, affected by the presence of all others. If a species such as ponderosa pine is eliminated from forest stands in the San Bernardino Mountains, significant ecosystem changes will take place. Species that depend on ponderosa pine for food and shelter must adapt to the new ecological order as ponderosa pine is replaced by more O_3-tolerant species. Significant ecological changes would also occur as a result of other forest declines described previously.

When a population of a species is subject to an environmental stress, it responds in typical Darwinian fashion. The most sensitive, or least tolerant, individuals die and are eliminated. If the stress continues, the species survives, but individuals tolerant to the stress dominate the population. As such, the effect on ecosystems may be limited. Such responses can be expected for many species subject to low-level exposure to pollutants such as O_3, atmospheric deposition pollutants such as sulfuric and nitric acids (H_2SO_4 and HNO_3), increased exposure to ultraviolet (UV) light associated with stratospheric O_3 depletion, and climate changes associated with greenhouse gases.

Individual species, as well as ecosystems of which they are members, are relatively resistant to environmental stressors. If the stress is limited, they may genetically adapt without significant ecological changes. If the environmental stress is sufficient to eliminate one or more species or provide a competitive advantage to other species, significant ecological changes may take place. Major ecosystem changes are characteristic of aquatic systems undergoing significant acidification as a result of atmospheric deposition. They affect all levels of food chains, including plants and animals.

With the exception of acidic deposition, the effects of atmospheric pollutants on ecosystems have been little studied. The potential exists for subtle to major ecosystem changes as a result of

1. Direct exposures to phytotoxic pollutants such as O_3
2. Responses to components of atmospheric deposition (in addition to strong acids) such as nitrate N, elemental mercury (Hg), and methyl mercury (CH_3Hg); pesticides; endocrine disruptors such as polychlorinated biphenyls (PCBs) and dioxins; and other pollutants that accumulate in watersheds and estuaries
3. Increased UV radiation over portions of the southern and northern hemispheres
4. Global warming

6.3.1 Phytotoxic Pollutants

With the exception of major forest declines linked to O_3 exposures, few studies have been conducted to evaluate potential changes in plant, animal, and microbial populations that may be occurring as a result of pollutant exposures on a regional level. This has been due, in part, to limited support for such studies and the inherent complexity of associating changes in species composition with exposure to atmospheric pollutants or other potential stress factors. For example, a progressive decline in flowering dogwood has been observed at Mammoth Cave National Park. It is not known whether such changes are natural or due to other factors such as phytotoxic air pollutants, acidic deposition, and a variety of other abiotic environmental stresses. A change is taking place, but the cause is unknown. Other unstudied changes in forest ecosystems are likely to be taking place as well.

6.3.2 Atmospheric Deposition

Atmospheric deposition poses significant ecological concerns. Major attention is given here to acidic deposition because of its well-recognized and documented environmental effects.

6.3.2.1 Acidic Deposition

Acidic deposition poses a threat to ecosystems that, because of local or regional geology (crystalline or metamorphic rock), have soils and surface waters that cannot adequately neutralize acidified rain, snow, and dry deposited materials. Such soils and waters contain little or no Ca or Mg carbonates and therefore have low acid-neutralizing capacity (ANC). Acid-sensitive ecosystems in North America are illustrated in Figure 4.8. These include mountainous regions of New York and New England; the Appalachian Mountain chain (including portions of Pennsylvania, West Virginia, Virginia, Kentucky, and Tennessee); upper Michigan, Wisconsin, and Minnesota; the Precambrian Shield of Canada, including Ontario, Quebec, and New Brunswick; and various mountainous regions of the North American west.

6.3.2.1.1 Aquatic Ecosystems

Chemical constituents that cause significant biological changes in aquatic ecosystems include H^+, aluminum (Al^{3+}), and calcium (Ca^{2+}) ions. Increased H^+ (decreased pH) and Al^{3+} cause a variety of adverse biological effects that may be exacerbated by low Ca^{2+} concentrations. As pH decreases to less than 5.5, Al^{3+} concentrations increase to what are often toxic levels.

Changes in water chemistry associated with chronic or episodic acidification affect aquatic organisms at all ecological levels. Because of their large size and recreational importance, significant scientific attention has been given to the effects of acidification on fish populations. Declines in fish populations in acidic and low ANC lakes in the Adirondacks have been directly linked to atmospheric deposition-induced acidification. More than 200 lakes are now known to be fishless. Declines in fish populations in lakes and streams in New England and mid-Atlantic mountains have also been reported. In Canada, thousands of lakes in Ontario, Quebec, and New Brunswick are fishless, apparently due to increased H^+ concentrations associated with acidic deposition.

Extinction of fish populations in freshwater lakes and decline of fish populations (and fish kills) in streams seem to be a result of episodic as well as chronic acidification. Sudden shifts in pH following spring snowmelt have reportedly caused fish kills. Long-term increases in H^+ and Al^{3+} concentrations result in recruitment failure; that is, few young fish reach maturity. Recruitment failure occurs because of the high mortality of eggs, larvae, and young fish that are particularly vulnerable to elevated H^+ and Al^{3+} concentrations. As a consequence, acidified waters show decreased population densities, with a shift of size and age toward larger, older fish. As acidification continues, a lake becomes fishless. Because acidification is a progressive phenomenon, the most sensitive fish species are affected first. Based on toxicity models developed from bioassay data, approximately 20% of lakes in the Adirondacks, 6% in New England, and 3% in the upper Midwest have H^+, Al^{3+}, and Ca^{2+} index values unsuitable for the survival of sensitive fish species such as the common minnow and rainbow trout. Almost 10% of lakes in the Adirondacks have H^+, Al^{3+}, and Ca^{2+} index values that make them unsuitable for the more acid-tolerant brook trout. In the mid-Atlantic coastal plain, an estimated 60% of upstream sites have H^+ and Al^{3+} levels during spring that would kill 50% of anadromous (ocean species that spawn in freshwater rivers and streams) fish species such as the larval stage of the blueback herring.

As with other stressed environments, species diversity reduction is one of the initial significant biological changes that occur in acidified waters. Although total biomass may initially remain unchanged, fewer species are present. Shifts in species composition occur at pH levels in the range of 6.0 to 6.5. Unicellular phytoplankton species decrease and certain filamentous species increase in abundance (and in severely acidified lakes, cover all surfaces with a thick mat). Acidification decreases the diversity and abundance of benthic (bottom-dwelling) invertebrates, which, because of their role as microdecomposers, are essential to nutrient cycling in aquatic systems. Reduced invertebrate populations result in fewer prey species for higher aquatic organisms.

Decreased decomposition by benthic organisms and bacteria results in the accumulation of organic debris, with less mineralization and release of organically bound nutrients. Strong acidity causes nutrient ions to bind to particles, decreasing their uptake by phytoplankton.

Acidification effects are profound, whether associated with acidic deposition or natural processes. As lakes become increasingly acidified, significant species changes and decreases in biomass production occur.

6.3.2.1.2 Terrestrial Ecosystems

Acidic deposition has the potential to directly and indirectly affect terrestrial plants and other organisms in natural ecosystems. Major scientific attention has been given to evaluating potential causal connections between acidic deposition and "declines" of Norway spruce and beech in Germany and other countries of central Europe, high-elevation spruce and fir forests in the northern and southern Appalachian Mountains of the eastern United States, low-elevation spruce and fir forests in the northeastern United States, southern pines in the southeastern United States, and sugar maple in the northeastern United States and southeastern Canada.

As indicated in Section 6.1.3, acidic deposition has been implicated as a contributing factor to forest declines of high-altitude red spruce in the Appalachian Mountains and apparent widespread growth reductions in southern Appalachian tree species. Acidic deposition has a significant potential to affect high-elevation eastern forest species because of acid cloud water (pH 2.8–3.8) exposures that occur upwards of 100 days/year. Such exposures result in alterations in plant nutrition, cold hardening of buds, and a variety of other physiological processes. Acidic deposition may cause reduced growth as a result of (1) foliar nutrient leaching; (2) reduced soil nutrient availability; (3) interference with normal root function associated with increased mobilization of Al^{3+} and inhibition of Ca^{2+}, magnesium (Mg^{2+}), and phosphorous (P^{2+}) uptake; (4) reduced carbon (C) levels due to reduced photosynthetic activity and increased respiratory C consumption; and (5) increased sensitivity to cold damage. Acidic deposition may also affect tree growth as a result of soil mobilization of Al^{3+} that causes direct toxic effects to roots. Acidification of forest soils results in interference with development of mycorrhizae, the symbiotic growth of roots and fungi essential for the normal growth of many forest species.

Direct effects of acidic deposition on forest species, as well as indirect effects resulting from changes in soil chemistry and microbiology, have the potential to cause significant changes in forest ecosystems. The magnitude of such effects over the long term (next half century) may be large.

6.3.2.2 Nitrogen

The atmosphere is a significant source of N input into terrestrial ecosystems, watersheds, and estuaries. Nitrogen is an important plant nutrient, as it is required for the synthesis of amino acids, proteins, nucleic acids, and chlorophyll. To a large extent, it governs the utilization of potassium (K^{+}), P^{2+}, and other nutrients.

6.3.2.2.1 Terrestrial Ecosystems

Until the industrial revolution, the availability of soil N was the limiting factor in the growth of trees and possibly other plants in forest ecosystems. Modern-day atmospheric inputs of N to forest environments is on the order of 5 to 20 times greater than that in the preindustrial era.

Nitrogen associated with atmospheric deposition serves as a plant nutrient in N-poor soils, particularly in forests. Such N inputs to forest ecosystems may result in increased growth of some species and altered species composition and diversity because of changes in competitive relationships.

Soil N increases can change the vegetative structure of an ecosystem. When N becomes readily available in environments previously characterized by low N availability, plants adapted to such an environment will be replaced by those capable of using now-increased N levels. Such effects have been reported for heathlands in the Netherlands, where nitrophilous (nitrogen-loving) grass species have replaced slower-growing species. Herbaceous forest plant composition in the Netherlands has shifted in the past several decades to species commonly found on N-rich soils.

In other European countries, N deposition has altered the composition of nonwoody plants in mixed oak forests. Twenty of 30 species closely associated with high N deposition have increased in abundance.

Excessive N deposition may adversely affect the growth of forest trees. Decline of some tree species in German forests has been attributed to Mg^{2+} deficiencies caused by excessive N levels. Chronic N addition studies indicate that excessive N levels may result in decreased forest tree productivity and increased mortality. Studies in Vermont suggest that coniferous forest stands undergoing decline are being replaced by fast-growing deciduous forests that cycle N more rapidly. Plant succession and biodiversity are being significantly affected by chronic inputs of N into historically low-N-input environments.

A major evolving environmental concern is "N saturation" that results when atmospheric inputs to what are normally background soil N levels exceed the capacity of plants and soil microorganisms to utilize and retain it. Excess N "runs off," resulting in increased stream, river, and lake nitrate levels. Nitrogen saturation has been reported in spruce–fir ecosystems in the Appalachians, eastern hardwood watersheds in West Virginia, and mixed conifer and chaparral forests in the Los Angeles Basin.

A variety of ecosystem changes resulting from N saturation have been suggested. These include (1) permanent increase in foliage N and reduced P^{2+} and lignin in wood, (2) reduced productivity due to physiological disruptions, (3) decreased root growth, (4) increased nitrification and nitrate $\left(NO_3^-\right)$ leaching, and (5) reduced soil fertility resulting from increased leaching.

Increased soil N can significantly alter nutrient cycling in terrestrial ecosystems. Processes affected include (1) plant nutrient uptake and allocation, (2) litter production, (3) release of N as NH_3 (ammonification) and conversion of NH_3 to NO_3^- (nitrification), and (4) leaching of NO_3^- and release of trace gases such as nitrous oxide (N_2O). In addition, under N saturation, soil microbial communities change from predominantly fungal (mycorrhizal) to those dominated by bacteria.

6.3.2.2.2 Aquatic Ecosystems

Nitrogen associated with atmospheric deposition changes the fertility, or nutrient condition, of lakes, rivers, streams, and estuaries (shallow water areas where rivers enter the sea). Increased plant nutrients in aquatic ecosystems significantly increase the growth of algae and aquatic weeds. This increase in plant productivity results in increased biomass production at all levels of the food chain, including fish, and increases the turbidity of lakes and streams. Plant nutrients (e.g., NO_3^-), when present at elevated levels, may cause population explosions of algae called algal blooms. As these massive algal populations die, their decomposition results in episodic oxygen (O_2) depletion, with significant adverse ecological consequences. Plant nutrients contribute to the process of eutrophication, wherein freshwater lakes mature and fill in more rapidly as a result of increased biomass production and organic sediment deposition.

The atmosphere is but one source of plant nutrients that significantly affects the ecology of lakes, rivers, streams, and estuaries. Other sources such as agricultural runoff and wastewater treatment dominate the N budget of such systems. Nevertheless, in some stream, river, and lake waters, the atmosphere may be the source of 40% of nitrate N. Atmospheric deposition of NO_3^- to estuaries, such as those of the Chesapeake Bay, may be more significant in terms of ecological effects than in freshwater aquatic ecosystems. In the latter, the limiting factor for plant growth is P^{2+}, not N. This is not the case in estuaries.

6.3.2.3 Ozone

Ozone (O_3) deposition has significant ecological effects, particularly on vegetation, forests, and sensitive ecosystems.

6.3.2.3.1 Impaired Plant Health and Growth

Elevated ozone (O_3) levels impair plant health and growth through a range of physiological and structural effects. Ozone can damage leaf tissue, causing visible symptoms such as stippling, bronzing, and necrosis. It also interferes with photosynthesis by disrupting chlorophyll production and stomatal function, ultimately reducing a plant's ability to produce energy efficiently. Chronic exposure suppresses overall plant development, limiting root expansion, decreasing leaf area, and reducing total biomass in both crops and native vegetation. Notably, sensitive species such as soybeans, wheat, black cherry, and ponderosa pine are particularly vulnerable to these effects. Long-term exposure to elevated ozone levels contributes to tree decline by weakening their physiological defenses, making them more vulnerable to drought stress, disease, and insect infestations. This chronic stress can lead to reduced growth and premature mortality in affected species. As ozone-sensitive trees decline, forest composition may shift, altering biodiversity and disrupting ecosystem structure and function. Such impacts have been well documented in forests across the eastern United States, California, and parts of Europe, where visible signs of ozone injury and significant reductions in tree growth have been observed.

6.3.2.3.2 Crop Yield Reductions

Ozone is considered the most damaging air pollutant to crops worldwide. Studies by the EPA and USDA estimate yield losses of 5%–15% for soybeans, 3%–10% for wheat and corn, and even greater reductions for more sensitive fruits and vegetables, highlighting ozone's significant impact on agricultural productivity.

6.3.2.3.3 Deposition to Water and Soil

While ozone's primary effects are on plant foliage, its deposition can indirectly influence aquatic and soil ecosystems. Altering leaf litter quality affects organic matter decomposition rates and disrupts trophic interactions within forest-floor food webs, leading to broader ecological imbalances.

6.3.2.4 Mercury and Other Trace Metals

Mercury deposition on land and water surfaces, and its accumulation in the muds of freshwater lakes, is an increasing environmental and public health concern. Although, in a relative sense, environmental Hg levels are low, their significance is greatly increased because Hg is biomagnified in food chains either as ionic Hg^{2+} or organic CH_3Hg. Methylmercury, produced by microorganisms as they transform Hg from its elemental form, tends to remain dissolved in water. As such, it moves upward through food chains and accumulates in fish tissues. This process is called biomagnification. It can result in CH_3Hg or Hg^{2+} levels in fish that are orders of magnitude higher than those found in water.

Because fish are at the top of most aquatic food chains, Hg levels in fish may be high. They are so high in some cases that the US government issues fish consumption advisories for many bodies of water, particularly in the Great Lakes Basin and Florida Everglades.

The primary environmental concern associated with Hg pollution has been the potential exposure of the fish-consuming public to unsafe Hg levels. Mercury is a very toxic substance in its elemental, ionic, and organic forms.

Emissions to the atmosphere from major Hg sources have been regulated under hazardous air pollutant standards since 1973, and significant limits have been placed on industrial water sources. Nevertheless, because of emissions from unregulated sources such as coal-fired power plants, Hg pollution of the environment is a major concern.

Although regulatory efforts have focused on public health, Hg poses an exposure/toxicity risk to a number of wildlife species that consume large quantities of fish. These include bald eagles, ospreys, loons, kingfishers, herons, and mink. It also poses a poisoning risk to sharks in near-shore waters. There is only limited evidence available on the chronic effects of Hg exposure on wildlife species. Significant neurological and reproductive system effects in populations of fish-eating birds and mammals may be occurring.

Atmospheric deposition contains a number of other trace metals that can accumulate in the environment. These include Zn, Cu, Pb, Ni, tin (Sn), chromium (Cr), cadmium (Cd), silver (Ag), Mb, and vanadium (V). Heavy metals tend to accumulate in forest soil humus or soil layers immediately below where root activity occurs. As a consequence, they may be toxic to roots and soil microorganisms and interfere with nutrient cycles. Heavy metal deposition has affected the root growth of plants in heavily contaminated soils around smelters. The effects of low-level heavy metal contamination associated with atmospheric deposition have received only limited study. Some invertebrates (e.g., earthworms that consume leaf litter) are known to accumulate Cd, Pb, Zn, and Ni.

6.3.2.5 Organochlorine Compounds

6.3.2.5.1 Pesticides

The era of widespread use of persistent organochlorine pesticides in the United States began coming to a close with a 1972 ban on the use of DDT; cancellation of the registrations of other persistent pesticides or restrictions on their use followed in the next decade.

Because of their low volatilities (~10^{-4} mmHg), organochlorine pesticides have never been major atmospheric pollutants of concern. Nevertheless, as semivolatile and very persistent compounds, they have rather remarkable environmental mobility. For example, they have been found in fat tissue of Antarctic penguins 6,450 km (4,000 mi) from the nearest site of application.

Pesticides move through the atmosphere by a continual process of volatilization and deposition on particles and other surfaces. As such, they are commonly found in atmospheric deposition samples, albeit at very low levels.

Organochlorine pesticides, such as Hg, are bioaccumulated. Consequently, they are biomagnified in food chains, with the highest levels in fish and predator birds. The effects of pesticides on predator birds in the period 1950 to 1975 were dramatic, resulting in a precipitous decline in such species in North America. With the ban of DDT and other organochlorine pesticides and restrictions on the use of other similar organochlorine compounds, pesticide levels in fish and birds began to decline, with the latter slowly recovering from its endangered species status (e.g., bald eagles, ospreys, peregrine falcons).

Although organochlorine pesticides are, for the most part, no longer used in the United States, Canada, and Europe, they are still used in many parts of the world. Because of their persistence (half-life = ~12 to 15 years) and continued atmospheric transport, organochlorine pesticides are commonly measured in North American precipitation samples.

The environmental significance of this low-level contamination of the environment and movement through food chains is unknown. These substances have been reported to be endocrine disruptors in that they mimic estrogen, a female sex hormone. Major concerns have been expressed relative to their potential effect on male reproductive organs and processes, such as those observed in alligators associated with a spill of a large quantity of DDT.

The organochlorine pesticides, because of their persistence, mobility in the environment, and passage through food chains, have very likely contaminated every living thing on the planet and will likely continue to do so for some time.

6.3.2.5.2 Polychlorinated Biphenyls, Dioxins, and Furans

Although their use in the United States was discontinued in the 1970s, PCBs remain widespread contaminants of the Earth's environment. Like organochlorine pesticides, they are semivolatile and persistent. They are also biomagnified in food chains, with concentrations in aquatic organisms orders of magnitude greater than in the surrounding water. They affect the liver and nervous and reproductive systems and are animal carcinogens. The effects of PCBs on wildlife are largely unknown. Exposure through aquatic food chains is believed to be responsible for genetic or congenital abnormalities observed among fish-eating birds (particularly cormorants) in the Great Lakes Basin. Like other organochlorine compounds, PCBs seem to be endocrine disruptors.

Dioxins and furans are chlorinated organic compounds widely present in the environment at very low concentrations. They are characterized by their very high mammalian toxicity (LD_{50} [dose required to kill 50% of animals under test] = 2 ppb w/w), persistence, movement, and accumulation in food chains. 2,3,7,8-Tetrachlorodibenzo-*p*-dioxin (TCDD) has been shown to cause endocrine disruption, weaken immune systems, and cause adverse reproductive changes in wildlife species.

6.3.3 Stratospheric Ozone Depletion

The primary environmental concern associated with stratospheric O_3 depletion is increased UV-B radiation over Antarctica and potential increases over the northern hemisphere. UV light is a plant growth regulator; it controls cell elongation in terrestrial plants. As a consequence, plants tend to grow "shorter" in full sunlight. As a result of increases in UV-B exposure, plants may not grow as tall, and therefore, regions subject to increased UV-B levels would not be as biologically productive. This could be a concern for annual crop plants and forest tree species.

Plants of species located at different latitudes are adapted to UV-B levels characteristic of their region. As such, it is likely that tolerant forms would, in a matter of a decade or so, begin to replace less tolerant forms in response to increased UV-B levels. Species would survive by adapting to the changed UV light conditions. Less tolerant or less genetically variable species may, however, be under severe environmental stress and be replaced by more tolerant species.

There is a significant potential for adverse biological changes in phytoplankton and other parts of the food chain in the upper zone of ocean water surrounding Antarctica. Exposures to UV-B can decrease photosynthesis and productivity in several marine species of algae and adversely affect various forms of larvae and other organisms. Algae or phytoplankton are the base of the Antarctic food chain that includes the very abundant krill (a small, shrimp-like animal) and higher-order species such as fish, birds, and sea mammals (e.g., seals and whales). As such, any major UV-B-related effects on phytoplankton could have major consequences for species in these food chains.

A worldwide decline in many amphibian species has been observed in the past several decades. The cause of this decline is not known. A few studies have shown that several amphibian species are very sensitive to near-surface UV-B levels. On this limited basis, a potential causal relationship between increased UV levels associated with stratospheric O_3 depletion and amphibian mortality has been proposed.

6.3.4 Climate Change

Atmospheric temperature is one of the most important environmental factors that affect the distribution of plant and animal communities around the planet. Temperature effects on individual plant or animal species and ecosystems are both direct and indirect (e.g., effects on water [H_2O] availability). Even slight changes in average atmospheric temperature can result in changes in the presence of a particular species in a given environment, its abundance, and its distribution. Limited evidence already exists that the near-surface planetary warming that has occurred in the past two decades has caused ecological changes. Projections of future global atmospheric warming, with associated effects in the polar regions (thinning of ice cover), coastal areas (sea level rises), and desert zones (increased desertification), as well as changes in precipitation patterns and migration of climatic zones, indicate that global warming will result in ecological changes, the likes of which the Earth has not experienced in thousands of years. These changes will affect, to varying degrees, most species that exist on our planet. The effects of global warming are expected to be particularly significant in polar regions and coastal areas.

6.3.4.1 Polar Regions

Based on climate models, polar regions will experience the greatest temperature increases (due to decreased albedo [see Chapter 1] associated with the melting of ice and snow). Significant evidence exists that changes in the Arctic and Antarctic environments have occurred. These include (1) satellite and ground-based analyses—summarized by NOAA, EPA, and the IPCC AR6—show that Arctic snow cover has declined at an average rate of 3.7% per decade for May and ~15% per decade for June, relative to the 1981–2020 baseline, (2) in Alaska, average temperatures have risen by roughly 2°C (3.6°F) over the past 60 years, with winter warming even greater (up to 6°F) since the 1950s, (3) the decline of lake ice duration in middle and high latitudes, (4) declines in the extent and thickness of Arctic sea ice, and (5) in West Antarctica, ice sheet thinning and coastal glacier retreat have accelerated in recent decades, contributing to rising sea levels, particularly from the Amundsen Sea sector of the West Antarctic Ice Sheet.

Sea ice is a major factor in regulating heat and moisture exchange and ocean salinity. It provides a habitat for major polar species such as polar bears, seals, and penguins. As it thins, coastal areas become subject to erosion, altering marine mammal habitat.

During the past two decades, the health of polar bears in Arctic regions has been declining. Canadian scientists have suggested that the breakup of sea ice 3 weeks sooner than normal reduces the hunting season for seals (bears' primary food source) on sea ice. A 50% decline in Adelie penguins on the Antarctic peninsula has been observed over the past 25 years, presumably because of regional changes in sea ice and snowfall related to global warming.

6.3.4.2 Coastal Zones

Climatic models predict significant sea level rise associated with thermal expansion of seawater and melting of polar ice caps. Sea level rises would inundate coastal wetlands, erode shorelines, and increase the salinity of estuaries, bays, and rivers. As coastal wetlands are inundated, previously dry areas would be flooded. Overall, the size of coastal wetlands would decrease significantly.

Coastal wetlands, which include estuaries, are among the most biologically productive environments in the world. Their loss would result in enormous changes in biological diversity and the abundance and distribution of individual species. Red tide refers to harmful algal blooms (HABs), typically caused by the rapid growth of toxic marine algae such as Karenia brevis. These blooms can discolor coastal waters and release potent toxins that harm marine life, trigger respiratory irritation in humans, and contaminate shellfish, posing public health risks. Red tides are often driven by warm temperatures, nutrient pollution, and calm ocean conditions and have become more frequent and intense in recent decades due to climate change and increased coastal runoff.

6.3.4.3 Plant and Animal Communities

Plant and animal communities are characterized by the major plant types found in them. These include forests, grasslands, deserts, savannahs, and tundra. The nature of these plant communities, called biomes, depends on the combination of environmental factors that exist in a given region or location. Global warming can be expected to result in significant changes in climatic zones; therefore, individual biomes will expand or contract, depending on environmental changes that occur.

Before the advent of modern humans, forests covered vast areas of the Earth's surface. Although a large percentage of forests has been cleared for agricultural purposes, many forested areas still exist. A projected 2°C (3.6°F) warming of the Earth's atmosphere is expected to shift the ideal range for many North American forest species northward by approximately 300 km (180 mi). As global warming would occur progressively, forest species would slowly begin to colonize more northerly locations and be replaced in their current southern range by species that are more tolerant of hotter, and in some cases drier, conditions. If the climate were to become wetter, forests would expand into what are now rangelands.

Increased carbon dioxide (CO_2) levels are expected to have a positive effect on forest growth as well as other plant species because CO_2 is required for photosynthesis; it also enables plants to use H_2O more efficiently. As is the case with N, some species would respond more favorably to increased CO_2 levels and thus have a competitive advantage. Consequently, changes in species composition and abundance are likely to occur.

Each biome, and the ecosystems that comprise it, is characterized by plant and animal species that have evolved over time. Some are endemic to a single region, whereas others migrate across climatic and biological zones, for example, migratory birds.

Climate changes associated with global warming can have both direct and indirect effects on plants and animals. As wetlands are lost, some species are likely to experience a significant loss of habitat and thus a significant reduction in population size. As temperatures warm, many species can be expected to expand their range poleward and up in elevation. Other expected changes include (1) increases or decreases in population density; (2) timing of events such as migration, breeding and nesting, and flowering; (3) changes in body size and behavior; and (4) shifts in genetic frequency.

Climate change–induced wildfires are increasingly having severe ecological effects in many regions. Rising temperatures, prolonged droughts, and changing precipitation patterns have led to more frequent, larger, and longer-lasting wildfires. These fires can devastate ecosystems by destroying vegetation, altering soil composition, and disrupting nutrient cycles. Wildlife habitats are either lost or fragmented, and recovery after a fire can be slow or incomplete, particularly in sensitive ecosystems. Moreover, increased wildfire smoke contributes to air pollution, impacting both ecological and human health. Over time, repeated fires can change the structure of ecosystems, favor fire-adapted species, and reduce biodiversity, which compounds the long-term effects of climate change on natural systems.

As indicated in Chapter 4, significant changes in species range and phenology (timing) in the direction expected, if global warming were occurring, have been documented for hundreds of plant and animal species. The average shift in spring phenology for temperate zone species has averaged

5.1 days earlier per decade. An example of this is the migration and nesting of birds in Michigan (2–3 weeks earlier for several species compared to 1980, with 20 species nesting up to 9 days earlier than in 1971). Notable expansion of species ranges poleward in North America includes the opossum, mockingbird, and armadillo.

6.4 EFFECTS ON MATERIALS

Gaseous and particulate-phase pollutants are known to have adverse effects on materials. Of particular importance are effects on metals, carbonate building stones, paints, textiles, fabric dyes, rubber, leather, and paper. In western Europe, a repository for many monuments of history and fine works of art, air pollutant–induced damage has been incalculable. Because these cultural treasures are irreplaceable, their preservation from destructive effects of airborne contaminants poses a significant challenge to their present guardians. In a sense, destruction of western European antiquities symbolizes the seriousness of the materials damage problem.

Materials can be affected by both physical and chemical mechanisms. Physical damage may result from the abrasive effect of wind-driven particles impacting surfaces and the soiling effect of dust deposition. Chemical reactions may occur when pollutants and materials come into direct contact. Absorbed gases may act directly on the material, or they may first be converted to new substances that are responsible for the observed effects. The action of chemicals usually results in irreversible changes. Consequently, chemical damage is a more serious problem than the physical changes caused by particulate dusts.

6.4.1 Metal Corrosion

Corrosion of metals in industrialized areas represents one of the most ubiquitous effects of atmospheric pollutants. Because corrosion is natural, it is difficult to assess the relative contribution of pollutants to this process. As the ferrous metals, iron (Fe) and steel, account for approximately 90% of all metal usage, their pollution-induced corrosion is of particular importance. When ferrous metals corrode, they take on a characteristic rusty appearance. Corrosion is rapid at first and then slows as a partially protective film develops. Acceleration of corrosion in industrial communities has been reported to be due, at least in part, to SO_2. The corrosive effects of SO_2 on Fe, steel, and steel alloys increase with increased SO_2 concentration and deposition rates. Deposition is facilitated by surface moisture, which promotes corrosion by electrochemical reactions. The corrosive effect of SO_2 seems to be reduced in atmospheres with elevated oxidant levels.

Nonferrous metals can also experience significant pollution-induced corrosion. For example, Zn, widely used to protect steel from atmospheric corrosion, will corrode when acidic gases destroy the basic carbonate coating that normally forms on it. The reaction of sulfur oxides (SO_x) with Cu results in the familiar green patina of copper sulfate ($CuSO_4$) that forms on the surface of Cu materials. This coating of $CuSO_4$ makes the metal resistant to further corrosion. The green discoloration of SO_x-induced Cu corrosion is most commonly observed on bronze statues in city parks and squares. Copper may also be discolored by exposures to hydrogen sulfide (H_2S), which blackens it.

Because nonferrous metals are used to form electrical connections in electronic equipment, corrosion of these connections can result in serious operational and maintenance problems for equipment users. The stress corrosion and breakage of Ni–brass spring relays that used to occur in the central offices of telephone companies in numerous major cities is such an example. Breakage occurred when Ni–brass connections were under moderate stress and positive electrical potential. Stress corrosion and breakage of the Ni–brass relay springs were caused by the high NO_3^- concentrations in dust that accumulated on these metal components. The problem was resolved by using Cu–Ni spring relays that were not susceptible to similar stress corrosion.

By forming thin insulating films over contacts, pollutants can damage low-power electrical contacts used in computers, communications, and other electronic equipment. Such films may result in open circuits, causing equipment to malfunction. They are produced by chemical reactions between the contact metals and contaminants such as SO_2 and H_2S. Equipment malfunction may also result from contamination of electrical contacts by PM. Particles may physically prevent closing or may result in chemical corrosion of contact metals.

6.4.2 Building Materials

Unlike metal corrosion, whose true cause is difficult to discern, pollutant effects on building surfaces are very evident. Soiling and staining of buildings in urban areas that have or have had high atmospheric aerosol levels is a particularly striking example. The "dirtiness" associated with many cities is due to the dark coating of particulate dusts deposited by the smoke of industries and coal-fired space heating over many years.

In addition to soiling, building materials such as marble, limestone, and carbonate cement are chemically eroded by SO_2 and acid aerosols. The reaction of SO_2 with carbonate building stones results in the formation of hydrated Ca salts, $CaSO_3 \cdot 2\ H_2O$ and $CaSO_4 \cdot 2\ H_2O$ (gypsum), both of which are H_2O soluble. The soluble salts produced may precipitate from solution and form encrustations; as they are washed away, the stone surface becomes more susceptible to pollutant effects. Stone damage is further accelerated by sulfur (S)-oxidizing bacteria and fungi.

Deterioration rates are affected by pollutant concentration, moisture content and permeability of stone, and deposition rates. Dry deposition of SO_2 between rain events has been reported as the major causal factor in carbonate building stone deterioration. Pollutant-induced erosion on a carbonate stone street monument can be seen in Figure 6.6.

Chemical erosion of priceless, irreplaceable historical monuments and works of art in western and southern Europe can be described as no less than catastrophic. Although primarily a European problem, similar destruction is occurring on monuments in the United States. In the late nineteenth century, the obelisk, Cleopatra's Needle, was moved from Egypt to a park in New York City. The monument, which had withstood the ravages of desert heat and sands for thousands of years without significant deterioration, has had hieroglyphics obliterated on two of its faces in the past half century. The other two faces have eroded less, as they were away from the direction of prevailing winds.

A significant concern is the Acropolis, located downwind of Athens, Greece. Although chemical erosion of marble structures in the Acropolis has been occurring for more than 100 years, rapid population growth and use of high-S heating oils since World War II have greatly accelerated their destruction.

In addition to the Acropolis, the Colosseum in Rome and the Taj Mahal in India are in various stages of dissolution. In many European cities, marble statues have had to be moved indoors. Cities such as Venice have enacted legislation to preserve and restore hundreds of damaged piazzi, churches, and historical buildings. The grand cathedrals of Europe (e.g., Notre Dame, Chartres, and Cologne) are constantly undergoing restoration of acid-ravaged limestone statuary and other features on their external façades.

6.4.3 Paints

Paints consist of two functional components: vehicles and pigments. The vehicle consists of a binder and additives that hold the pigment to a substrate surface. The principal materials used as vehicles include hydrocarbon solvents, glycols, and resins, which may be natural, modified natural, or synthetic. Pigments, classified as white or colored, provide aesthetic appeal, hiding power, and durability. White Pb, titanium dioxide (TiO_2), and zinc oxide (ZnO) have been widely used white pigments. Colored pigments are a variety of mineral, metal, and organic compounds. The function of paints is to provide an aesthetically appealing surface coating that protects underlying material from deterioration.

FIGURE 6.6 Erosion of carbonate column in Maltese square. (Courtesy of DHD Multimedia Gallery, http://aj.hd.org.)

The appearance and durability of paints are affected by natural weathering processes. Paint weathering is accelerated by environmental factors such as moisture, temperature, sunlight (particularly UV light), and fungi. Paint appearance and durability may also be affected by air pollutants such as PM, H_2S, SO_x, NH_3, and O_3.

Air pollutants affect the durability of painted surfaces by promoting chalking, discoloration, loss of gloss, erosion, blistering, and peeling. They may also decrease paint adhesion and strength and increase drying time. Sulfur dioxide can increase the drying time of some paints by reacting with drying oils and competing with auto-oxidative curing reactions responsible for cross-linking the binder. Erosion of oil-based house paint is reportedly enhanced by exposure to SO_2 and high relative humidity. Apparently, SO_2 reacts with extender pigments such as calcium carbonate ($CaCO_3$) and ZnO. Particles can also damage painted surfaces by serving as carriers for corrosive pollutants or by staining or pitting painted surfaces.

One of the more obvious effects of pollutants on paints is the dirty appearance that results from accumulation of aerosol particles. These may cause chemical deterioration of freshly applied paints that have not completely cured. Particles may also serve as wicks that allow chemically reactive substances to migrate into underlying surfaces, resulting in corrosion (if the underlying material is metallic) and subsequent paint peeling.

Exterior paints pigmented with white Pb may be discolored by reaction of Pb with low atmospheric levels of H_2S. Blackening of the paint surface by formation of lead sulfide (PbS) can occur at H_2S exposures as low as 0.05 ppmv for several hours. The intensity of discoloration is related not only to the concentration and duration of H_2S exposure, but also to Pb content of the paint and the presence of moisture.

There have been numerous reports over the last several decades of damage to automobile paints and coatings. Damage occurs on horizontal surfaces of freshly painted vehicles and appears as irregularly shaped and permanently etched areas. It is easily observed on dark-colored vehicles and appears after evaporation of moisture droplets. The consensus in the automobile industry is that this paint damage problem is due to environmental "fallout." The most likely cause is believed to be acid rain, as chemical analyses of exposed test panels have shown elevated levels of sulfate.

6.4.4 Textiles and Textile Dyes

Textiles consist of a basic fiber component and additives such as dyes, water repellents, and finishes. Each of these is subject to pollutant damage.

Exposures to atmospheric pollutants may result in significant deterioration and weakening of textile fibers. Fabrics such as cotton, hemp, linen, and rayon are particularly sensitive to acid-forming gases and aerosols. Acids chemically break their cellulose chains, producing water-soluble products that have low tensile strength. Synthetic fibers such as nylon may also incur significant acid damage as well as undergo oxidation by NO_2, which reduces the affinity of nylon fibers for certain dyes.

Several pollutants can react with fabric dyes by oxidation or reduction reactions, causing them to fade. Fading of textile dyes associated with NO_x exposure, particularly NO_2, has had a long history. For example, NO_x emissions from open electric-arc lamps and incandescent gas mantles caused fading of stored woolen goods in Germany prior to World War I. Nitrogen oxides emissions from gas heaters were also believed to be responsible for a serious fading problem on acetate rayon fabrics during the 1930s. During the 1950s, NO_x emissions from home gas dryers caused fading of colored cotton fabrics. Nitrogen dioxide–induced fading has also been reported for fabrics exposed to urban environments.

Reports of NO_2-induced fading of textile materials from ambient exposures were soon followed by observations that ambient O_3 levels were also a major fading cause. In the early 1960s, O_3 fading was reported for polyestercotton permanent press slacks stored in warehouses or on retail shelves. Some manufacturers suffered significant economic losses. Fading of nylon carpets was a problem along the Gulf Coast from Florida to Texas. Blue dyes were particularly sensitive to a combination of ambient O_3 exposure and high humidity. This Gulf Coast fading was overcome by using O_3-resistant dyes and modifying nylon fibers to decrease exposure to O_3. Pollutant-induced fading of fabric dyes has posed significant challenges for textile manufacturers. Consequently, the industry had to develop resistant dyes and chemical inhibitors to mitigate such problems.

6.4.5 Paper and Leather

Papers that came into use after 1750 are very sensitive to SO_2 because of its conversion to H_2SO_4 by the metallic impurities in paper. Sulfuric acid causes paper to become brittle, decreasing its service life. This embrittlement is a major concern to libraries and museums in cities with elevated SO_2 concentrations, as it makes the preservation of historical books and documents more difficult.

Historical book preservation involves more than the paper component. Air pollution damage to leather-bound books is also a major concern. Damage to leather is initially seen as cracking. Cracking results in increased leather exposure, thus accelerating deterioration. Leather eventually loses its resiliency and turns to a reddish brown dust.

6.4.6 Rubber

Ozone can induce cracking in rubber and plastic compounds, most notably those under stress. The depth and nature of cracking depend on O_3 concentration, rubber or plastic formulation, and degree of stress. Unsaturated natural and synthetic rubbers, for example, butadiene-styrene and butadiene-acrylonitrile, are especially vulnerable to O_3 cracking. Ozone reacts with the double bonds in such products. Saturated compounds such as thiokol, butyl, and silicon polymers are relatively resistant to O_3 cracking. Rubber compounds are often formulated by the addition of antioxidants to reduce cracking. This is particularly important in rubber tires.

Rubber cracking was one of the first effects of smog observed in the Los Angeles area. During the 1950s, it was used to monitor O_3 concentrations. Under standardized conditions, O_3 concentrations could be directly related to the depth of O_3-induced cracks.

6.4.7 Glass

With the exception of F, glass is resistant to most air pollutants. Fluoride reacts with silicon (Si) compounds to produce a familiar etched appearance. Etching of glass has been observed in the vicinity of major F sources such as phosphate fertilizer mills, Al smelters, enamel frit plants, and even steel mills. Continuous exposure to F may render glass opaque.

6.4.8 Economic Losses

Economic losses due to air pollution–induced damage to materials are difficult to assess; it is very difficult to distinguish between natural deterioration and that caused or accelerated by air pollutants. Only limited efforts have been made to quantify the costs associated with air pollution damage to materials. It is probable, however, that such costs are in the tens of billions of dollars.

6.5 ODOR POLLUTION

The welfare effects described in this chapter generally do not affect people directly. Odor, on the other hand, does. Although exposure to malodorous substances may cause symptoms in some individuals, the problem of odor, or odor air pollution, is usually viewed as being one of annoyance.

The term *odor*, as used in everyday speech, commonly carries with it an unpleasant connotation. The term defined by scientists is neutral; it connotes only that some volatile or semivolatile substance is sensed by the human olfactory system.

The ability to smell chemical substances is common among members of the animal kingdom. In many species, the olfactory sense plays an important role in locating food, attracting individuals of the opposite sex for breeding, and warning of approaching danger. In humans, the olfactory sense plays a lesser and apparently less obvious role.

The olfactory function in humans consists of two different organs found in the nose. The olfactory epithelium (found in the highest part of the nose) consists of millions of receptor cells that connect directly to the olfactory bulbs of the brain. The free endings of trigeminal nerves distributed throughout the nasal cavity serve as a secondary organ of smell. Trigeminal nerves respond to odoriferous substances that cause irritation, tingling, or burning. The chemical senses that correspond to these two olfactory organs are not easily separated, with many odoriferous substances or odorants stimulating both systems. They stimulate different parts of the brain and therefore have different effects. The major function of the trigeminal nerve system is to stimulate reflex actions such as sneezing or the interruption of breathing when the body is exposed to potentially harmful substances.

6.5.1 Odor Measurement

Odor perception is a psychophysiological response to the inhalation of odoriferous chemicals. Odors cannot be chemically measured. However, some sensory attributes of odors, for example, detectability, intensity, character (quality), and hedonic tone (pleasantness or unpleasantness), can be measured by exposing individuals in controlled environments.

The limit of detection (the odor threshold) may be characterized in two ways. It may be the concentration of a substance that is detectably different from the background. Alternatively, it may be defined as the first concentration at which an observer can positively identify an odor. The former is the detection threshold; the latter, a recognition threshold.

Although the olfactory sense in humans is not as acute as it is in many animals, it nevertheless has the ability to detect many substances at very low concentrations. Odor thresholds for a variety

TABLE 6.2
Odor Thresholds and Characteristics of Selected Chemical Compounds

Chemical	Odor Threshold (ppmv)	Odor Character
Acetaldehyde	0.21	Green, sweet
Acetone	100.0	Chemical sweet, pungent
Dimethyl amine	0.047	Fishy
Ammonia	46.8	Pungent
Benzene	4.68	Solvent
Butyric acid	0.001	Sour
Dimethyl sulfide	0.001	Vegetable sulfide
Ethanol	10.0	Sweet
Ethyl mercaptan	0.001	Earthy, sulfide-like
Formaldehyde	1.0	Hay, straw-like, pungent
Hydrogen sulfide	0.00047	Eggy sulfide
Methanol	100.0	Sweet
Methylethyl ketone	10.0	Sweet
Paracresol	0.001	Tar-like, pungent
Perchloroethylene	4.68	Chlorinated solvent
Phenol	0.047	Medicinal
Sulfur dioxide	0.47	Sulfury
Toluene	2.14	Mothballs, rubbery

of chemicals are summarized in Table 6.2. Note that humans can detect H_2S at approximately 0.5 ppbv.

Odors differ in their character or quality. This odor parameter allows us to distinguish odors of different substances by prior odor associations. The characters of a variety of selected chemicals are summarized in Table 6.2. For example, dimethyl amine is described as fishy, phenol as medicinal, paracresol as tar-like, and so on.

The olfactory sense is highly subjective, with some odors being pleasant to some individuals and unpleasant to others. The difficulty of defining odor problems lies in part with the lack of unanimity as to what is unpleasant.

Unpleasantness may also be related to odor intensity. Some otherwise unpleasant odors may elicit a pleasant sensation in an individual when they are in relatively low to moderate concentrations. The olfactory response to an odorant decreases nonlinearly as the concentration decreases. For a substance such as amyl butyrate, the perceived odor intensity decreases by 50% for a tenfold reduction in concentration. This logarithmic relationship is common for most odorants.

The perceived intensity of an odor rapidly decreases after initial exposure. Within a few minutes, it may not be perceived at all. This phenomenon is called olfactory fatigue. Sensitivity to the odor is restored within minutes after exposure ceases.

A person may become habituated to unpleasant odors ("get used to it" or become "more tolerant"). Habituation operates over much longer periods than olfactory fatigue.

Unpleasant odors may affect our sense of well-being. Responses to a variety of malodors (bad odors) can include nausea, vomiting, headache, coughing, sneezing, shallow breathing, disturbed sleep, appetite disturbance, sensory irritation, annoyance, and depression. Effects may be physiological, psychological, or both.

6.5.2 Odor Problems

It is likely that malodors from nearby sources are responsible for more complaints to regulatory agencies than any other form of air pollution. Particularly notable sources of malodors (and, in many instances, citizen complaints) are rendering plants, soap-making facilities, petrochemical plants, refineries, pulp and kraft paper mills, fish-processing plants, diesel exhaust, sewage treatment plants, and agricultural operations (including feedlots, poultry houses, and hog confinements). Malodors associated with such sources include a variety of amines, sulfur gases (such as H_2S, methyl and ethyl mercaptan, and carbon disulfide [CS]), phenol, NH_3, aldehydes, fatty acids, and the like.

In addition to the annoyance caused by malodors, the presence of a continuous source may result in decreases of property values as prospective homebuyers avoid such areas. Nevertheless, as urban–suburban populations expand, many new homes are built in once-rural areas where malodorous agricultural operations exist. Not surprisingly, new residents often find barnyard odors objectionable. In many cases, the nature of the "barnyard" is changing. Many, if not most, agricultural livestock operations today involve large numbers of animals. These "factory farms" produce enormous quantities of agricultural wastes and, of course, stench. In agricultural states, such farms may be exempt from air pollution regulations and even zoning requirements.

REFERENCE

OECD, *The Economic Consequences of Outdoor Air Pollution*, OECD Publishing, Paris, 2016. http://dx.doi.org/10.1787/9789264257474-en.

READINGS

Adams, R.M., Hamilton, S.A., and McCord, B.A., An assessment of economic effects of O_3 on U.S. agriculture, *JAPCA*, 35, 938, 1985.

Alscher, R.G., and Wellburn, A.R., Eds., *Plant Responses to the Gaseous Environment: Molecular, Metabolic, and Physiological Aspects*, in Third International Symposium on Air Pollutants and Metabolism, Blackburg, VA, Chapman & Hall, London, 1992.

Ambosht, R.S., Ed., *Modern Trends in Ecology and Environment*, Backhuys Publishers, Leiden, Netherlands, 1998.

Butlin, R.N., Effects of air pollutants on buildings and materials, *Proc. R. Soc. Edinburgh*, 97B, 255, 1991.

Graedel, T.E., and McGill, R., Degradation of materials in the atmosphere, *Environ. Sci. Technol.*, 20, 1093, 1986.

Hellman, T.M., and Small, F.H., Characterization of the odor properties of 101 petrochemicals using sensory methods, *JAPCA*, 24, 979, 1974.

Irving, P.M., Ed., *Acidic Deposition: State of Science and Technology*, Vols. I–IV, The U.S. National Acid Precipitation Assessment Program, Washington, DC, 1991.

Jacobson, J.S., and Hill, A.C., Eds., *Recognition of Air Pollution Injury to Vegetation: A Pictorial Atlas*, Air Pollution Control Association, Pittsburgh, PA, 1970.

Johnson, D.W., and Taylor, G.E., Role of air pollution in forest decline in eastern North America, *J. Water Air Soil Pollut.*, 48, 21, 1989.

Krupa, S.R., *Air Pollution, People, and Plants*, APS Press, St. Paul, MN, 1997.

Legge, A.H., and Krupa, S.V., Eds., *Air Pollutants and Effects on Terrestrial Ecosystems*, John Wiley & Sons, New York, 1986.

Lepp, N.W., Effect of heavy metal pollution on plants, in *Metals in the Environment*, Vol. 2, Lepp, N.W., Ed., Applied Science Publishers, London, 1981.

Lucier, A.A., and Haines, S.G., Eds., *Mechanisms of Forest Response to Acidic Deposition*, Springer-Verlag, New York, 1990.

Mackenzie, J.J., and El-Ashry, M.T., Eds., *Ill Winds: Airborne Pollution's Toll on Trees and Crops*, Yale University Press, New Haven, CT, 1989.

Manion, P.D., and Lachance, D., *Forest Decline Concepts*, APS Press, St. Paul, MN, 1992.

National Research Council, Committee on Biological Effects of Atmospheric Pollutants, *Fluorides*, National Academy of Sciences, Washington, DC, 1971.

Olszyk, D.M., Bytnerowicz, A., and Takemoto, B.K., Photochemical oxidant pollution and vegetation: Effects of gases, fog and particles, *Environ. Pollut.*, 61, 11, 1989.

Prokop, W.H., Odors, in *Air Pollution Engineering Manual*, 2nd ed., Davis, W.T., Ed., Air and Waste Management Association, Pittsburgh, PA, 2000, p. 136.

Schutze, E.D., Air pollution and forest decline in a spruce *(Picea abies)* forest, *Science*, 244, 776, 1989.
Skiffington, R.A., and Wilson, E.J., Excess nitrogen: Issues for consideration, *Environ. Pollut.*, 54, 159, 1988.
Smith, W.H., Ed., *Air Pollution and Forests: Interactions between Air Contaminants and Forest Ecosystems*, 2nd ed., Springer-Verlag, New York, 1990.
Tasdemiroglu, E., Costs of materials damage due to air pollution in the United States, *Renew. Energy*, 1, 639, 1991.
Taylor, G.E., Johnson, D.W., and Andersen, C.P., Air pollution and forest ecosystems: A regional to global perspective, *Ecol. Appl.*, 4, 662, 1994.
US EPA, *Air Quality Criteria for Ozone and Related Photochemical Oxidants*, EPA/600/AP-93/004AF-CF, EPA, Washington, DC, 1996.
US EPA, *Air Quality Criteria for Particulate Matter*, EPA/600/P-99/002C, EPA, Washington, DC, 1999.
US EPA, *Nitrogen Oxides: Impacts on Public Health and the Environment*, 1997. http://ttncaaal/t/ria.html.
Winner, W.E., Mooney, H.A., and Goldstein, R.A., Eds., *Sulfur Dioxide and Vegetation: Physiology, Ecology and Policy Issues*, Stanford University Press, Palo Alto, CA, 1985.
World Health Organization, *Nitrogen Oxides*, 2nd ed., Environmental Health Criteria 188, WHO, Geneva, Switzerland, 1997.

QUESTIONS

1. Describe major injury types caused by phytotoxic pollutants.
2. Indicate pollutants that cause the following symptoms in plants: (a) interveinal necrosis, (b) stipple and chlorotic mottle, (c) undersurface glazing.
3. What are the distinguishing characteristics of forest declines?
4. What are the likely causes of the following forest declines: (a) ponderosa pine in California, (b) red spruce in Vermont, (c) forest tree species in Germany?
5. How does acidic deposition affect forest trees?
6. In many developing countries, lead is smelted in facilities without any emission controls. What welfare concern(s) does this pose?
7. Describe adverse effects of lake acidification as pH progressively decreases from 6.5 to 5.0.
8. Acidic deposition has adverse effects on aquatic organisms as a result of the mobilization of metals. Describe how this occurs.
9. What are the potential adverse effects of nitrogen deposition on forest ecosystems?
10. Why is nitrate deposition in estuaries an important environmental concern?
11. What is biomagnification and what is its significance relative to mercury and PCB contamination of the environment?
12. What is the environmental significance of dioxin emissions from incinerators?
13. What are some of the potential ecological effects associated with stratospheric O_3 depletion?
14. Describe potential or existing effects of global warming on Arctic ecosystems and wildlife.
15. What effect would global warming have on the distribution of forest trees in North America?
16. Describe pollutant effects on metals and building stones.
17. Describe pollutant effects on paper, leather, and rubber.
18. Describe malodors as an air quality problem.
19. Considering the various welfare effects discussed in this chapter, what pollutant or pollutants have the potential to cause the most environmental harm?
20. In 2001, the US EPA began the process of developing limits on mercury emissions from coal-fired power plants. Why?

7 Air Quality and Emissions Assessment

The quality of air in our communities, countryside, and even remote locations changes from hour to hour, day to day, and over longer timescales. Concentrations of pollutants depend on the magnitude of emissions from individual sources, source emission density, topography, and state of the atmosphere.

Air quality can be defined qualitatively. It is poor when pollutants (1) cause a reduction in visibility, (2) soil building surfaces and damage materials, (3) damage crops and other plants, or (4) cause adverse health effects. It is deemed good when the sky appears clean and no adverse environmental effects are evident.

Qualitative assessments of air quality, although indicative of the nature of the atmosphere relative to pollution concerns, cannot be used to support regulatory programs designed to protect the environment. Air quality must therefore be characterized quantitatively. As such, it is defined in the context of concentrations of specific target pollutants and observed environmental and public health effects.

Protection of human health and the environment from adverse effects of pollutants is the primary goal of all air pollution control programs. Protection and enhancement of air quality require that good data on ambient concentrations of major pollutants and emissions from individual sources and groups of sources be available to regulatory authorities. Systematic efforts to monitor ambient pollutant concentrations both temporally and spatially, as well as to characterize and quantify emissions, are vital to the success of all air quality protection and pollution control programs. These efforts include ambient air quality monitoring, source emissions assessment, and modeling.

7.1 AIR QUALITY MONITORING

Pollutant levels in North America are assessed using systematically conducted, long-term ambient air quality monitoring efforts. In the United States, monitoring provides data to

1. Determine compliance with National Ambient Air Quality Standards (NAAQSs) for seven pollutant categories (criteria pollutants) in air quality control regions (AQCRs) or portions thereof
2. Determine long-term trends
3. Determine human exposures
4. Support the Air Quality Index (AQI) program
5. Support emission reduction programs
6. Determine effectiveness of emission control programs
7. Support environmental assessments such as visibility impairment and degradation of watersheds
8. Support research efforts designed to determine potential associations between pollutant levels and adverse health and environmental effects

Air quality monitoring efforts are important in identifying new pollutant-related environmental problems (e.g., mercury [Hg] deposition) and tracking changes in tropospheric and stratospheric concentrations of ozone (O_3)-destroying chemicals (as well as O_3 itself) and atmospheric concentrations of greenhouse gases such as carbon dioxide (CO_2), methane (CH_4), nitrous oxide (N_2O), and chlorofluorocarbons (CFCs).

DOI: 10.1201/9781003664109-7

7.1.1 Monitoring Considerations

Air quality monitoring involves numerous measurements of individual pollutants over time at several locations in organized, systematic programs. Concentration measurements are made from samples of small volumes of ambient air.

7.1.1.1 Sampling

Because of the vastness and dynamic nature of the atmosphere, it is not possible to determine the infinite number of concentration values typical of individual pollutants. As in other instances where a population of values is potentially large, pollutant concentrations must be inferred from a relatively limited number of samples. Consequently, they must be determined from fixed-site monitors whose locations reflect the objectives of air quality monitoring programs.

Concentrations are determined by collecting pollutants in or on a sampling medium or in automated continuous systems, where they are drawn through a sensing device. In manual methods, sampling and analysis are discrete events, with sample analysis occurring days to weeks after collection. In automated systems, sampling and analysis are simultaneous or near-simultaneous events and concentrations are measured in real time.

In manual methods, a sufficient amount of pollutant must be collected to meet the lower limit of detection (LOD) requirements of the analytical procedure used. This quantity will depend on the atmospheric concentration and sample size (the volume of air that passes through or comes in contact with the sampling medium). Sample size is a direct function of sampling rate and duration; samples are collected intermittently.

Sampling rates for individual pollutants using manual methods reflect collection efficiencies and instrument limitations. Collection efficiency is the ratio of pollutant collected to the actual quantity present in the air volume sampled. It depends on the sampling technique employed and on the chemical and physical properties of individual pollutants. For gases, collection efficiency decreases as flow rate increases above the optimum value. Optimum collection efficiencies are often achieved at relatively low sampling rates for gas-phase contaminants (< 1 L/min). Because of low flow rates or low collected pollutant mass, sampling duration may be extended to collect a sufficient quantity of pollutant for analysis. For some gas-phase substances, refrigeration or use of preservatives is necessary to minimize pollutant loss before samples are analyzed.

In continuous real-time sampling, the analytical technique is usually sensitive enough that the sample size or volume needed to accurately detect and quantify specific pollutants is relatively small (typically milliliters). As a consequence, flow rates used in such instruments are also relatively low (milliliters per minute), and optimal rates of flow are used to achieve the sensitivity and accuracy for which the instruments are designed.

7.1.1.2 Averaging Times

Collection and analytical limitations associated with manual, intermittent methodologies often require extended sampling durations; thus, concentration values are integrated or averaged over the sampling period. Sampling durations required to collect samples and the intended use of the data determine the averaging time. For example, 24-hour averages are appropriate for long-term trends.

The real-time air quality monitoring instruments used for more than four decades provide a continuous record of fluctuating concentrations. Although such data are deemed more useful than integrated data obtained from intermittent samples, real-time data are so voluminous that they are, for the most part, uninterpretable. Therefore, data from real-time monitoring instruments are integrated to provide hourly average concentrations or concentrations reflective of the needs of NAAQSs. For pollutants such as O_3, in which peak levels occur for a limited period, 8-hour averaging times are used. Differences in average concentrations associated with real-time data are illustrated in Figure 7.1.

Averaging times are specified for regulated air pollutants. For example, the averaging times for particulate matter with aerodynamic diameters of less than or equal to 2.5 μm ($PM_{2.5}$) are 24 hours

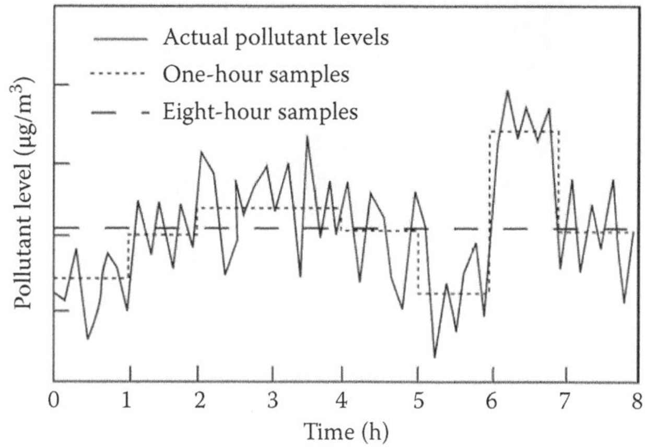

FIGURE 7.1 Pollutant concentrations associated with different averaging times applied to real-time data.

and 1 year, but for those with diameters of less than or equal to 10 μm (PM_{10}), it is 24 hours; for carbon monoxide (CO) it is 1 hour and 8 hours; for ozone, 8 hours; for sulfur dioxide (SO_2), 1 hour and 3 hours; for nitrogen dioxide (NO_2), 1 hour and 1 year; and for lead (Pb), 3 months. Concentrations averaged over the entire year may be based on a continuous record of measurements or, in the case of PM_{10}, NO_x, and Pb, a more limited number of samples.

7.1.1.3 Sampling Techniques

The principal objective of sampling is to collect a pollutant or pollutants for subsequent analysis or provide a sensing environment for real-time measurements. Both require a system whereby gases or particles are drawn to the surface of a collecting medium or into a sensing environment. These functions are accomplished by sampling trains that include a vacuum pump, flow regulator, and collecting device or sensing unit. Sampling trains for gases may also utilize filters to prevent particles from entering the collection unit, scrubbers for interfering gases, and the like.

Based on the type of information desired, as well as the collection and analytical limitations, sampling may be conducted by passive, intermittent, or continuous procedures. Such procedures provide air quality data representing a range of averaging times, from the instantaneous values of continuous systems to the 7-day average employed for some passive samplers.

Because of low cost and ease of use, passive sampling techniques were widely used in the early years of pollution control. These early techniques included the capture of particles on sticky paper, collection of settleable particles in dustfall jars or buckets, and measurement of sulfation rates on lead dioxide–impregnated gauze-covered cylinders (Pb candles). These techniques were relatively crude and abandoned with the advent of dynamic sampling methods.

New developments in passive sampling technology in the past three decades have made passive sampling a widely accepted and used methodology for personal exposure monitoring in occupational environments, measurement of indoor air contaminants such as formaldehyde (HCHO) and radon (Rn), and air quality and health research studies. Such sampling devices include small badge-type devices, tubes, canisters, and the like (Figure 7.2). Because of its low cost, passive sampling techniques would be appropriate for use in ambient monitoring activities of developing nations.

In grab sampling, air samples are collected in a relatively short time (seconds or minutes) using a variety of devices. Grab sampling is typically used in problem environments (e.g., chemical spills) as a means of identifying pollutants and their concentrations present at the time samples were collected. Grab sampling using evacuated canisters for collection and identification of volatile organic compounds (VOCs) is an integral part of Photochemical Assessment Monitoring Stations (PAMS) network activities.

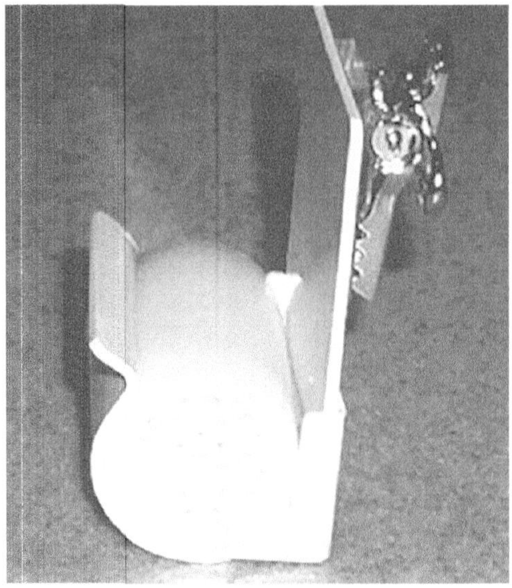

FIGURE 7.2 Passive ozone sampler. (Courtesy of Ogawa & Company, USA Inc.)

Intermittent sampling using dynamic sampling methods (i.e., manual methods) came into widespread use in air quality monitoring networks in the United States in the 1950s through the early 1970s, both for PM and gases such as SO_2 and nitrogen oxides $(NO)_x$. Intermittent sampling has the advantage of relatively low cost and provides reasonably useable data. It has the disadvantage of relatively long averaging times (typically 24 hours), historically viewed by regulators as undesirable for gas-phase pollutants. Intermittent sampling continues to be the sampling method of choice for PM_{10}, $PM_{2.5}$, Pb, and atmospheric deposition networks. Because of its relatively low cost, intermittent sampling is often used to supplement continuous monitoring systems. Intermittent sample collections at multiple locations may provide more useful data than a continuous record at a few locations. It is particularly useful where resources for pollutant monitoring are limited.

Continuous monitoring devices provide instantaneous or near-instantaneous sampling results. Because data are acquired in real time, continuous monitors are the systems of choice in air quality monitoring programs. Their major disadvantages are high capital equipment and system maintenance costs and a need to reduce data to a manageable form.

7.1.1.3.1 Gas-Phase Pollutants

Manual collection of gas-phase substances may be conducted by employing the principles of absorption, adsorption, condensation, or capture. Of these, absorption has been the most widely used. It involves drawing contaminated air through a liquid reagent. The contaminant is collected when it dissolves in the absorbing reagent contained in a bubbler or impinger.

Adsorption is a process wherein gas-phase substances are collected on a solid surface. In adsorption, gases and vapors are physically attracted to a solid that has a large surface area, such as activated carbon (C), silica gel, molecular sieve, and others. Collected gases may be desorbed from the collecting medium and subsequently analyzed. Adsorption sampling is common for a variety of organic gases. Because adsorbents are not very specific, a mixture of gases is collected, making analysis more involved. Gases may be collected by condensation or freezing. In such sampling, air is drawn through a collection vessel maintained at subambient temperatures. The low temperature causes vapors to condense or freeze and be retained in the collecting vessel. The sample must be removed and subsequently analyzed. Such cryogenic, or freeze-out, sampling has been used to condense and isolate a variety of organic gases.

In grab sampling, air is collected or captured for subsequent laboratory analysis using gas-tight syringes, gas sampling bags, or evacuated tubes, bottles, or canisters. Sample volumes vary from less than 1 mL using a gas syringe to 100 L in a gas sampling bag.

7.1.1.3.2 Particulate-Phase Pollutants

Particulate-phase pollutants may be collected by gravitational settling, filtration, inertial impaction, electrostatic precipitation, or thermostatic deposition. In some continuous sampling devices, they are drawn through a chamber that contains an optical sensor. For the most part, ambient air quality monitoring for PM is conducted using filtration or inertial impaction.

Filtration has been the primary PM collection technique used in air quality monitoring. To determine concentrations gravimetrically (i.e., by determining the weight of exposed and clean filters), large volume flow rates are used to collect sufficient particle mass. A typical sampling duration is 24 hours.

The standard sampling device for monitoring atmospheric PM concentrations in the United States until 1988 was the high-volume sampler (commonly referred to as a Hi-Vol). A glass fiber filter was located upstream of a heavy-duty vacuum cleaner–type motor operated at a rate of 1.13 to 1.17 m^3/min (40–60 cfm). The sampler was mounted in a protective shelter with the filter parallel to the ground. It efficiently collected suspended particles in the size range of 0.3 to 100 μm. Concentrations were reported as micrograms per cubic meter (μg/m^3).

The Hi-Vol collected particles in a wider size range than was appropriate for a health-based air quality standard. As such, its use was superseded by size-selective devices that collect particles that can easily enter the human respiratory system. Hi-Vol use in the United States is now limited to sampling airborne particles in conformance with Pb monitoring requirements.

Various devices are used to collect particles in distinct size ranges. These include those in which a size-selective inlet module is mounted atop a Hi-Vol–type device. The module removes particles that are more than 10 μm by impaction on a greased surface. Particles with cutoff diameters of less than or equal to 10 μm are then collected on a quartz glass fiber filter. A PM_{10} sampling device is illustrated in Figure 7.3.

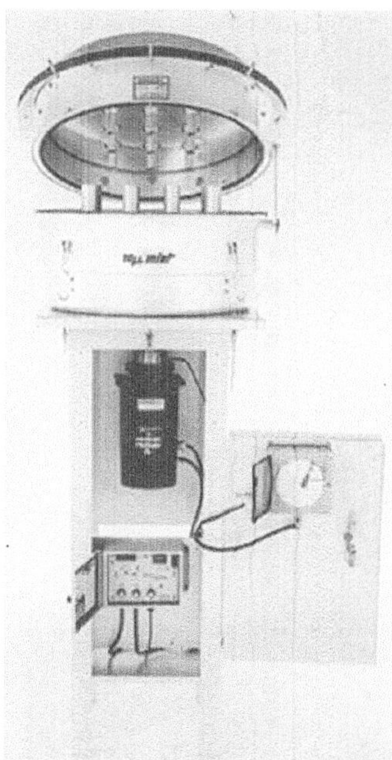

FIGURE 7.3 PM_{10} sampler with size-selective inlet. (Courtesy of Thermo Andersen.)

Cascade impactors are fractionating devices that collect particles on filter surfaces by impaction using orifice plates with smaller and smaller diameter holes. They are designed to collect particles using up to six fractionating stages. Air is drawn through orifice plates that deflect it from its flow path. Due to inertial forces, particles strike a filter surface where they are collected.

Dichotomous impactors are designed to collect particles in two size ranges: those with aerodynamic diameters less than 10 µm but greater than 2.5 µm, and those with diameters less than 2.5 µm. Both fractions are collected on individual membrane filters. Although dichotomous, or virtual, impactors operate like typical inertial units, large particles are impacted into a void rather than onto an impervious surface.

Various techniques are used to measure particle concentrations continuously in automated systems. These devices collect particles on filter tape, with concentrations determined hourly or integrated over 24 hours by β-particle attenuation or tapered element oscillating monitors (TEOMs). An automated dichotomous sampler can be seen in Figure 7.4.

TEOMs measure particle concentrations on a continuous basis. Particle mass collected on an exchangeable filter cartridge is determined from frequency changes in a tapered element. As particle mass collects on the exchangeable filter, the element's natural oscillation frequency changes. Because a direct relationship exists between the element's change in frequency and particle mass, this methodology provides PM measurements of high accuracy and precision.

7.1.2 Reporting Pollutant Concentrations

In the United States, concentrations of gas-phase substances in air are commonly expressed as mixing ratios and reported in parts per million by volume (ppmv) or whole number fractions thereof, such as parts per hundred million (pphmv), parts per billion (ppbv), or parts per trillion (pptv).

One part per million by volume is equal to a volume of gas mixed in a million parts of air. A microliter of a gas mixed in a liter of air would be equal to 1 ppmv:

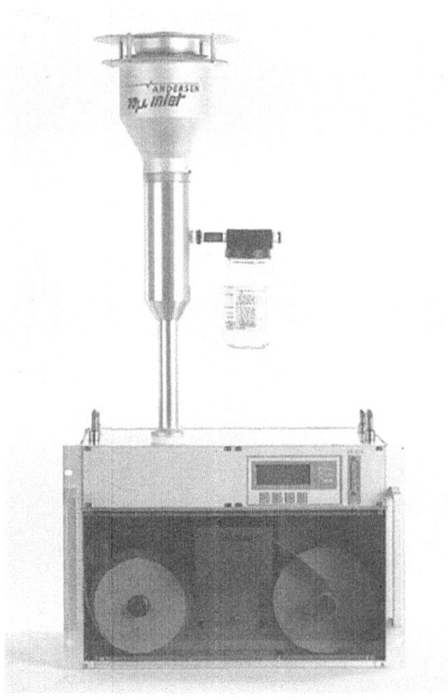

FIGURE 7.4 Automated dichotomous sampler. (Courtesy of Thermo Andersen.)

$$1\,\text{ppmv} = 1\,\mu\text{L gas} / 1\,\text{air} \tag{7.1}$$

Mixing ratios commonly used to express air concentrations must not be confused with those used for weight–volume mixtures (milligrams per liter) for water (H_2O) and weight–weight ratios (micrograms per gram, milligrams per kilogram) for solids. Although all are expressed as parts per million, they do not express equal quantities.

Concentrations of gas-phase substances mixed in air are, in many countries, expressed in metric units as mass per unit volume (micrograms per cubic meter [$\mu g/m^3$], milligrams per cubic meter [mg/m^3]), in conformance with international practice. However, air concentrations in metric-using countries are increasingly being expressed as mixing ratios. This reflects the fact that concentrations across the chemical spectrum can be more easily compared when mixing ratios are used; it also reflects the effect that American science has on the world community.

Concentrations can be converted from one form of expression to another by application of the gas laws. Assume that a manual instrument sampled SO_2 at an average flow rate of 0.2 L/min for 24 hours. Concentration is determined from the mass of SO_2 (in this case, 40 μg) collected in the sample solution using the following equation:

$$\text{ppmm} = \left[(m)(24.45)\right] / MV \tag{7.2}$$

where *m* is the mass of collected gas (μg), *V* is the sample volume (L), *M* is the molecular weight of the gas (e.g., $SO_2 = 64$), and 24.45 is the volume (in microliters) of a gas produced from a micromole of a substance at reference temperature and pressure (25°C and 760 mmHg).

Therefore,

$$\text{ppmv} = \left[(40\text{g})(24.45\text{L})\right] / \left[(64)(288\text{L})\right]$$
$$\text{ppmv} = 0.053$$

The ppmv concentration of a gas in air can be converted into its respective International System of Units (SI) concentration using Equation 7.3, derived from Equation 7.2:

$$\text{mg}/\text{m}^3 = (M / 24.45)\,\text{ppmv} \tag{7.3}$$

For purposes of illustration, let us convert the ambient air quality standard for O_3 (0.075 ppmv, 8-hour average) to its corresponding concentration in $\mu g/m^3$:

$$\text{mg}/\text{m}^3 = (48 / 24.45)(0.075)$$
$$= 0.147\text{mg}/\text{m}^3$$
$$= 147\text{g}/\text{m}^3$$

If the concentration of a gas in milligrams per cubic meter is known, it can be converted to parts per million by volume using Equation 7.4:

$$\text{ppmv} = (24.45 / M)\left(\text{mg}/\text{m}^3\right) \tag{7.4}$$

For purposes of illustration, let us convert the 1-hour standard for SO_2 expressed in SI units as 0.196 mg/m^3 to ppmv:

$$\text{ppmv} = (24.45 / 64)(0.196)$$
$$= 0.075\,\text{ppmv or } 75\,\text{ppbv}$$

Concentrations of gas-phase substances are commonly reported for reference conditions of temperature and pressure, 25°C and 760 mmHg. A measured volume of air may be standardized (V_s) with the following equation:

$$V_s = V_m\left[(P_1 / P_2)(T_2 / T_1)\right] \tag{7.5}$$

where V_m is the measured volume, P_1 is the field atmospheric pressure (mmHg), P_2 is the reference pressure (mmHg), T_1 is the field ambient temperature (K), and T_2 is the reference temperature (K).

Assuming that field sampling conditions were 22°C and 740 mmHg and V_m = 288 L, the standardized volume for the SO_2 sample above would be

$$V_s = 288\left[(740/760)(295/295)\right]$$
$$V_s = 283.3$$

This will change the concentration calculated from Equation 7.2 from 0.053 to 0.054 ppmv, an approximate 2% difference.

Particle concentrations in air quality monitoring programs are always reported as weight per unit volume, that is, in micrograms per cubic meter.

7.1.2.1 Accuracy, Precision, and Bias

Measured concentrations of pollutants must reliably reflect actual concentrations. Three concepts are important in determining such reliability: accuracy, precision, and bias.

Accuracy is the relative agreement between a measured value and an accepted reference or true value. Although 100% accuracy is desirable, such accuracy is difficult to achieve because all sampling and analysis techniques are subject to errors associated with the sampling instrument, analytical method, personal errors by those conducting measurements, and extending a measurement system beyond its limits.

Sampling and analytical techniques used to measure concentrations of the same chemical species vary in accuracy. As such, their utility and acceptability are determined by meeting an accuracy performance objective, typically within approximately 10% of the reference or true value.

Accuracy is determined and reported in the context of absolute or relative error. Absolute error is defined as:

$$E = O - A \tag{7.6}$$

where O is the observed value and A is the actual value.

If, in making a series of measurements of a known concentration (1 ppmv) of SO_2, the observed average value were 0.925 ppmv, then

$$E = 0.925 - 1.0$$
$$= -0.075$$

Absolute error is −0.075. The measured concentration in this case is less than the actual concentration. Consequently, the error is negative. Had the measured concentration been 1.075 ppmv, the error would have been positive. The relative error (E_r) determined as a percentage can be calculated from the following equation:

$$E_r = \left[(\mathrm{E})(100)\right] / \mathrm{ppmv} \tag{7.7}$$
$$= -7.5\%$$

Accuracy is commonly reported as a percentage of the true value. In this case, the sampling or analytical procedure has an accuracy of 92.5%. If the measured value were 1.075 ppmv, it would have an accuracy of 107.5%.

Error and accuracy values are based on average values determined from a series of measurements. In the preceding case, the mean of 10 measurements (0.850, 1.000, 0.950, 0.960, 0.870, 0.900, 0.880, 0.920, 0.940, and 0.990 ppmv) from a 1 ppmv reference concentration of SO_2 was 0.925 ppmv. Variation of measured values around the mean value describes the scientific/statistical concept of precision, a measure of the reproducibility of results. (Note: Dictionaries define precision as the exactness of a result, and as such, concepts of accuracy and precision are indistinguishable and are commonly used interchangeably.)

The precision of a sampling/analytical technique is determined from calculations of the mean of repeated measurements and its standard deviation. In a normally distributed population, approximately 67% of sample values are within approximately 1 standard deviation (σ) of the mean; 95% are within 2σ values.

Standard deviation can be calculated from the following equation:

$$\tilde{A} = \left[\sum_{i=1}^{n} \left(X_i = \bar{X} \right)^2 / N \right]^{0.5} \tag{7.8}$$

where $\bar{X}$ is the mean, $\bar{X}_i$ is the individual measured value, and N is the total number of measurements.

By using a programmable calculator or spreadsheet software, 1σ is determined to be 0.048. Therefore, the precision of our sampling results is more or less 0.048 ppmv. Precision, however, is usually reported as a percentage by calculating the coefficient of variation (CV), where $\bar{X}$ is the sample mean:

$$\begin{aligned} \text{CV} &= \left(\sigma / \bar{X} \right)(100) \\ &= 5.2\% \end{aligned} \tag{7.9}$$

Using approximately 1σ as a reference, the precision of our measurements was more or less 5.2%.

In this case, our sampling/analytical technique had high relative accuracy (92.5%) and good precision (low variability). Sample precision values within more or less 10% of the measured value are considered to be good.

Differences in the concepts of accuracy and precision can be seen in Figure 7.5. In the first target (a), values are scattered and do not come close to the target's center. In the second (b), all values are close to the middle of the target. In the former, accuracy is relatively low, with poor precision. In the latter, results are both accurate and precise.

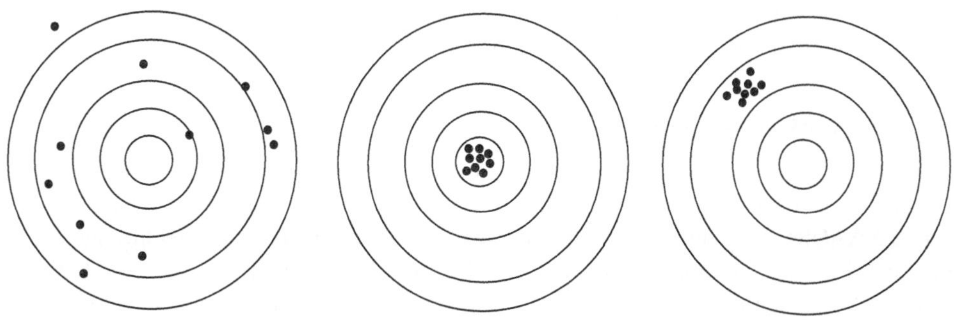

(a) (b) (c)

FIGURE 7.5 Graphical illustration of accuracy, precision, and bias.

The third target (c) illustrates the concept of bias. Results show good to excellent precision but deviate from the true value in a systematic way. Bias occurs as a result of calibration errors (the instrument may be out of calibration), but it may also be due to hardware problems associated with instrument electronics or systematic software problems.

Accuracy in the context of air quality monitoring activities includes a combination of random (precision) and systematic (bias) errors. The term *accuracy* is used when these two types of error cannot be separated. Therefore, accuracy values are stated in the same format as precision. Reported accuracy values include precision determinations. For most ambient air quality monitoring activities, accuracies (which take precision into account) within more or less 10% of the true value are considered acceptable. For passive sampling techniques, accuracy values within approximately 25% are considered acceptable.

7.1.2.2 Calibration

All sampling and monitoring instruments must be periodically calibrated to reduce bias-associated errors and ensure data reliability. Calibration is a process in which measured values are compared with standard reference values (in the case of pollutant concentrations) or volume airflows are measured using standard airflow measuring techniques and devices. When bias-type errors are identified, instrument adjustments are made to reduce observed errors, that is, bring the instrument back into calibration.

In some manual sampling methods, instrument calibration is limited to measuring airflow rates. Such calibration may be conducted using flow-measuring instruments based on primary or secondary standards. A primary standard is one that is directly traceable to the National Institute of Standards and Technology (NIST). A primary standard would be, for example, a bubble meter (gas burette and soap solution used for low flows) or volumetric flask. In the latter case, the volumetric flask may be used to calibrate wet or dry test meters that may be used to calibrate rotometers. Wet and dry test meters and rotometers are secondary standards. Because of their portability and ease of use, field instruments are commonly calibrated with secondary standard devices (such as rotometers). Based on US Environmental Protection Agency (US EPA) regulations, airflow rate measurements must be made using instruments traceable to an authoritative volume (e.g., NIST).

Monitoring instruments for gas-phase pollutants such as O_3, NO_2, CO, and SO_2 must be periodically calibrated to determine whether measured values are consistent with reference values. Concentration calibrations are made using gas standards traceable to a NIST reference material or a NIST-certified gas manufacturer's internal standard.

Standards for gases such as CO, SO_2, NO_2, and nitric oxide (NO) are available in gas cylinders or permeation tubes. Gas permeation tubes contain a sealed volume of calibration gas that passes through a membrane at a known rate. Permeation tubes can be inserted into a gas sampling stream at a constant flow rate to obtain standardized concentrations for calibration.

Due to its high reactivity, gas standards are not available for O_3 calibration. Ozone monitors must therefore be calibrated by producing test atmospheres using an O_3 generator and gas dilution system. Some O_3 monitors are also used to calibrate other O_3 monitors.

7.1.3 Federal Monitoring Requirements

Under regulations promulgated by the US EPA, ambient air quality monitoring programs must be developed and operated by states as part of control programs for assessing and determining compliance with NAAQSs. They must meet minimum design and quality assurance requirements.

State and local air monitoring stations (SLAMS) must have sites that monitor (1) highest pollutant concentrations, (2) representative concentrations in areas of high population density, (3) the effect of major emission sources, and (4) regional background concentrations. Recently, the US EPA added two new objectives for air quality monitoring efforts: determination of (1) extent of pollutant transport among populated areas, and (2) welfare-related effects in more rural and remote areas.

In siting air quality monitoring stations, states are required to consider spatial scales consistent with federal monitoring objectives. Spatial scales are estimates of the sizes of areas around monitoring locations that experience similar pollutant concentrations.

Spatial scale categories are (1) microscale, (2) middle scale, (3) neighborhood scale, (4) urban scale, and (5) regional scale. The microscale ranges from a few meters to 100 m (~300 ft), the middle scale from 100 m to 0.5 km (~0.3 mi), the neighborhood scale from 0.5 to 4.0 km (~2.5 mi), the urban scale from 4 to 50 km (~30 mi), and the regional scale from tens to hundreds of kilometers.

Spatial scales for the highest concentration or source impact are micro, middle, neighborhood, and, less frequently, urban scales. Appropriate spatial scales for high population densities and representative concentrations are middle, neighborhood, and urban. Neighborhood or regional scales are appropriate for background levels. Urban and regional scales are also appropriate for the determination of pollutant transport and welfare-related effects in rural remote areas.

States that have AQCRs that exceed air quality standards for O_3 must also establish a PAMS network in conformance with federal requirements.

All federally mandated air quality monitoring programs must use monitoring techniques based on federal reference methods (FRMs) or approved equivalents, meet minimum quality assurance requirements, and report air monitoring results to the US EPA annually.

7.1.3.1 Air Quality Monitoring Networks

Air quality monitoring networks are systems of instruments that measure various ambient air pollution concentrations and meteorological data at fixed intervals across a broad geographic area and that are used to study trends in air pollution impacts and emission, transport, and deposition. Several networks make up the US EPA Air Quality System (AQS).

Operation of SLAMS and PAMS networks is supported in part by federal grant monies. They are operated in conjunction with federal monitoring activities that include National Air Monitoring Stations (NAMS). The NAMS program is an urban-area, long-term air monitoring network that provides a systematic, consistent database for air quality comparisons and trends analysis. The NAMS are, in most cases, similar to SLAMS. The SLAMS program, however, has more flexibility in siting and operation to reflect state, local, and tribal needs. In many cases, NAMS are operated by SLAMS programs under a US EPA contract.

The NAMS and SLAMS data, as well as those of other sites (e.g., industrial monitors), are reported annually to the US EPA's Aerometric Information Retrieval System, where data are compiled and published.

The PAMS network is designed to monitor O_3 and photochemical air pollutants in areas with persistently high ground-level O_3 concentrations. Sites are located along the east coast from Portsmouth, NH, to Washington, DC, and in California from San Diego to Sacramento. They are also in Phoenix, AZ; El Paso, Houston, and Dallas–Ft. Worth, TX; Baton Rouge, LA; Atlanta, GA; and the Lake Michigan lakefront from Milwaukee to northwest Indiana. The PAMS sites monitor NO_x, total nonmethane organic compounds, target VOCs, and upper air meteorology during the summer months.

The Interagency Monitoring of Protected Visual Environments (IMPROVE) network was set up in the mid-1990s. It represents a cooperative effort between the US EPA and other federal resource management agencies such as the National Park Service. Its focus is monitoring $PM_{2.5}$ and visibility and assessing the relationship between them. IMPROVE consists of 110 federally maintained monitoring stations and several state and organization sponsored stations which monitor concentrations and compositions of fine and coarse particulate matter. 24-hour samples are collected every three days.

Other federal monitoring networks are associated with intensive research-oriented studies and atmospheric deposition. Included in the former are the North American Research Strategy for Tropospheric Ozone, $PM_{2.5}$ supersites, and PM health centers.

The Clean Air Status and Trends Network (CASTNET) is this country's primary atmospheric data source for rural O_3 levels and dry atmospheric deposition. It includes 79 monitoring stations operated by the US EPA or the National Park Service, which reports weekly average concentrations of sulfate SO_4^{2-}, nitrate NO_3^-, ammonium NH_4^+, SO_2, and nitric acid (HNO_3), as well as hourly O_3 concentrations. The National Oceanic and Atmospheric Administration operates a smaller dry deposition network called the Atmospheric Integrated Monitoring Network, which focuses on research questions related to atmospheric deposition.

Other atmospheric deposition networks include the Integrated Atmospheric Deposition Network (IADN) and the National Atmospheric Deposition Program/National Trends Network (NADP/NTN). The NADP began in 1978 as part of the National Acid Precipitation Assessment Program in an effort to assess the problem of atmospheric deposition and its effects on aquatic and terrestrial ecosystems, as well as agricultural crops. It has been a cooperative program between federal and state agencies, universities, and utilities and other industries to measure concentrations of hydrogen ion (H^+), NH_4^+, SO_4^{2-}, NO_3^-, chloride (Cl^-), calcium (Ca^{2+}), magnesium (Mg^{2+}), and potassium (K^+) in wet deposition. Its mission has been expanded to include Hg. At the time of this writing, the Mercury Deposition Network (MDN) is located at 40 NADP sites in 16 states and two Canadian provinces. Its focus is to measure Hg levels in wet deposition to assess seasonal and annual fluxes into sensitive ecosystems.

The IADN network was established in 1990 as a joint effort between the United States and Canada to conduct air and precipitation monitoring in the Great Lakes Basin. It measures target substances in wet and dry deposition and gas-phase organic vapors. Of particular concern are semivolatile organic compounds such as polychlorinated biphenyls (PCBs), dioxins, polycyclic aromatic hydrocarbons (PAHs), and a variety of organochlorine pesticides. In the year 2000, the network included one master station on each of the Great Lakes plus 14 satellite stations.

The SLAMS program, and less so the NAMS program, has been undergoing a significant reevaluation and reports to the AQS. Considerable progress has been made in controlling emissions of Pb, SO_2, and CO. In many AQCRs, levels of these pollutants as well as NO_x are well below NAAQSs and have been for a long time. As such, opportunities exist to discontinue monitoring activities for these pollutants at selected sites to redirect resources for evolving pollutant concerns (e.g., $PM_{2.5}$ and air toxics).

7.1.3.2 Federal Reference Methods

All air quality monitoring activities conducted at NAMS, SLAMS, and PAMS sites for criteria pollutants (those regulated under NAAQSs) must use methodologies approved by the US EPA as reference methods or their equivalent. All reference and equivalent methods are subject to intensive performance evaluations before approval.

Federal reference methods for the seven criteria pollutants are summarized in Table 7.1. Although the FRM has primary regulatory status, it is not uncommon for one or more equivalent methods to be more widely used in day-to-day monitoring activities.

TABLE 7.1
Federal Reference Methods for Criteria Pollutants

Pollutant	Reference Method
Sulfur dioxide	Spectrophotometry (pararosaniline method)
Nitrogen dioxide	Gas-phase chemiluminescence
Carbon monoxide	NDIR photometry
Ozone	Chemiluminescence
PM_{10}	Performance-approved product
$PM_{2.5}$	Performance-approved product
Lead	TSP sampling; analysis by wet chemistry, atomic absorption spectrometry

Note: FID, flame ionization detection.

7.1.3.2.1 Sulfur Dioxide

The FRM for SO_2 is the spectrophotometric pararosaniline method. In its manual application, SO_2 is collected by bubbling air through a potassium tetrachloromercurate solution to form a $HgCl_2SO_3^{2-}$ complex. Subsequent reactions with HCHO and colorless pararosaniline hydrochloride produce a red–violet product that can be measured spectrophotometrically and converted to an SO_2 concentration using a standardized concentration curve.

Manual collection and analysis is both cumbersome and limiting. As a consequence, the FRM is more often used in automated continuous monitoring devices. Even as an automated procedure, such monitors require considerable maintenance. As such, many air quality monitoring stations use a US EPA-approved equivalent method.

In the year 2000, a multigas long-path monitoring system was approved as an automated equivalent method for SO_2. This device can provide remote sensing data over an open path (total path length up to 1 km) using a Fourier transform infrared (FT-IR) spectrometer. In these systems, a source generates a beam of infrared energy across the air medium to be measured by a retroreflector (Figure 7.6) in a monostatic system or a detector at the other end of the path in a bistatic system. Infrared energy absorbed at specific wavelengths can be directly related to the total concentration of a target gas (e.g., SO_2) along the path length.

Other equivalent SO_2 monitoring methods include conductimetry (collection of SO_2 in a hydrogen peroxide [H_2O_2] solution and measurement of increased conductance of sulfuric acid [H_2SO_4] formed), amperometry or electrochemistry (measurement of an electric current produced on SO_2's reaction with the collecting solution), ultraviolet (UV) fluorescence detection, flame photometry, and gas chromatography.

7.1.3.2.2 Nitrogen Oxides

The FRM for measurement of NO_x is gas-phase chemiluminescence. Chemiluminescence occurs as a result of the emission of light from electronically excited chemical species formed

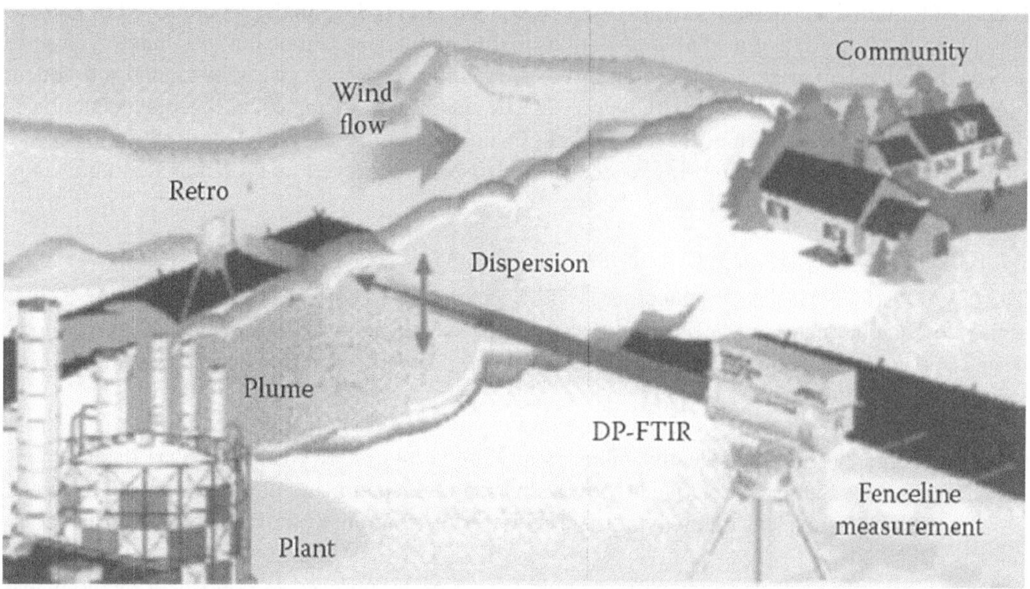

FIGURE 7.6 Schematic of FT-IR spectrophotometer use in the field. (Courtesy of US EPA.)

in chemical reactions. The reaction of O_3 with NO produces an electronically excited NO_2 molecule:

$$NO + O_3 \rightarrow NO_2 \cdot + O_2 \tag{7.10}$$

When $NO_2\cdot$ returns to its ground state, it emits light in the 600 to 3,000 nm range. Emitted light is amplified by a photomultiplier tube; the intensity of emitted light is proportional to the NO concentration.

Measurement of NO_2 is based on its conversion to NO and subsequent detection and measurement by chemiluminescence. Such analyses are subject to interference by other nitrogen (N)-containing compounds (e.g., peroxyacyl nitrate), resulting in higher measured NO_2 concentrations than may actually be present.

The manual equivalent method (including an automated version) uses a sodium arsenite reagent to collect NO_2, which, upon analysis, forms a colored solution that is measured spectrophotometrically.

The open-path FT-IR long-path monitoring system has also been approved for use as an automated reference method for NO_2.

7.1.3.2.3 Carbon Monoxide

The FRM for CO is nondispersive infrared (NDIR) spectrometry. Measurement is based on the ability of CO to strongly absorb infrared energy at certain wavelengths. NDIR spectrophotometric devices utilize two cylindrical cells (usually 100 cm [~39 in.] long), a sample and a reference cell, in which a source emits CO-absorbing infrared wavelengths. As CO passes through the sample cell, it absorbs infrared energy in direct proportion to its concentration. Differences in infrared energy (or heating) in the two cells cause a slight movement in a diaphragm, which is detected and recorded. Concentrations can be measured with a relative accuracy of approximately 5% in the optimum range (0–150 ppmv).

7.1.3.2.4 Ozone

The FRM for O_3 is chemiluminescence. It is based on the measurement of light emissions produced upon the reaction of O_3 with ethylene (C_2H_4). Because of C_2H_4's flammability, the FRM was amended in 1979 to replace C_2H_4 with Rhodamine B dye embedded in a disk. Because Rhodamine B does not attain a stable baseline rapidly after exposure to O_3, UV analyzers approved as equivalent methods are used in almost all ambient air quality monitoring stations in the United States.

The measurement principle of UV analyzers is based on absorption of UV by O_3 and subsequent use of photometry to measure the reduction of UV energy at a detector at 254 nm; reduction is dependent on the path length of the sample cell, O_3 concentration, and wavelength of UV light.

The FT-IR long-path monitoring system has also been approved as an automated equivalent method for ambient O_3 monitoring.

7.1.3.2.5 Particulate Matter: PM_{10} and $PM_{2.5}$

The NAAQSs for PM include standards for PM_{10} and $PM_{2.5}$. The FRMs for PM_{10} and $PM_{2.5}$ are defined by a combination of design and performance specifications. Consequently, manufacturers design and fabricate samplers to meet reference method requirements. Each approved sampling instrument can be designated as an FRM.

Manual reference samplers have included instruments that (1) collect PM on a filter medium or collecting surface using size-selective inlets on modified types of Hi-Vols, and (2) are dichotomous virtual impactors. Samples are collected over a period of 24 hours and concentrations determined gravimetrically. Equivalent methods must provide results that are comparable in performance to reference methods.

Several automated commercial PM_{10} samplers have been approved as equivalent methods and will likely be approved as equivalent methods for $PM_{2.5}$. These devices collect particles on filter tape, with concentrations determined hourly or integrated over 24 hours by β-particle attenuation or TEOMs.

7.1.3.2.6 Lead

Both the FRM and equivalent methods for Pb require manual collection of total suspended particulate (TSP) mass by a Hi-Vol particulate collector, operated at a flow rate of 1.12 to 1.70 m^3/min (40–60 cfm), on glass fiber filters for 24 hours. Filter samples are collected, dried, weighed, and analyzed for Pb. Lead analysis reference methods are based on wet chemistry and spectrophotometry.

Commonly used equivalent analytical techniques include flame atomic absorption spectrometry, graphite furnace atomic absorption spectrometry, inductively coupled plasma emission spectrometry, and energy-dispersive x-ray fluorescence spectrometry.

7.1.3.3 Quality Assurance Programs

All environmental monitoring and measurement programs must have a quality assurance and quality control (QA/QC) program in place that provides for the acquisition of both valid and reliable air quality monitoring data. All data submitted to the US EPA under SLAMS must meet minimum QA/QC program requirements.

Quality assurance and quality control are distinct and interrelated quality system functions. Quality assurance is a program that sets policy and oversees management controls that include planning, review of data collection activities, and data use. It also involves setting data quality objectives, assigning responsibilities, conducting reviews, and implementing corrective actions.

Quality control, on the other hand, focuses on the technical aspects of data quality programs. It includes the implementation of specific QC procedures such as calibrations, checks, replicate samples, routine self-assessment, and audits.

Under federal rules, QA/QC efforts must be documented so that users can evaluate data quality before any conclusions are drawn from collected data. Under SLAMS programs, federally required quality system-monitoring efforts must meet a well-defined need; take cost into consideration; and implement a QA program whose policies, procedures, standards, and documentation provide data of acceptable quality with minimal data loss.

All monitoring methods used in SLAMS must be assessed periodically to quantitatively determine accuracy, precision, and bias independent of routine instrument calibrations for airflow and concentration determinations. These technical system audits involve on-site inspection of SLAMS programs to assess compliance with requirements that govern collection, analysis, validation, and reporting of air quality data. Audits are typically conducted at specified frequencies to establish the consistency of data collection activities and comparability of data across a monitoring network.

7.1.3.4 Data Summarization and Characterization

Air quality monitoring programs generate an enormous number of individual pollutant concentration values. After collection, data must be summarized or treated in such a way that their regulatory and environmental significance can be evaluated.

Data summarization and characterization efforts are designed to reduce a large number of measured values to meaningful summary tables, graphical representations, or statistics. Air quality monitoring data are commonly summarized using graphical techniques and statistics.

7.1.3.4.1 Graphical Techniques and Displays

Air quality monitoring data are continuous (i.e., potentially include an infinite number of values), even if they have been integrated over a particular averaging time. Continuous data are often summarized using frequency distributions.

Frequency distributions are prepared by identifying an appropriate number of frequency classes or class intervals. A frequency class is a range of data values that represent a part of the collected data. In standard statistical practice, data are summarized in as few as 10 to as many as 20 or more frequency classes. Frequency class data are graphically presented as histograms. A histogram for TSP data is shown in Figure 7.7. Note that the bar indicating frequency for each class interval touches the adjoining bar; this indicates that data are continuous. Histograms should not be confused with the bar graphs that are used to summarize discreet data. (Note that popular spreadsheet and data manipulation software such as Excel commonly produce bar graphs under the histogram function.)

The histogram is an important tool in evaluating how data are distributed. Data that are normally distributed will be seen as a bell-shaped curve, with an equal number of values to the left and right of center of the data distribution. The symmetry of the distribution is described by its skewness. A normal distribution has a skewness of 0. In Figure 7.7, data appear skewed to the right (positive skewness values). As such, the data are not normally distributed. With some data sets, the frequency curve will be skewed to the left (negative skewness values).

The form of the frequency distribution curve must be known so that appropriate statistical procedures are used in data analysis. If data are not normally distributed, they cannot, in theory, be analyzed by application of parametric statistics (the most widely used statistical procedures). Such procedures can be applied if data, upon transformation, fit a normal distribution. Not uncommonly, environmental data are log normally distributed (after log transformation and plotting on semilogarithmic graph paper).

Histograms have been widely used to summarize continuous data (including environmental data). Increasingly, stem and leaf plots and box plots are used to summarize pollution data. Stem and leaf diagrams and box plots show the range of the data (with minima and maxima), as well as medians and quartiles. The median locates the 50th percentile and sorts data into halves. Quartiles divide each half into quarters. The upper quartile locates the 75th percentile, and the lower the 25th percentile.

Frequency distribution data can be summarized using stem and leaf plots. In such distributions, actual values truncated to two or three digits are plotted, with the first value the stem and succeeding

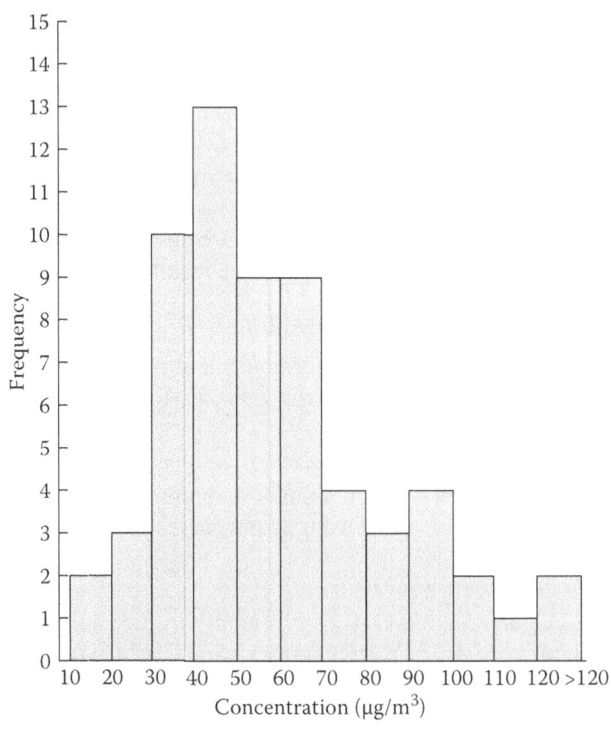

FIGURE 7.7 Histogram plot of TSP data.

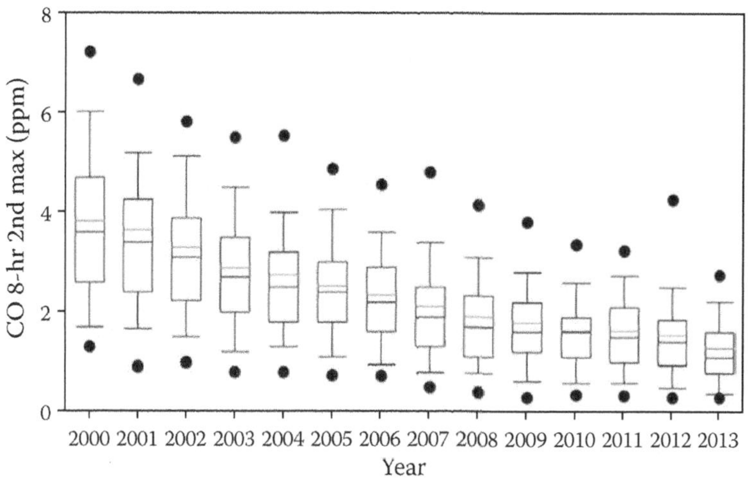

FIGURE 7.8 Box plot of the national trends in second 8-hour average CO concentrations over the United States from 2000 to 2013. (Calculated from US EPA Air Quality Statistics Report for the core based statistical areas [CBSA], 2013.)

numbers the leaves. Stems and leaves are separated by a vertical channel. Numbers in the channel indicate upper and lower quartiles.

A box plot of national trends in the second 8-hour average CO concentrations over the United States from 2000 to 2013 can be seen in Figure 7.8. The black line in the box indicates the location of the 50th percentile, or median. Each box stretches between the 25th and 75th percentiles. Whiskers stretch out from the rest of the box to span the remainder of the data except for outliers. The upper adjacent represents the largest value below the upper inner fence and the lower adjacent is the smallest value above the lower inner fence. Data lying outside the probability distribution described by the box plot are called outliers, which are shown as black solid points in the figure.

7.1.3.4.2 Statistical Summarization

Data can be summarized using measures of central tendency such as mean, median, and mode. In normally distributed data, the mean lies in the center, corresponding to the peak of the normal distribution curve. It is calculated simply by summing all data values and dividing the sum by the total number of values (N).

The median is the middle number when data are arranged from the lowest to the highest values. It is the average of the two middle values when N is an even number. The mode is the most commonly observed value in the data set. If data are normally distributed, the mean, median, and mode are all the same value.

The latest NAAQS for 8-hour average carbon monoxide is 9 ppm. Because the values starting from 2000 have already decreased to less than 9 ppm, the standard is therefore not plotted here.

The median is the appropriate measure of central tendency when data sets are not normally distributed. In such cases, the arithmetic mean does not accurately reflect where most of the values lie in a frequency distribution. An example is the use of the median to report family income by the Bureau of Labor Statistics. Median income is reported because approximately 5% of Americans earn 40% of the total national income. If the mean were reported, 70% or so of American households would appear to have below-average incomes.

In the TSP data summarized in Figure 7.7, median and mean values were 48.8 and 59.9 μg/m^3, respectively. In this case, a relatively few high values skewed the mean 10 μg/m^3 higher than the median.

Although the median is commonly used to describe central values in data sets that are not normally distributed, in most instances, it cannot be used to calculate the data variation required for statistical analyses. In such cases, it is usually desirable to log transform data and calculate the geometric mean (GM). The GM can be calculated using the following equation:

$$GM = \text{antilog}\left(\log x_1 + \log x_2 \cdots + \log x_n\right) / N \tag{7.11}$$

Applying such calculation to the values used in the histogram in Figure 7.7, the GM is found to be 50.8 μg/m^3, a value relatively close to the median.

The σ describes the distribution of data around the mean. It can be determined by using Equation 7.8. Sixty-seven percent of values in a data set lie within approximately 1σ of the mean. The larger the σ, the more variable the data; the smaller, the less variable.

Air quality standards are based on arithmetic means. When PM concentrations were based on TSP measurements, the TSP standard was based on the GM.

Environmental monitoring data are typically reported as arithmetic numbers. In some cases, for example, pH values in monitoring acidic deposition and decibel values in monitoring noise (see Chapter 12), data are in geometric form. In both cases, it is inappropriate to sum values and divide by N to calculate the mean.

Because pH is the negative log of H^+ concentration, pH values must be converted to their equivalent molar concentrations. In monitoring acidic deposition weekly over 6 weeks, the following pH values were reported: 4.5, 5.0, 3.8, 4.0, 4.2, and 4.8. Their corresponding molar H^+ concentrations, respectively, are $1 \times 10^{-4.5}$, 1×10^{-5}, $1 \times 10^{-3.8}$, 1×10^{-4}, $1 \times 10^{-4.2}$, and $1 \times 10^{-4.8}$ M. Summing and dividing equivalent molar concentrations by N, the average molar concentration is 0.000061 M or $1 \times 10^{-4.20}$. By taking the negative log of this value, the average pH is 4.20. Had each of the pH values been summed and divided by N, the average would have been 4.38.

Although these values do not seem to be significantly different from each other, in fact, they are. The actual difference can be determined by dividing the molar concentration of pH 4.20 (0.000061) by the molar concentration of pH 4.38 (0.000042; the lower H^+ concentration):

$$0.000061M / 0.000042M = 1.45$$

A pH value of 4.20 is seen to have an H^+ concentration 1.45 times greater than a pH of 4.38. Similarly, if our average pH value of measured precipitation samples were compared with the theoretical clean air value of 5.65, it would be 28 times more acidic.

7.1.3.5 Air Quality Index

Results of air quality monitoring can be used to provide guidance to public health and environmental officials as well as individuals relative to community air pollution exposures. For this purpose, the US EPA has developed and implemented an AQI program to provide timely and easy-to-understand information on local air quality and related health concerns.

The AQI is used to report daily air quality in metropolitan areas for five pollutants: O_3, PM (PM_{10} and $PM_{2.5}$), CO, SO_2, and NO_2. Since CO, SO_2, and NO_2 have fallen below their standards and are no longer included in AQI, the current AQI are mainly based on ozone and particle pollution. It uses a scaled approach according to the six levels of health concern in Table 7.2. The index varies from 0 to above 300. An AQI value of 100 corresponds to the air quality standard for a single pollutant. AQI values at less than 100 indicate good to moderate air quality, whereas values of 101 to 500 indicate increased public health risk; values of more than 300 are considered hazardous. Sensitive groups in the third AQI category (Table 7.2) include children and adults with respiratory disease, adults with cardiovascular disease, and children and adults who are active outdoors.

The AQI is calculated for each of the five pollutants in an AQCR. The reported value is based on the pollutant with the highest concentration relative to its standard. An AQI value can also be

TABLE 7.2
Air Quality Index Values, Levels of Concern, and Color Code

AQI Values	Level of Health Concern	Color Code
0–50	Good	Green
51–100	Moderate	Yellow
101–150	Unhealthy for sensitive groups	Orange
151–200	Unhealthy	Red
201–300	Very unhealthy	Purple
301+	Hazardous	Maroon

Source: US EPA, *Air Quality Index (AQI) Basics*. https://www.airnow.gov/aqi/aqi-basics/.

TABLE 7.3
Air Quality Index Values for Fine Particle Pollution, micrograms per cubic meter (μg/m³) in 2024

AQI Values	Previous AQI Breakpoints	Updated AQI Breakpoint	What Changes
0–50 (Good)	0.0–12.0	0.0–9.0	Due to new $PM_{2.5}$ updated annual standard of 9 μg/m³
51–100 (Moderate)	12.1–35.4	9.1–35.4	Due to new PM2.5 updated annual standard of 9 μg/m³
101–150 (Unhealthy for Sensitive Groups)	35.5–55.4	35.5–55.4	No change, because 24-hour standard of 35 μg/m³
151–200 (Unhealthy)	55.5–150.4	55.5–125.4	EPA updated the upper-end breakpoints for the Unhealthy to Hazardous categories based on new scientific evidence, and combined the two Hazardous category ranges into one.
201–300 (Very Unhealthy)	150.5–250.4	125.5–225.4	Same as above.
301+ (Hazardous)	250.5–350.4 and 350.5–500	225.5 +	Same as above.

Source: US EPA, Final Updates to the *Air Quality Index (AQI) for Particulate Matter*. 2024. https://www.epa.gov/system/files/documents/2024-02/pm-naaqs-air-quality-index-fact-sheet.pdf.

reported for a single pollutant, for example, O_3, when this is the primary pollutant of concern. An AQI value of 100 for O_3 corresponds to an 8-hour average concentration of 0.070 ppmv. For PM_{10}, it corresponds to a 24-hour average concentration of 150 μg/m³; for $PM_{2.5}$, a 24-hour average concentration of 35 μg/m³. Air quality indices of 100 for CO and SO_2 are based on an 8-hour average concentration of 9 ppmv and a 1-hour average concentration of 75 ppbv, respectively. Because no health effects are known to occur below an AQI of 200 for NO_2, an AQI is not calculated below 201. An NO_2 AQI of 201 corresponds to an NO_2 level of 0.65 ppmv averaged over 24 hours. AQI for fine particle pollution has been revised in Table 7.3 based on the new annual average standard of 9 μg/m³.

Relatively good air quality (i.e., reported AQIs of < 100) is common in many metropolitan areas in the United States. However, values between 101 and 150 are reported in many communities several times a year, particularly the summertime high-O_3 season. AQI values for O_3 of more than 101 commonly occur in the Los Angeles Basin and Houston, TX, where smog is a significant problem.

Values that are higher than 300 in the United States are extremely rare. They are likely to be common, however (if the AQI were calculated), in many other countries such as China since 2012, the European Union since 2006, India since 2014, Mexico since 2006, the United Kingdom since 1998, on different scales. The Air Quality Health Index is used in Canada and Hong Kong to indicate the impact of air pollution on human health.

The AQI for PM_{10} and $PM_{2.5}$ is based on manual sampling methods that use a 24-hour averaging time. Therefore, the AQI for PM is of limited usefulness. As a consequence, efforts have been made by the US EPA and others to utilize automated equivalent PM_{10} and $PM_{2.5}$ sampling methods to provide hourly values that are strongly correlated (using regression models) with 24-hour values.

7.2 EMISSIONS ASSESSMENT

Assessment of emissions of regulated pollutants from stationary and mobile sources is a vital part of federal, state, and local air pollution control efforts. Emissions data are needed to (1) identify sources that are primarily responsible for ambient pollutant concentrations in a region, (2) formulate emissions control policy, (3) comply with permitting requirements, (4) comply with emission reporting requirements for regulated target sources and pollutants, (5) compile national annual emission inventories, and (6) provide a database for ambient air quality modeling.

Emissions assessments of individual sources are conducted by sampling stack gases using manual one-time sampling procedures (stack testing) or continuous emission monitors (CEMs). Such assessments utilize the US EPA reference methods. Although emissions assessment by stack sampling is preferred by air pollution control agencies, most sources of regulated pollutants in the United States have never been required or volunteered to conduct stack testing to determine emissions under real-world conditions.

Most emission assessments conducted to meet regulatory reporting requirements have been based on emission factors (see Section 7.2.3).

7.2.1 Source Sampling and Monitoring

Source sampling is typically conducted on waste gas streams that flow to the atmosphere through stacks that elevate emissions above the source. Such stacks may vary from a meter (~3 ft) in diameter and 10 to 20 m (33–66 ft) in height to 10 m in diameter and 300 to 400 m (~990 to 1,300 ft) in height. For small industrial sources, source sampling can also be conducted on local exhaust ventilation systems.

Manual stack or source sampling is typically conducted by a team of trained technicians using the US EPA reference sampling and analytical methods. Because of the height of some stacks and the height at which representative waste gas flow occurs, sampling teams must often construct scaffolding to collect representative samples. Sample collection in many cases will require several days to a week of on-site activity.

Manual source sampling consists of introducing a probe into the waste gas stream to draw a sample volume into a sampling train. Sampling trains differ somewhat for individual pollutants and sampling methodologies; a sampling train for US EPA Method 5 for PM is illustrated in Figure 7.9. It includes a (1) probe inserted into the stack, (2) temperature sensor, (3) pitot tube, and (4) two-module sampling unit, which includes, in series, a heated compartment, an ice-bath compartment, a series of collecting impingers, and a vacuum pump, dry gas meter, and flow-regulating orifice. This sampling train can be easily modified to sample sulfur oxides (SO_x) and H_2SO_4 mist.

Because particles are subject to inertial forces, samples must be collected at the same rate of flow as stack gases. Such sampling is described as isokinetic.

Stack sampling is conducted in airflows and temperatures that differ considerably from ambient conditions, particularly if the waste stream is produced as a result of combustion or other high-temperature processes. Stack gas temperature and pressure must be measured to standardize the sample

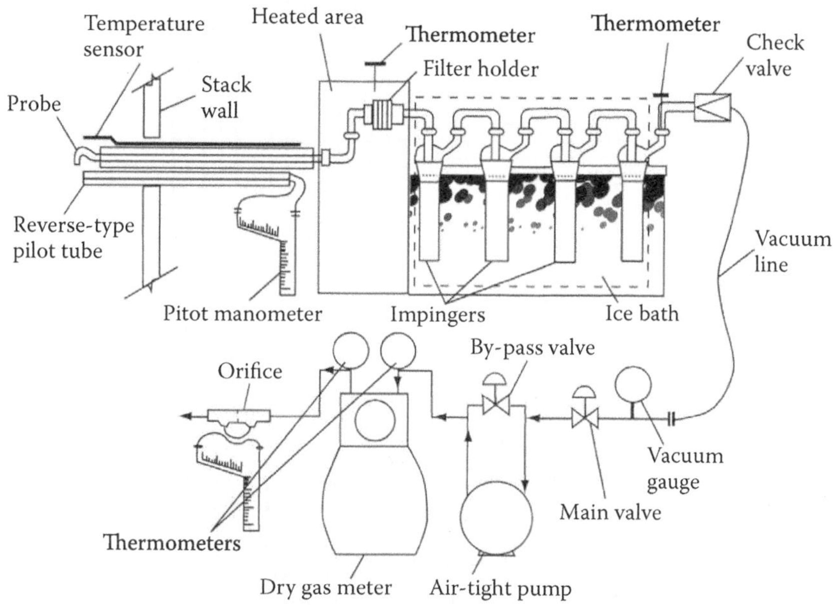

FIGURE 7.9 US EPA reference method for PM stack sampling. (From CFR 40, Part 60, Appendix 5, Method 5, p. 625, July 1, 1989.)

volume to reference conditions (i.e., 20°C, 760 mmHg). Gas velocities determined from pressure measurements using a pitot tube are used to calculate stack gas flow rates. A dry test meter is used to determine flow rates (uncorrected to reference conditions) through the sampling train.

A number of utilities and industrial sources are required to measure stack emissions continuously. The CEM systems must meet US EPA performance specifications for the measurement of SO_2, opacity, CO_2, total reduced sulfur (S), and hydrogen sulfide (H_2S). These systems may be *in situ* or extractive. *In situ* systems are, for the most part, no longer used because they cannot be calibrated using standard calibration gases. Extractive systems are configured for wet-based or dry-based measurements. Dry extractive systems deliver samples of stack gases to an analyzer cabinet through heated lines. Particles are collected at the sampling probe on a filter. Moisture is removed by a combination of refrigeration and condensation. This keeps the analyzer dry and eliminates interference that may be caused by water vapor. Moisture content is determined to convert the volumetric concentration into a mass emission rate.

Most CEM systems utilize dilution extraction. By diluting stack gas samples with clean, dry, instrument air, these systems eliminate the need for heating and conditioning extracted samples. A dilution extraction system allows measurement of the entire sample, including H_2O vapor.

7.2.2 Opacity

New source performance standards (NSPSs; see Chapter 8) for a number of source categories regulate plume opacity. Plume opacity regulations for PM emission sources have been in place at state and local levels for decades.

Opacity standards for particles require a different form of emission assessment than gas-phase pollutants or conventional emission standards for PM. Opacity describes the relative darkness of a plume and its ability to transmit light. It can be determined by *in situ* CEMs that use open-path optical sensing devices. Opacity can also be determined by trained "smoke readers" who assess plume optical density by comparing the color of a plume to Ringelmann opacity charts (Figure 7.10).

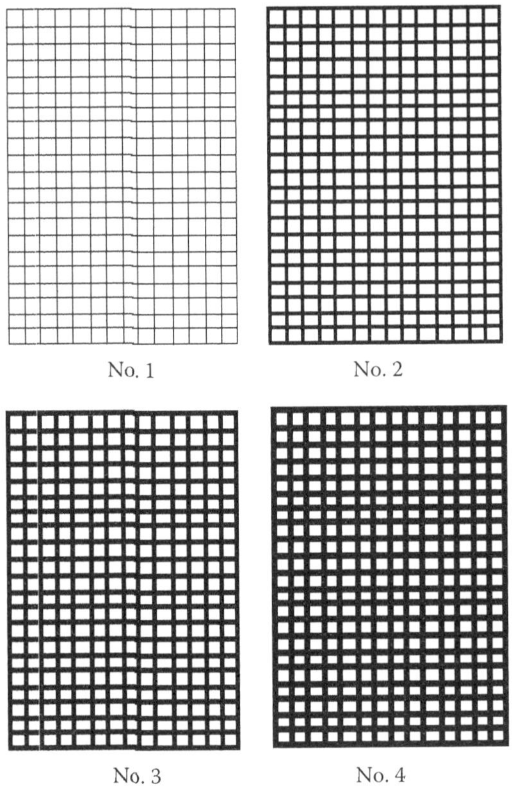

FIGURE 7.10 Ringelmann opacity charts.

Ringelmann measurements have the advantage of simplicity, low cost, and a long history of legal acceptance.

7.2.3 Emission Factors

Emission factors are tools used to estimate source pollutant emissions. They are expressed as pollutant mass divided by weight, volume, heat input, or duration of the activity and are expressed in such units as pounds per ton (lb/ton, kg/metric ton [mton]) of fuel or refuse burned or pounds per million British thermal units (MBtu) of heat input.

Emission factors can be used to estimate source emissions using the following equation:

$$E = \left[A(\mathrm{EF})(1 - \mathrm{CE}/100)\right] \tag{7.12}$$

where E is emissions, A is activity rate, EF is uncontrolled emission factor, and CE is control efficiency.

Published emission factors represent averages of data of acceptable quality that are assumed to be representative of long-term averages for facilities in a source category. They have been compiled by the US EPA and published in a continuously updated document called AP-42, which is available in hard copy, CD-ROM, and online. It contains Volume I, which addresses hundreds of stationary and area sources, and Volume II, which addresses mobile sources. Emission factors in AP-42 are obtained from source tests, material and mass balance evaluations, and engineering estimates. They are rated from A to E to provide an indication of their appropriateness for estimating average emissions for a source activity. An emission rating of A indicates high reliability with comparability equal to a CEM. On the other end, an E rating indicates a highly qualitative "engineering judgment."

Emission factors are particularly appropriate for developing emission inventories. Such inventories may be used in dispersion modeling, control strategy development, and screening for compliance. Because they represent average emissions for a source category, use of emission factors is appropriate when source-specific information is not available. Users are advised to be aware of their limitations in representing facility emissions. The uncertainty of emission estimates based on emission factors is increased when estimates of peak or short-term (hourly, daily) emissions are needed for regulatory purposes. Short-term emission fluctuations occur as a result of variations in process operating conditions, process materials used, control device operating conditions, ambient conditions, and others. Emission factors are generally based on normal operating conditions and represent long-term average conditions.

Many emission factors are determined from material and mass balance evaluations. For some sources, a material–mass balance approach may provide a better estimate of actual emissions than emission tests, as it would also account for fugitive emissions. Such mass balance estimates are particularly appropriate for processes in which a high percentage of a pollutant is lost (e.g., S from coal and fuel oil, solvent emissions).

Emission factors in AP-42 are given for three pollutant classifications. These are criteria pollutants and their precursors, hazardous air pollutants (HAPs), and greenhouse gases. Ammonia (NH_3) and O_3-destroying pollutants are also included. Because criteria pollutants were the original focus of AP-42 and major regulatory efforts, they are extensively treated. Emission factors for HAPs are increasingly being added to AP-42.

Most emission factors published in AP-42 have been determined from source testing using US EPA-published reference methods. Such reference methods (40 Code of Federal Regulations [CFR] Part 60, Appendix A, and Part 50, Appendix M) exist for PM and PM_{10}, SO_2, NO_2, CO, inorganic Pb, and VOCs. Methods for determining HAP emissions have also been published (40 CFR 61, Appendix, and US EPA document SW-846).

Emission factors can be accessed through the Factor Information and Retrieval of Emissions data system, which contains the US EPA's recommended criteria and HAP emission estimation factors.

Emission data from AP-42 for controlled-air medical incinerators for pollutants of both regulatory and special concern are summarized in Table 7.4. Note the relatively high emissions of hydrogen chloride (HCl) and that medical incinerators are a source of dioxins.

TABLE 7.4
Emission Factors (No Control Devices) for Selected Pollutants from Controlled-Air Medical Waste Incinerators

	Emission Factor		
Pollutant	**lb/ton**	**kg/mg**	**Emission Factor Rating**
NO_x	3.56	1.78	A
CO	2.95	1.48	A
SO_2	2.17	1.09	B
PM	4.67	2.33	B
HCl	33.50	16.80	C
Mercury	0.107	0.54	C
PCBs	4.65×10^{-5}	2.33×10^{-5}	E
Total chlorinated dibenzyl-p-dioxins	2.13×10^{-5}	1.07×10^{-5}	B

Source: US EPA, AP-42, EPA, Washington, DC, 2002.

For purposes of obtaining a permit under Title V of the Clean Air Act (CAA) Amendments of 1990, let us assume that we have to estimate emissions for a multichamber municipal incinerator that burns 50 tons of waste per day. The published emission factor is 17 lb PM/ton of refuse burned. Assuming an emission factor of 17 lb/ton and an average daily incineration of 50 tons of refuse, the emission rate per day, hour, and year, using Equation 7.13, can be calculated:

$$E = A(\text{EF}) \tag{7.13}$$

where A is the waste-burning rate and EF is the published emission factor. Then:

$$E = (50\,\text{tons/day})(17\,\text{lb/ton}) = 850\,\text{lb/day}\,(386\,\text{kg/day})$$

$$= (850\,\text{lb/day})(1\,\text{day}/24\,\text{h}) = 35\,\text{lb/h}\,(15.9\,\text{kg/h})$$

$$= (850\,\text{lb/day})(365\,\text{days/year})(1\,\text{ton}/2{,}000\,\text{lb}) = 155\,\text{tons/year}\,(141\,\text{mtons/year})$$

If the emission limit were 1.5 lb/ton, then a dust cleaning system would have to be installed with a minimum collection efficiency (CE) of:

$$\begin{aligned} \text{CE} &= [(\text{emission rate} - \text{emission limit})/\text{enission limit}](100) \\ &= [(17\,\text{lb/ton} - 1.5\,\text{lb/ton})17\,\text{lb/ton}](100) \\ &= 91\% \end{aligned} \tag{7.14}$$

Using Equation 7.12, the emission rate would be:

$$\begin{aligned} E &= (50\,\text{tons/day})(17\,\text{lb/ton})(1 - 91/100) \\ &= 76.5\,\text{lb/day}\,(34.8\,\text{kg/day}) \end{aligned}$$

7.3 AIR QUALITY MODELING

Because of resource limitations and practical considerations, it is desirable and even necessary to utilize approaches other than air quality monitoring to determine the effect of pollution sources on air quality in a region (particularly under worst-case conditions). Use of air quality models to predict the effect of new and existing sources on ground-level concentrations of emitted pollutants or pollutants produced as a result of atmospheric reactions addresses, in many cases, these limitations and constraints.

Air quality models are important tools for determining the environmental effect of pollution sources. Their use provides a relatively inexpensive and reliable means of determining compliance with NAAQSs and the extent of emission reductions necessary to achieve standards in an area not in compliance with a NAAQS. They are widely used by regulatory authorities as surveillance tools to assess the effect of emissions on ambient air quality. Models are also used to evaluate permit applications associated with permissible increments under the prevention of significant deterioration requirements and new source review programs.

Models consist of one or more mathematical formulae that include parameters that affect concentrations of pollutants at various distances downwind of emission sources. Models have been developed for several pollutants and time periods. Short-term models are used to calculate concentrations of pollutants over a few hours or days. They require a good understanding of the chemical reactions and atmospheric processes that occur over a period of minutes to hours. Short-term models can be employed to predict worst-case episode conditions and are used by regulatory agencies as a basis for control strategies. Long-term models are designed to predict seasonal or annual average concentrations, which may prove useful in studying atmospheric deposition as well as potential adverse health effects associated with pollutant exposures.

Models may be described according to chemical reactions. So-called nonreactive models are applied to pollutants such as CO and SO_2 because of the simple manner in which their chemistry can be represented. Reactive models address complex multispecies chemical reactions common to atmospheric photochemistry and are used for pollutants such as NO, NO_2, and O_3.

Models can be classified according to the type of coordinate system used. In grid-based models, the region of interest is subdivided into an array of cells that may range in horizontal size from 1 to 2,000 km^2 (0.38–769 mi^2), depending on the scale desired. The average concentration of each pollutant is then calculated for each grid cell. This approach is commonly used by regulatory agencies to determine compliance with NAAQSs. Trajectory models, unlike the fixed-grid approach described previously, follow parcels of air as they move downwind. These are simpler than fixed-grid models because concentrations need not be calculated for every point in the region.

Models can be described as simple or advanced based on the assumptions used and the degree of sophistication with which important variables are treated. Advanced models have been developed for photochemical air pollution, dispersion in complex terrain, and long-range transport. Gaussian models are widely used to predict the effect of emissions of relatively unreactive gas-phase substances such as SO_2 and CO, as well as PM in air quality downwind of point sources.

7.3.1 Gaussian Models

Gaussian models have been used in the United States since the mid-1960s. They are based on the assumption that plume spread, and dispersion of pollutants within it, results from molecular diffusion, and because of diffusion, pollutant concentrations in both the horizontal and vertical plume dimensions are distributed normally (bell-shaped curve). These models use the σ values of the Gaussian distribution in horizontal and vertical transects to represent the dispersion characteristics of a plume at various distances downwind of a source. Plume shape, as well as the σ values of pollutant concentrations, varies in response to different atmospheric meteorological conditions. The coordinate system in Gaussian models defines the x axis as downwind of the source, the y axis as horizontal to the x axis, and the z axis as the vertical dimension. Overlapping the two single distributions in each of the coordinate directions provides a double Gaussian distribution. Projecting this downwind gives the volume of space containing the plume (Figure 7.11).

As the plume moves downwind, the lowest edge will come into contact with the ground. Assuming no absorption or deposition, that portion of the plume is reflected upward. This reflection results in an increase in ground-level concentrations that must be factored into the model.

In Figures 7.12 and 7.13 (later in the chapter), σ_y and σ_z are standard deviations of Gaussian concentration distributions in plumes. They are described as horizontal and vertical dispersion coefficients. Their values increase with time and distance. The rate of increase in σ_y and σ_z depends on atmospheric conditions, particularly atmospheric stability. The rate of increase is greatest for stability class A and lowest for stability class F. In general, σ_y is greater than σ_z because σ_y is not influenced by inversion conditions. As both σ_y and σ_z increase in magnitude, pollutant concentrations along the centerline of the plume decrease.

Dispersion coefficients in Figures 7.12 and 7.13 are given for each of six atmospheric stability classes (Table 7.5). Stability classes are based on solar radiation, surface wind speed, and cloud cover. Stability classes A through C only occur in daylight hours; E and F, only at night. Classes E and F are characterized by stable air that minimizes thermal turbulence. Classes A, B, and C represent unstable conditions, with increasing stability from A to C. Class D is neutral, with moderate winds and uniform atmospheric mixing. Stability class A can result in pollutants being brought to the ground relatively close to the emission source, with resultant high concentrations. Stability class

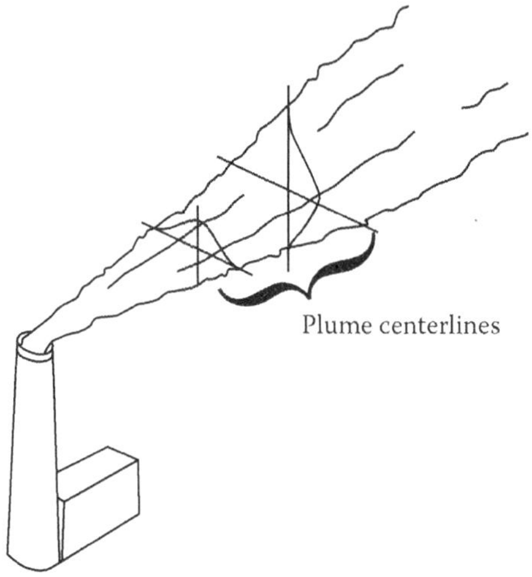

FIGURE 7.11 Gaussian plume spread associated with plume movement downwind. (From US EPA, EPA/450/2-81-0113, EPA, Washington, DC, 1981.)

TABLE 7.5
Atmospheric Stability Classes

	Day			Night	
	Incoming Solar Radiation			Thinly Overcast	
Wind Speed at 10 m (m/s)	**Strong**	**Moderate**	**Slight**	**≥ 4/8 Cloud**	**≤ 3/8 Cloud**
< 2	A	A–B	B	–	–
2–3	A–B	B	C	E	F
3–5	B	B–C	C	D	E
5–6	C	C–D	D	D	D
> 6	C	D	D	D	D

F allows the plume to be carried longer distances before it reaches the ground; therefore, it results in lower ground-level concentrations (except when topography is a factor). Both rough terrain and urban heat islands can change the stability classification under ambient conditions (usually upward by one class, i.e., more stable).

Under stable atmospheric conditions or unlimited vertical mixing, ground-level concentrations can be calculated from Equation 7.15:

$$C_x = \left(Q / \pi \tilde{A}_y \tilde{A}_z \bar{u}\right) e^{-1/2\left[\left(H/\tilde{A}_z\right)\left(H/\tilde{A}_z\right)\right]} e^{-1/2\left[\left(y/\tilde{A}_y\right)\left(y/\tilde{A}_y\right)\right]} \tag{7.15}$$

where C_x is the ground-level concentration at some distance x downwind (g/m^3), Q is the average emission rate (g/s), $\bar{u}$ is the mean wind speed (m/s), H is the effective stack height (m), σ_y is the standard deviation of horizontal distribution of plume concentration (m), σ_z is the standard deviation of vertical distribution of plume concentration (m), y is the off-centerline distance (m), and e is the natural log equal to 2.71828.

Equation 7.15 reduces to Equation 7.16 when $y = 0$:

$$C_x = \left(Q / \grave{A}\tilde{A}_y \tilde{A}_z \bar{u}\right) e^{-1/2\left[\left(H/\tilde{A}_z\right)\left(H/\tilde{A}_z\right)\right]} \tag{7.16}$$

If the emission source is at ground level and there is no effective plume rise, Equation 7.16 reduces to:

$$C_x = \left(Q / \pi \sigma_y \sigma_z \bar{u}\right) \tag{7.17}$$

Plume rise is calculated from equations developed in the early 1960s. They are used to calculate H (effective stack height) in Equation 7.15 and Equation 7.16. The H is the sum of the physical stack height and plume rise. For neutral and unstable atmospheric conditions and for buoyancy fluxes less than 55 m^4/s^3, H can be estimated from Equation 7.18:

$$H = h + 21.45F^{3/4} / \bar{u} \tag{7.18}$$

where h is the physical stack height (m) and F is the buoyancy flux (m^4/s^3). When $F \leq 55\ m^4/s^3$:

$$H = h + 38.71F^{3/5} / \bar{u} \tag{7.19}$$

The buoyancy flux, F, is calculated from the following equation:

$$F = g v_s d^2 \left(T_s - T\right) / 4T_s \tag{7.20}$$

where g is the gravitational constant (9.8 m/s^2), v_s is the stack gas velocity (m/s), d is the internal diameter of the top of the stack (m), T_s is the stack gas temperature (K), and T is the ambient temperature (K).

The horizontal distance (x_{pr}) from the stack to where the final plume rise occurs is assumed to be $3.5x^*$, where x^* is the distance in kilometers to where atmospheric turbulence begins to dominate entrainment. In Equation 7.18, $x_{pr} = 0.049\ F^{5/8}$, and in Equation 7.19, $x_{pr} = 0.119\ F^{2/5}$.

Taking factors into account such as multiple eddy reflections from the ground and bottom of the inversion layer and a σ_z value that is greater than 1.6 of the mixing height (m), an equation for determining maximum ground-level concentrations (C_{max}) can be derived:

$$C_{max} = 2Q\tilde{A}_z / \grave{A}\bar{u}eH^2\tilde{A}_y \tag{7.21}$$

This equation is only applicable when the ratio σ_z/σ_y is constant with distance. The distance to the maximum concentration is the distance at which $\sigma_z = (H/2)^{1/2}$.

Ground-level concentration (C_x) at some distance downwind (x) of an industrial boiler burning 12 tons (10.9 mtons) of 2.5% sulfur coal per hour with an emission rate of 151 g/s can be determined from Equation 7.16. Assume the following: $H = 120$ m, $\bar{u} = 2$ m/s, $y = 0$, it is 1 hour before sunrise, and the sky is clear.

From Table 7.4, the stability class for this atmospheric condition is F. From Figure 7.12, the horizontal dispersion coefficient σ_y for a downwind distance of 10 km for F atmospheric stability class is approximately 270 m; the vertical dispersion coefficient σ_z is 45 m (Figure 7.13).

Therefore, using Equation 7.16:

$$C_x = \left[151 / (3.14)(270)(45)(2)\right] e^{-1/2\left[(120/45)(120/45)\right]}$$

$$= 1.98 \times 10^{-3} e^{-1/2(7.1)}$$

$$= \left[1.98 \times 10^{-3}\right] / e^{1/2(7.1)}$$

$$= \left[1.98 \times 10^{-3}\right] / 34.8$$

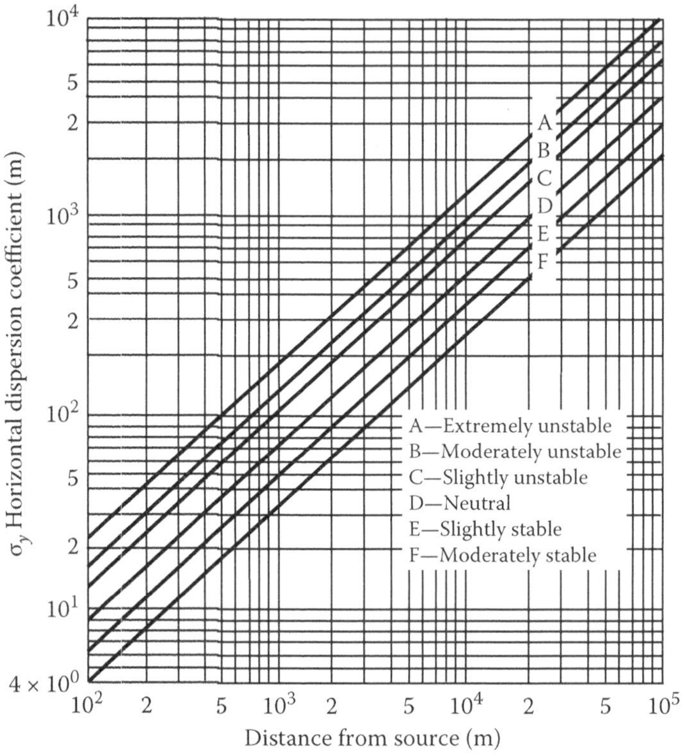

FIGURE 7.12 Pasquill's horizontal dispersion coefficients. (From Turner, D.B., *Workbook for Atmospheric Dispersion Estimates*, EPA Publication AP-26, EPA, Washington, DC, 1969.)

$$= 5.7 \times 10^{-5}\,\text{g/m}^3$$

$$= 57\,\text{g/m}^3$$

If emissions are from a ground-level source with $H = 0$, $\bar{u} = 4$ m/s, $Q = 100$ g/s, and a stability class of B, we can use Equation 7.17 to calculate downwind concentrations at various receptor distances (e.g., 200 and 1,000 m). For a receptor distance of 200 m:

$$\begin{aligned} C_x &= 100/\left[3.14(36)(20.5)(4)\right] \\ &= 10.8 \times 10^{-3}\,\text{g/m}^3 \\ &= 10.8\,\text{mg/m}^3 \end{aligned}$$

For a receptor distance of 1,000 m:

$$\begin{aligned} C_x &= 100/\left[3.14(155)(110)(4)\right] \\ &= 4.67 \times 10^{-4}\,\text{g/m}^3 \\ &= 467\,\text{g/m}^3 \end{aligned}$$

In the calculations above, H was given. Plume rise must be calculated using the model from Equations 7.18 and 7.19. Using these equations, we can calculate H for an 80 m high source (h) with

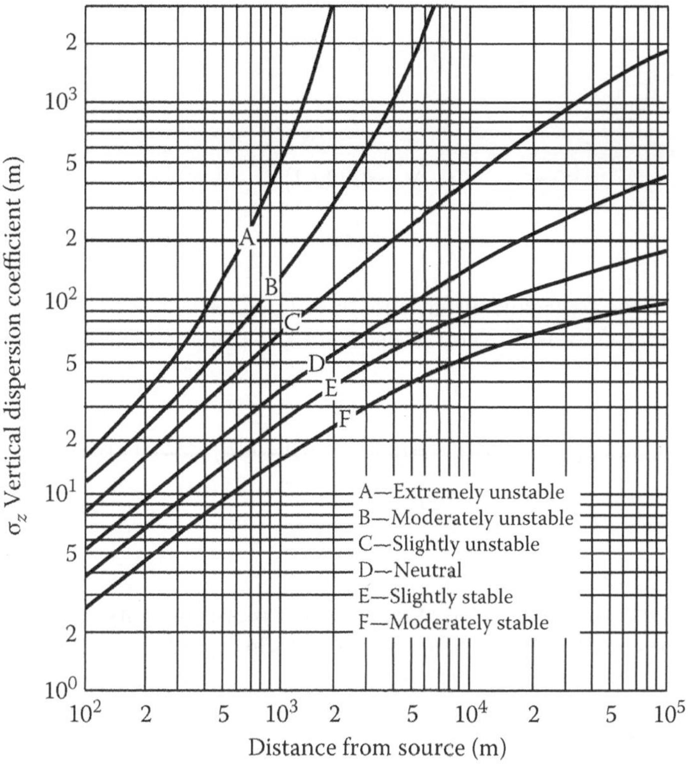

FIGURE 7.13 Pasquill's vertical dispersion coefficients. (From Turner, D.B., *Workbook for Atmospheric Dispersion Estimates*, EPA Publication AP-26, EPA, Washington, DC, 1969.)

a stack diameter of 4 m, a stack velocity of 14 m/s, a stack gas temperature of 90°C (363 K), an ambient temperature of 25°C (298 K), a $\bar{u}$ value at 10 m of 4 m/s, and a stability class of B.

The buoyancy flux (due to thermal characteristics of the plume) must first be calculated (Equation 7.20) to determine effective stack height:

$$F = 9.8 v_s d^2 (T_s - T)(4T_s)$$
$$= 9.8(14)(16)(65)/[4(363)]$$
$$= 98.27$$

Because $F \leq 55$ m^4/s^3, we use Equation 7.19:

$$H = h + (38.71)(98.27)^{3/5}/4$$
$$= 80 + 152$$
$$= 232$$

Using Equation 7.22, the maximum ground-level concentration can be calculated if the ratio σ_z/σ_y is constant with distance. The distance to the maximum concentration is the distance at which:

$$\tilde{A}_z = H/(2)^{1/2} \tag{7.22}$$

which, in the case above:

$$\tilde{A}_z = (232/2)^{1/2}$$
$$= 164\,\text{m}$$

From Figure 7.13, the downwind distance for this dispersion coefficient is approximately 1.4 km (0.87 mi) for stability class B; the corresponding σ (from Figure 7.12) is approximately 210 m (687 ft). Therefore:

$$\begin{aligned} C_{\max} &= [2(100)(164)]/[3.14(4)(2.718)(232)^2(210)] \\ &= 85\times10^{-6}\,\text{g}/\text{m}^3 \\ &= 85\text{g}/\text{m}^3 \end{aligned}$$

Gaussian dispersion modeling is based on a number of assumptions. These include (1) steady-state conditions (constant source emission strength); (2) wind speed, direction, and diffusion characteristics of the plume are constant; (3) no chemical transformations take place; and (4) wind speeds are greater than or equal to 1 m/s. In addition, Gaussian models are limited to predicting concentrations no greater than 50 km downwind.

7.3.2 Model Options

Dispersion models may be Gaussian, numerical, statistical, or physical; Gaussian-based models are the most widely applied. Their use is based on the ease of application and the conservative estimates they provide.

In its efforts to promote consistency in applying models, the US EPA has developed a document entitled "Guidelines on Air Quality Models" in which it lists models based on several comparative analyses and suggests best applications. Those that perform well for a general set of conditions are classified as Appendix A models (Table 7.6). They are classified by whether they are screening or refined models and their application to (1) simple or complex terrains; (2) rural or urban areas, or both; and (3) specific source concerns, pollutant production, and regions.

7.3.3 Photochemical Models

Air quality models simulate the physical and chemical processes governing the transport, dispersion, and secondary formation of air pollutants such as ozone and particulate matter using emissions and meteorological inputs. These models can be used to develop and assess regulations, conduct research, provide air quality forecasts, and evaluate control and mitigation strategies. Air quality models can range in complexity from simple box models to complex multidimensional Eulerian models. In an Eulerian model, the domain under investigation is divided into a network of grid cells in which principles of conservation of mass are utilized to mathematically determine the generation, loss, and transport of pollutants into and out of each cell at specific points in time. In contrast, Lagrangian models instead approach the domain as a non-stationary parcel of air that can travel and change with time as shown in Figure 7.14.

Most air quality models utilize a Eulerian framework and include models such as CMAQ, CAMx, and WRF-Chem. CMAQ is the Community Multiscale Air Quality Model developed and maintained

TABLE 7.6
Preferred US EPA Air Quality Dispersion Models

Type/Terrain	Mode (Urban/Rural)	Model
Screening		
Simple	Both	SCREEN 3
Simple	Both	AERSCREEN
Simple	Both	TSCREEN
Simple	Urban	RAM
Complex	Rural	COMPLEX I
Complex	Urban	SHORTZ
Complex	Rural	RTDM 3.2
Complex	Rural	Valley
Complex	Both	CTScreen
Line	Both	BLP
Refined		
Simple	Urban	RAM
Simple	Both	AERMOD
Simple	Urban	CDM 2.0
Complex	Both	CTDMPLUS
Line	Both	BLP
Line	Both	CALINE-3
Ozone	Both	CMAQ/CAMx
Coastal	Both	OCD
Particulate matter	Both	CMA/CAMx

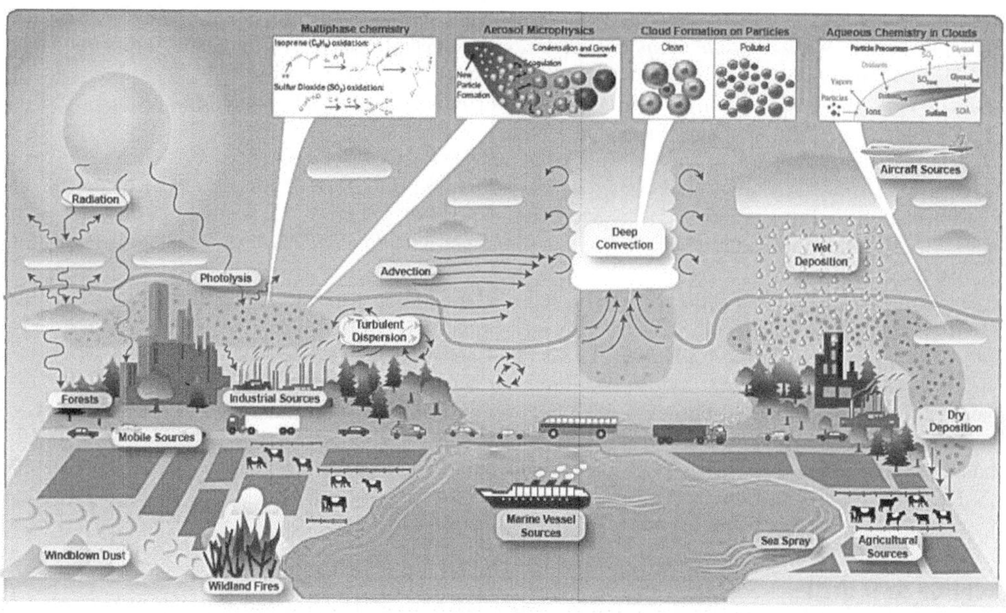

FIGURE 7.14 Overview of science processes in photochemical models. (*Source*: US EPA.)

by the US EPA as a regulatory model. CMAQ is an opensource model and is used by regulatory agencies and research communities worldwide and can simulate air pollution transport and deposition and small local or regional scales or at the hemispheric scale and fine time scales. WRF-Chem, the Weather Research and Forecasting Chemistry model, is developed and maintained by the National Center for Atmospheric Research. WRF-Chem is a global scale coupled meteorological and chemical model that can simulate the complex chemical interactions of air pollutant species and their transport and deposition as influenced by weather patterns. WRF is also used to provide meteorological inputs for CMAQ and other models or can be coupled directly with CMAQ to provide more realistic realtime meteorological inputs. The Comprehensive Air Quality Model with Extensions (CAMx) is a commercially developed open-source model and can be used as a regulatory model similar to CMAQ.

READINGS

American Conference of Governmental Industrial Hygienists, *Air Sampling for Evaluation of Atmospheric Contaminants*, 8th ed., ACGIH, Cincinnati, OH, 1995.

Axelrod, H.D., and Lodge, J.P., Jr., Sampling and calibration of gaseous pollutants, in *Air Pollution*. Vol. III: *Measuring, Monitoring and Surveillance of Air Pollution*, 3rd ed., Stern, A.C., Ed., Academic Press, New York, 1977, p. 145.

Boubel, R.A., Fox, D.L., Turner, D.B., and Stern, A.C., *Fundamentals of Air Pollution*, 3rd ed., Academic Press, San Diego, CA, 1994.

Briggs, G.A., Plume rise predictions, in *Lectures on Air Pollution and Environmental Impact Analysis*, Haugen, D.A., Ed., American Meteorological Society, Boston, 1975.

Bucholtz, F., Ed., *Environmental Monitoring and Instrumentation*, Optical Society of America, Washington, DC, 1997.

Couling, S., Ed., *Measurement of Airborne Pollutants*, Butterworth Heineman, Oxford, 1993.

Harrison, R.M., Measurements of concentrations of air pollutants, in *Air Pollution and Health*, Holgate, S.T., Samet, J.M., Koren, H.S., and Maynard, R.L., Eds., Academic Press, San Diego, CA, 1999, p. 63.

Harrison, R.M., and Young, R.J., Eds., *Handbook of Air Pollution Analysis*, 2nd ed., Chapman & Hall, London, 1986.

Jacobson, M.Z., *Fundamentals of Atmospheric Modeling*, Cambridge University Press, Cambridge, UK, 1998.

Krzyzanowski, M., and Schwela, D., Patterns of air pollution in developing countries, in *Air Pollution and Health*, Holgate, S.T., Samet, J.M., Koren, H.S., and Maynard, R.L., Eds., Academic Press, San Diego, CA, 1999, p. 105.

Lodge, J.P., Jr., Ed., *Methods of Air Sampling and Analysis*, 3rd ed., Intersociety Committee, Lewis Publishers, Chelsea, MI, 1988.

McGregor, G.R., Basic meteorology, in *Air Pollution and Health*, Holgate, S.T., Samet, J.M., Koren, H.S., and Maynard, R.L., Eds., Academic Press, San Diego, CA, 1999, p. 51.

Noll, K.E., *Air Monitoring Survey Design*, Ann Arbor Science Publishers, Ann Arbor, MI, 1976.

Ozkaynak, H., Exposure assessment, in *Air Pollution and Health*, Holgate, S.T., Samet, J.M., Koren, H.S., and Maynard, R.L., Eds., Academic Press, San Diego, CA, 1999, p. 115.

Powals, R.J., Zaner, L.J., and Sporek, K.F., *Handbook of Stack Sampling and Analysis*, Technomic Publishing, Westport, CT, 1978.

Turner, D.B., *Atmospheric Dispersion Estimates*, 2nd ed., Lewis Publishers/CRC Press, Boca Raton, FL, 1990.

US EPA, *Air Quality Index: A Guide to Air Quality and Health*, EPA-454/R-00-005, EPA, Washington, DC, 2000.

Wight, G.D., *Fundamentals of Air Sampling*, Lewis Publishers/CRC Press, Boca Raton, FL, 1994.

Willeke, K., and Baron, P.A., Eds., *Aerosol Measurement*, Van Nostrand Reinhold, New York, 1993.

Winegar, E.D., and Keith, L.H., Eds., *Sampling and Analysis of Airborne Pollutants*, Lewis Publishers/CRC Press, Boca Raton, FL, 1993.

Zannetti, P., *Air Pollution Modeling: Theories, Computational Methods and Available Software*, Van Nostrand Reinhold, New York, 1990.

ONLINE SOURCES

1. https://www.epa.gov/outdoor-air-quality-data—Access to Air Pollution Data
2. https://www.airnow.gov/aqi-and-health/—Air Quality Index

3. https://www.epa.gov/air-emissions-factors-and-quantification/ap-42-compilation-air-emissions-factors—Access to AP-42 with Emission Factors and Inventory Groups
4. https://www.epa.gov/air-emissions-inventories/air-emissions-inventory-improvement-program-eiip—Emissions Inventory Improvement Program
5. https://www.epa.gov/haps—Air Toxics Emission Factors and Inventory
6. https://www.epa.gov/air-emissions-modeling—Emissions Modeling Clearinghouse
7. https://www.epa.gov/air-emissions-factors-and-quantification/basic-information-air-emissions-factors-and-quantification—Emissions Factor and Inventory resource material
8. https://www.epa.gov/scram/air-quality-dispersion-modeling—Air Quality Dispersion Modeling

QUESTIONS

1. Describe differences between qualitative and quantitative assessments of air quality.
2. What is air quality monitoring?
3. Why is air quality monitoring conducted in the United States?
4. What is the relationship between sampling and air quality monitoring?
5. Describe passive, grab, intermittent, and continuous sampling and monitoring methods.
6. What is the significance of the following in conducting air sampling?
   a. Sample size
   b. Sampling rate
   c. Sampling duration
7. What averaging times are used for the following pollutants: O_3, PM_{10}, CO, and SO_2? Why is each averaging time used?
8. Describe the sample collection principles of
   a. Absorption
   b. Adsorption
   c. Impaction
   d. Filtration
9. What are grab samples? Why and when are they used?
10. Describe the use of mixing ratios for expressing air pollutant concentrations.
11. The primary 1-hour NAAQS for CO is 35 ppmv. What is its SI equivalent concentration in milligrams per cubic meter at reference conditions of 25°C and 760 mmHg?
12. The primary annual average NAAQS for NO_2 in SI units is 100 μg/m^3. What is the equivalent concentration in parts per million by volume at reference conditions of 25°C and 760 mmHg? In parts per billion by volume?
13. A PM_{10} sample was collected under ambient average temperature and pressure conditions of 5°C and 740 mmHg at a measured flow rate of 17 ft^3/min (0.48 m^3/min) over a period of 24 hours. What was the volume standardized to reference conditions of 25°C and 760 mmHg?
14. Describe differences between accuracy and precision. How is each determined?
15. What is bias? What is its primary cause?
16. What is calibration? Why must one periodically calibrate sampling and monitoring instruments?
17. Describe differences between primary and secondary standards.
18. What criteria are used to locate individual monitoring sites in SLAM network designs?
19. Describe spatial scales used in identifying monitoring locations.
20. Describe pollutants monitored and the purpose of the following air quality monitoring networks:
   a. NAMS
   b. PAMS
   c. AQS

   d. NADP
   e. IMPROVE
   f. CASTNET
21. Describe the concept of federal reference and equivalent methods as it applies to air quality monitoring in the United States.
22. What are the federal reference methods for SO_2, NO_x, and PM_{10}?
23. Distinguish between quality assurance and quality control as they are applied to air quality monitoring activities.
24. What is the purpose of data summarization procedures and efforts?
25. What information can one obtain from evaluating histogram and stem and leaf plots?
26. What is the usefulness of box plots in summarizing pollutant data?
27. Describe when each of the following measures of central tendency should be used in data analysis:
   a. Mean
   b. Median
   c. Geometric mean
28. Determine the mean, median, and geometric means of the following PM_{10} values: 55.2, 74.5, 102.2, 40.3, 45.6, 20.5, 62.4, 80.7, 35.8, 65.8, and 48.9 μg/m^3.
29. Weekly samples of wet acid deposition were collected over a 2-month period. Calculate the average pH and determine how many times more acidic it is relative to the theoretical clean air value of 5.65. Measured values were 3.4, 4.2, 5.0, 3.8, 4.0, 4.4, 4.8, and 3.9.
30. Describe how the Air Quality Index is derived. Describe its usefulness.
31. Describe the nature of, and need for, stack sampling.
32. Describe what emission factors are, how they are derived, and their use and limitations.
33. A municipal solid waste incinerator has an emission factor of 20 lb PM/ton (10 kg/mton) of waste. Assume that the maximum 2-hour permissible emission level is 0.18 g/m^3 (corrected to 12% CO_2). If 80 dry standard cubic feet (2.26 m^3) of flue gas are produced per pound (0.45 kg) of waste combusted, determine the control equipment collection efficiency needed to meet the emission limit.
34. Use emission factors from Table 7.4 and assume that a medical waste incinerator burns 10 tons (9 mtons) of mixed hospital wastes per month. What would the daily emission rate (lb/day, kg/day) of hydrogen chloride be? The annual emission rate of dioxins (lb/year, kg/year)?
35. Describe two procedures that can be used to monitor plume opacity.
36. Why are continuous emission monitors used?
37. Describe how you would determine the emission factor for PM from sewage sludge incinerators.
38. By analyzing/evaluating Equation 7.15, what would be the expected effect of increasing wind speed on downwind pollutant concentrations?
39. What factors affect plume rise and therefore effective stack height?
40. From Figures 7.12 and 7.13, what is the effect of the following on dispersion coefficients?
   a. Stability class
   b. Increasing distance downwind
41. Using Gaussian dispersion equations, the stability class table, and dispersion coefficient data (Figures 7.12 and 7.13) for a source emitting SO_2 at a rate of 100 g/s on a sunny summer day with a wind speed at anemometer height of 4 m/s and effective stack height of 36.4 m, determine the stability class and dispersion coefficients for receptor distances of 300 and 1,000 m and ground-level concentrations at each distance in parts per billion by volume.
42. Describe model types used in regulatory programs.
43. What are the common photochemical models?

8 Regulation and Public Policy

Nominal regulatory efforts to control air pollution in the United States began in the late nineteenth and early twentieth centuries when several large cities approved ordinances to control "smoke" produced by a variety of industrial activities. Although largely ineffectual because they lacked enforcement authority, these smoke ordinances nevertheless served as forerunners of present-day regulatory programs.

In response to the increasingly worsening smog problem in its southern coastal area, California began serious efforts to control air pollution in the 1950s. As scientists unraveled the chemistry of smog, regulatory efforts changed from controlling emissions from stationary sources to those from automobiles. As a consequence, by the late 1960s, both California and the federal government began in earnest their significant and continuing regulatory efforts to reduce emissions from motor vehicles.

By the 1960s, there was increasing recognition at all governmental levels that air pollution was a serious environmental problem. Many state and local governments began to form agencies and develop regulatory frameworks to control air pollution. These efforts (in retrospect) were relatively timid and, for the most part, not very effective.

As the modern environmental movement evolved, the premise that polluted cities and countrysides were a price of progress was no longer accepted. A major change occurred in American attitudes. The economic well-being that provided Americans the so-called good life was seen to diminish the quality of the environment in which fruits of economic prosperity were to be enjoyed. The desire of Americans to prevent further deterioration and improve environmental quality gave rise to the significant regulatory efforts to control air pollution and other environmental problems that have taken place since 1970.

While other countries have undergone regulatory efforts, the World Health Organization (WHO) has also developed air quality guidelines in its effort to protect the health of individuals worldwide. WHO reported that 91% of the world's population in 2016 was living in places where its 2005 air quality guidelines had not yet been met and that 4.2 million premature deaths had occurred in 2016 due to ambient (outdoor) air pollution. WHO has guidelines for PM_{10}, $PM_{2.5}$, O_3, NO_2, and SO_2 with an update of its guidelines in 2021. In general, US National Ambient Air Quality Standards (NAAQSs), to be presented in more detail in this chapter, are equal to or more stringent than WHO guidelines.

8.1 NONREGULATORY LEGAL REMEDIES

Before modern regulatory air pollution control programs, individuals or groups of citizens were limited in what they could do to abate the pollution adversely affecting them. They could ask corporate officers and others responsible for managing sources to be better neighbors and appeal to their sense of civic responsibility. Such appeals (if many of them were indeed made) would, for the most part, have gone unheeded. Pollution abatement requires the allocation of what many economists and corporate financial officers describe as unproductive resources, that is, money that does not add to product value.

In the absence of strong governmental regulatory programs, a citizen or group of citizens has a right to file suit against a source employing common law legal principles. Although significant air pollution control regulations have been in place for more than five decades in the US, its citizens continue to have this right.

Individuals claiming injury from emissions produced by a source can file a civil suit against the source in an effort to recover damages and abate the problem. Such suits are called torts. Tort

DOI: 10.1201/9781003664109-8

law involves a system of legal principles and remedies that attempts to address harm to individuals or their property. Historically, both nuisance and trespass principles of tort law have been used to address claims of harm associated with air pollution.

8.1.1 Nuisance

A nuisance is an intentional or negligent act that results in an unreasonable interference with the use and enjoyment of one's property. It may be private (affecting some individuals and not others) or public (affecting everyone equally). Under the private nuisance concept, the plaintiff, or injured party, must convince a judge or jury that the action is unreasonable. If the suit is successful, he or she is entitled to recover damages. To abate the problem, however, the plaintiff must seek an injunction against the source.

In deciding whether to grant such an injunction, the court must balance the equities; that is, it must take into account the following: (1) extent of damage, (2) cost of pollution abatement, (3) whether the pollution source was on the scene first, and (4) the economic position of the polluting source in the community. If the dispute is between a corporation with a large local workforce and a single party, or even a group of landowners, courts invariably balance the equity in favor of the polluting source and refuse to grant an injunction. In such cases, courts reason that injunctive relief, if granted, will result in great economic harm to the local community. Although damages are awarded to the affected party or parties, injunctive relief is only rarely provided.

If all members of a community are equally affected, then no individual citizen can sue. In such cases, a public authority, such as a local prosecutor, would have to file a public nuisance suit against the pollution source that is causing "an unreasonable interference with a right common to the general public." As in private suits, courts must assess the seriousness of alleged harm relative to the social utility of the source's conduct. Public authorities have historically been reluctant to file such actions.

8.1.2 Trespass

The trespass concept applies to cases in which there is a physical violation of a landowner's right to exclusive possession of his or her property. Particulate-laden smoke could be considered to violate trespass provisions of common law. Historically, in determining liability, courts have not required that the invasion be proven unreasonable or that damages be proven.

8.2 REGULATORY STRATEGIES AND TACTICS

Successful abatement of air pollution and protection of public health and welfare requires the selection of appropriate regulatory strategies and tactics to implement them. A variety of air quality regulatory strategies may be employed. These include air quality management (air quality standards), emission standards, and economics-based approaches. Air quality management and emission standard strategies are the mainstay of pollution control efforts in the United States and other developed countries. Several economics-based pollution control approaches have been proposed and adopted. Some have been incorporated into air pollution control laws and regulatory practice.

8.2.1 Strategies

8.2.1.1 Air Quality Management

Air quality management strategies are based on the widely accepted toxicological principle that pollutant exposures below threshold (lowest dose at which toxic effects are observed) values are relatively safe, and therefore some level of atmospheric pollution is acceptable and thus legally permissible. Pollutants identified as posing potentially significant public health risks and welfare concerns may be regulated by the promulgation of ambient air quality standards.

Air quality standards are legal limits on atmospheric concentrations of regulated pollutants. Promulgation of an air quality standard for an ambient air pollutant is typically characterized by intensive review of scientific studies, recommendations on acceptable levels, and ultimate decision making by a regulatory authority.

Setting ambient air quality standards is a difficult and arduous process. In many cases, the relationship between pollutant levels, adverse health effects, and the level of protection necessary is subject to some degree of uncertainty. In the United States, the Environmental Protection Agency (US EPA) is required to promulgate air quality standards that provide an "adequate margin of safety," with special consideration for those individuals who may be the most sensitive, that is, asthmatics, children, the aged or infirm, those with existing cardiovascular disease, and the like. As a consequence, ambient air quality standards are much more stringent than occupational standards designed to protect nominally healthy working adults.

Because air quality management strategies are based on the premise that a safe level of exposure (a threshold) exists, they are generally not applied to contaminants not known to have threshold values. It has been widely accepted in both the scientific and regulatory communities that there is no generally safe level of exposure to carcinogens.

Once an air quality standard for a pollutant or pollutant category has been promulgated, regulatory authorities at federal, state or provincial, and local levels are responsible for developing and implementing control and management practices (tactics) that will ensure that air quality standards are not exceeded or will not be exceeded beyond a certain date as defined in the standard.

Air quality management, although generally assumed to be a cost-effective approach to achieving air quality goals, is a relatively complicated undertaking. Its complexity involves both technical and political dimensions. Regulatory authorities must select and implement tactics that, in their best judgment, will achieve each air quality standard. In the case of ozone (O_3), control efforts have been confounded because of uncertainties associated with the contributions of anthropogenic and biogenic sources of hydrocarbons (HCs) to tropospheric O_3 chemistry, as well as other factors. In efforts to achieve particulate matter (PM) standards, regulatory authorities attempted to employ a control tactic that made both technical and economic sense but was not politically popular. Such was the case with prohibitive bans on leaf burning in autumn. Although relatively cost-effective, it has been politically unacceptable to some communities. On paper, air quality management is a highly practical and cost-effective control strategy. However, as is the case with many things in life, "the devil is in the details."

8.2.1.2 Emission Standards

In an emission standard strategy, limits are placed on the maximum quantity (typically mass per unit time or thermal input) of a pollutant or pollutants that can be emitted from specific sources. The same emission standard is applied equally to all sources in a source category; that is, all members of a category must comply with the emission limit without regard to existing air quality. The same emission limits apply to sources in "clean" or "dirty" air regions.

With the exception of hazardous or toxic pollutants in the United States, emission standards, when used as a strategy, are governed by practicability, that is, emission standards reflect the highest degree of pollution reduction achievable taking cost of control into consideration.

The "best practicable means" or "good practice" approach is widely used for the development of emission standards in the United States and other developed countries. The term *practicable* indicates economic, technical, and political practicality. The "best practicable means" in practice may be the degree of emission reduction achieved by the best industrial plants in a source category, or technology that can be reasonably applied by borrowing from other industries. Implicit in this approach is that emission standards may become more stringent (at least for new sources) as control technology improves and becomes more affordable. "Best practicable means," the traditional approach to air pollution control in Europe, would be equivalent to what is described in the United States as "reasonably available control technology" (RACT). Such requirements are generally applied to new or significantly modified existing sources under new source performance standards (NSPSs), to be discussed later in this chapter, and to achieve air quality standards in some air quality control regions (AQCRs).

An emission standards strategy may also require very stringent emission limits. For pollutants that pose unique health hazards, for example, mercury (Hg), beryllium (Be), benzene, radioactive isotopes, and asbestos, emission reduction requirements reflect "best-available control technology" (BACT). The objective of BACT standards is to achieve the highest degree of emission reduction that technology is capable of, with a more limited consideration of capital and operating costs. New coal-fired utilities in Class I prevention of significant deterioration (PSD) regions (see Section 8.4.6) are required to use BACT.

The 1990 Clean Air Act (CAA) Amendments require setting maximum achievable control technology (MACT) standards for toxic or hazardous air pollutants (HAPs). This concept, in theory, suggests a higher degree of emission control than BACT. This is because emission reductions can be achieved by a variety of control practices other than treating just waste gas streams.

Emission standards are usually specified in some numerical form. In the case of coal-fired power plants licensed under the 1980 NSPS for sulfur dioxide (SO_2), the standard requires 90% removal and a maximum allowable emission rate of 1.2 lb of SO_2/MBtu (million British thermal unit) or 70% removal and a maximum allowable emission rate of 0.6 lb of SO_2/MBtu in the case of lower sulfur coals. Numerical standards can be based on heat input (lb/MBtu), unit of time (lb/h), weight per unit air volume (grains/standard cubic foot), or weight of process material (lb/ton). They may also be prescribed in metric units; in some cases, mixed units (US and metric) are used (e.g., g/mi).

Emission standards have the advantage of simplicity. There is no need for extensive and expensive air quality monitoring networks, modeling to determine compliance, and complex technical and public policy decisions. All sources in the source category must meet the same requirements regardless of existing air quality. It is a highly attractive control strategy from an administrative standpoint. As used in the United States, it has the advantage of being applied to special pollution concerns such as new sources and pollutants believed to pose uniquely hazardous or toxic exposure problems.

Use of emission standards as a control strategy also has limitations or problems. Although costs are taken into consideration in setting emission limits, sources must comply regardless of existing air quality. In areas with relatively good air quality, sources may be required to control emissions beyond that needed to protect public health and possibly welfare as well. In "dirty" air regions, emission limitations influenced by cost considerations may not be sufficiently stringent to protect public health and the environment.

8.2.1.3 Economics-Based Approaches

Several economics-based approaches have been proposed and adopted to achieve air quality objectives. They assume that command-and-control regulations tend to force industry and other sources to implement control measures and practices that do not take cost and economic efficiencies into consideration. They also presume that command-and-control regulation freezes the development of new technologies and practices that would result in higher levels of control and provides few or no financial incentives for sources to reduce emissions below control targets. Other terms used include "cap and trade" policies and pollutant "tax" policies. The proposed "carbon tax" is an example of a policy intended to curtail the emission of CO_2 emissions from fossil fuels. Such proposed policies, even in the proposal stage by a state or federal agency, may result in decisions being made by sources (e.g., electric utilities) to retire existing older plants in favor of less-emitting sources (i.e., a shift from coal to natural gas) or zero-emitting renewable energy sources such solar, wind, and water. Once such decisions are made, the utility may or may not reverse its direction, even when the proposed policies fail to be implemented due to its commitment to protecting the environment or its anticipation of future similar policies.

Economics-based approaches proposed or actually adopted include cost–benefit analysis and market-based policies.

8.2.1.3.1 Cost–Benefit Analysis

A control strategy based on cost–benefit analysis has been advocated by free-market economists. It would require the quantification of all damage costs associated with atmospheric pollutants and costs of control. Damage costs would be limited to the tangible, that is, the dollar value of health care and mortality, crop losses, reductions in property values, and effects on materials. Intangible values such as loss of friends and loved ones, changes in ecosystems, and diminished aesthetic quality would not be included. The primary focus of a cost–benefit strategy would be to select one or more pollution control options that minimize damage and control costs.

Theoretical costs of pollution effects on public health and welfare are compared with control costs in Figure 8.1. As damage costs decrease, control costs increase. Note the steep increase in control costs as damage costs decrease toward zero. This illustrates a major problem in pollution control; that is, there is a disproportionate increase in control costs relative to environmental benefits (decreased damage costs) achieved. In economic theory, it would be desirable to implement the degree of control that provides the greatest reduction of damages (or provides the greatest benefit) per unit cost. Damage and control costs would be "optimized" where the two lines intersect.

Use of cost–benefit analysis is limited (in fact, rendered unusable) by a variety of practical and theoretical problems. First, it assumes that both damage and control costs are easily quantified. Although true for control costs, the uncertainty of the nature and extent of costs and damages associated with the effects of pollutants on humans and the environment makes their quantification very difficult. Second, it assumes that damage costs to public health and welfare above the point of optimization are socially and politically acceptable. A cost–benefit-based control strategy would pose a public policy nightmare (more so than is already the case) for regulatory officials who would have to make decisions on who or what is to be protected.

The cost of control versus risks to humans and the environment has been a continuous source of contention between the regulated community and regulatory authorities. Under clean air legislation in the United States, control costs must be taken into consideration when promulgating NSPSs but not air quality standards and standards for HAPs. In 1995, pro-industry legislation that would have required cost–benefit analyses for most new environmental regulations was introduced into the US Congress. Despite considerable congressional support, the measure failed to be enacted. In the late

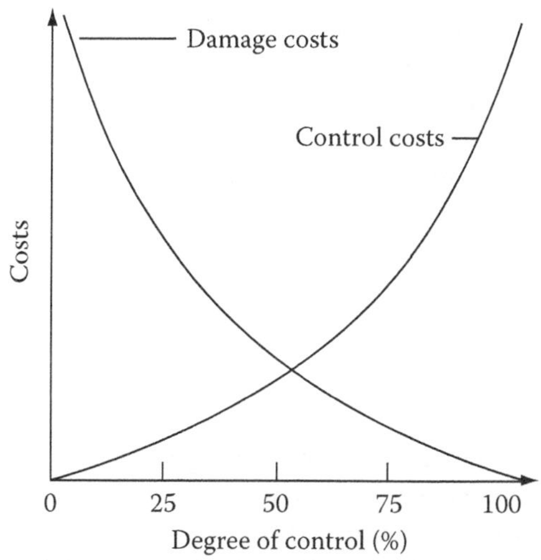

FIGURE 8.1 Relationship between control costs and pollutant environmental damage costs.

1990s, the US EPA's revised PM and O_3 standards were challenged (but upheld) in federal courts, in part because the US EPA did not conduct a cost–benefit analysis before the promulgation of these standards. Control cost considerations are a continual demand of industrial stakeholders and their supporters.

8.2.1.3.2 Economics-Based and Market-Based Policies

Several economics-based and market-based policies have been implemented because of their acceptance by regulators, the regulated community, and environmental groups. These include use of the bubble concept, emission trading, and allowances. Use of allowances and free-market trading of allowances were mandated in the 1990 CAA Amendments.

The bubble concept was the first economics-based regulatory innovation adopted by the US EPA. Under this policy, existing plants and groups of plants may increase (or not decrease) their emissions at one or more sources by decreasing emissions to a more significant degree at other sources within a facility. As a consequence, plant environmental managers have the flexibility to implement more cost-effective means of complying with air quality requirements. To receive regulatory approval, a bubble must result in emission reductions equivalent to or better than baseline emission limits relative to their environmental effect and enforceability and comply with all relevant provisions of clean air legislation. Excess emission reductions may be banked or traded.

Emission reductions beyond those legally required under the bubble concept, other regulatory requirements, and voluntary emission reduction efforts may be stored in US EPA-approved emission credit banks for later use in emission transactions. These emission reduction credits (ERCs) may be sold or traded to other corporations to meet regulatory requirements. ERCs only apply to the same criteria pollutants (i.e., pollutants for which an air quality standard has been promulgated) and may not be used to meet technology-based emission reduction requirements.

The use, trading, and selling of allowances to achieve objectives of Title IV: Acidic Deposition Control of the 1990 CAA Amendments are excellent examples of a market-based approach to achieving major air quality objectives. Under an allowance system, a source is allowed to emit so many units of a regulated pollutant. This allowance is some fraction of baseline emissions. Sources can "live within their allowance" by employing those emission reduction approaches that they deem to be cost-effective. Emission reductions that exceed allowances may be banked or traded. The allowance system used in the acidic deposition control program facilitated a 9 million metric tons (mtons)/year (10 million tons/year) reduction of SO_2 emissions from utilities without significant increases in the cost of electricity, a remarkable accomplishment.

Pollution charges as a pollution control strategy have received limited attention. Pollution charge systems attempt to reduce emissions by imposing fees or taxes on emissions of regulated pollutants. Pollution charges are seen as a mechanism for providing incentives to corporations to adopt improved control technologies and operating practices. The imposition of pollution charges based on emissions rather than pollution-generating activities allows regulators the flexibility to encourage sources to reduce emissions when marginal control costs are equal to pollution charge rates. The challenge for regulators is to identify the desirable charge level. If it is too low, insufficient pollution control will result; if it is too high, costs may be unreasonably burdensome.

Pollution charges have had limited use as a means of achieving emission reductions. Examples include charges on (1) SO_2 emissions from large sources in high-pollution areas in Japan, (2) combustion sources with 750 MW or greater electrical generating capacity in France, (3) oil sulfur (S) content in Norway, and (4) chlorofluorocarbons (CFCs) during the CFC phaseout in the United States.

The pollution charge concept has had a perception problem. Environmental groups have perceived pollution charges as providing industry with the "right to pollute," whereas conservative groups view them as another tax to oppose.

Despite their limited past use, pollution charges may play a significant role in future international efforts to reduce emissions of carbon dioxide (CO_2) and curb predicted greenhouse gas effects on global climate. The use of broad-based carbon (C) fees to limit emissions of greenhouse gases such as CO_2 has been proposed although current regulations are emission standard based. Their primary effect would be to promote energy conservation and energy-conserving technologies.

8.2.2 Tactics

Once control strategies have been selected, regulatory agencies must develop a plan or plans to achieve control objectives. Specific actions taken to implement a control strategy can be described as tactics.

Tactics employed to reduce emissions from stationary sources have included setting emission standards for specific sources in a control region, restrictions on fuel use (e.g., prohibiting the use of coal or oil with a S content of > 1%), use of tall stacks, and prohibitive bans on open burning or use of certain types of combustion equipment (e.g., apartment house incinerators).

The tactics used to control emissions from or reduce air quality problems associated with motor vehicles include (1) control requirements for blowby gases, evaporative losses, refueling, and cold starts; (2) exhaust emission standards; (3) inspection and maintenance; (4) fuel-additive standards; (5) special fuel use requirements; (6) use of alternative power systems; and (7) transportation plans.

As can be seen, emission standards can be used as both an overall control strategy and as a tactic to achieve air quality standards. Various economics-based approaches may also be used as tactics.

Permit systems are a major component of all air pollution control programs. A permit extends a legal right to a source to emit pollutants into the atmosphere. By requiring emission sources to obtain permits, regulatory authorities can impose emission limits on the source or require other actions that reduce emissions. Permits are important tools for obtaining information about emissions from a source and requiring compliance with emission limits. In the broad context, permitting can be viewed as a major air pollution control tactic.

8.3 FEDERAL LEGISLATIVE HISTORY

At the end of the nineteenth and through the first half of the twentieth century, air pollution in the United States was seen as a local problem. This began to change in the 1950s as states, led by California, began a variety of regulatory initiatives. A federal role began to evolve at about the same time. The legislative history of federal air pollution control efforts is described next.

8.3.1 1955 Clean Air Legislation

Legislation enacted in 1955 authorized the Public Health Service, of the Department of Health, Education, and Welfare (DHEW), to conduct air pollution research and training programs and provide technical assistance to state and local governments. It affirmed that state and local governments had the fundamental responsibility for air pollution control. Amendments to this law in 1960 and 1962 called for special studies relating to health effects associated with motor vehicle–related pollutants.

8.3.2 1963 Clean Air Act

In response to the deteriorating quality of air in postwar United States, Congress passed the CAA of 1963. The CAA broadened the federal role in abating air pollution. Specifically, it provided for (1) grants for program development and improvement of state and local air pollution control efforts; (2) accelerated research, training, and technical assistance; (3) federal enforcement authority to abate interstate air pollution problems; and (4) federal research responsibility for automobile and sulfur oxides (SO_x) pollution. Although the federal role was considerably expanded, state and local governments were primarily responsible for air pollution control.

8.3.3 Motor Vehicle Air Pollution Control Act

In 1965, Congress amended the 1963 CAA. These amendments were described as the Motor Vehicle Air Pollution Control Act. It authorized the secretary of DHEW to promulgate emission standards for motor vehicles. As a consequence, federal emission standards were established for all 1968 model light-duty motor vehicles. The 1965 amendments also established the National Air Pollution Control Administration (NAPCA) within DHEW to provide regulatory leadership in the nation's efforts to control air pollution.

8.3.4 Air Quality Act of 1967

Although some progress was made under the CAA of 1963, control programs at all governmental levels remained relatively weak and air pollution problems continued, for the most part, unabated. In 1967, Congress attempted to strengthen this country's air pollution control efforts by enacting the Air Quality Act. Many of the regulatory concepts developed in the 1967 legislation continue to provide the framework for current control efforts. Of special significance were the institution of a regional approach to air pollution control and the development and implementation of air quality standards.

The Air Quality Act of 1967 required NAPCA to designate AQCRs and develop and issue air quality criteria and control technique information. After the federal agency completed these mandated requirements, individual states were to develop state air quality standards and plan for their implementation on a fixed time schedule. Air quality criteria documents were to serve as the scientific basis for development of air quality standards.

The Air Quality Act also provided for the continuation of the federal role in air pollution research and development, training programs, and grants-in-aid to state and local air pollution control programs. Although federal authorities were given significant new powers to initiate lawsuits whenever interstate violations occurred, states maintained primary responsibility for enforcement.

Progress under the Air Quality Act was slow and limited. States could not promulgate air quality standards until the NAPCA designated AQCRs and issued air quality criteria and control technique documents for specific pollutants. These tasks took considerable time; for example, air quality criteria documents for SO_x and PM were not issued until 1969; for HCs, carbon monoxide (CO), and oxidants, 1970; and for nitrogen oxides (NO_x), 1971. The process of designating AQCRs was also very slow. Publication of control technique documents was carried out more expeditiously.

8.3.5 1970 Clean Air Act Amendments

In 1970, public concern for environmental quality had reached its zenith. In response to limited progress under the Air Quality Act of 1967 and public support for strong legislation, Congress enacted tough new CAA Amendments. The NAPCA was eliminated and its air pollution control functions were transferred to the US EPA, an independent federal agency created by executive order of the president.

The CAA Amendments of 1970 sought to remedy deficiencies of earlier legislation. Most importantly, they completed federalization of air pollution control efforts, provided significant federal enforcement authority, and placed air pollution control on a rigorous timetable.

Specific provisions of the 1970 CAA Amendments included (1) setting uniform national ambient air quality standards (NAAQSs), (2) immediately designating AQCRs, (3) requiring state implementation plans (SIPs) to achieve NAAQSs, (4) setting stringent new automobile exhaust emission standards and standards for fuel additives, (5) setting emission standards for new or significantly modified sources and HAPs, (6) allowing the right of citizen suits, and (7) providing for federal enforcement authority in air pollution emergencies and interstate and intrastate air pollution violations.

8.3.6 1977 Clean Air Act Amendments

The goal of the 1970 CAA Amendments was to achieve clean air for all regions of the country by July 1, 1975. Although significant progress for some pollutants was made, many AQCRs did not achieve one or more NAAQSs by the congressionally mandated deadline.

Recognizing that air quality goals set in 1970 could not be met without major economic disruptions, Congress amended the CAA in 1977. These amendments would best be described as a "midcourse correction." Specifically, they (1) postponed compliance deadlines for national primary (health-based) air quality standards; (2) postponed, and in some instances modified, federal automobile emission standards; (3) included a PSD provision into the language of the act; (4) provided a mechanism by which the US EPA could have flexibility in allowing some growth in "dirty" air regions; and (5) gave the US EPA authority to regulate stratospheric O_3-destroying chemicals. Regulation of O_3-destroying chemicals was the major new initiative in the 1977 Amendments.

8.3.7 1990 Clean Air Act Amendments

Although it was the intent of Congress to reauthorize and update by amendment all major federal legislation on a 5-year schedule, disagreements over policies related to the control of acidic deposition contributed to an 8-year hiatus in amending the CAA. In 1990, major agreements on CAA Amendments were reached by Congress and the president. The 1990 CAA Amendments represented a significant political achievement. Major new or expanded authorities included (1) changes in timetables for achieving air quality standards in nonattainment areas, (2) regulation of emissions from motor vehicles, (3) regulation of hazardous and toxic air pollutants, (4) acidic deposition control, (5) stratospheric O_3 protection, (6) permitting requirements, and (7) enforcement.

8.3.8 Federalization

In the evolution of federal legislation, the most significant trend was increasing involvement of the federal government in the problem, and ultimate primacy in its solution. This federalization of air pollution control efforts was not unique; it was consistent with the centralization of responsibility at the federal level for many problems that beset post–World War II American society. During this period, the principle of "states' rights" slowly eroded, with the recognition that problems such as air pollution could only be effectively dealt with by a coordinated approach at all levels of government.

8.4 AIR POLLUTION CONTROL UNDER THE 1970, 1977, AND 1990 CAA AMENDMENTS

Prior to enactment of the CAA Amendments in 1970, the NAPCA and Congress considered several different control strategies. The outcome of their deliberations was an ambient air pollution control strategy based on setting NAAQSs, with the proviso that states develop implementation plans to achieve them. Additionally, national emission standards would be established for major new sources and for existing and new sources of HAPs. As a consequence, the nation's ambient air pollution control program has been based primarily on strategies employing both air quality management and emission standards concepts.

8.4.1 Air Quality Standards

Under the 1967 Air Quality Act, states were required to set air quality standards for AQCRs within their jurisdiction. These standards were to be consistent with air quality criteria developed and issued by the NAPCA. Development of state air quality standards under the 1967 legislation was slow, due in part to the slowness of the NAPCA to publish air quality criteria and designate AQCRs.

In addition, states were concerned that the adoption of stringent standards would result in their industries relocating to states with weaker standards. This concern created a great deal of uncertainty, confusion, and ultimately inaction.

Under the CAA Amendments of 1970, the federal government was charged with the responsibility of developing uniform NAAQSs. These were to include primary standards designed to protect health and secondary standards to protect public welfare. Primary standards were to be achieved by July 1, 1975, and secondary standards "at a reasonable time thereafter." Implicit in this dual-standard requirement was the premise that pollutant levels protective of public welfare were to be more stringent than those for public health and that achievement of primary health-based standards had immediate priority. In 1971, the US EPA promulgated NAAQSs for five air pollutants (CO, NO_2, photochemical oxidants, PM, and SO_2). In 1978, an air quality standard was promulgated for lead (Pb) as the sixth criteria pollutant. Also, in 1978, the 1971 photochemical oxidant standard was revised to an O_3 standard, with an increase in permissible levels from 0.08 ppmv to 0.12 ppmv. In 1987, the suspended PM standard was revised and designated a PM_{10} (PM with an aerodynamic diameter of $\leq$ 10 μm) standard and included a 24-hour (150 μg/m^3) and an annual average (50 μg/m^3) standard. This revision reflected the need for a particle size-based PM standard that better reflected respiratory health risks. In 1997, the US EPA promulgated a new 8-hour standard for O_3 (0.08 ppmv) and a $PM_{2.5}$ standard (in addition to the existing PM_{10}). The $PM_{2.5}$ standard also consisted of a 24-hour (65 μg/m^3) and an annual (15 μg/m^3) standard. The latest NAAQSs (including the most recent changes to the SO_2 and NO_2 1-hour standards in 2010, the change in the $PM_{2.5}$ annual standard in 2012 to 12 μg/m^3 and the change in the O_3 standard to 0.070 ppmv in 2015) and the averaging times and forms of the standards are summarized in Table 8.1. The table and detailed summaries of the changes in the standards since their establishment as criteria pollutants can be found on the US EPA website at www.epa.gov/criteria-air-pollutants/naaqs-table.

8.4.2 Air Quality Criteria

Promulgation of NAAQSs must be preceded by publication of air quality criteria. These are issued in document form and summarize all relevant scientific information on the health and welfare effects of individual pollutants. The principal function of the air quality criteria process is to determine the relationship between pollutant concentrations and health and welfare effects so that air quality standards are supported by strong scientific evidence. These documents are reviewed on a periodic basis to determine if there is a basis for adjusting the standard based on new scientific evidence. As examples, the CO standard (established in 1971) was reviewed and retained based on comprehensive reviews conducted in 1985, 1994, and 2011 and the Pb standard was established in 1978, revised in 2008, and reviewed and retained in 2016.

8.4.3 Air Quality Control Regions

The concept of a regional approach to air pollution control was first introduced in the Air Quality Act of 1967. This approach recognized that air pollution was a regional problem that did not respect political boundaries. Initially, the regional concept was only applied to interstate problems. With enactment of the 1970 CAA Amendments, Congress mandated that both interstate and intrastate AQCRs be established for the purpose of achieving NAAQSs.

8.4.4 State Implementation Plans

After promulgation of NAAQSs by the US EPA or a revision to the NAAQS, areas within the AQCRs are determined to either be in attainment or in nonattainment for the specific pollutant. If in nonattainment, individual states are required to develop and submit plans for their implementation, enforcement, and maintenance. These SIPs can be approved by the US EPA in whole or in part. If a

TABLE 8.1
National Ambient Air Quality Standards

Pollutant	Primary/Secondary	Averaging Time	Level	Form
Carbon monoxide (CO)	Primary	8 hour	9 ppmv	Not to be exceeded more than once per year
		1 hour	35 ppmv	
Lead (Pb)	Primary and secondary	Rolling 3-month average	0.15 μg/m^3 [a]	Not to be exceeded
Nitrogen dioxide (NO_2)	Primary	1 hour	100 ppbv	98th percentile of 1-hour daily max, averaged over 3 years
	Primary and secondary	Annual	53 ppbv[b]	Annual mean
Ozone (O_3)	Primary and secondary	8 hour	0.070 ppmv[c]	Annual fourth-highest daily maximum 8-hour concentration, averaged over 3 years
Particle pollution ($PM_{2.5}$)	Primary	Annual	9 μg/m^3	Annual mean, averaged over 3 years
	Secondary	Annual	15 μg/m^3	Annual mean, averaged over 3 years
	Primary and secondary	24 hour	35 μg/m^3	98th percentile, averaged over 3 years
Particle pollution (PM_{10})	Primary and secondary	24 hour	150 μg/m^3	Not to be exceeded more than once per year on average over 3 years
Sulfur dioxide (SO_2)	Primary	1 hour	75 ppbv[d]	99th Percentile of 1-hour daily maximum concentrations, averaged over 3 years
	Secondary	Annual	10 ppbv	Annual mean, averaged over 3 years

[a] *In areas designated* nonattainment for the Pb standards prior to the promulgation of the current (2008) standards, and for which implementation plans to attain or maintain the current (2008) standards have not been submitted and approved, the previous standards (1.5 μg/m^3 as a calendar quarter average) also remain in effect.

[b] The official level of the annual NO_2 standard is 0.053 ppmv, equal to 53 ppbv, which is shown here for the purpose of clearer comparison to the 1-hour standard.

[c] Final rule signed October 1, 2015, and effective December 28, 2015. The previous (2008) O_3 standards are not revoked and remain in effect for designated areas. Additionally, some areas may have certain continuing implementation obligations under the prior revoked 1-hour (1979) and 8-hour (1997) O_3 standards.

[d] The previous SO_2 standards (0.14 ppmv 24-hour and 0.03 ppmv annual) will additionally remain in effect in certain areas: (1) any area for which it is not yet 1 year since the effective date of designation under the current (2010) standards, and (2) any area for which an implementation plan providing for attainment of the current (2010) standard has not been submitted and approved and which is designated nonattainment under the previous SO_2 standards or is not meeting the requirements of a SIP call under the previous SO_2 standards (40 CFR 50.4(3)). A SIP call is an EPA action requiring a state to resubmit all or part of its State Implementation Plan to demonstrate attainment of the required NAAQS.

portion of an SIP is disapproved, the state must revise it and seek approval for revisions. If the plan, in whole or in part, is not acceptable, the US EPA is empowered to develop an SIP for the state, which the state must then enforce.

In developing implementation plans, states formulate policies for each air pollutant for which NAAQSs have been promulgated. In this process, they are required to determine and report on the nature and quantity of emissions from existing sources and require, at a minimum, RACT to limit emissions. They negotiate compliance schedules with major sources and indicate the effects that emission reductions will have on air quality. Implementation plans are directed to pollutants in AQCRs where levels are above the NAAQSs.

8.4.4.1 Episode Plans

State plans must include control strategies for air pollution episodes or emergencies. Episodes occur as a result of stable atmospheric conditions associated with stagnating, migratory, or semipermanent high-pressure systems. The poor dispersion associated with these atmospheric conditions may result in ambient pollutant levels that pose increasingly greater health risks. States are required to develop response criteria and plans. An example of the pollutant concentrations triggering various episode stages from advisory, to alert, to emergency level is shown in Table 8.2, taken from the current State of Illinois episode plan in 2020. Note that each stage or level reflects a higher degree of pollution severity. Control actions at each stage are to be sufficient to prevent the next stage from occurring. Episode plans must provide for a reduction of emissions on a prearranged schedule and should therefore, in theory, assist in protecting public health from worsening air quality during the episode.

8.4.4.2 Legal Authority

To gain approval for its SIP, a state must have adequate legal authority to (1) adopt and enforce emission regulations; (2) implement episode control plans; (3) regulate new or modified sources; (4) require stationary sources to provide emission data; and (5) gather information to determine compliance with regulations, including the right to inspect and test. Although federal law is "the law of the land," it cannot be enforced by state and local officials. State and local governmental regulatory agencies must enact enabling legislation and ordinances to enforce provisions of federal law. Although these reflect local circumstances and sentiments, they must be consistent with provisions of federal clean air legislation.

8.4.4.3 State Implementation Plan Approval

In practice, the process of SIP development and approval has been a demanding experience for federal, state, and local officials. Plans had to be submitted within 9 months of primary standard promulgation. In many cases, data on emissions (emission inventory) were scant and manpower for

TABLE 8.2
Episode Plan Showing Concentrations That Trigger Each Episode Stage ()

Pollutant	Averaging Time	Advisory	Yellow Alert	Red Alert	Emergency
SO_2	2-hour	0.3	-	-	-
(ppmv)	4-hour	-	0.30	0.35	0.40
PM_{10}	2-hour	420	-	-	-
(µg/m^3)	4-hour	-	350	420	500
CO	2-hour	30	-	-	-
(ppmv)	8-hour	-	15	30	40
O_3	1-hour	0.12	0.20	0.30	0.50
(ppmv)					
NO_2	2-hour	0.40	-	-	-
(ppmv)	1-hour	-	0.60	1.20	1.60
	24-hour	-	0.15	0.30	0.40

Source: State of Illinois Administrative Code, Title 35, Subtitle B, Chapter 1, Subchapter 1, Part 244: Episodes, approved, 1992.

plan development insufficient. Many of these early hastily prepared plans were subject to numerous revision requests before federal approval could be granted. Additional revision requirements were necessitated by federal court decisions resulting from citizen suits filed against the US EPA relative to requirements of SIPs. As a result of these decisions, states in the early 1970s had to revise their implementation plans to provide for maintenance of air quality once standards were achieved, and PSD of air quality in regions where air quality was better than the standards. Since then, SIPs have had to be revised in response to changes in standards and regulatory policies initiated by the US EPA or mandated by the 1977 and 1990 CAA Amendments.

8.4.5 Nonattainment of Air Quality Standards

As the 1975 statutory deadline for attainment of NAAQSs was approaching, it was evident that many AQCRs would not be in compliance with one or more standards. Congress recognized that an extension of deadlines was necessary. Under the 1977 CAA Amendments, attainment deadlines for SO_2 and PM were extended to 1982, and deadlines for automobile-related pollutants were extended to 1987.

Because many AQCRs were not in compliance with one or more air quality standards, the US EPA classified these as nonattainment areas. Nonattainment status had considerable significance because the introduction of new major sources in a dirty air region would be expected to aggravate threats to public health and would therefore be prohibited. The prospect of limitations on economic development where economic growth would naturally occur created major political and economic concerns. In response, the US EPA administratively developed a "compromise" emission offset policy that would allow growth as long as new sources did not produce a net increase in emissions. Under the offset policy, a new industrial facility would have to identify and use emission reductions from existing facilities to offset emissions associated with the new source. In cases where there was to be an expansion of an existing facility, sufficient emission reductions had to be made in the existing facility to qualify as offsets.

The 1977 CAA Amendments mandated that the US EPA offset policy be incorporated into the SIP process. The 1977 CAA Amendments also provided an alternative; that is, a major new source could be approved for a nonattainment region if it would attain the "lowest achievable emission rate" (LAER). To be eligible under this provision, an industrial source having other facilities within a state would have to be in compliance with all relevant emission reduction programs for all its facilities. In practice, LAER has required the use of BACT on such new sources.

The problem of nonattainment was addressed in a major way in the 1990 CAA Amendments. Deadlines were again postponed, and the US EPA was given increased flexibility in setting deadlines in nonattainment areas. In general, attainment dates were those that could be achieved as expeditiously as practicable, but no later than 5 years after designation. This could be further extended for a period no longer than 10 years if the US EPA deemed it appropriate, considering the severity of the nonattainment problem and the availability and feasibility of control measures.

Attainment dates for O_3, CO, and $PM_{2.5}$ were to be determined based on the classifications in Table 8.3. The ozone standard changed from a 1-hour standard (0.120 ppm) to an 8-hour standard of 0.080 ppm in 1997 to an 8-hour standard of 0.075 in 2008 to 0.070 in 2015. The classifications for these are shown in the table. As can be seen, the more severe the problem and the more difficult its control, the more time a state had/has to bring an AQCR into compliance with primary air quality standards. The passage of a more stringent standard, such as is the case for ozone, prior to a state having achieved attainment, creates a complicated matrix. The authors have tried to capture this complexity by adding the 2015 ozone information to the table that was previously published on the US EPA website for the 2008 standards.

8.4.6 New Source Performance Standards

The 1970 CAA Amendments required that new or significantly modified existing sources (that result in increased pollution emissions) comply with NSPSs. Emission standards established under

TABLE 8.3
Classification and Attainment Dates for Nonattainment Areas under the 1990 CAA Amendments

Area Class	Design Value (ppmv)	Attainment Date
Ozone, 1-hour (revoked)		
Marginal	0.121–0.138	3 years after enactment
Moderate	0.138–0.160	6 years after enactment
Serious	0.160–0.180	9 years after enactment
Severe	0.180–0.280	15 years after enactment
Extreme	≥ 0.280	20 years after enactment
Ozone, 8-hour std, 1997		
Marginal	0.085–0.092	
Moderate	0.092–0.107	
Serious	0.107–0.120	2008–2012[a]
Severe 15	0.120–0.127	2009–2012[a]
Severe 17	0.127–0.186	2012[a]
Extreme	≥ 0.186	
Ozone, 8-hour, 2008		
Marginal	0.076–0.086	July 20, 2018
Moderate	0.086–0.100	July 20, 2018[a]
Serious	0.100–0.113	July 20, 2018[a]
Severe 15	0.113–0.119	July 20, 2018[a]
Severe 17	0.119–0.175	July 20, 2018[a]
Extreme	≥ 0.175	3 years after designation
Ozone, 8-hour, 2015	0.071–0.081	6 years after designation
Marginal	0.081–0.093	9 years after designation
Moderate	0.093–0.105	
Serious		
Severe	0.105–0.163	15 or 17 years after designation
Extreme	≥ 0.163	20 years after designation
Carbon monoxide		
Moderate	9.1–16.4	December 31, 1995
Serious	≥ 16.5	December 31, 2000
$PM_{2.5}$		
All NA areas classified as moderate		December 31, 2015

[a] Exact dates varied by area.

NSPSs were to be applied to all new sources in designated source categories uniformly, without regard to whether they were being constructed in clean or dirty air regions. These performance or emission standards were to reflect the degree of emission limitation achievable through application of the best system of emission reduction, taking cost into consideration. The purpose of these emission limitations was to prevent new pollution problems by requiring the installation of control measures during construction, when they would be least expensive. As time has passed, many of the NSPSs have been amended to include existing sources that were in existence prior to passage of the initial NSPS for that source category.

NSPSs are applied to specific source categories, the definition of which is the responsibility of the US EPA administrator. In 2020, there were more than 70 NSPSs listed at the website (www.epa.gov/stationary-sources-air-pollution/new-source-performance-standards). A sample of source categories and pollutants regulated under NSPSs is presented in Table 8.4. The majority of these source categories were also required to meet National Emission Standards for Hazardous Air Pollutants (NESHAPs). These latter standards have incorporated more of the noncriteria pollutants, some of which were included in the initial NSPSs. For example, NSPSs have required emission limits on fluoride from phosphate and aluminum (Al) plants, acid mist from sulfuric acid (H_2SO_4) plants, and total reduced S from paper mills. NSPSs can be applied to new sources for pollutants not regulated under NAAQSs or NESHAP standards. Although the formulation and promulgation of NSPSs for any category of sources and pollutants is a federal responsibility, the US EPA delegates enforcement authority to states with acceptable SIPs.

In June 2014, following the adoption of the Climate Action Plan by the Office of the President in 2013, the US EPA issued a proposed NSPS related to carbon pollution from electric utility generation units (EGUs). The final rule (Standards of Performance for Greenhouse Gas Emissions from New, Modified or Reconstructed EGUs) was published in October 2015. For newly constructed coal-fired and gas-fired EGUs, the standard would have required CO_2 emission limits of 1,400 lb/MWh and 1,000–1,030 lb/MWh, respectively, with lesser restrictive limits for modified or reconstructed EGUs. These levels would have reduced carbon pollution by encouraging reliance on efficient units like natural gas-fired stationary combustion turbines, integrated gasification combined cycle (IGCC) systems, and CO_2-controlled fossil-fired utility boilers using technologies such as partial carbon capture and storage (CCS). The proposed standards would have required 40% to 50% capture and storage of CO_2 emissions from conventional coal-fired power plants typical of those in current operation. Given that the carbon content of coal is approximately 50%–80% of coal by weight, and CO_2 (MW of 44) is 3.7 times that of C (MW of 12), the CO_2, if captured, would have been approximately equal 1 to 1.5 times the total weight or tonnage of coal being burned, and thus represented a significant materials handling as well as technological challenge.

In December 2018, the US EPA proposed revisions to the greenhouse gas NSPS for EGUs that would allow adoption of the best system of emission reduction (BSER) that has been adequately demonstrated. The proposed BSER reduced the stringency of the NSPSs for new coal-fired EGUs to 1,900 lb/MWh and would not require carbon capture and storage. The primary reason for this proposed revision was stated to be the high costs and limited geographic availability of CCS. The implementation of the NSPS for greenhouse gases remained uncertain as of early 2020. While it has not been implemented, the likelihood of the aforementioned or more stringent future greenhouse gas emission standards has been a driving force in decisions made by many utility companies to retire older coal-fired EGUs in favor of less carbon-intensive options. While not a part of the CAA, the United States established an energy tax credit for both industry and EGUs who voluntarily capture CO_2 emissions. The tax credit and its impact on emissions is discussed in more detail in Chapter 10, Section 10.4.8.1.

8.4.7 Prevention of Significant Deterioration

Although the principle of PSD in areas where air quality was better than the standards was not specifically mentioned in the CAA Amendments of 1970, environmental groups maintained that nondegradation of air in clean air areas was implicit in the wording of the preamble of the act, "to protect and enhance," and reflected congressional and federal agency intent documented in congressional hearings before its passage. As a result of a lawsuit filed by environmental groups, the US EPA was compelled to develop regulations to protect air quality in clean air areas.

In 1975, the US EPA developed a PSD policy with a classification system that would allow some economic development in clean air areas and still protect air quality from significant deterioration. To eliminate legal uncertainty, the classification system and provisions for its implementation were incorporated into the 1977 CAA Amendments. Clean air areas were divided into three classes. Very little deterioration would be allowed in Class I regions, that is, all international parks, national wilderness areas with more than 5,000 acres, national monuments with more than 5,000 acres, and

TABLE 8.4
Pollutants Regulated under NSPSs in Selected Source Categories

Sources	Pollutants
Electric utility steam generation	PM, opacity, SO_2, NOx, Hg
Coal preparation plants	PM, opacity
Municipal waste combustors	PM, metals, organics, acid gases, NO_x
Primary lead, zinc, and copper smelters	PM, opacity, SO_2
Secondary lead smelters	PM, opacity
Secondary brass and bronze smelters	PM, opacity
Primary aluminum production	Fluorides, opacity
Iron and steel plants	PM, opacity
Ferroalloy production facilities	PM, opacity, CO
Steel plants—electric arc furnaces	PM, opacity
Sulfuric acid plants	SO_2, acid mist, opacity
Nitric acid plants	NO_x
Portland cement manufacturing	PM, opacity, NO, SO_2
Hot mix asphalt facilities	PM, opacity
Sewage treatment plant incineration	PM, opacity
Petroleum refineries	PM, SO_x, CO, reduced sulfur
Kraft pulp mills	PM, opacity, reduced sulfur
Grain elevators	PM, opacity
Lime manufacturing plants	PM
Phosphate fertilizer industry	Fluorides

national parks with more than 6,000 acres. These included 155 areas in 36 states and one international park. The US EPA was required to develop and implement regulations that would protect visibility in these pristine air areas. Class II areas would allow moderate air quality deterioration, and Class III areas would allow air quality deterioration up to the secondary standard. Maximum allowable increases above baseline values for each of the three PSD classes are summarized in Table 8.5.

Under the PSD program, all areas in attainment for SO_2 and PM are designated Class I, II, or III for that pollutant. Additional pollution allowed in these areas is an increment beyond a baseline level established at the time of the location of the first major emitting facility. Similar requirements have also been established for NO_2. PSD increments were established for $PM_{2.5}$ and became effective on October 20, 2011. All areas of an individual state that meet NAAQSs are initially designated as Class II except for those areas specifically designated by the CAA as Class I.

Special permitting requirements must be met by major new facilities in PSD areas. If they have the potential to affect visibility in Class I areas, they are subject to a visibility review. They must use BACT. Major existing stationary sources in federal Class I areas that adversely affect visibility are required to install best available retrofit technology (BART) to minimize visibility impairment.

Although there was considerable congressional debate over visibility protection prior to the passage of the 1990 CAA Amendments, actual PSD changes were relatively limited. Most importantly, additional emphasis was placed on regional haze issues. The US EPA was required to work with several western states to establish a commission to address visibility in the Grand Canyon. The agency promulgated regional haze rules (1999) subsequent to the Grand Canyon Visibility Transport Commission's report of 1996.

These regulations are designed to improve visibility in 156 Class I areas. Because fine particles (< 2.5 μm) are transported hundreds of miles, most states, including those that do not have Class I regions, have to participate in planning, analyses, and, in many cases, emission reductions. States

TABLE 8.5
Allowable Increments for PM_{10}, $PM_{2.5}$, SO_2, and NO_2 in PSD-Regulated Regions

Pollutant	Standard	Class	Allowable Increment (μg/m³)
PM_{10}	Annual mean	I	4
		II	17
		III	34
	24-hour maximum	I	8
		II	30
		III	60
$PM_{2.5}$	Annual mean	I	1
		II	4
		III	8
	24-hour maximum	I	2
		II	9
		III	18
SO_2	Annual mean	I	2
		II	20
		III	40
	24-hour maximum	I	5
		II	91
		III	182
	3-hour maximum	I	25
		II	512
		III	700
NO_2	Annual mean	I	2.5
		II	25
		III	50

were required to develop long-term strategies with periodic goals for improving visibility in Class I areas. They were also required to develop implementation plans that included measures enforceable by 2008 and identify facilities that would have to install BART controls. Haze rules considerably broaden the scope of existing PSD review requirements that primarily focus on new sources and use existing baseline air quality as a reference for making decisions.

8.4.8 Hazardous and Toxic Air Pollutants

Under the 1970 CAA Amendments, Congress authorized the US EPA to regulate pollutants deemed to be more hazardous than those regulated under NAAQSs. As a result, the US EPA initially promulgated National Emissions Standards for Hazardous Air Pollutants for Hg, Be, and asbestos. By 1990, NESHAPs had also been promulgated for vinyl chloride, inorganic arsenic (As), benzene, and radioactive isotopes.

Over a period of 20 years, the US EPA designated and regulated only seven HAPs. Many states had gone their own way in regulating what they described as air toxics. By 1990, state air pollution control agencies had established emission standards for approximately 800 "toxic" pollutants.

The relatively slow pace of designating and regulating HAPs reflected policy uncertainties associated with the NESHAP provision. The US EPA interpreted this provision as requiring the setting of emission limits that would provide an ample margin of safety. Because many HAPs are carcinogenic, the US EPA had at various times interpreted the statutory language as requiring an emission standard of zero for carcinogenic substances. It was therefore reluctant to regulate emissions of economically important substances that were potential human carcinogens, as such regulation would have required a total ban on their production.

Regulation of hazardous or toxic air pollutants received special attention under the 1990 CAA Amendments. In Title III, Congress listed 182 HAPs for which the US EPA was to identify source categories and promulgate emission standards by the year 2000. The US EPA was also authorized to add substances to this list if there were adequate data to show that emissions, ambient concentrations, bioaccumulation, or deposition could reasonably be anticipated to cause adverse human health or environmental effects. EPA considers per- and polyfluoroalkyl substances (PFAS) are potentially harmful as HAPs. PFAS are a group of manufactured chemicals that can be released into the air. Per- and polyfluoroalkyl substances are widely used, long-lasting chemicals that break down very slowly over time. Due to their persistence and widespread use, PFAS have been detected in the blood of people and animals worldwide, as well as at low levels in food, water, air, soil, and fish.

Scientific research has linked exposure to certain PFAS with potential health risks in both humans and animals. With thousands of PFAS compounds found in a wide range of consumer, commercial, and industrial products, evaluating their health and environmental impacts remains a significant challenge. Two species of PFAS are perfluorooctanoic acid (PFOA) and perfluorooctanoic sulfonic acid (PFOS), which are human-made chemicals used in various consumer products. PFAS started to be observed and identified in their emissions. They will be defined as a group of HAPs in the near future.

One of the unique features of Title III is that hazards are not limited to human health. By reference to adverse environmental effects, Congress indicated its intent to protect wildlife, aquatic life, and other natural resources, including populations of endangered or threatened species, and prevent the degradation of environmental quality over broad areas. Also unique is the use of technology-based standards rather than standards based on known health effects.

The US EPA was required to (1) publish a list of categories and subcategories of major sources (> 10 tons/year single pollutant, > 25 tons/year aggregate) of HAPs and area sources for pollutants listed, and (2) promulgate emission standards for each in accordance with a congressionally mandated schedule. In promulgating technology-based emission standards, the US EPA was given the flexibility to distinguish among classes, types, and sizes of sources. It was to require the maximum degree of reduction in emissions, taking into account the cost of achieving such emission reductions, any non-air quality health and environmental effects, and energy requirements. Maximum achievable emission reduction could be achieved by (1) process changes, substitution of materials, or other modifications; (2) enclosing systems or processes; (3) collecting or treating pollutants released from a process, stack, storage, or fugitive source; or (4) design, equipment, work practice, or operating standards. Maximum achievable control technology standards are used to describe these emission reduction requirements for air toxics and HAPs.

As of 2020, there were approximately 140 different sources/categories of industrial and municipal sources that are subject to NESHAPS and MACT standards, and revisions occur to one or more standards every year. An up-to-date listing of categories and standards can be found on the US EPA website at www.epa.gov/stationary-sources-air-pollution/national-emission-standards-hazardous-air-pollutants-neshap-9.

There is often an overlapping relationship between the NESHAPs and the NSPSs for a given source category as well as political debate related to the economic benefits of the standards. This is illustrated by the activities in the 2011 to 2013 period in which the US EPA (under the Obama Administration) proposed and promulgated the new Mercury and Air Toxics Standards (MATS) under NESHAP. The MATS established new or more stringent emission standards for Hg, filterable PM and HCl from new and existing coal-fired and integrated gasification combined cycle units, and PM, HCl, and HF standards for new and existing oil-fired electric utility generating units, all of which were already under the NSPS. The more restrictive PM standards served as surrogates for non-Hg-based HAPs, whereas the HCl and HF standards served as surrogates for acid gases. Alternative emission standards were also included in the MATS for the specific non-Hg HAPs and acid gases in cases where the EGU might choose to control the individual pollutants for which PM and HCl served as surrogates. The proposed MATS NESHAP created a need to revisit the utility NSPS, resulting in revisions to both that were finalized in 2013. Altogether, it was estimated that 1,400 EGUs were affected by MATS—1,100 existing coal-fired and 300 oil-fired at 600 power plants. The cost (in 2007 USD) to comply, at the time of passage of MATS, was estimated to be \$9.6 billion a year but would provide \$37 billion to \$90 billion per year in public health benefits. The MATS was projected to reduce mercury emissions from EGUs by 90%, acid gases by 88%, and SO_2 by 41% compared to 2011 levels. In February 2019, the US EPA (under the Trump administration) issued a proposed rule, partially in response to a US Supreme Court ruling in *Michigan vs. EPA*, that "after considering the cost of compliance relative to the HAP benefits of regulation, the EPA proposes to find that it is not 'appropriate and necessary' to regulate HAP emissions from coal- and oil-fired EGUs, thereby reversing the Agency's prior conclusion under CAA." While the US EPA stated that it would not remove EGUs from the CAA under NSPS, it did state further,

> We are soliciting comment, however, on whether the EPA has the authority or obligation to delist EGUs from CAA section 112(c) and rescind (or to rescind without delisting) the National Emission Standards for Hazardous Air Pollutants (NESHAP) for Coal- and Oil-Fired EGUs, commonly known as the Mercury and Air Toxics Standards (MATS).

While MATS was still in effect in early 2020 (a presidential election year), the proposal has been the subject of substantial political, economic, and public health debate.

8.4.9 Stratospheric Ozone Depletion

The 1977 CAA Amendments provided the US EPA statutory authority to regulate substances, practices, processes, and activities that could reasonably be anticipated to damage the O_3 layer. Exercising this authority, CFC use for aerosol propellants was banned in 1978. US emissions of CFCs declined by 35% to 40% of their preban levels as a result. The 1977 amendments provided the US EPA the authority it used to meet the emission reduction targets of the 1987 Montreal Protocol and subsequent amendments to it.

The 1990 CAA Amendments repealed the O_3 protection section of the 1977 CAA Amendments by adding Title VI: Stratospheric O_3 Protection. These amendments updated US EPA authority and added specific requirements that reflected more recent needs for protecting the O_3 layer and policies to achieve such protection.

Major provisions of Title VI include a listing of Class I and Class II substances (see later), monitoring and reporting requirements, phaseout of Class I and Class II substances, accelerated phaseout authority, a national recycling and emission reduction program, servicing of motor vehicle air conditioners, regulating nonessential products containing CFCs, labeling requirements for Class I and II substances, a policy for the development of safe alternatives, and international cooperation.

Classification of chemicals into Class I and II substances reflects differences in O_3 depletion potential and a recognition that Class II substances, used as substitutes for CFCs, can also destroy O_3, albeit this potential is at least an order of magnitude lower. Under substantial strengthening of the Montreal Protocol in 1990, production of Class I substances, including CFCs, three types of bromine (Br)-containing halons, and carbon tetrachloride (CCl_4), was to be banned by the year 2000, and methyl chloroform by 2002. However, several countries did not ban these until 2010, and worldwide bans are an enforcement challenge. As late as 2019, scientists observed emissions of CFC-11 coming from certain regions in Asia. The countries involved have reportedly now ceased production. Class II compounds, which include hydrochlorofluorocarbons, were to be banned by 2020 (2030 for developing countries), with a phaseout that began in 2015.

A critical review of the literature and the latest findings on stratospheric ozone can be found in the June 2013 critical review of stratospheric ozone conducted by the Air and Waste Management Association referenced in the section on additional readings at the end of the chapter.

8.4.10 Acidic Deposition

For many years, congressional efforts to amend the CAA were stymied by contentious debate over control of acidic deposition. Congress put that debate to rest and enacted the 1990 CAA Amendments with specific authority and language addressing acidic deposition.

The purpose of Title IV: Acidic Deposition Control was to minimize the adverse effects of acidic deposition by reducing annual emissions of SO_x and NO_x by 10 and 2 million tons, respectively, from 1980 emission levels.

Mindful of potentially significant economic costs and dislocations that may have occurred as a result of traditional "command-and-control" regulatory approaches, Congress developed an economics-based approach that it believed to be both more flexible and cost-effective. Emission reductions to meet the goals of the acidic deposition program would be achieved by establishing a program of emission allowances. An allowance under the Act was equal to 1 ton of SO_x.

Emission reduction requirements for SO_x, using the allowance provision, were applied to large fossil fuel–fired electric boilers specifically identified in the Act. These were to be achieved using a phased approach, with Phase 1 commencing in 1995 and Phase 2 in 2000. These requirements reduced total annual emissions in the 48 contiguous states and the District of Columbia to approximately 8 million metric tons/year (8.9 million tons/year).

In the allowance program, each unit of each affected source was allocated allowances that were determined from total emissions and emission reductions required to achieve Phase 1 and Phase 2 goals. Each source had to develop a plan that described a schedule and the methods to achieve compliance. Unused allowances at one unit or source could be used by utility operators to offset emissions at other units or sources. They could be traded or banked for use in future years.

Title IV of the 1990 CAA Amendments repealed the SO_2 NSPS (requiring use of scrubbers) prescribed for coal-fired power plants under the 1977 CAA Amendments. Congress mandated that the US EPA promulgate an NSPS for coal-fired power plants that required the best system of emission reduction, taking into account the costs of achieving the reduction, any non-air quality health and environmental effect, and energy requirements.

Nitrogen oxide emission reductions were to be implemented in two phases, in 1996 and 2000. Although the NO_x program utilized some of the same emission reduction principles, for example, emissions trading and flexibility in how emission reductions were to be achieved, it did not cap NO_x emissions or utilize an allowance system. Compliance was to be achieved by using an individual emission rate for each boiler or averaging emission rates over two or more units to meet an overall emission rate reduction requirement.

Title IV allowed regulated sources to determine their own compliance approaches. These have included switching to cleaner-burning fuels, reassigning some energy production from relatively high emission-producing units to cleaner units, and adopting conservation or efficiency measures. Prior to Title IV requirements, most utilities used what was described as least-cost dispatching. Under such a system, generating units that produced electrical power at the least cost were used more than higher-cost units. Not surprisingly, lower-cost units had the highest SO_x emissions. In recent years, as summarized in Chapter 2, emissions of SO_2 have been further reduced due to concerns over greenhouse gas emissions (specifically CO_2) and the retiring of older coal-fired EGUs in preference for less carbon-emitting sources such as natural gas and renewable sources (wind, solar, and hydro).

8.4.11 Motor Vehicle Emission Control

The pioneering atmospheric chemistry studies of Dr. Arie Haagen-Smit in the 1950s provided a fundamental understanding of the role of automobile emissions in photochemical smog development. As seen in Tables 2.1 through 2.3, automobiles have been the major anthropogenic source of CO, HCs, NO_x, and Pb. They are the major source of precursor molecules for the production of photochemical oxidants such as O_3. It is for this reason that automobile emissions have been subject to significant regulatory requirements.

Automobile air pollution has been of greater concern in southern California than in any other location in the world. This is due to a culture heavily dependent on automobile usage, abundant sunshine, topographical barriers to air movement, and frequent elevated inversions (Chapter 2) during summer months. Consequently, California has led the nation and the world in controlling automobile emissions. California emission requirements have typically preceded those required nationally and have, in most cases, been more stringent.

8.4.11.1 Vehicle Emissions

Emissions from light-duty motor vehicles have included blowby gases, exhaust gases, volatile nonmethane hydrocarbon (NMHC) losses from the carburetor and fuel tank, and Pb. Control of motor vehicle emissions has included regulatory actions to reduce crankcase emissions, exhaust gases, and evaporative losses. Fuel composition standards have also been necessary to facilitate the use of catalytic systems, phase out the use of Pb, reduce the sulfur content of fuel, and control fuel volatility (Reid vapor pressure) to achieve the emission reductions from vehicles. Similar programs have been implemented for heavy-duty vehicles, including a variety of mobile sources including on-road and off-road vehicles, railroads, and aircraft. Here, we will provide a summary primarily of the development of standards for light-duty vehicles, which now encompasses both passenger vehicles as well as light-duty trucks.

8.4.11.2 Regulatory History

The regulatory history of control requirements mandated for motor vehicles sold in California and nationwide is summarized in Table 8.6. Because of the relatively simple and low-cost technology (Chapter 9) required, crankcase emissions were effectively controlled in the early 1960s, whereas evaporative emissions were controlled in the early 1970s. Control of exhaust gases, which account for the largest percentage of emissions (~90%), has had a complex history of regulatory requirements, technology development, cost, and public acceptance. Table 8.6 illustrates the overlapping complexity of establishing regulations, each of which had a promulgation date, followed by a phasing in of the standards for future model years, and in some cases, new clean air amendments that superseded the previous standards prior to their complete phase-in.

8.4.11.3 Exhaust Gas Standards

Emission standards for exhaust gases were first required on 1966 model, light-duty vehicles sold in California and in 1968 models nationwide. Emission standards were applied only to NMHCs and CO, as these were more easily controlled than NO_x. Table 8.7 illustrates the values of and the dates on which the first California and national standards were established. California established standards for NO_x (1971), PM (1984), and formaldehyde (HCHO, 1994). The US EPA Tier 1 standards, set in 1991 after passage of the 1990 Clean Air Act Amendments, tightened the standards for light-duty vehicles including LDV (passenger cars) and LDT (trucks) with separate standards for each vehicle type for NO_x, NMHCs, and CO. NMHCs were subsequently redefined as nonmethane organic gases. California also adopted its very ambitious Low-Emission Vehicle Program (LEV) in 1990, which tightened emission standards for light- and medium-duty vehicles. The LEV program (now referred to as LEV I) established tiers of emission standards for increasingly more stringent categories of low-emission vehicles; a mechanism requiring each manufacturer to phase in progressively cleaner vehicles each year with an option of banking and trading emissions; and a requirement that a certain percentage of vehicles be zero-emission vehicles (ZEV). The program had a series of five progressively more stringent vehicle categories and associated standards, including TLEV (transitional LEV), LEV, ULEV (ultra-low), SULEV (super-ultra-low), and ZEV. The latter category is primarily electric vehicles and hydrogen-fueled vehicles. While the ZEV has no emissions, it does derive its energy from other energy sources that are necessary to produce stored electrical energy and/or hydrogen. In 1998,

TABLE 8.6
Regulatory History of Control Requirements on New Light-Duty Motor Vehicles

Model Year	Control Actions
1961	California—crankcase emission controls
1963	US—crankcase emission controls
1966	California—exhaust emission controls for CO and NMHCs
1968	US—exhaust emission controls for CO and NMHCs
1970	California—evaporative control systems; new, more stringent emission standards for CO and NMHCs; required reduction of NO_x emissions for the first time
	US—Congress passes CAA Amendments; mandates stringent new motor vehicle exhaust emission standards for CO and NMHCs to be achieved by 1975 model year; mandates stringent NO_x emission standard to be achieved by 1976 model year; provides for fuel composition standards; US EPA imposes more stringent CO and NMHC standards
1971	US—evaporative control systems
1972	California—emission standards for NO_x are tightened
	US—US EPA imposes tighter standards for CO and NMHCs; first federal emission standard for NO_x promulgated
1974	California—stringent emission standards mandated for 1975 model cars are postponed; California imposes interim standards for all new cars sold in California
	US—statutory emission standards for CO, NMHCs, and NO_x mandated in 1970 legislation are postponed; US EPA imposes interim standards for CO and NMHCs
1977	California—tightens NMHC and NO_x emission standards and added PM standards for the first time in 1984
	US—US EPA imposes interim emission standards for NO_x mandated by Congress in 1974 legislation; CAA Amendments passed; interim standards for CO and NMHCs continued through 1979 model year; 1975 statutory emission requirements for NMHCs required in 1980, CO in 1981; NOx standard relaxed with 1981 deadline
1990	CAA Amendments passed; established Tier 1 standards (1991); significant additional reduction in NMHCs required; emission limits for CO and NO_x under 1970 CAA Amendments to be met in the period 1993–1995
	California—adopted a new Low-Emission Vehicle (LEV) program that tightened light- and medium-duty vehicle standards from 1994 to the present, including adoption of zero-emission vehicle programs
1999	US—Tier 2 standards set in Dec. 1999 tightening the NMHC, NO_x, and CO standards and adding PM and HCHO standards with standards to be phased in 2004 to 2009
2014	US—Tier 3 standards set with standards to be phased in 2017 to 2025

California introduced an additional category, the partial zero-emissions vehicle (PZEV) which was used to describe a vehicle with zero evaporative emissions from its fuel system, with at least a 15-year (or 150,000 mi) warranty on emission components, and that meets the SULEV emission standards.

In late 1999, with the establishment of the US EPA Tier 2 standards, all five of the pollutants previously addressed in the California standards were included. Tier 2 and subsequent standards are the same for both LDV and LDT. The US EPA Tier 3 standards were finalized in 2014 with the intended phase-in to occur between 2017 and 2025. All emission standards are given in grams per mile (g/mi). The California LEV classifications that correspond to the US EPA bin numbers are shown in Table 8.7 for the Tier 3 standards. Note that methods of measuring emissions, as well as expressing their concentrations, have changed several times since emission limits were initially mandated, so comparisons between early standards and Tier 1–3 standards provide approximate indications of progress in tightening of standards and emissions to the environment.

The Tier 2 and Tier 3 standards included smog rating standards, also referred to as bins, in which automakers choose to certify their vehicles. The automakers' overall fleet of vehicles was

TABLE 8.7
History of Light-Duty Motor Vehicle Exhaust Emission Standards

Standard	NO_x (g/mi)	NMOG (g/mi)	CO (g/mi)	PM (g/mi)	HCHO (g/mi)
Precontrol	4.1	10.6	84		
First Regulations					
California—1966		6.3	51		
US—1968		6.3	51		
California—1971	4.0				
US EPA—1973	3.0				
California—1984				0.60	
California—1994					0.150
Tier 1 (Finalized by US EPA in 1991; Phased in 1994 to 1997)					
LDV	0.60[c]	0.31	4.2		
LLDT[a]	0.60[c]	0.31	4.2		
LLDT[b]	0.97	0.40	5.5		
	Emission Limits at Full Useful Life[g]				
Tier 2 (Finalized by US EPA in December 1999; Phased in 2004 to 2009)					
Bin 1	0	0	0	0	0
Bin 2	0.02	0.010	2.1	0.01	0.004
Bin 3	0.03	0.055	2.1	0.01	0.011
Bin 4	0.04	0.070	2.1	0.01	0.011
Bin 5	0.07	0.090	4.2	0.01	0.018
Bin 6	0.10	0.090	4.2	0.01	0.018
Bin 7	0.15	0.090	4.2	0.02	0.018
Bin 8[a]	0.20	0.125	4.2	0.02	0.018
Fleet average[d]	0.07	na	na	na	na
Tier 3 (Finalized by US EPA in 2014; To be phased in 2017 to 2025)					
Bin 1	0		0	0	0
Bin 20 (SULEV)	0.02[h]		1.0	0.003	0.004
Bin 30 (SULEV)	0.03		1.0	0.003	0.004
Bin 50 (ULEV)	0.05		1.7	0.003	0.004
Bin 70 (ULEV)	0.07		1.7	0.003	0.004
Bin 125 (ULEV)	0.125		2.1	0.003	0.004
Bin 160 (LEV)	0.16		4.2	0.003	0.004
Fleet average	0.03[e]		na	0.003[f]	na

[a] light-duty trucks with < 3,750 lb gross vehicle weight rating (GVWR)
[b] light-duty trucks with > 3,750 but < 8,500 lb
[c] gasoline standard; diesel engines = 1.25 g/mi
[d] manufacturer's fleet average
[e] fleet average is lowered each year to meet 0.03 g/mi by 2025
[f] fleet average is lowered each year to meet 0.003 by 2021
[g] useful life has varied from 100,000 to 150,000 miles, depending on pollutant and tier
[na] not applicable

required to meet the overall fleet average by certain dates, whereas vehicles within each bin were restricted to the value designated for that bin. The Tier 3 standards also combined the sum of the g/mi of NO_x and NMOG into a single category to provide more flexibility to the automaker, as both are participants in photochemical smog formation. As an example, a vehicle that is to be certified under Tier 3 standards for Bin 30 must not exceed the 0.03 g/mi sum of NO_x and NMOG, 1.0 g/mi of CO, 0.003 g/mi of PM, and 0.004 g/mi of HCHO throughout its useful life. Table 8.7 clearly illustrates the substantial progress that has been made in reducing emissions since the 1960s and the continuing effort to reduce new model vehicle emissions even further. The reader is cautioned, however, that the table is an oversimplification of the standards. The original standards were based on a vehicle operating life of 5 years or 50,000 mi. The Tier 2 standards were typically for 100,000 miles or 10 years, whichever came first, and Tier 3 standards apply for the useful life of 10 years or 120,000 miles. Intermediate life standards (e.g., 50,000 miles, and in some cases up to 150,000 miles) are also embedded in the multiple standards for certain conditions, the complexity of which is beyond the scope of this textbook. It is likely that many of the Tier 3 vehicles manufactured in 2025 will also meet the standards adopted by California for PZEVs, including the 15-year/150,000-mile warranty. It was reported in 2018 that at least five major auto manufacturers were selling PZEV vehicles—still gasoline powered, but with very low emissions.

The US EPA maintains multiple website locations related to vehicle emissions (e.g., https://www.epa.gov/emission-standards-reference-guide/all-epa-emission-standards). While many of those sites contain information with posted dates, many do not provide a specific date that the site information was posted, so it is often challenging to determine whether the information is the latest update.

Throughout the history of federal involvement in regulating emissions from motor vehicles, California alone had been permitted to regulate emissions more stringently than the federal government. This "California exemption" recognized the state's leadership role in motor vehicle emissions control and its unique motor vehicle–related air pollution problems. Under the 1990 Amendments, other states were allowed to adopt emission standards and requirements similar to, but no more stringent than, those of California. With the implementation of the Tier 3 emission standards, the differences between California and federal standards are smaller.

The emission standards shown in Table 8.7 only apply to new vehicles, so it takes time for newer vehicles to be infused into the vehicle mix and for older vehicles to be retired from the mix. Figure 8.2 illustrates the percentage of the emissions of CO, HC, and NO_x contributed by each model year in 2018 for Denver, Chicago, and West Los Angeles. Different pollutants were chosen for each city to illustrate the similarities between cities. In 2018, the percentage of new vehicles (MY2018) was less than 2% for each city, and the median model year was between 2011 and 2013 as shown by the open circle on the dashed line, which represents the percentage of each model year in the 2018 fleet (right axis). Per US EPA methodology, gas trucks includes SUVs, as well as light- and medium-duty trucks. Taking West Los Angeles as an example, the largest percentage of NO_x emissions in 2018 came from MY2002 vehicles (16-year-old vehicles!). The percentage of MY2002 vehicles was only about 3%, but MY2002 vehicles contributed about 8% of the diesel emissions, 7.5% of the gas truck emissions, and 4% of the passenger car emissions. Older model years contributed less (as there were fewer on the road), and newer vehicles contributed less because they had to meet more stringent emission standards. The data in Figure 8.2 clearly illustrate the time lag that occurs between implementing a new standard and actually achieving reduced emissions into the environment.

While the light-duty vehicle standards have been highlighted herein, the standards for other mobile sources, including on-road heavy-duty trucks and non-road mobile sources, have followed similar paths of increasingly restrictive standards, although generally in a time frame later than that of the lightweight vehicles. The reader is encouraged to review the emission standards of these other sources and their development by going to the sites referred to herein and by utilizing the search engine on the US EPA website and other public available sites. Each of the combustion source categories (on-road and off-road mobile and stationary combustion) has benefited greatly from the technologies developed by the others. It has also been said by many that the establishment of emission standards, and particularly the tactic of the phase-in of standards, has in many cases driven the technology, but all toward the goal of protecting the environment and our health. The development and current status of greenhouse gas emission standards is discussed in this chapter under Section 8.6.3.

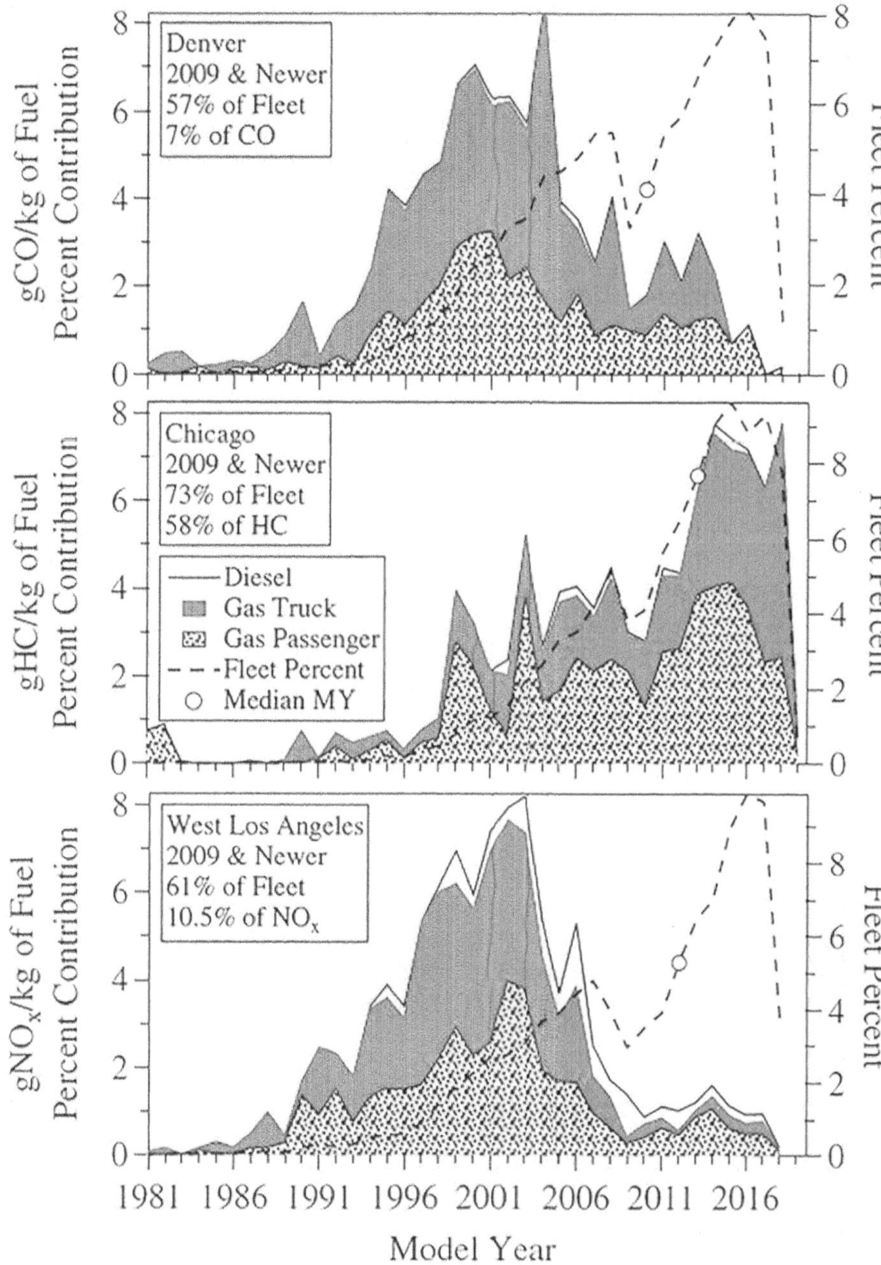

FIGURE 8.2 Percent emissions contribution by fuel and vehicle type (left axis) for CO in Denver, HC in Chicago, and NO_x in West Los Angeles by model year for 2018. (*Source*: Bishop, G.A., EM—*The Magazine for Environmental Managers*, A&WMA, April 2020. Used with permission.)

8.4.11.4 Fuel Additives and Cleaner Fuels

Anticipating that catalytic converters were to be used on 1975 and later model vehicles (see Chapter 9), Congress authorized the regulation of fuel additives. Lead additives, which had been universally used to increase octane ratings of gasolines, were anticipated to coat catalytic materials, rendering them ineffective. Using this authority, the US EPA began a phasedown in leaded gasoline use. Other regulatory actions included the prohibition of (1) misfueling, that is, the introduction of leaded gasoline into a vehicle labeled unleaded gasoline only; (2) use of gasoline or diesel fuel with a S concentration of more than 0.05% by weight (500 ppm by weight); (3) sale and use of high-volatility (Reid vapor pressure of > 9 $lb/in.^2$ [psi] [465 mmHg]) gasoline during the high-O_3 season; and (4) sale and highway use of Pb additives after 1995. The sulfur content of gasoline was further restricted to an average of 120 ppm by mass (maximum of 300) in 2004, to 30 ppm (maximum of 80) in 2006 and to 10 ppm (maximum of 80) in 2017, to avoid the gradual contamination of catalysts used to control NO_x and NMOG emissions in the Tier 3 program. For on-road diesel fuel, the sulfur content was reduced from 500 ppm to 15 ppm in 2006. Fuel for non-road diesel engines was also limited to 500 ppm in 1993 and further to 15 ppm beginning in 2007 and expanded to locomotives in 2010. The 15 ppm sulfur fuel is referred to as ultra-low sulfur diesel (ULSD).

Specific provisions of the CAA Amendments addressed the issue of reformulated gasoline and oxygenated fuels. The US EPA was to promulgate regulations establishing requirements for reformulated gasoline to be used in motor vehicles in specified O_3 and CO nonattainment areas. These were to require significant reductions in emissions of O_3-forming volatile organic compounds (VOCs) during the O_3 season as well as toxic air pollutants. Toxic pollutants included in these regulations were benzene, 1, 3-butadiene, polycyclic organic matter, acetaldehyde, and HCOH. In CO nonattainment areas, oxygenated gasoline containing no less than 2.7% O_2 by weight was to be sold in high-emission seasons.

Gasolines described as reformulated have had their compositions changed to reduce the reactivity of exhaust products and decrease emissions of NMHCs, CO, and NO_x. Reformulated gasolines can be used in conventional internal combustion engines (ICEs) with no modification of the propulsion system. They are produced by refining modifications and adding oxygenated compounds. They have lower olefinic and aromatic HC contents and lower Reid vapor pressures (RVPs). They improve the efficiency of combustion and reduce CO emissions. Commonly used oxygenated HC additives included methyl-*t*-butyl ether (MTBE), ethyl-*t*-butyl ether (ETBE), tertiary-amyl-methyl ether (TAME), CH_3OH, and C_2H_5OH (ethanol). MTBE, while used early in the program, was banned by many states and finally phased out due to concerns over health effects and potential contamination of groundwater. Ethanol, however, is now widely used across the United States as an oxygenated additive. All gasoline engines can use up to E10, which is gasoline 10% ethanol. Ethanol has about 34% less energy than gasoline, so E10 vehicles suffer about a 3% loss in fuel economy.

8.4.11.5 Emission Test Cycles

Exhaust emission standards must reflect actual emissions in normal driving. Emissions vary, however, as automobiles are operated under different conditions, for example, long trips, short trips, stop-and-go urban traffic, and superhighway travel. In addition, individual drivers operate vehicles differently under similar driving conditions. Because it is not possible to duplicate all these driving factors in specifying emission requirements, standard emission assessment procedures are used to represent a driving cycle that is responsible for emissions in areas where the most severe motor vehicle-related air pollution occurs.

As a result, both California and US test procedures utilize a dynamometer test cycle that simulates early-morning driving conditions in an urban area. These test cycles ensure that each manufacturer's vehicles are tested under the same conditions and that the quality of the test programs is maintained. The primary test protocol currently used by the US EPA is the federal test protocol (FTP or FTP75), which involves dynamometer testing in which the driver and vehicle undergo an urban driving pattern typical of early morning traffic on the dynamometer. The test length is 11.04 miles with speeds ranging from at rest to approximately 56.7 mph (average speed of 21.19 mph), including a number of accelerations and decelerations with the driver staying within plus or minus 2 mph of the required speeds as the vehicle is taken through the cycle. The cycle includes a cold start, a cold-stabilized period, a hot soak (engine off), and a hot-start stabilized period. A number of other test procedures (US06, SC03, and the Urban Dynamometer Driving Schedule [UDDS]) are utilized in vehicle testing that account for the effects of air conditioning and alternative urban simulations. The UDDS protocol (also referred to as FTP72 and LA4) is similar to FTP75 except that it is only for the cold start and the cold-stabilized periods of the cycle. It is also used in several other countries including Sweden and Australia under different titles. The federal procedures for the US EPA test protocols address the number of vehicles to be tested, the precise dynamometer specifications, the exact protocol for each test, the required configuration of sampling (gases and PM), the instrumentation specifications, and the quality assurance protocols required. These requirements are contained in and updated daily in the Code of Federal Register, Title 40, Chapter 1, Subchapter C, Part 86. To illustrate the complexity of the overall process, Part 86, entitled *Control of emissions from new and in-use highway vehicles and engines* (February 25, 2020) contained 1,321 pages outlining the specifications/requirements for testing.

8.4.11.6 Inspection and Maintenance

The Clean Air Act Amendments of 1970 authorized the US EPA to provide oversight over automobile inspection and maintenance programs. Approximately 70 US cities and several states had automobile emission inspections and maintenance (I&M) programs prior to 1990. The 1990 Amendments expanded I&M requirements to 40 additional metropolitan areas, many in the northeastern United States, who were nonattainment for ozone or carbon monoxide. Specific enhanced requirements were mandated for new and existing I&M programs.

I&M programs were deemed to be an essential element of pollution control programs designed to achieve NAAQSs. Monitoring studies have shown that in some major metropolitan areas, approximately 50% of emissions of regulated pollutants from motor vehicles are associated with approximately 10% of the motor vehicle population. It was these polluting vehicles with poorly functioning or malfunctioning emission control systems that I&M programs were designed to identify and have repaired and properly maintained. In many states, this led to specific counties or parishes that were nonattainment being required to implement I&M programs while adjacent areas were not required to do so, even though vehicles traveled daily between the areas.

Congress, in 1990, also mandated that the US EPA promulgate regulations that would require manufacturers to install onboard diagnostic (OBD) systems on 1994 and later model vehicles. Such systems were to detect emission-related failures and alert drivers to the need for potential repair. They include dashboard warning lights and trouble code storage that helps mechanics identify malfunctions and possible causes. In 2001, the I&M programs were required to begin incorporating vehicle OBDs as part of I&M, with full phase-in for all 2006 and later vehicle models. As vehicle emission standards have continued to become more stringent for newer vehicles, as illustrated in Table 8.7, many metropolitan areas have questioned whether I&M and its requirements are still worth the expense to the agencies and the public, given the full life standards. However, the I&M programs became a regulatory part of each state's SIP and any change to a particular I&M program requires a change in the SIP. The argument for continuation is that despite OBDs and the more stringent standards, a large portion of emissions occur from a small percentage of poorly maintained vehicles; without I&M, those would go unchecked.

8.4.11.7 Tampering Prohibitions

Tampering (willful disabling of emission control systems) has been a continuing problem since emissions controls were required on motor vehicles, particularly early on in the program when the disabling of the control system was deemed to improve the engine performance. Tampering prohibitions were included in both 1970 and 1977 CAA Amendments, with a focus on motor vehicle dealers, repair shops, and others. In 1990, tampering prohibitions were extended to individuals who manufacture products designed to disable or interfere with the proper functioning of emission control systems. They prohibited everyone, including "the private individual," from tampering with motor vehicle emission control systems.

8.4.11.8 Evaporative Controls and Onboard Vapor Recovery Systems

A major concern with mobile sources has been the accidental loss of fuel in the handling and delivery of fuels to the end user as well as the evaporative loss of fuel from the vehicle itself. While the losses represent only a small fraction of the fuel burned, these volatile organic compounds are not only potentially hazardous when breathed, they also play a major role in the formation of other pollutants such as ozone. A major component of those losses occurred through the slippage of unburned fuel and incomplete combustion products by the pistons of the vehicle engine into the crankcase. These gases were released into the atmosphere and commonly referred to as blowby emissions. They contained both CO and unburned hydrocarbons. As shown in Table 8.6, they were controlled in the early 1960s by installing a recycle line from the crankcase with positive crankcase ventilation of the blowby gases back into the engine through a positive crankcase ventilation (PCV) valve.

Regulatory actions took two separate paths with respect to control of evaporative emissions from fuels. The first was to reduce evaporative emissions in the delivery and storage of gasoline. The second was to reduce evaporative emissions from the vehicle. During the late 1970s and early 19809s, the US EPA issued an NSPS for the petroleum industry related to the control of evaporative VOC emissions associated with both storage of gasoline at terminals and tanker truck loading and unloading operations. While several techniques were used, one of the primary techniques was referred to as vapor balancing. During the loading or unloading of gasoline, whether at the terminal or from a truck to a service station, the vapor within the tank being filled had to be expelled from the tank as the new gasoline was transferred to the tank. Vapor balancing was accomplished by modifying the transfer procedures, so that exiting vapor was returned to the tank being emptied. This reduced, and in some cases almost eliminated, the emissions during transfer. When the technology was applied to service stations (i.e., gasoline dispensing facilities or GDFs), it became known as Stage I vapor control. By 1982, more than 50% of GDFs had installed Phase I controls, and they were soon required nationally. In a similar manner, and primarily in ozone nonattainment areas, a similar Stage II technology was implemented in which the VOC emissions at the gasoline pump were captured during refueling of vehicles. The typical Stage II dispenser utilized a double-hosed nozzle. The inside hose delivered fuel into the fuel tank, while the outer hose formed an approximate pressure seal around the gas cap and captured the fumes, which were then pulled by vacuum back into the fuel pump and captured or recirculated into the GDF fuel storage tank. As a result of the CAA of 1990, all ozone nonattainment areas that were defined as Serious, Severe, or Extreme NA areas were required to implement Stage II controls. As with I&M programs, these were only required in Serious and above nonattainment areas, although some Moderate NA areas adopted I&M and Stage II as part of their SIP to achieve required emission reductions. By 2012, the US EPA reported that there were about 30,600 GDFs with Stage II control with an annual operating/maintenance cost of about $91 million/year. However, that changed rather abruptly shortly after 2012 due to other technological developments with onboard vehicle evaporative controls.

Even though vehicles had installed the PCV technology in the 1970s, evaporative losses still occurred on a continuous basis due to the changing vapor pressure above the fuel in vehicle fuel tanks due to diurnal temperature changes and the need to relieve the pressure through the gas cap. The early evaporative loss technology provided a tighter gas cap seal and utilized a passive activated carbon adsorption cannister through which the expanding fuel vapors were expelled. The cleaned air was then released into the atmosphere (see Chapter 9). During vehicle operation, the engine pulled air back through the cannister into the engine where the adsorbed VOCs were combusted. The technology was later expanded to an active system, referred to as onboard refueling vapor recovery (ORVR), which also captured the vapors during refueling in an expanded activated carbon cannister. Beginning with the 1998 model year, ORVR was phased in to incorporate nearly all new vehicle models by 2006. Due to the phasing in of the new models and the gradual reduction in non-ORVR models, the percentage of light-duty vehicles on the highway without ORVR had decreased to about 26% in 2012 with a projected further decrease to less than 8% by 2020. In 2012, the US EPA made the decision to allow states to modify their SIPs and eliminated mandated Stage II controls.

8.4.11.9 Clean-Fuel Vehicles

The US EPA was mandated to promulgate standards for clean-fuel vehicles under the 1990 CAA Amendments. So-called clean fuels include any fuel, reformulated gasoline, natural gas, liquefied petroleum gas (LPG), hydrogen (H_2), or power source (including electricity) used in vehicles that complied with clean-fuel vehicle standards.

8.4.11.10 Clean Fleets

The 1990 CAA Amendments placed new requirements on centrally fueled fleet operations in certain O_3 nonattainment areas. Thirty percent of new vehicles purchased were to use clean fuels and meet exhaust emission standards (0.075 g/mi NMHC, 3.4 g/mi CO) that were lower than those required for general passenger vehicles; these went into effect in 1998, with purchase requirements increased to 70% in 2000. This clean fleet program affected 22 metropolitan areas in 19 states.

8.4.12 Enforcement

Early pollution control programs were largely ineffective because agencies had limited enforcement authority. The 1970 and subsequent CAA Amendments provided considerable federal enforcement authority and mandated similar authority for states under SIP provisions. Since 1970, the administrator of the US EPA has had the authority to take enforcement actions when provisions of SIPs, NSPSs, NESHAPs, and other regulatory requirements are violated.

Under current statutory language, the US EPA can, in response to violations of SIPs and other major provisions of the CAA, issue administrative compliance orders. The order specifies the time by which a source must comply with emission limits. The US EPA may commence civil actions in federal district courts should the affected party not respond to administrative orders. In such actions, the US EPA can seek a permanent or temporary injunction or civil penalties. If violations are willful, criminal actions may be taken. The enforcement value of US EPA administrative actions was significantly enhanced in 1990 when the US EPA was given authority to impose civil penalties (subject to formal adjudicatory hearings) administratively. The US EPA may also issue noncompliance penalties. These are assessed for certain broad categories of violations in an amount no less than the economic value derived from delayed compliance.

The 1990 CAA Amendments provided authority for tough criminal sanctions. These include fines of upward of a million dollars and prison terms of up to 15 years. Such criminal sanctions may be imposed for "knowing" violations of the law.

The US EPA may file suit to abate any air pollution problem that presents "an imminent and substantial endangerment to the health of persons," even if there has been no violation of the law. It may also issue administrative orders to protect public health if it is not practical to initiate a court action.

Although the US EPA has significant enforcement authority, most enforcement occurs at state and local levels. Federal enforcement authority is only used when state and local agencies are reluctant to enforce provisions of their own SIPs and other programs (e.g., NSPSs). It is applied to violations of regulatory requirements for which the US EPA has direct responsibility, that is, motor vehicle emissions, O_3-destroying chemicals, and acidic deposition.

8.4.13 Permit Requirements

Permit systems are effective tools for achieving compliance with emission reduction requirements. Permit requirements put the burden of proof of compliance on emission sources. Permits may be required for existing sources to discharge any pollutants into the atmosphere and to install and operate pollution control equipment. They are required for new or significantly modified existing sources before construction commences. Legally, a source cannot operate, that is, emit pollutants, without a permit. To be granted a permit, a source must comply with all existing emission reduction requirements.

It is not possible in many cases for a source to immediately comply with permit conditions due to time requirements for designing, ordering, and installing emission control systems. For a source to operate in violation of emission regulations, a legal exception or variance is granted for a specified period, for example, 6 months. These variances provide a reasonable period for a source to install the control equipment necessary to comply with emission limitations.

The role of permitting in achieving air quality goals was given significant new attention in the 1990 CAA Amendments. Under Title V, the US EPA promulgated regulations that established minimum requirements for air pollution permit programs. Among these were (1) permit applications, including a standard application form and criteria for determining completeness of an application; (2) monitoring and reporting; (3) an annual fee assessed to permittees; (4) personnel and funding to administer the permits program; and (5) adequate authority to issue permits and assure compliance.

Major stationary sources, sources subject to NSPSs, regulated hazardous pollutant sources, and sources subject to PSD requirements must have permits. Permits contain enforceable emission limitations, monitoring and reporting requirements, and in some cases a compliance plan that includes an enforceable schedule.

8.4.14 Citizen Suits

The 1970 CAA Amendments gave citizens the right to sue both private and governmental entities to enforce air pollution requirements. Prior to 1970, citizens did not, in most cases, have legal standing to seek court review of federal agency decisions. The premise of the citizen's suits provision is to make the federal enforcement agency (US EPA) accountable when its actions are deemed to conflict with clean air legislation and regulatory requirements. These provisions have been used by environmental groups, such as the Natural Resources Defense Council, Environmental Defense Fund, Sierra Club, and others when they believed that the US EPA was not complying with its statutory obligations. Successful citizen suits have required the US EPA to promulgate regulations for air quality maintenance, indirect source review, PSD, and the air quality standard for Pb.

One of the principal effects of citizen suits has been to require the US EPA to implement provisions or concepts embodied in clean air legislation when it was reluctant to do so. The reasons for such reluctance include practical and economic considerations, availability of workforce, administrative priorities, and bureaucratic inertia. An attendant result of citizen suits has been the legal clarification of the language of clean air legislation.

Although citizen suits have, for the most part, been used to require the US EPA to develop and implement new policies, citizen legal action can also be taken against a pollution source to force compliance with emission reduction requirements. Under the 1990 CAA Amendments, citizens were given a legal right to file lawsuits to seek civil penalties against parties in violation of the CAA.

In the early years of air pollution control, citizen suits were commonly brought against the US EPA by environmental groups. Over time, a number of citizen groups sympathetic to industry interests were formed. Such groups commonly file suits in response to US EPA regulatory actions, often contending that they are arbitrary, capricious, and not supported by "good science."

8.4.15 Pollution Prevention

Congress passed the Pollution Prevention Act (PPA) in 1990. Under this Act, the scope of federal policy to cover releases of pollutants to all media was expanded, including accessibility of the public to the national Toxics Release Inventory (TRI). The focus of the Act was to establish new policies for waste reduction (interpreted broadly). The guiding principle is that source reduction is fundamentally different and more desirable than waste management and pollution control. It focuses on source reduction as the primary means of reducing waste production. A source reduction strategy deemphasizes command-and-control regulation and emphasizes voluntary cross-media pollution prevention activities. The PPA mandates the US EPA to promote governmental involvement in overcoming institutional barriers by providing grants to state technical assistance programs under a new federal pollution prevention program. Pollution prevention is given a new emphasis under the 1990 CAA Amendments. The Amendments required that the US EPA conduct an engineering research and development program to develop, evaluate, and demonstrate nonregulatory technologies for air pollution prevention. In establishing emission standards for HAPs, the US EPA is required to consider cross-media effects, substitution of materials or other modifications to reduce the volume of HAPs, the potential for process changes, and eliminating emissions entirely.

8.5 STATE AND LOCAL AIR POLLUTION CONTROL FUNCTIONS

All states and many municipalities had some kind of air pollution legislation or ordinances prior to the enactment of the 1970 CAA Amendments. These state and local programs varied in scope, enforcement authority, administration, and public support. All were limited in their ability to deal adequately with the complex problems of achieving control objectives.

Passage of the 1970 CAA Amendments, with their attendant requirements for federal approval of SIPs, imposed significant homogeneity on state and local programs. As a consequence, many state and local control programs strongly resemble each other. Despite this homogeneity, differences exist because they reflect unique state and local control concerns.

Most stationary source control functions are carried out at the state or local level. Federal clean air legislation provides the fundamental framework for effective pollution control at these levels. This framework is provided in the US EPA regulations promulgation process, the SIP approval process, grants-in-aid to state and local agencies, and backup federal enforcement authority should state and local agencies fail to carry out their responsibilities.

The federal posture in working with state and local agencies is a combination of carrot and stick. State and local programs that meet federal requirements receive limited monies to maintain air pollution control programs. Noncompliance with federal requirements can result in loss of federal support for pollution control activities as well as highway trust funds.

Most state and local programs operate under the philosophy that local authorities best know how to deal with state and local problems, that is, which control tactics should be used and how they are to be applied. In theory, this is true. It is also true that political pressure to hamper or slow pollution control efforts is the most intense at state and local levels. As a result, a federal presence is necessary to enforce, if necessary, the implementation of air pollution control regulations should state or local agencies choose not to enforce them. In most instances, state and local agencies prefer not to invite "federal interference."

8.5.1 Boards and Agencies

Regulation of air pollution is carried out by state and local authorities through the "police powers" delegated to states under the US Constitution. Utilization of these police powers requires enactment of state legislation and local ordinances to provide a legal framework for air pollution control. Because legislation and ordinances cannot deal with or anticipate all the individual circumstances of source control, they must provide for the establishment of an administrative agency to carry out the provisions embodied in them. These agencies may have quasi-legislative powers to adopt rules and regulations and recommend appropriate legal action. In some instances, these powers are vested in separate pollution control boards or commissions.

Air pollution control agencies have the day-to-day executive responsibilities of planning to achieve NAAQSs, source data collection, air quality monitoring, development of emission standards, development of source compliance schedules and permit review, surveillance of sources to determine compliance status, and recommendations for enforcement actions in cases of noncompliance.

As previously discussed, states are required to submit SIPs for US EPA approval that outline in detail how NAAQSs are going to be met in each AQCR under their jurisdiction. States may delegate responsibility to local agencies to develop local control plans. These local plans must be compatible with the SIP because they will be incorporated into it.

Once it has been determined that ambient levels of a specific pollutant exceed one or more NAAQSs, it may be necessary to promulgate emission standards for sources in the AQCRs. These standards may be based on computer models that use emission source data and dispersion characteristics of the atmosphere to predict ambient concentrations. Once needed emission reductions are determined from these models, compliance schedules are negotiated with industrial sources and incorporated into each state's SIP.

8.5.2 Effectiveness of State and Local Programs

The effectiveness of state and local air pollution control programs varies from state to state, and from one municipality to another. Among the many factors determining the effectiveness of air pollution control efforts are (1) a legal framework for regulation promulgation and enforcement, (2) the availability of economic resources, (3) sufficiently trained workforce, (4) the commitment of pollution control officials to enforcement of control requirements, (5) the cooperation of public officials, and (6) public acceptance of needed pollution control measures. Although each of these is important, the most vital factor is the acceptance of pollution control requirements by private citizens, public officials, and sources. Without this support, air pollution control objectives may not be achieved.

8.6 PUBLIC POLICY ISSUES

Clean air legislation has provided a regulatory framework for controlling ambient air pollution. Within that framework, the US EPA is required by law to implement provisions of the CAA in a manner protective of public health and the environment.

Regulatory needs change as new problems are identified, the advancement of science indicates a need for changes in existing standards and requirements, and court decisions mandate new actions or changes in proposed or existing regulatory requirements. Therefore, air pollution regulations, and the policies that they embody, are continuously evolving to reflect current realities. Policy considerations may involve using existing regulatory authority or seeking expanded authority through congressional action. Policy issues may also involve negotiation, ratification, and implementation of international treaties.

Environmental policy initiatives, or implementation of new regulatory initiatives, commonly result in a measure of controversy. This is particularly the case when proposed policies and regulations have real or perceived major economic implications for those to be regulated. On the other hand, environmental organizations may not accept that regulatory policies are sufficiently stringent.

The nature of air pollution control engenders a continual public policy development effort that involves proposal of new regulations, participation of the regulated community and other publics, and regulation promulgation and subsequent implementation. Examples of major contemporary public policy issues include new source review, MATS, climate change, the Kyoto Protocol, corporate average fuel economy standards, and greenhouse gas emission standards. For the reader, searches of the US EPA website (http://www.epa.gov) and the broader internet provide immediate online access to the many fact sheets, discussions, and controversial aspects related to the regulatory and public notice aspects of proposed changes to existing and new regulations.

8.6.1 New Source Review

New or significantly modified existing sources are required under Title I of the CAA to comply with all applicable NSPSs before they can be granted a permit to operate. They may also be subject to (1) additional emission limitation requirements if they are to be located in nonattainment areas and (2) PSD requirements in attainment areas (Chapter 8, Section 8.4.7). These requirements can be described as NSR.

An NSR is triggered when a new source's emissions exceed a threshold range of 10 to 100 tons/year (9–90 metric tons/year) in a nonattainment area and 250 tons/year (225 metric tons/year) for a PSD area. An NSR can also be triggered by major modification of an existing source, defined by the CAA as a physical change or change in the method of operation that results in an increase in emissions or emissions of a new pollutant. Existing regulations exempt routine maintenance, repair, or replacement of equipment, increases in hours of operation or production rate, or changes in ownership. If there is no physical change or change in the method of operation that results in significant net increased emissions of a regulated pollutant, then an NSR is not required.

NSR regulations establish significance levels that define major modifications that vary by pollutant and the attainment status of the AQCRs. These regulations allow for net emission increases if the source offers past or future emission decreases at its other units to counterbalance an increase from the proposed change. However, NSR only applies if net emissions increase above the significance level.

In 1998, the US EPA launched a major new NSR enforcement initiative in which it notified several coal-fired utilities that they were in violation of CAA provisions as a result of what the agency described as "maintenance" activities that were significantly greater than what would be described as "routine." In the utilities' view, the US EPA significantly narrowed its historical interpretation of what constituted routine maintenance and what triggers NSR.

Enforcement actions were taken by the US EPA and several states against the owners and operators of coal-fired power plants, alleging that their facilities had been modified without NSR permits. Modifications in individual cases included the construction of entire coal-fired, steam-generating units without incorporating modern pollution control equipment, increases in hourly capacity, redesign and modification of original design defects, and life extension projects. In all such cases, the US EPA maintained that net emission increases occurred above significance levels and should have been subject to NSR. Settlements were reached with several power companies and others put on hold as a new presidential administration took office.

Electric utilities cited for violations under NSR claimed that their projects were limited to routine maintenance and thus exempt from regulatory requirements. They and the rest of their industry maintained that the Clinton administration erroneously interpreted NSR requirements. They also maintained that the new NSR interpretation was preventing them from expanding production to meet energy supply needs.

In June 2002, a new, more industry-friendly presidential administration announced major proposed changes in enforcement rules governing older coal-fired power plants and refineries. These included (1) the promotion of pollution prevention projects, (2) expanded limits on the applicability of NSR designed to provide facilities greater flexibility to modernize without increasing emissions, (3) a clean unit provision that encourages the installation of state-of-the-art pollution controls, (4) changes in calculating emissions increases and establishing emissions baselines, (5) clarification of the definition of routine maintenance, and (6) clarification of how NSR applies to multiple projects implemented in a short time when a company modifies one part of a facility in such a way that output in other parts of a facility increases. Changes in NSR proposed by the US EPA in 2002 were claimed by proponents as efforts to increase energy efficiency and encourage emission reductions. To opponents, it was a major relaxation of NSR rules that would preclude future federal legal action except in the most flagrant cases.

In 2011, the US EPA issued PSD and Title V permitting guidance for new and/or modified sources that emit greenhouse gases (GHG) on the basis of their finding in December 2009 that elevated atmospheric concentrations of six GHGs endangered both public health and welfare. The guidance was issued in anticipation of future regulations that would require control or reduction of GHG emissions. The nonattainment component of NSR was not addressed in the guidance since no NAAQSs had (and still have not) been established for GHG.

8.6.2 Global Climate Change, the Kyoto Protocol, and the Paris Agreement

As indicated in Chapter 4, an increasing consensus has developed among atmospheric scientists and world leaders that greenhouse gas emissions are having a significant effect on global climate, with even more significant warming and other climatic effects to come. As a result of concerns, an international treaty, United Nations Framework Convention on Climate Change (UNFCCC), was developed and opened for signature at the 1992 Earth Summit in Rio de Janeiro. The objective of the treaty was to stabilize the concentration of GHG in the environment at a level that would prevent dangerous anthropogenic interference with climate systems. The conference has met annually since that time. The treaty is widely accepted and has more than 190 parties who have accepted it. The framework set up a process whereby protocols or agreements could be established to accomplish its objective. The most significant of the protocols was the Kyoto Protocol where more than 100 countries (including most industrialized nations) negotiated a treaty in Kyoto, Japan, to reduce greenhouse gas emissions to target levels.

The Kyoto Protocol, signed in December 1997, called for a worldwide reduction of greenhouse gas emissions by an average of 5.2% below 1990 levels by 2012 (an actual 33% reduction based on projected increases in emissions). Different targets were adopted by different countries. Members of the European Union committed themselves to an 8% reduction; the United States, 7%; Japan, 6%; and Russia and the Ukraine agreed to stabilize at 1990 levels. Sensitive to their economic development needs, no greenhouse gas reductions were required of developing countries.

The Kyoto Protocol provided considerable flexibility in how emission targets were to be achieved. Policy options included emission trading, enhancement of energy efficiency in relevant economic sectors, protection and enhancement of sinks and reservoirs, sustainable agriculture promotion, increased use of renewable energy sources, removal of subsidies and exemptions that serve to increase greenhouse gas emissions, and limitation and reductions in methane (CH_4) emissions.

While the United States, under the Clinton administration, signed the Kyoto Protocol in 1997, the US Senate signaled, by an overwhelming vote (on a resolution), that it would not ratify it, and the following president (Bush) also refused to sign it. Many political leaders claimed the treaty was flawed because it exempted large C emitters such as China and India. Because of its perceived potential adverse effect on the American economy, the United States, by an act of the president, withdrew from the treaty. The treaty was subsequently renegotiated in 2001 (without US participation) and went into effect after being ratified by countries whose total C-based emissions exceeded 55% of the emission reductions mandated by the treaty. Even with the protocol in place, and with 36 countries in the protocol reducing their emissions, global greenhouse gas emissions increased substantially between 1990 and 2012, and the 2012 commitments expired. Figure 2.5 shows that global CO_2 emissions increased by almost 60% in that period.

In December 2012, the Kyoto Protocol, referred to as the Doha Amendment, was amended, calling for further reductions in GHG emissions. It established emission reductions that needed to occur between 2012 and 2020. While the amendment was adopted, it required that 144 parties submit instruments of acceptance prior to its being entered into force. As of February 2020, 137 parties had submitted the required instruments of acceptance. Canada withdrew its commitment to the Kyoto Protocol in December 2012 and the United States never signed the agreement.

In late 2015, recognizing that the Soha Amendment had not yet entered into force because the required minimum of 144 party ratifications had not been reached, the 196 parties to the UNFCCC met in Paris, France, at their annual conference. The outcome was the adoption by consensus of a new agreement (not under the Kyoto Agreement) now referred to as the "*Accord de Paris*" or Paris Agreement. It was entered into force on November 4, 2016. As of February 2020, all signatories to UNFCCC had signed the agreement, including Canada, China, the EU, India, and the United States. The United States also ratified the agreement.

On August 4, 2017, the United States (under the Trump administration) provided notice of intent to the United Nations to withdraw from the Paris Agreement, followed by its notice of withdrawal on November 4, 2019. The withdrawal became effective as of November 4, 2020. The basis was that it would undermine the US economy and place the country at a permanent disadvantage. After the 2020 presidential election, President-elect Joe Biden pledged to return the United States to the Paris Agreement on his first day in office. On January 20, 2021, shortly after his inauguration, he signed an executive order to rejoin the agreement, and the United States formally resumed its membership on February 19, 2021, 107 days after the withdrawal had taken effect. On January 20, 2025, shortly after beginning his second term, President Trump signed an executive order initiating a second U.S. withdrawal from the agreement. A summary of the withdrawal, rejoining, and subsequent second withdrawal, including global responses can be found at the website https://en.wikipedia.org/wiki/United_States_withdrawal_from_the_Paris_Agreement.

Economic growth is a powerful factor in politics. As such, requirements of the Kyoto Protocol, as well as the Paris Agreement, are viewed as much, if not more, in political as in economic terms. The political hurdles are considerable. Most, but not all, business leaders believed at that time that emission reduction requirements under the Kyoto Protocol were unrealistic, and the actual growth in worldwide CO_2 emissions, shown in Figure 2.5, provides a fairly strong indicator suggesting reductions such as those included in the Kyoto Protocol were unrealistic, given the short time frame included to meet those goals. While the United States did not sign the Kyoto Protocol and intends to withdraw from the Paris Agreement, its annual CO_2 emissions have either been relatively flat or decreasing each year since its peak of emissions in 2007 with an overall reduction of approximately 11% that occurred by 2018. Both China's and India's emissions have grown steadily during that period. On a very positive note, emissions in the European Union have been declining annually since about 1980 with over a 30% decrease since 1980. Their per capita CO_2 emissions are also only 44% of the per capita emissions of the United States.

Since the United States consumes approximately 16%–18% of the world's energy production and emits about 14% of the world's CO_2, and its per capita emissions are higher than any other country, its participation in C emission reduction programs is vital to any international effort to reduce greenhouse gas emissions, atmospheric accumulation, and related climate impacts. Progress in reducing greenhouse gas emissions from major sources such as the electric utility industry and motor vehicles is occurring, however, at least from the perspective of the emissions from individual new sources.

8.6.3 CAFE and Greenhouse Gas Emission Standards

In 2011, the US EPA (via the CAA) and the Department of Transportation's National Highway Traffic Safety Administration (NHTSA) (via the Energy Policy and Conservation Act), under the Obama administration, announced a joint proposal to reduce greenhouse gases and improve fuel economy for vehicles. New corporate average fuel economy (CAFE) standards were set in 2012 with the intent of substantially improving fuel economy for model years (MY) 2017 through 2021 light-duty vehicles (passenger cars, light-duty trucks, and medium-duty passenger vehicles). The agency also put forth "augural" or anticipated CAFE standards for MY 2022–2025 that were consistent with the US EPA's standards for those later years, as Congress prohibits setting CAFE standards for more than 5 years at a time. The CO_2 emission standards were also established for MY2017–2025, with the MY2022–2025 standards being subject to a required "mid-term evaluation" by mid-2018. Table 8.8 provides a summary of the standards for passenger cars and trucks and the combined anticipated average for the fleet. The CAFE and CO_2 emission standards were projected to increase for each model year for both passenger vehicles and trucks. For light-duty vehicles (combined passenger vehicles and trucks), the changes were to be from 35.5 mpg and 250 g/mi of CO_2 for the base MY2016 to 54.5 mpg and 163 g/mi of CO_2 for FY2025—a 54% increase in fuel economy (or about 5%/year) and a 35% decrease in emissions. The standards for trucks were less stringent, considering the need for trucks to retain hauling capacity. Figure 8.3 shows the actual trend of CO_2 emissions from light-duty vehicles and the average fuel economy for the period 1975 to 2018, showing that substantial improvements have occurred in both CO_2 emissions and in fuel economy since initiation

TABLE 8.8
Projected Fleet-Wide Emissions Compliance Targets under the Greenhouse Gas CO_2 Standards (US EPA) and Corresponding Fuel Economy Standards (NHTSA)

Vehicle Type	Vehicle Model Year									
	2016 Base	2017	2018	2019	2020	2021	2022	2023	2024	2026
Passenger cars (g/mi)	225	212	202	191	182	172	164	157	150	143
Light Trucks (g/mi)	298	295	285	277	269	249	237	225	214	203
Combined Cars and Trucks (g/mi)	250	243	232	222	213	199	190	180	171	163
Combined Cars and Trucks (mpg)	35.5	36.6	38.3	40.0	41.7	44.7	46.8	49.4	52.0	54.5

Source: US EPA, EPA-420-F-12-051, 2012.

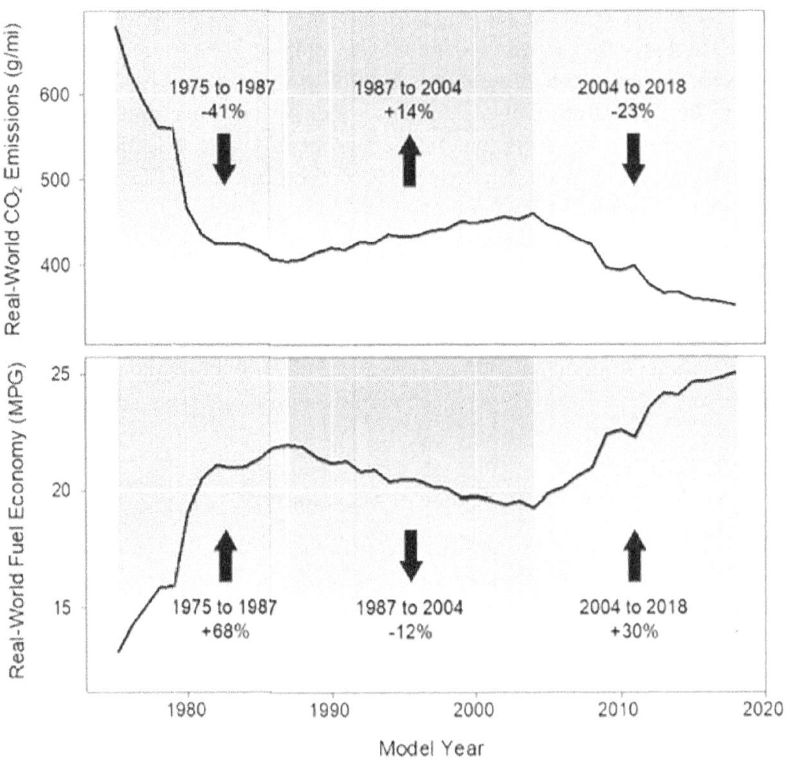

FIGURE 8.3 Estimated CO_2 emissions and fuel economy for operating light-duty vehicles. (From US EPA, *The 2019 EPA Automotive Trends Report: Greenhouse Gas Emissions, Fuel Economy, and Technology since 1975*, EPA-420-R-20-006, March 2020.)

of the standards. The latest MY2018 data showed that the fuel economy was at an all-time high of 25.1 miles per gallon and CO_2 emissions were at an all-time low of 353 g/mi.

Vehicle manufacturers are able to use a combination of CO_2 equivalent credits (i.e., reduced leakage of refrigerants), improved fuel economy, and electric and alternative fueled vehicles that reduce or eliminate CO_2 emissions (i.e., electric vehicles, EV; plug-in hybrids, PHEV; fuel cell, FC; and natural gas, NG vehicles). The values shown in Figure 8.3 include the effect of all light-duty vehicles, including electric vehicles. While progress is being made, it is important to note that for electric vehicles, where the battery must be charged by external sources instead of by regenerative braking (i.e., hybrid vehicle), the source of energy may still be a fossil fuel–based EGU that has CO_2 emissions.

As stated earlier, NHTSA and the US EPA were required to revisit and review the standards for MY2021 through MY2025. In August 2018, NHTSA and the US EPA (under the Trump Administration) issued a notice of proposed rulemaking (NPRM) entitled the Safer Affordable Fuel Efficient (SAFE) Vehicles rule with the intent to amend and eliminate the anticipated increasing stringency of the CAFE and greenhouse gas emission standards. It was proposed to freeze the requirements for passenger cars to 43.7 mpg and 204 g/mi, trucks to 31.3 mpg and 284 g/mi, and the combined light-duty vehicles to approximately 36.9 mpg and 241 g/mi for MY2021–2026. These values are approximately equal to the goals that had previously been set for vehicles in the FY2018–2020 time frame. The NPRM, on the basis that future emission standards and CAFE standards should be under one national program, included the intent to revoke California's authority both to set more stringent light-duty vehicle emission standards than the national programs and to require an increasing number of zero-emission vehicles to the vehicle mix. California and 22 other states challenged the administration's intent to revoke its authority and threatened to sue if the administration finalized the new requirements. In September 2019, the agencies published the final rule entitled the SAFE Vehicles Part One: One National Program. The final rule made it clear that federal law preempts state and local tailpipe GHG emission standards as well as zero-emission vehicle mandates, and it withdrew the CAA preemption waiver that was granted to California in January 2013 related to GHG and ZEV programs. The agencies were to continue to work on finalizing the GHG and CAFE standards that would replace the tentative ones shown in Table 8.8 for FY2021–2026.

On March 31, 2020, the US EPA (under the Trump administration) finalized a rollback of the CAFE standards that effectively reduced the approximate 5% growth/year in fuel economy standards to about 1.5%/year growth through 2026. It was reported that the rollback would achieve a fleet average fuel economy of about 40.4 mpg rather than the 54.5 mpg value shown in Table 8.8. As stated by the US EPA administrator,

> We are delivering on President Trump's promise to correct the current fuel economy and greenhouse gas emissions standards. . . . Our final rule puts in place a sensible one national program that strikes the right regulatory balance that protects our environment, and sets reasonable targets for the auto industry. This rule supports our economy, and the safety of American families.

It was reported that the rollback would reduce the price of future vehicles by approximately $1,000, thus helping the average vehicle owner. This does not take into account that the savings in gasoline for the average consumer driving a 54.5 mpg vehicle versus a 40.4 mpg vehicle would be approximately $200/year. The aforementioned actions clearly illustrate the tension and difference of opinions that exist over the urgency of reducing greenhouse gases to minimize climate impact and the political and economic impacts related to these issues.

Other programs and options have been discussed by states, industry, corporations, environmental groups, and citizens at large as possible ways of reducing greenhouse gas emissions, given those groups' conviction that impacts on the climate are occurring and that the time for action is "now." Some of those initiatives would require national programs, while others could be implemented by individual states or voluntarily by companies and groups. While a thorough summary is beyond the scope of the textbook, these include a potential carbon tax on emissions and/or adoption of goals to achieve a certain percentage of emission reductions, zero-carbon emissions or net-negative carbon emissions in a specific location (e.g., city, state, company, school system, building). The internet and other new sources are filled with these types of programs and will likely continue to be into the future. Examples of some of them are included as follows to illustrate the point.

A group of 16 large corporations, 10 energy corporations, and three nongovernmental organizations (NGOs) founded the Climate Leadership Council. Following a full-page ad in the *Wall Street*

Journal in 2017 and the addition of more members, it updated its bipartisan concept of instituting a carbon tax/dividend program. The plan is outlined on their website (clcouncil.org) in a document updated in September 2019 entitled "The Four Pillars of Our Carbon Dividends Plan." The plan's first pillar includes a gradually rising economy-wide carbon fee on CO_2 emissions that start at $40/ton (in 2017 USD) and increase by 5% per year above inflation. It was projected by economists that this would be the most effective climate policy solution and would halve US emissions by 2035, allowing the country to far exceed the goals of the Paris Agreement. The second pillar is that net proceeds from the carbon fee would be returned to the citizens, with a family of four receiving about $2,000 in carbon dividend payments in the first year. The dividend would grow as the carbon fee increased. The third pillar is that the concept would streamline and eliminate complex regulations as the fee would replace the need for regulations in many sectors of the economy and for various source categories. The fourth pillar would provide border carbon adjustments to address carbon-intensive exports and imports to enhance the competitiveness of US firms that are more energy efficient than their foreign competitors. It is uncertain at this time as to whether a carbon fee–based solution could be implemented, and it has been the subject of much debate in the media and among congressional members.

Another initiative, the Transportation and Climate Initiative (transportationandclimate.org) is a

> regional collaboration of 12 Northeast and Mid-Atlantic states and the District of Columbia that seeks to improve transportation, develop the clean energy economy and reduce carbon emissions from the transportation sector. The participating states in the TCI are: Connecticut, Delaware, Maine, Maryland, Massachusetts, New Hampshire, New Jersey, New York, Pennsylvania, Rhode Island, Vermont, and Virginia.

If successful, the initiative would take effect in 2022 and would be focused on clean vehicles and fuels, sustainable communities, and freight efficiency.

Similarly aggressive programs are being proposed by many individual companies and municipalities, some of which are also engaged in the CLC and/or TCI. A few brief examples from various media sources are included here for illustration, although the authors have not confirmed or explored the information in detail. It was reported that Berkeley, CA, had banned the use of natural gas in new construction effective as of January 2020 as a means of reducing GHG; however, a trade group had filed a suit as this would unduly affect the ability of restaurants to prepare food (*Houston Chronicle*, November 21, 2019). On that same date, an Associated Press release reported that the city of Brookline, MA, had adopted a new bylaw at their town meeting requiring the use of all-electric appliances for heating, hot water, and other uses to reduce their local emission of greenhouse gases from oil- and gas-fueled appliances. It was reported that gas and oil groups would likely challenge the bylaw. In January 2020, Microsoft's president announced,

> While the world will need to reach net zero, those of us who can afford to move faster and go further should do so. That's why today we are announcing an ambitious goal and a new plan to reduce and ultimately remove Microsoft's carbon footprint. . . . By 2030 Microsoft will be carbon negative, and by 2050 Microsoft will remove from the environment all the carbon the company has emitted either directly or by electrical consumption since it was founded in 1975.
>
> (https://news.microsoft.com/2020/01/16/)

Other companies have made similar commitments.

The political and economic aspects of regulations related to greenhouse gases, emission standards, and carbon capture and sequestration will be the subject of continued concern from regulatory (state and federal), industrial, and environmental perspectives for years to come. It represents one of the most, if not the most, challenging problems of the twenty-first century. Setting politics and economics aside, science indicates that a failure to adequately address greenhouse gas emissions from fossil fuels will result in substantial impacts on the climate, and therefore, the physical world in which we live.

READINGS

Ackerman, B.A., and Stewart, R.B., Reforming environmental law: The democratic case for economic incentives, *Columbia J. Environ. Litigation*, 13, 171, 1988.

Anderson, S.O., Halberstadt, M.L., and Borgford-Parnell, N., Stratospheric ozone, global warming and principle of unintended consequences—an on-going science and policy success story, *JAWMA*, 63, 607–647, 2013.

Beaton, S.B., Bishop, G.A., Zhang, Y., Ashbaugh, L.L., Lawson, D.R., and Stedman, D.H., On-road vehicle emissions: Regulations, costs, and benefits, *Science*, 268, 991, 1995.

Bishop, G.A., Are electric vehicles a panacea for reducing ozone precursor emissions, *EM—The Magazine for Environmental Managers*, A&WMA, 2020.

Burke, R.L., *Permitting for Clean Air: A Guide to Permitting under Title V of the CAA Amendments of 1990*, Air and Waste Management Association, Pittsburgh, PA, 1992.

Calvert, J.G., Heywood, J.B., Sawyer, R.F., and Seinfeld, J.H., Achieving acceptable air quality: Some reflections on controlling vehicle emissions, *Science*, 261, 37, 1993.

Cannon, J.A., The regulation of toxic air pollutants: A critical review, *JAPCA*, 36, 562, 1986.

Committee for Economic Development, *What Price Clean Air? A Market Approach to Energy and Environmental Policy*, Committee for Economic Development, Washington, DC, 1993.

DeNevers, N.H., Neligan, R.E., and Stater, H.H., Air quality management, pollution control strategies, modeling, evaluation, in *Air Pollution*. Vol. V: *Air Quality Management*, 3rd ed., Stern, A.C., Ed., Academic Press, New York, 1977, p. 4.

Grant, W., *Autos, Smog and Pollution Control: The Politics of Air Quality Management in California*, Edward Elgar, Brookfield, VT, 1995.

Lee, B., Highlights of the clean air act amendments of 1990, *JAWMA*, 41, 48, 1991.

Leonard, L., *Air Quality Permitting*, Lewis Publishers/CRC Press, Boca Raton, FL, 1997.

Lloyd, A.C., Lents, J.M., Green, C., and Nemeth, P., Air quality management in Los Angeles: Perspectives on past and future emission control strategies, *JAPCA*, 39, 696, 1989.

Mintzer, I.M., *A Matter of Degrees: The Potential for Controlling the Greenhouse Effect*, World Resources Institute Research Report 5, Washington, DC, 1987.

National Research Council, *On Prevention of Significant Deterioration of Air Quality*, National Academy Press, Washington, DC, 1981.

National Research Council, *Ozone Depletion, Greenhouse Gases and Climatic Change*, National Academy Press, Washington, DC, 1989.

National Research Council, *Rethinking the Ozone Problem in Urban and Regional Air Pollution*, National Academy Press, Washington, DC, 1991.

Office of Technology Assessment, *Acid Rain and Transported Air Pollutants: Implications for Public Policy*, OTA-O-205, OTA, Washington, DC, 1984.

Padgett, J., and Richmond, H., The process of establishing and revising ambient air quality standards, *JAPCA*, 33, 13, 1983.

Roberts, J.J., Air quality management, in *Handbook of Air Pollution Technology*, Calvert, S., and Englund, H., Eds., John Wiley & Sons, New York, 1984, p. 969.

Schneider, S.H., The greenhouse effect: Science and policy, *Science*, 243, 771, 1989.

Schulze, R.H., The 20-year history of the evolution of air pollution control legislation in the U.S.A., *Atmos. Environ.*, 27B, 15, 1993.

Stavins, R.N., and Whitehead, B.W., Market-based incentives for environmental protection, *Environment*, 34, 7, 1992.

Stenvaag, J.M., *Clean Air Act Amendments: Law and Practices*, John Wiley & Sons, New York, 1991.

Stern, A.C., Ed., *Air Pollution*. Vol. V: *Air Quality Management*, Academic Press, New York, 1977.

Stern, A.C., Prevention of significant deterioration: A critical review, *JAPCA*, 27, 440, 1977.

Stewart, R., Stratospheric ozone protection: Changes over two decades of regulation, *Nat. Res. Environ.*, 7(2), 24–27, 53–54, 1992.

Tietenberg, T.H., *Emissions Trading: An Exercise in Reforming Pollution Policy*, Resources for the Future, Washington, DC, 1985.

US EPA, *The 2019 EPA Automotive Trends Report: Greenhouse Gas Emissions, Fuel Economy, and Technology since 1975*, EPA-420-R-20-006, March 2020.

US EPA, *Draft Guidance for PM2.5 Permit Modeling*, OAQPS, RTP, NC, March 4, 2013.

US EPA, *EPA and NHTSA Set Standards to Reduce Greenhouse Gases and Improve Fuel Economy for Model Years 2017–2025 Cars and Light Trucks*, EPA-420-F-12-051, 2012.

US EPA, Implementing the 1990 clean air act: The first two years, in *Report of the Office of Air and Radiation to Administrator William K. Reilly*, EPA 400-R-92-013, EPA, Washington, DC, 1992.

US EPA, National emission standards for hazardous air pollutants from coal and oil-fired electric utility steam generating units and standards of performance for fossil-fuel-fired electric utility, industrial-commercial-institutional, and small industrial-commercial-institutional steam generating units, *Fed. Reg.*, 77, 32, Rules and Regulations, February 16, 2012.

US EPA, and NHTSA, The Safer Affordable Fuel-Efficient (SAFE) vehicles rule part one: One national program, *Fed. Reg.*, 84, 188, Rules and Regulations, September 27, 2019.

Watson, A.Y., Bates, R.R., and Kennedy, D., Eds., *Air Pollution, the Automobile, and Public Health*, Health Effects Institute, National Academy Press, Washington, DC, 1988.

WHO, *Air Quality Guidelines for Particulate Matter, Ozone, Nitrogen Dioxide and Sulfur Dioxide*, 2005. who.int/phe/health_topics/outdoorair/outdoorair_aqg/en/.

QUESTIONS

1. Despite enactment of clean air legislation and regulations, some property owners may wish to sue a source that is alleged to be depositing particulate matter on their property. What would they reasonably expect to gain from filing a civil suit, and what legal principle might they use?
2. What are air pollution control strategies and how do they differ from tactics?
3. What air pollution control strategies are used under federal clean air laws? What is their relative significance in air pollution control?
4. Describe the theoretical and practical uses of cost–benefit analyses, the bubble concept, and pollution charges in achieving pollution control objectives?
5. Describe the principle of using allowances to achieve the objectives of acidic deposition control.
6. What is the primary advantage of controlling air pollution by using air quality standards? Emission standards?
7. What tactics are used to achieve air quality standards?
8. Describe differences between reasonably available and best-available control technologies. When is each used?
9. Describe a brief history of local, state, and federal efforts in controlling air pollution in the United States.
10. What enforcement tools does the US EPA have available under the 1970, 1977, and 1990 CAA Amendments?
11. Under the CAA Amendments of 1970, Congress required that air quality standards be set and applied to every AQCR in the country. What agency or authority has responsibility for (a) setting air quality standards and (b) enforcing air quality standards?
12. What is the importance of the following in achieving air quality standards: (a) AQCRs, (b) state implementation plans, and (c) emission standards?
13. The deadline for achieving air quality standards in all AQCRs in the United States was July 1, 1975. It is many years later, and some AQCRs have yet to attain these standards. Why is this the case?
14. What options does an industry have if it wishes to build a new facility or expand an existing one in an AQCR where the NAAQS for a pollutant it emits has not been attained?
15. Both NSPS and NESHAP provisions of clean air legislation employ the "emission standard" concept of controlling air pollution. In practice, what differences are there in the application of NSPSs and NESHAPs?
16. Prior to the 1990 CAA Amendments, seven pollutants were regulated as hazardous pollutants. Under the Air Toxics provision of the 1990 CAA Amendments, an additional 182

substances were to be regulated under this concept. These additional substances are regulated using technology-based MACT standards. Describe what MACT standards are.

17. What are the responsibilities of federal and state agencies relative to air quality standards, NSPSs, NESHAPs, and the regulation of motor vehicle emissions?
18. Describe the purpose and statutory requirements of PSD.
19. Describe regulatory authorities and actions used by the US EPA to protect the environment from O_3-destroying chemicals.
20. In the 1990 CAA Amendments, Congress required that a uniform permit system be developed. Describe the use and significance of permits in achieving air quality objectives.
21. Since 1990, more than a 90% (20 million ton/year) reduction in SO_x emissions has been achieved in the United States. Describe the various programs through which this has been accomplished.
22. Briefly describe the statutory history of air pollution control requirements on light-duty motor vehicles.
23. The US EPA has the authority to control fuel additives. Why and how has this authority been used to reduce the lead content of gasoline?
24. In addition to requiring emission reductions from motor vehicles, Congress has required petroleum refiners to produce gasolines that result in emission reductions. Explain these requirements and their significance.
25. Despite the fact that most new cars and trucks have emissions that are a small fraction of those of motor vehicles built 35 years or more ago, motor vehicle air pollution is still a major problem. Why?
26. Describe how state boards and agencies function in implementing the requirements of federal legislation.
27. What are the Kyoto Protocol and Paris Agreements? Why are they so controversial in the US?
28. How could provisions of the Kyoto Protocol, as they would have applied to the US, be achieved without a significant economic impact?
29. New air quality standards for $PM_{2.5}$ and O_3 were promulgated in 1997 but were not immediately implemented. Why was this the case?
30. What is new source review and how does it apply to coal-fired power plants?
31. What are PFAS, PFOA, and PFOS?

9 Control of Motor Vehicle Emissions

As seen in Chapter 2, transportation (highway vehicles) in the United States was the major emission source category of three primary pollutants: volatile organic compounds (VOCs), carbon monoxide (CO), and nitrogen oxides (NO_x) in 1970. As a result of regulatory and technological efforts, VOCs, CO, and NO_x emissions from motor vehicles were reduced by 90%, 90%, and 74%, respectively, in the period from 1970 to 2018. Although substantial reductions have occurred, highway vehicles in 2018 were still the largest emission category for CO and NO_x. The reduction in emissions occurred despite a nearly 200% increase in annual vehicle miles traveled (VMT) from 1.1×10^{12} in 1970 to 3.2×10^{12} as reported by the Federal Highway Administration in 2018. The annual compound growth in VMT was about 2.2%/year for the 48-year period, although it has slowed to about 0.8% based on the last 10 years.

In addition to primary pollutants, transportation also utilizes approximately 32% of the total energy obtained from fossil fuels (Figure 2.1) and is thus a major source category for CO_2 emissions in the United States and other parts of the world. The emissions of CO_2 have been addressed in more recent years, as summarized in Chapter 8 through the use of CAFE and greenhouse gas emission standards. These have placed emphasis on improving the efficiency of engines and the use of lighter-weight materials for vehicles as well as encouraging a transition to hybrid and electric vehicles.

Motor vehicle pollutants such as nonmethane hydrocarbons (NMHCs) or VOCs and NO_x serve as precursor molecules for the production of increased tropospheric ozone (O_3) levels, and the CO_2 emissions contribute to global warming. Within the context of primary pollutants and subsequent formation of secondary pollutants such as O_3, other smog and haze constituents, and climate change, it is not surprising that motor vehicle emission control has been a major regulatory priority in the United States for more than five decades and an increasing priority in other developed and developing countries.

Another dynamic that relates to the emissions from vehicles is human behavior and desire. In 1975, the most popular vehicle on the road in the light-duty category (including sedans, car SUVs, truck SUVs, minivans/vans, and pickups) was the sedan, which held 80.6% of the market share. The average sedan had a fuel economy of 13.5 mpg and emitted 660 g/mi of CO_2 as well as primary pollutant emissions. This was followed by trucks (13.1%). While truck share has remained relatively constant since 1975, the sedan market share has steadily decreased, and car SUVs and truck SUVs have increased. Preliminary 2019 estimates show that the market share for sedans has steadily declined and continues to do so, with only 38.5% of the market. Both car SUV and truck SUV market shares have steadily increased from 0.1% and 1.7%, respectively, in 1975, to 11.3% and 33.1% in 2019, or 44% of the market. While these are heavier vehicles than the sedan, the good news is that all of these have substantially higher fuel efficiencies and less CO_2 emissions in 2019 than in 1975 as follows: sedans (30.8 mpg, 283 g/mi CO_2), car SUVs (27.0 mpg, 327 g/mi) and truck SUVs (23.7 mpg, 375 g/mi). This is largely due to improvements in engine design and efficiency, use of lighter-weight materials in the chassis and interior, and inclusion of hybrid vehicles with energy recovery through regenerative braking. While the change in driver desires has slowly moved toward more efficient, yet larger, vehicles, the emissions of both primary pollutants and greenhouse gases have decreased. The US Energy Information Administration (US EIA), in its 2020 base case, projects that the fuel economy of passenger cars and light-duty trucks will increase by another 55% by 2050.

It is possible that the future reduction in emissions of both types of gases may not be as much from engine design as from a potentially rapid shift to electric vehicles powered by electric motors (i.e., source reduction). The worldwide media (2019–2020) are full of reports from both large and

DOI: 10.1201/9781003664109-9

small automotive manufacturers, electric utility generation units (EGUs), economic forecasters, and think-tank groups projecting a potential rapid transition to electric vehicles in the next decade. Vehicles are commonly available that have features such as lane tracking, adaptive cruise control, cylinder deactivation, and stop/start. The age of 5G, the Internet of Things (IoT), and artificial intelligence (AI) all enhance the ability to move forward with advanced technologies, including the electric vehicle. Companies worldwide have announced multibillion-dollar investments in electric vehicle batteries/manufacturing and the conversion or development of new electric vehicle manufacturing facilities, and the number of new EV models is at an all-time high. The cost of batteries (primarily lithium-ion) has also dropped 50% in the last 5 years to a new low of approximately $135/kW-hour. The Federal Highway Administration (FHWA) reported in 2018 that there are 273 million vehicles registered in the United States, the vast majority of which are powered by gasoline internal combustion engines. While highway vehicles with internal combustion engines will be around for many years due to their current dominance in the overall fleet, some have predicted that EVs (with a current car market share of approximately 2%) could dominate new car sales by as early as 2030.

Other predictions are not as optimistic, tempered by concerns related to costs, acceptability of lesser vehicle range, lack of adequate on-highway charging stations, and possible limitations in the supply chain related to batteries. The US EIA, in its 2020 Annual Energy Outlook, projected that electric vehicles would reach annual sales of approximately two million battery-powered vehicles (about 12% of the market) by 2050, and hybrid or plug-in hybrid vehicles would approach one million (about 6% of the market). Despite the uncertainties of market share, most EGUs in the United States have begun to prepare for the likely increase in electrical energy supply needed to meet the increased charging demand of electric and plug-in hybrid vehicles.

Regulatory efforts to control motor vehicle emissions in the United States (Chapter 8) have a history dating back to the early 1960s. Emission reduction requirements over that time have become more stringent, requiring motor vehicle manufacturers to continuously develop and employ new emission control techniques. The following sections present an overview of the technologies and concepts being employed but are not intended to get into the complexity of design.

9.1 MOTOR VEHICLE ENGINES

Two types of engine systems are commonly used in motor vehicles: (1) Otto cycle spark ignition (SI) reciprocating internal combustion engines (ICEs), and (2) diesel cycle compression ignition (CI) ICEs. The former are widely used in cars and light-duty trucks, motorcycles, boats, and other small-horsepower consumer products. The latter are used in utility vehicles; for example, trucks, earth-moving and mining equipment, farm tractors, and ships, where high-torque, low-speed performance is required. Of the 218 million highway vehicles registered in the United States in 2014, only 4% were diesel CI vehicles, with the majority being medium- to heavy-duty trucks. With respect to passenger car and light truck sales in 2014, less than 1% of sales were diesel CI engines. This is compared with Europe, where 50% of vehicle sales are diesel. While diesel engines have lower CO_2 emissions due to better fuel economy, they emit higher quantities of NO_x and PM. As a result of the dominance of gasoline ICEs in the United States, this chapter focuses more on the SI ICE, although the basic principles of combustion and many of the technologies for controlling emissions are the same. Figure 9.1 illustrates the components of a typical light-duty vehicle, including its fuel-injected SI ICE engine and exhaust system. Many of the components shown will be referenced in this chapter.

9.1.1 Otto Cycle Spark Ignition Engines

9.1.1.1 Engine Characteristics

Otto cycle SI engines, like the one shown in Figure 9.1, may be four-stroke or two-stroke systems. The former are used in cars and light-duty trucks; the latter, in motorcycles and lower-horsepower-requiring vehicles.

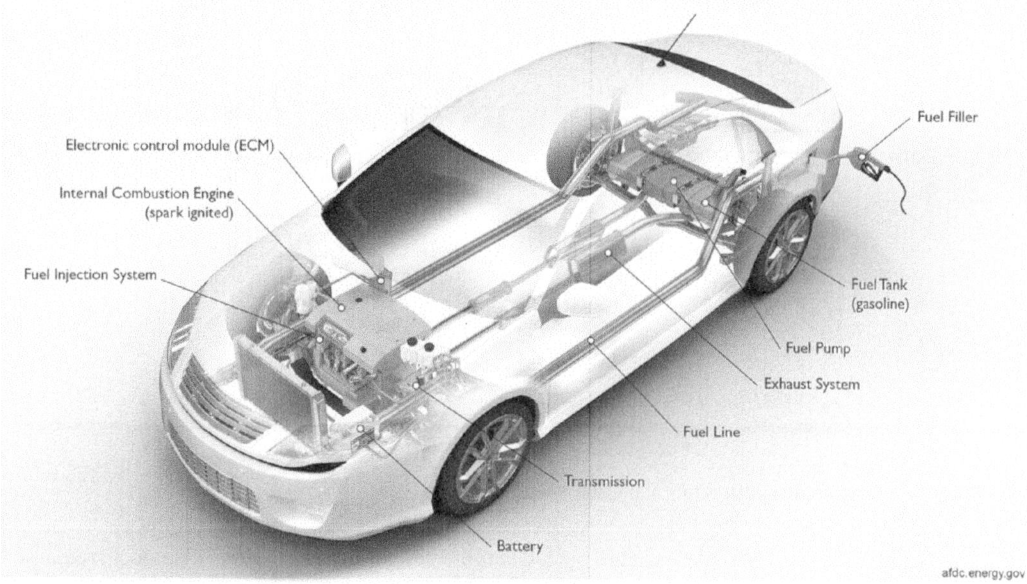

FIGURE 9.1 Vehicle illustrating the components of a gasoline-powered ICE. (*Source*: US DOE EERE, Alternative Fuels Data Center, 2020, https://afdc.energy.gov/vehicles/how-do-gasoline-cars-work.)

The operation of a four-stroke SI engine is illustrated in Figure 9.2. The cycle begins as a mixture of air and fuel is drawn or injected into the combustion chamber (i.e., the volume described by the top of the piston at the beginning of the compression stroke, cylinder walls, and intake valves); this is the intake stroke. In the compression stroke, the valves close and the piston moves up through the cylinder to top dead center, near where ignition takes place. The piston returns to its original position (power stroke) before it moves upward again to purge exhaust gases (exhaust stroke). As a consequence, there is one power stroke per two piston revolutions.

Ignition of the fuel–air mixture drives the piston downward, providing power to the drive shaft that propels the vehicle. In newer hybrid vehicles, the engine may simply provide the power to electric motor(s) that propel the vehicle. Residence time (the time it takes to complete the power stroke) in a 3,000 rpm engine is on the order of 10 ms.

In a two-stroke engine, the basic processes (intake, compression, heat induction, expansion, heat exhaust, gas exhaust) are essentially the same. They occur in half the cycle period of a four-stroke engine. This has the advantage of increasing the power-to-weight ratio and the disadvantage of allowing a shorter period for each process. This results in a substantial fuel penalty (up to 25%) and higher HC emissions.

9.1.1.2 Emissions

Exhaust gases from motor vehicles powered by SI four-stroke ICEs include the normal products of complete combustion (CO_2, H_2O, N_2), products of incomplete combustion (HCs, CO, unburned carbon as PM), and other pollutants that might be formed during combustion (i.e., NO_x, sulfur-based compounds).

In early vehicles, blowby emissions and evaporative fuel losses also occurred, although these have been substantially reduced or eliminated in the modern vehicle. Some unburned HCs "blew by" the piston rings, rather than being combusted in the pressurized combustion chamber, and accumulated in the crankcase. In the uncontrolled vehicle, these would have been vented to the atmosphere

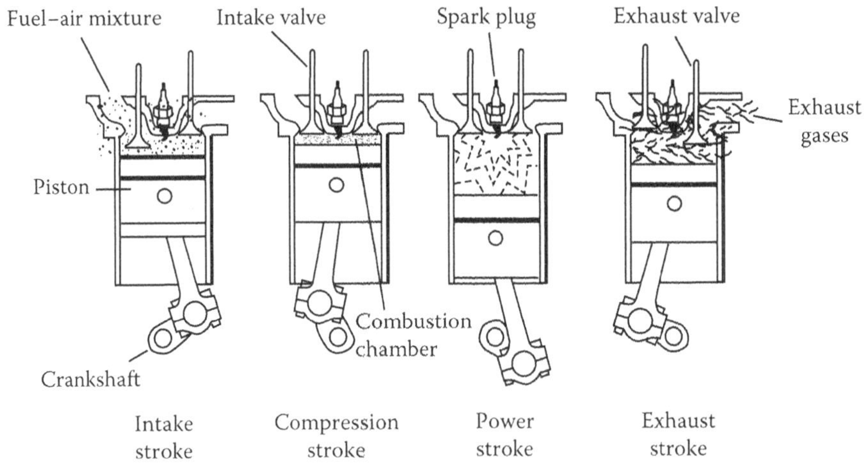

FIGURE 9.2 Four-stroke spark ignition engine operation.

through a small tubular crankcase exhaust port. Gasoline evaporative emissions also occurred associated with the motor vehicle fuel tanks and engine, referred to as diurnal, hot soak, and operating losses. Diurnal emissions occurred when fuel tanks of parked vehicles cooled at night, drawing air in and "exhaling" vapor-containing air upon heating during the day. This fuel tank's "breathing" produced evaporative HC emissions, particularly on hot days. Hot soak emissions occurred just after the engine was turned off and residual fuel evaporated from the warm vehicle. Operating, or running, losses occurred as vapors were driven from the fuel tank when the vehicle was being operated and the fuel warmed. Operating losses were significant at elevated ambient temperatures or when the fuel system became hot during vehicle operation. Evaporative losses may also be associated with vehicle refueling/spillage. The blowby and evaporative emissions control concepts were presented in Chapter 8, Section 8.4.11.

9.1.1.3 Formation of Combustion Pollutants

A variety of by-products are produced in ICE gasoline and diesel combustion. Because of significant emission concentrations and effects on the environment, CO, HCs, NO_x, and CO_2 are of particular importance.

Gasoline and diesel fuels consist of a variety of HC species. They have an average composition that can be described by the formula C_xH_y. For gasoline, x has a value of approximately 7 to 8, and y has a value of 13 to 18. Diesel fuel is typically represented as $C_{12}H_{24}$ and is a heavier molecule weight fuel. For illustrative purposes, values such as C_7H_{13} and C_8H_{18} (octane) have been used to describe the combustion of gasoline in an engine. Using C_7H_{13}, for example, gasoline would require about 10.25 molecules of oxygen (O_2) to be fully oxidized to carbon dioxide (CO_2) and water (H_2O) vapor. In combustion balances, it is common to assume that air is about 79% N_2 and 21% O_2 or a mole ratio of 3.76, so 10.25 moles of O_2 would also have 38.54 moles of N_2, since air is used as the oxidizer. The following equation illustrates the stoichiometric or complete combustion on a molar basis:

$$C_7H_{13} + 10.25O_2 + 38.54N_2 \rightarrow 7CO_2 + 6.5H_2O + 38.54N_2 \tag{9.1}$$

On a mass basis (i.e., multiplying the moles of each constituent by its molecular weight), 97 g of fuel would require 1408 g of air (328 g of O_2 plus 1080 g of N_2) at the stoichiometric condition. The air to fuel ratio (A/F) on a mass basis would be approximately 14.5:1. For diesel fuel and octane, the stoichiometric ratios would have been slightly higher at 14.7 and15.0, respectively.

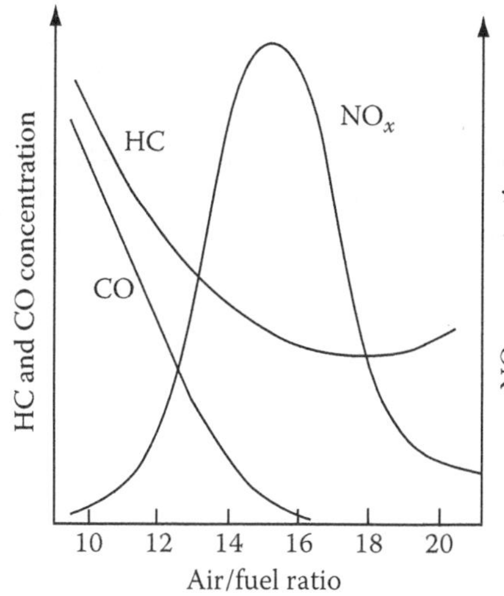

FIGURE 9.3 Relationships between air/fuel ratios and pollutant emissions.

The inverse of the A/F ratio, F/A, is often used in the automotive industry to calculate the equivalence ratio, ϕ:

$$\phi = \frac{(F/A)\text{actual}}{(F/A)\text{stoichiometric}} \tag{9.2}$$

If the equivalence ratio is greater than one, it means the engine is operating with more fuel than is stoichiometrically required (i.e., excess or rich fuel). When an HC fuel is oxidized, CO is initially formed and then further oxidized to CO_2 by O_2. If O_2 is present in less than stoichiometric proportions (i.e., the A/F ratio is less than stochiometric and ϕ is greater than one—a fuel-rich condition), combustion will be incomplete, and CO will begin to increase as shown in Figure 9.2. As the A/F ratio increases above stoichiometric (ϕ decreases to less than 1.0—a fuel-lean condition), very little CO is produced due to its reaction with the excess oxygen.

As can be seen in Figure 9.3, the A/F ratio also affects emissions of HCs. If the A/F ratio is too low, unburned HCs and CO are emitted as a result of the incomplete combustion. If the A/F is substantially higher than the stoichiometric condition (excess air), the HCs begin to increase as the ideal combustion temperature cannot be maintained. However, significant emissions can occur even when sufficient air is present due to quenching (cooling) phenomena within the combustion chamber. A temperature gradient occurs near the internal walls of the combustion chamber as a result of the large quantity of heat that is transferred from combustion gases to cylinder walls and then to the engine's cooling system. Near cylinder walls, temperatures may be less than required to maintain a flame. The flame is "quenched," leaving a small zone of partially combusted fuel. These "quench" zones exist in many locations in combustion chambers. Manufacturers have focused on improved engine designs that minimize the quench zones, as a quench zone of 125 μm could represent 1.5% of the chamber volume. "Crevice" zones of unburned HCs are also formed. These are spaces that are confined on several sides. They include the upper ring between the piston and wall, behind the rings, between the upper rings, within the spark plug screw threads, where the gasket has a small mismatch with the engine block and head, and where the different shapes of valves and seats form a small gap. Crevice zones can contain as much as 8% gasoline by mass and 3% of the cylinder clearance volume but have been minimized by modeling and engine design.

Major engine operating parameters that affect HC emissions include equivalence and compression ratios (chamber volume remaining at the end of the compression stroke compared with that at the end of the intake stroke), engine speed, and spark timing. Unburned HC emissions increase as the compression ratio increases because crevice regions, being of constant volume, make up a larger portion of the compressed volume. Emissions of unburned HCs decrease with increased engine speed as the associated increase in turbulence enhances combustion and postcombustion mixing and oxidation. They also decrease as the spark is retarded because more combustion occurs and exhaust gases are hotter.

Formation of NO_x, which includes nitric oxide (NO), nitrogen dioxide (NO_2), and nitrous oxide (N_2O), is highly dependent on combustion chamber temperatures which are directly related to A/F ratios. As can be seen in Figure 9.3, peak NO_x formation occurs where the A/F ratio is just slightly above the stoichiometric condition and where combustion chamber temperatures are still near their highest (2,000–3,000 K) values due to the combined high temperature and slight excess O_2 effect.

9.1.1.4 Emission Control Techniques and Systems

Both engine modifications and specific control technologies have been used to control motor vehicle emissions. These have included engine operation and design (including valvetrain), engine-based control systems, and exhaust gas control systems. Early control efforts were directed at reducing blowby gases, evaporative emissions, HCs, CO, and NO_x from exhaust systems. Control of HC and NO_x emissions reflected US Environmental Protection Agency (US EPA) policies that focused on reducing tropospheric O_3 production. More recently, the focus has been broadened to include the reduction in overall engine CO_2 emissions to address climate change.

9.1.1.4.1 Engine Operation and Design

Significant reductions in emissions of CO, HCs, and NO_x have been achieved in SI light-duty motor vehicles by engine operation and engineering design changes. These include, but are not limited to, use of lean-burn combustion, changes in spark timing, changes in fuel delivery systems/valvetrains, and engine temperature control.

9.1.1.4.1.1 Lean-Burn Combustion When NMHC and CO exhaust emission standards were first required on 1966 model cars in California, and in 1968 model cars sold nationwide, most automobile manufacturers complied by employing improved combustion systems. The most important combustion modification was to use lean air/fuel mixtures (above the stoichiometric ratio). Carbon monoxide emissions were reduced significantly as A/F ratios were increased from approximately 12:1 to 16:1. NMHCs were also reduced, but less effectively than CO. Although lean A/F ratios increase combustion efficiency, they reduce engine performance and vehicle drivability and increase emissions of NMHCs (because of ignition failures).

To achieve optimal engine power production, the A/F or equivalence ratio should be set at approximately 10% rich. Because vehicle power was a major consumer concern in the 1950s and 1960s, vehicles were designed to operate at rich equivalence ratios, or low A/F, ratios. Lean-burn modifications resulted in significant reductions in engine power. Although lean-burn combustion significantly reduced CO and HC emissions, the associated increase in peak combustion temperatures resulted in optimum conditions for NO_x production (Figure 9.3).

9.1.1.4.1.2 Spark Timing Emission reductions can be achieved by changing ignition timing, an important performance variable. Optimally, it is set to maximize efficiency to what is described as minimum advance for best torque or maximum brake torque (MBT). Optimum efficiency results when peak engine pressure occurs at 10° to 15° of top dead center of the compression stroke. Although actual spark timing varies considerably under different operating conditions, it may be retarded from the optimum to minimize emissions. It was one of the earliest engine operation adjustments used for emission control because of its ease of adjustment.

When the spark is retarded, exhaust temperatures are hotter and both CO and HCs continue to be oxidized in the exhaust system. Once the piston commences its descent (expansion or power stroke), the increased volume in the combustion chamber reduces the proportion of the A/F mixture in quench zones.

Unfortunately, spark retardation reduces engine efficiency. It was a major cause of fuel consumption increases associated with emission controls before the use of catalytic converters. After catalytic converters were introduced, ignition timing was able to be adjusted to optimum settings.

9.1.1.4.1.3 Fuel Delivery Systems and Valvetrains In the 1970s and early 1980s, most SI engines used suction-type carburetors to deliver a mixture of air and gasoline to the engine. These have been replaced with electronic fuel injection/ignition systems which are controlled by an electronic engine control unit (ECU) or as part of the powertrain control module (PCM). The ECU controls the fuel mixture, the idling speed, the ignition or spark timing, and, more recently, the valve timing. These injection systems have become progressively "smarter" at controlling the engine and have transitioned from throttle body injection systems (TBI) to multiple port fuel injection systems (PFIs or MPIs) to gasoline direct injection systems (GDI) where the fuel is injected directly into the engine rather than into the air being supplied to the engine.

Fuel injection systems are readily adaptable to feedback control from measurements of exhaust parameters. The multiport systems improve fuel distribution between cylinders and its timewise distribution to each cylinder. The multiport injection systems, and particularly GDI, have a number of advantages: (1) they increase engine power and performance; (2) the intake manifold can be better designed (because it is for air alone); (3) less heating of induction air is required; and (4) less performance loss occurs due to abnormal A/F ratios in cylinder-to-cylinder fuel distribution.

Valvetrain design is also another key aspect of engine design and comes as a result of the ability to electronically control the engine. The cylinders within the engine have a set of valves that allow for either air or the air/fuel mixture to enter the engine and for the combusted gases to exit the engine as shown in an elementary way in Figure 9.2 in the intake and exhaust strokes. The typical engine now has four valves per cylinder (two for air intake and two for exhaust), and variable valve timing (VVT) has been added to vehicles in recent years. The US EPA now reports annually on the progress and trends of all the major automotive manufacturers. The report provides a thorough review of the many technologies and the annual progress in those technologies and is downloadable at https://www.epa.gov/automotive-trends/download-automotive-trends-report?utm_source=chatgpt.com. Figure 9.4, taken from the 2019 report, illustrates the progress made by the automotive industry in adopting the aforementioned technologies and the resulting simultaneous improvements in specific power and fuel economy (as measured by a decrease in fuel consumption per horsepower of engine) from 1975 to 2018.

There is also a trend toward the use of turbochargers to increase engine power. The turbocharger's turbine, driven by the engine's exhaust gas, compresses the air at the intake of the engine and allows for more fuel to be injected into the engine, increasing the overall power of the engine. Turbochargers are often combined with VVT and MPI or GDI to create an engine with more power or conversely to decrease the size (and thus weight) of the engine. The latter, referred to as turbo-downsizing, has a positive effect on fuel consumption. This has also led to the public's greater acceptance of the "turbocharged" four-cylinder engine as a replacement for the six- and eight-cylinder engines of the past that were known for their greater power.

Other features that have been added to some vehicles in recent years to reduce emissions include cylinder deactivation and stop/start features. In cylinder deactivation, the onboard control module is programmed to sense when less engine power is needed during steady driving speeds and deactivates some of the cylinders, essentially creating a smaller engine and fewer emissions. The start/stop feature senses when the vehicle is not moving and shuts the engine off until it senses a need for additional power. In the hybrid vehicle, this can include delays in starting the engine, even after the vehicle begins to move as the power can be derived from the battery.

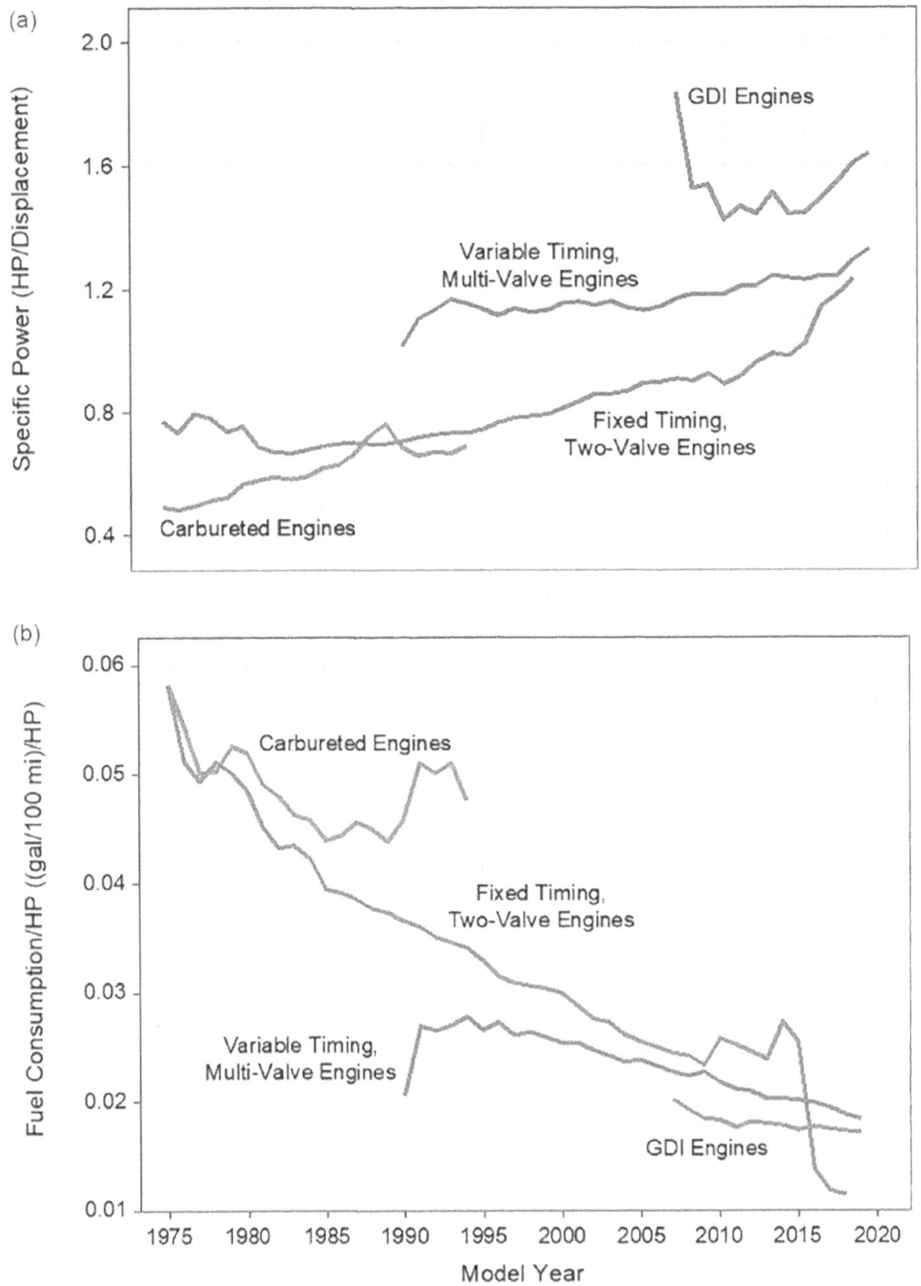

FIGURE 9.4 Improvements in (a) engine specific power and (b) fuel consumption from 1975 to 2019. (*Source*: US EPA, *The 2019 EPA Automotive Trends Report*, 2020.)

9.1.1.4.1.4 Engine Temperature Control Engine temperature in the cylinders and induction system is an important parameter because a high proportion of emissions are produced when the engine is cold or only partially warmed. As a consequence, it is desirable to bring engines up to normal operating temperatures as quickly as possible. This is achieved by careful adjustment of cooling system thermostats and directing the flow of intake air across the engine. It is now common to use heat from the exhaust manifold to warm cool or cold intake air using a control valve.

9.1.1.4.2 *Engine-Based Control Systems*

Motor vehicle emissions are also treated or prevented from being formed by engine-based control systems. These include (1) crankcase ventilation to control HC-rich blowby gases, (2) evaporative controls, and (3) exhaust gas recirculation to control NO_x emissions.

9.1.1.4.2.1 Crankcase Ventilation Blowby, or crankcase, emissions are controlled by a relatively simple and inexpensive technology known as positive crankcase ventilation (PCV) through a PCV valve. Gases slipping by the piston rings into the crankcase are returned to intake airflow by negative pressure in the intake manifold. There, they are mixed with the incoming air–fuel charge and drawn into the combustion chamber and combusted. The one-way PCV valve regulates this flow, ensuring that it only moves from the crankcase and is not excessive. In modern emission control systems, ambient air is also routed to the crankcase so that it is continually purged.

9.1.1.4.2.2 Evaporative Emissions Evaporative emissions from the engine fuel intake (previously, the carburetor) and fuel tank are controlled passively when the engine is off by adsorbing fuel vapors in an activated carbon bed in a small canister connected to the fuel system. The evaporative emission control (EVAP) system is passive in that the increased vapor pressure due to diurnal heating pushes the vapor through the cannister after which the cleaned vapor is then vented to the atmosphere. During engine operation, the adsorbed vapors are desorbed from the sorbent by siphoning air needed for combustion through the cannister prior to its being directed into the engine where the desorbed HCs are combusted. The efficiency of HC collection varies with the fuel components. High-volatility, low-molecular-weight substances are less likely to be sorbed and retained than lower-volatility, higher-molecular weight substances. Newer vehicles utilize this canister to also control refueling emissions at the gas cap, called the onboard refueling vapor recovery (ORVR) system, eliminating the need for Stage II evaporative controls at the pump. The ORVR system can control vapor emissions to about 99%.

9.1.1.4.2.3 Exhaust Gas Recirculation Emissions of NO_x can be reduced to varying degrees by introducing a diluent into the combustion chamber that absorbs heat. Diluents must be readily available and neutral relative to other engine emissions; they should not decrease engine performance. Exhaust gases meet these requirements as they are cooler than the gases in the combustion chamber and consist primarily of CO_2, H_2O, and N_2 as shown in Equation 9.1.

Heat sorption by the recycled exhaust gases reduces peak combustion temperatures and, consequently, NO_x production. Recirculation of a portion of exhaust gases is accomplished by passing a small fraction of the exhaust gas through an exhaust gas recirculation (EGR) control valve that is controlled by the powertrain control module (PCM) and its engine control module (ECM). The effectiveness of EGR in reducing NO_x emissions depends on the quantity of exhaust gases used. An EGR value of 10% may reduce NO_x emissions by 30% to 50%. Because combustion instability occurs above 15% to 20% EGR, this range represents its practical limit.

9.1.1.4.3 *Exhaust Gas Control Systems*

Exhaust gas emissions can be controlled by devices located downstream of the combustion chamber referred to as catalytic converters.

9.1.1.4.3.1 Oxidizing Catalytic Systems Oxidizing catalytic converters were first used on 1975 model cars in the United States to control emissions of HC and CO. Catalytic systems have the advantage of being able to operate efficiently at moderately low exhaust temperatures and optimal compression ratios and spark timing. Oxidation efficiencies of more than 90% can be achieved

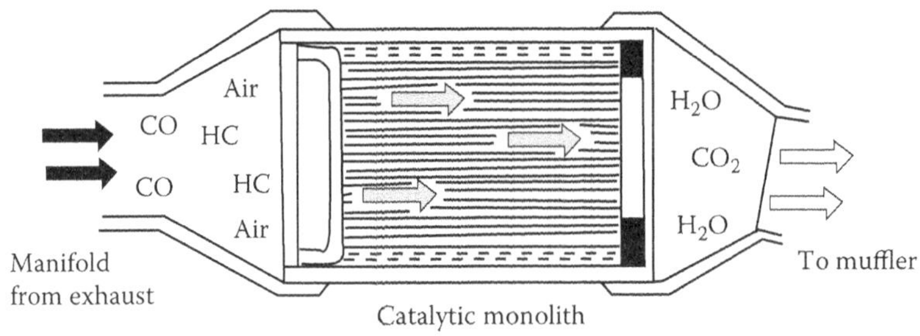

FIGURE 9.5 Movement of exhaust gases through a catalytic chamber.

at temperatures of 300°C (572°F) for CO and 350°C to 370°C (662°F–698°F) for HCs. Catalysts promote desired oxidation reactions at temperatures much lower than would otherwise be required.

Noble metals such as platinum (Pt), palladium (Pd), and rhodium (Rh) are preferred catalysts for catalytic converters. Most catalytic converter systems use some combination of these metals applied to the surface of small, 3 mm diameter pellets, a ceramic honeycomb structure, or a metal matrix. Pellettype systems have been largely replaced by honeycomb or matrix types (because of high pressure drops). Catalyst thickness on substrate surfaces is on the order of 20 to 60 μm. Gas flow through the converter is usually axial (Figure 9.5), but in some cases it is radial.

Oxidation of CO and HCs occurs best using lean A/F fuel mixtures because more O_2 is available for oxidation. Air pumps must be used to provide additional air with richer A/F mixtures. Control efficiencies on the order of 90% for HCs and 95% for CO can be achieved with ϕ values that are stoichiometric or on the lean side of stoichiometric.

The same catalytic materials promote the reduction of NO_x to nitrogen (N_2) and O_2. Reduction of NO_x is facilitated by removing O_2 (from NO_x) by reaction with CO and hydrogen (H_2). Rhodium is particularly effective in reducing NO_x, but Pt and Pd also work relatively well. Peak conversion of NO_x (90%) can be achieved at ϕ values that are approximately 0.5% richer than stoichiometric. Under leaner conditions, conversion efficiency declines significantly. For A/F mixtures 1.5% lean of stoichiometric, conversion efficiency decreases to 20%. Therefore, ϕ values that are optimal for oxidation reactions are less than optimal for NO_x reduction.

For much of their history of use, catalytic converters have been oxidizing systems. This reflects the fact that (1) two of the three major pollutants in auto exhaust are subject to oxidation, and (2) oxidation catalysts obviate the detrimental effects of emission controls on fuel consumption and vehicle operability. Consequently, NO_x emission reduction had to be achieved through other means. EGR was the control technique of choice when oxidizing catalytic systems were used.

9.1.1.4.3.2 Reducing Catalytic Systems Separate catalytic systems to reduce NO_x to N_2 and O_2 can be used in conjunction with oxidizing systems. A reducing catalyst must be placed upstream of the oxidation catalyst because exhaust gases passing through it need to be fuel-rich and contain CO as a reductant. Because the system is fuel-rich, additional air must be pumped into the exhaust gas stream. Platinum has been the catalyst of choice for reduction systems. Copper ion-exchange zeolites can also be used to reduce NO_x under lean conditions. These systems were often referred to as a dual-bed catalyst due to the nature of having both a reducing and an oxidizing catalyst bed.

9.1.1.4.3.3 Three-Way Catalytic Systems In these systems, oxidation and reduction occur in a single catalytic unit. The three-way catalyst is now the most widely used exhaust gas control system, as all three major pollutants are controlled in one bed. Its use was facilitated by the addition of the ECM as a component of the operation and control of the engine and its exhaust system.

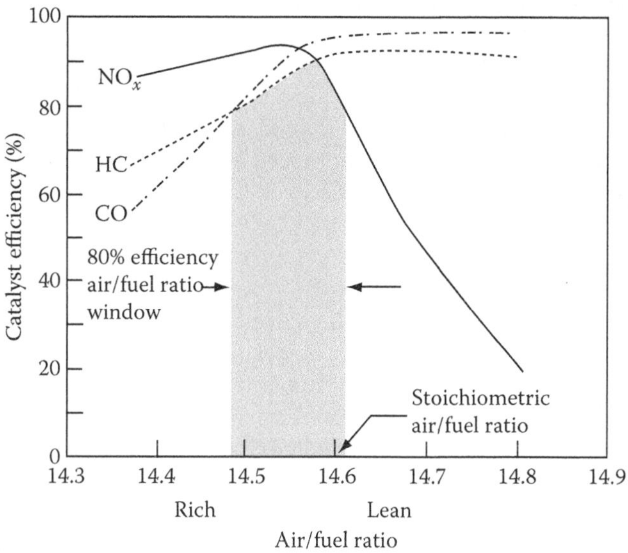

FIGURE 9.6 Relationship between air/fuel ratios and performance of catalytic converters. (From Sher, E., Ed., *Handbook of Air Pollution from Internal Combustion Engines*, Academic Press, San Diego, 1998. With permission.)

For oxidizing catalysis, an A/F ratio on the lean side of the stoichiometric ratio provides the highest efficiency, whereas for reduction catalysis of NO_x, an A/F ratio that is stoichiometric to slightly rich provides the highest efficiency. However, at ϕ values of 0.995 to 1.008, conversion efficiencies approaching 90% are possible for each of the three major pollutants. The effect of A/F ratios on catalyst efficiency in reducing emissions can be seen in Figure 9.6. These systems typically use a catalyst coated with platinum, rhodium, and cerium oxide. The three-way catalyst system utilizes a lambda sensor (often called an oxygen sensor) whose voltage can be used to rapidly alternate the exhaust gas in the catalyst between slightly lean and slightly rich conditions. The platinum oxidizes the HCs and CO, the rhodium reduces the NO_x, and the cerium oxide releases and stores oxygen to optimize the process as it goes back and forth between a reducing and oxidizing environment.

Moderately high temperatures (> 300°C, 572°F) must be attained and maintained for efficient catalytic system performance. Such temperatures are only attained several minutes after engine operation begins. Under cold-start conditions, catalytic converter performance is relatively poor, resulting in higher exhaust emissions. Catalytic systems must therefore be constructed of materials that have a low specific heat but high thermal conductivity to quickly warm to operating conditions. Catalytic materials are mixed with aluminum oxide (Al_2O_3) to facilitate rapid warming.

9.1.1.4.3.4 Deterioration of Catalyst Performance Catalytic system performance can be easily degraded. Degradation results from overheating and contamination of catalytic surfaces by substances in fuel (e.g., sulfur [S], phosphorous [P], and lead [Pb]). Lead, S, P, and possibly other substances are said to "poison" catalysts. In fact, they simply coat the surface of catalytic materials, preventing contact with exhaust pollutants.

Gasoline is formulated to minimize concentrations of P, S, and Pb to prevent contamination of catalytic materials. Phosphorous and S levels can be reduced or nearly eliminated at refineries. Lead was a significant concern, as it was added to gasoline to increase octane ratings and engine performance. Motor vehicles equipped with catalytic systems must therefore be operated on Pb-free gasoline.

Overheating occurs when engine malfunctions allow excessive fuel to pass into the exhaust system. Ignition failure for as little as 20 seconds may produce sufficiently elevated temperatures to totally destroy the catalytic system.

Catalytic converters can convert S to sulfur dioxide (SO_2) and hydrogen sulfide (H_2S), thereby producing malodors. Refiners are now allowed to produce gasoline with a range of sulfur levels as long as their annual corporate average does not exceed 10 parts per million (ppm) in the Gasoline Sulfur Program as presented in more detail in Chapter 8, Section 8.4.11.4.

9.1.2 Compression Ignition Engines

Emission reduction concerns associated with CI engines vary across the developed and developing world. In the United States, they are primarily associated with light-duty and heavy-duty trucks and commercial vehicles such as buses and constitute less than 4% of the registered vehicles and less than 1% of new vehicle sales. In Europe, diesel cars have a reputation as fuel-efficient, durable vehicles. Lack of consumer acceptance in the United States has been attributed to many different things, including tradition (i.e., gasoline engines gained popularity with US manufacturers and the public), fuel tax structure (i.e., higher taxes per gallon of diesel than gasoline), noise (i.e., the historical chatter of the diesel engine), and odors and soot particles. While many of these are no longer valid, the diesel engine has never been accepted, as measured by the annual sales, to the same extent as the gasoline engine, even though many countries push for greater acceptance of diesel engines due to the lower climate impact.

CI engines are naturally aspirated (air is inducted), turbocharged (inlet air is compressed by an exhaust-driven turbine–compressor combination), or supercharged (air is compressed by a mechanically driven pump or blower). Both turbocharging and supercharging increase engine power output by increasing airflow and, therefore, fuel flow. Like SI engines, CI engines use either four-stroke or two-stroke cycles. Fuel is injected under high pressure into a combustion chamber or bowl in the top of the piston toward the end of the compression stroke. The atomized fuel vaporizes, mixes with high-temperature air and, because of the high compression ratios used (12 to 24:1, depending on engine type), spontaneously ignites. Once combustion is initiated, additional fuel mixes with air and is combusted. Emissions of HC and CO are low because combustion is nearly 100% complete and the engine operates with excess air. Because the temperatures produced are high, NO_x emissions are also high. In addition, the fuel-mixing process produces C particles in fuel-rich regions where fuel is sprayed. Some of this elemental C passes through the combustion process unburned and sorbs high-molecular-weight HCs and sulfate S from the fuel and lubricating oil.

Because of lean engine operation, NO_x emissions cannot be controlled using three-way catalysts. Reduction of NO_x has been achieved by careful control of inlet air temperatures and injection retardation to delay most of the combustion process to the early expansion stroke. Nitrogen oxide emission reductions have been on the order of 50% to 65%.

Elemental C, or soot, formation has been reduced by 65% to 80% as a result of modifications to the combustion process. These have included the use of fuel injection equipment with very high injection pressures and carefully matching the geometry of the bowl-in-piston combustion chamber, air motion, and spray. Lubricant control has reduced the concentration of high-molecular-weight substances sorbed onto elemental C particles formed during combustion. Oxidation catalysts may also be used to reduce soluble organic components of particulate-phase emissions. Oxidation catalysts reduce emissions of both HC and CO by 80% to 90% and particulate matter (PM) by 50% to 60%. These catalysts are fitted to all European diesel light-duty motor vehicles to meet European stage 2 emission standards.

Particulate-phase emissions can also be reduced by filtration systems. A number of design approaches have been used in the last decade and a half. Soot filters are located on the tailpipe. They are subject to filter regeneration and avoidance of excessive backpressure. Soot filters typically work by removing and storing PM until a certain level of resistance is detected. Filters are then regenerated *in situ*.

9.2 AUTOMOTIVE FUELS

Different fuel types are used in SI and CI engines. In the former, HC mixtures called gasoline are used; in the latter, diesel fuel.

9.2.1 Gasolines

Gasoline or gasoline products are produced in crude oil refining. Due to different oil sources and refining procedures, gasoline products are complex mixtures of HC species that include paraffins, cycloparaffins, olefins, and aromatic HCs. Consequently, the composition of HC constituents varies from one refiner to another. It also varies as gasolines are formulated to provide optimum engine performance under different driving and climatic conditions. Individual constituents vary in ignition temperature and other combustion characteristics.

Gasolines may contain small quantities of non-HC substances. These include normal constituents of petroleum such as P and S or contaminants introduced in the gasoline production process (such as H_2O). Phosphorous and S reduce the effectiveness of catalytic converters; as such, additional refining steps are needed to reduce their levels. Water negatively affects engine performance.

Gasolines may contain additives such as Pb, Mn, or oxygenates. The use of Pb in gasolines has been restricted in many countries to minimize damage to catalytic converters and to safeguard public health. Nevertheless, it is still used in several developing countries. A Mn-containing compound, methylcyclopentadienyl manganese tricarbonyl (MMT), is used in some countries as an octane booster. Refiners have promoted its octane-boosting potential and requested its approval in the United States. Because of public health concerns, the US EPA has declined to approve its use. Oxygenates are used in the United States to boost gasoline octane ratings and decrease emissions of CO.

9.2.1.1 Octane Rating

Gasolines are formulated to prevent knock, the noise transmitted through the engine structure when spontaneous ignition of the end-gas (fuel, air, residual gas) mixture occurs before the propagating flame. Such ignition produces high local pressures and pressure waves with substantial amplitude. Although knock can be severe enough to cause major engine damage, in most cases, it is just an objectionable source of noise.

The presence (or absence) of knock is a function of the antiknock quality of gasoline. A gasoline's resistance to knock is defined by its octane number. The higher the octane number, the greater the knock resistance. The octane scale is based on two HCs, heptane and isooctane. The former has an octane rating of zero; the latter, 100. Blends of these define the knock resistance of octane numbers between 0 and 100.

The antiknock quality of gasolines is determined by two different methods. These produce what are called motor octane numbers (MONs) and research octane numbers (RONs). The antiknock quality of a gasoline is determined by averaging its MON and RON values.

From 1923 until the advent of catalyst-equipped motor vehicles, Pb alkyls were added to gasoline to increase octane numbers. Their use allowed an increase in antiknock quality to be achieved at lower cost than modifying fuel composition by additional refining steps. Lead additives used were tetraethyl lead ($[C_2H_5]_4Pb$) and tetramethyl lead ($[CH_3]_4Pb$). As previously indicated, a Mn antiknock agent, MMT, has also been used.

Octane ratings can be increased by adding oxygenates such as ethanol (C_2H_5OH), methanol (CH_3OH), tertiary butyl alcohol, and methyl-tertiary-butyl-ether (MTBE), although MTBE is no longer used. Octane ratings can also be increased by increasing the aromatic HC content of gasolines. Aromatic HCs such as benzene, toluene, ethylbenzene, and xylene (when combined, commonly referred to as BTEX) significantly enhance antiknock performance. The composition of aromatic HCs in unleaded gasolines has averaged approximately 30%, but has been as high as more than 40%. Because of health concerns (it is regulated as a hazardous pollutant), benzene concentrations in gasoline have recently been limited to 1%.

9.2.1.2 Gasoline Composition and Emissions

Motor vehicle emissions are significantly affected by fuel constituents and formulation. A prime example of this is Pb. Its use was the major source of Pb in the environment and human exposure to it. Emissions are also affected by the presence of low-molecular-weight HCs such as butane that have a high Reid vapor pressure (RVP), or volatility. Fuels with high RVPs have increased evaporative, running, and refueling emissions even in evaporative emissions-controlled systems. Increases in what is described as midrange volatility lead to reduced HC and CO emissions and increased NO_x emissions.

The olefin content of gasoline is an important emissions concern because of the reactivity of olefins and their role in producing increased O_3 levels in urban–suburban areas. High olefin contents are therefore undesirable.

The aromatic HC content of gasolines affects emissions of CO, HCs, benzene, polycyclic aromatic hydrocarbons (PAHs), and NO_x. As aromatic HC concentrations in fuels increase, emissions of CO, HCs, benzene, and PAHs also increase; NO_x emissions, on the other hand, decrease.

9.2.2 Alternative Fuels

A variety of fuels have been evaluated or are being used as lower-emission alternatives to conventional gasolines. While E10 is the most common one used, others include other alcohol–gasoline blends, CH_3OH, liquefied natural gas (LNG), liquefied petroleum gas (LPG) and hydrogen. Such fuels have the potential to improve air quality, particularly in urban areas. Reduction in emissions and improvement in air quality depend on the type of fuel used and on other factors. Alternative fuels may improve air quality by reducing mass emission rates from motor vehicles or reducing emissions of photochemically reactive HC compounds, or as in the case of hydrogen, eliminating the emissions of HCs and CO_2. As with HC-based fuels, hydrogen, if combusted at high temperature, does not eliminate NO_x emissions as air, containing nitrogen, is the source of oxygen.

9.2.2.1 Alcohol Fuels

There has been considerable interest in using CH_3OH, C_2H_5OH, and alcohol–gasoline blends as motor vehicle fuels. Methanol is a colorless liquid with low vapor pressure and high heat of vaporization. Both properties contribute to lower emissions under warm conditions but tend to make pure CH_3OH-operated vehicles difficult to start when cold (thus resulting in increased emissions). Additives such as gasoline (5%–15% by volume) are blended with CH_3OH to increase vapor pressure and improve starting under colder weather conditions. The most common blend, containing 85% CH_3OH by volume, is described as M85. Flex-fuel vehicles/engines have been developed to run on M85, E15, and E85 vehicles, however their market share remains very small.

The major emission products (on a mass basis) from CH_3OH-fueled vehicles are CH_3OH, HCOH, gasoline-like NMHC components, and NO_x. Of greatest concern is HCOH. Even after catalytic reduction, M85 vehicles may emit approximately 48 to 64.5 mg of HCOH/km (30–40 mg/mi), about three to six times more than conventional vehicles. Formaldehyde has high photochemical reactivity. Limited studies of flexibly fueled vehicles indicate that NO_x may be slightly reduced. However, dedicated CH_3OH-fueled vehicles use higher compression ratios and thus produce higher NO_x emissions. Evaporative emissions are lower and CH_3OH is less reactive than many constituents of gasoline.

9.2.2.2 Compressed and Liquefied Gases

Natural gas (which is > 90% methane, CH_4), propane (C_3H_8), and hydrogen (H_2) can be used to power motor vehicles. Natural gas is compressed and stored at pressures of 4,500 psi (pounds per square inch; 2.32×10^5 mmHg) or liquefied for use as an automotive fuel. Hydrogen-fueled vehicles are typically compressed and stored at 5,000–10,000 psi. Because LNG must be cryogenically cooled and stored, its use has been relatively limited compared with compressed natural gas (CNG). Natural gas vehicles (NGVs) emit primarily CH_4. Because of its low photochemical reactivity, CH_4 has low O_3-forming potential. This benefit is reduced by CNG impurities such as ethane (C_2H_6) and C_3H_8, which may produce up to 25 times more O_3 than CH_4. The presence of olefins may further reduce the benefits of CNG use. Although combustion by-products such as aldehydes are quite reactive, mass emission rates are relatively small. Carbon monoxide emissions from the lean-burn operating condition of NGVs are very low—90% less than gasoline-powered vehicles. Because of lean-burning conditions, CNG vehicles produce lower engine-out (released prior to the catalytic converter) emissions of NMHCs, NO_x, and CO, with higher fuel economy. Evaporative emissions from CNG vehicles are limited and contribute little to O_3 formation. In the United States, CNG is used in fleet vehicles by some corporations and institutions. It is not normally available to consumers at US service stations, but it is available at "petrol" stations in some countries.

LPG consists primarily of C_3H_8, a by-product of petroleum refining. It has many of the attributes of CNG but several disadvantages. These include limited supply and higher exhaust reactivity than CNG. However, it has a higher energy output per unit volume and thus can be stored in smaller tanks.

Hydrogen-fueled vehicles are also available in selected areas within the United States. Most of the vehicles are fuel cell vehicles (FCVs) that convert hydrogen into water, producing electricity that drives the electric motor propulsion system. Some experimental hydrogen-fueled vehicles have been developed that oxidize hydrogen in an internal combustion engine (HICE). The HICE typically has NO_x emissions that must be controlled due to the high temperature and oxidizing air (containing nitrogen) in the engine. The fuel for both is typically produced by steam-methane reforming or by electrolysis of water to hydrogen, both of which could have secondary CO_2 emissions, unless the electricity is produced by a zero-carbon energy source.

9.3 LOW-EMISSION AND ZERO-EMISSION VEHICLES

9.3.1 Low-Emission Vehicles

The SI ICE has been the dominant propulsion system used in light-duty motor vehicles for over a century. It is reliable and economical and gives excellent performance. Because of pollutant emissions associated with SI and CI engines, a number of regulations and technological improvements have been made to gasoline- and diesel-powered vehicles, as presented in Chapters 8 and 9. Examples of these lower-emission vehicles that still utilize an IC engine are the various levels of LEVs. Manufacturers have also developed alternative propulsion systems, including the stratified charge, gas turbine, Wankel, and Rankine cycle engines, all of which combust a fuel to generate the power needed to propel the vehicle. Some of these have only had limited success as alternatives due to the popularity of the standard Otto cycle engines.

The 1990 Clean Air Act (CAA) Amendments called for the development of low-emission and zero-emission vehicles in air quality control regions where achieving the O_3 air quality standard has been, and will continue to be, difficult (e.g., the south coast of California). In the 1990 CAA Amendments, light-duty motor vehicles were to meet more stringent emission limits that were to be phased in starting in 1994. These were referred to as Tier I standards. California was given the authority to develop stricter requirements to address the more serious problems that it faces in urban areas. As a consequence, the US EPA approved California's low-emission vehicle I (LEV I) program in 1993, which was in place from 1994 to 2003. The LEV II regulations were in place from 2004

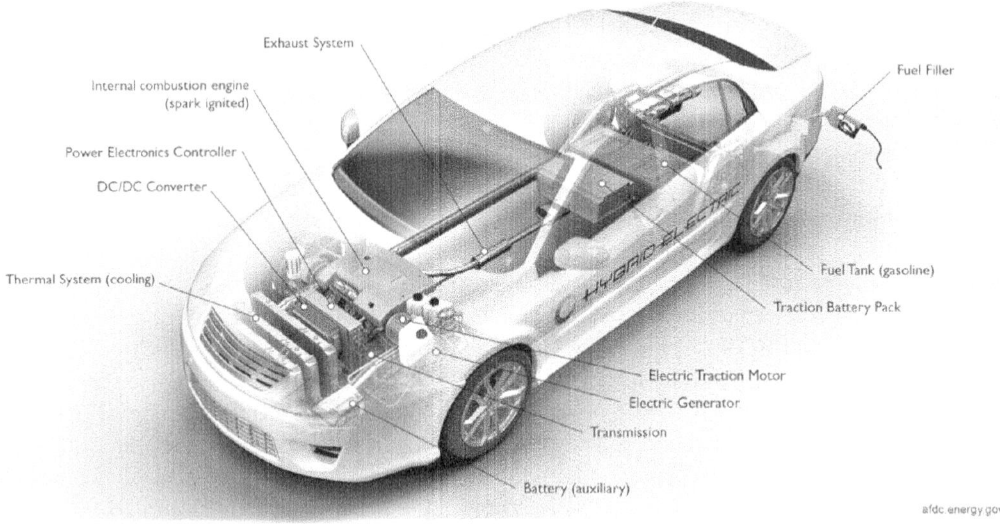

FIGURE 9.7 A hybrid vehicle illustrating the components of a gasoline-powered ICE. (*Source*: US DOE EERE, Alternative Fuels Data Center, 2020, https://afdc.energy.gov/vehicles/how-do-gasoline-cars-work.)

to 2010 to continue progress in emission reductions. In 2012, LEV III was adopted, to be phased in during the period 2015 to 2025. California's LEV program required the development of LEVs that would be phased into the vehicle population over a given period. These included LEVs, ultra LEVs (ULEVs), super ultra-low emission vehicles (SULEVs), and zero-emission vehicles (ZEVs). These would reduce (HC + NO_x) emissions from Tier I levels by 50%, 85%, 94%, and 100%, respectively; as well as CO emissions by 0%, 60%, 76%, and 100%, respectively.

The US EPA also established a voluntary national LEV program under which manufacturers agreed to produce cleaner vehicles for sale in eight northeastern states and the District of Columbia beginning in 1999, and nationwide in the year 2000.

9.3.2 Hybrid Electric Vehicles

A component of the low-emission programs was the hybrid electric vehicles (HEVs), which combined the benefits of both ICEs and electric vehicles (EVs). The combination addressed the limited driving range of the all-electric vehicle and reduced the emissions from the ICE-based vehicle by taking advantage of the regenerative braking/battery storage of the electric vehicle.

HEVs combine a power unit, a vehicle propulsion system, and an energy storage unit. They are typically configured as shown in Figure 9.7 with the engine and electronic circuitry in the front and the battery in the rear. The plug-in hybrid is similar except that the motor is generally smaller, and the battery pack is larger, since it depends more heavily on the electric components. Power units may include SI engines, CI direct injection engines, gas turbines, or fuel cells. In the case of the fuel cell, the gasoline engine and fuel tank are replaced by the fuel cell and a high-pressure hydrogen storage tank. Energy storage may be achieved using batteries, ultracapacitors, or flywheels, although the lithium-ion battery has been the primary choice of storage to date. Vehicles can be designed in a series, parallel, or series/parallel configuration. In the true series design, the electric motor propels the vehicle (a true electric vehicle), deriving its energy from a battery pack. In this mode, there is usually no connection between the engine and drivetrain, as it is used as a generator that recharges the energy storage when needed. Examples of these include the BMWi3 and the Chevrolet Volt, although the Volt does have the ability for the engine to propel the vehicle once the battery has been depleted. In the parallel mode design, the engine and electric motor are integrated to allow either the engine or the electric motor to propel the vehicle. The first-generation Honda Insight is an example. Most HEVs utilize a combination of series and parallel operation through a clutch or gear system that provides a variable range of operation of the engine and electrical motor(s) from zero up to 100% engine to 100% electrical motor, depending on the operating condition of the vehicle (i.e., starting and stopping, queuing, interstate speed).

In 2003, there were only three HEV models being marketed in the United States—the Toyota Prius, Honda Insight, and Honda Civic. By 2020, there were more than 30 HEV models available, including sedans, crossover vehicles, SUVs, and pickups, with many manufacturers offering multiple models. In the United States, federal and state incentives have been applied to partially offset the higher purchase cost of HEVs. In addition, the warranty related to storage batteries often extends to 8–10 years (or an equivalent mileage warranty).

HEVs have the potential to receive wider acceptance in the future because they meet the stringent emission limits for HCs, CO, and NO_x and have about 17%–24% higher fuel economy (and 15%–19% lower CO_2 emissions) than their non-hybrid counterparts. In most cases, and not considering any federal or state incentives, the savings of fuel cost would allow the owner to recover the increased cost of the hybrid vehicle within 8–10 years. Whether HEVs are just a transitional vehicle that will eventually be replaced by all-electric vehicles remains to be seen.

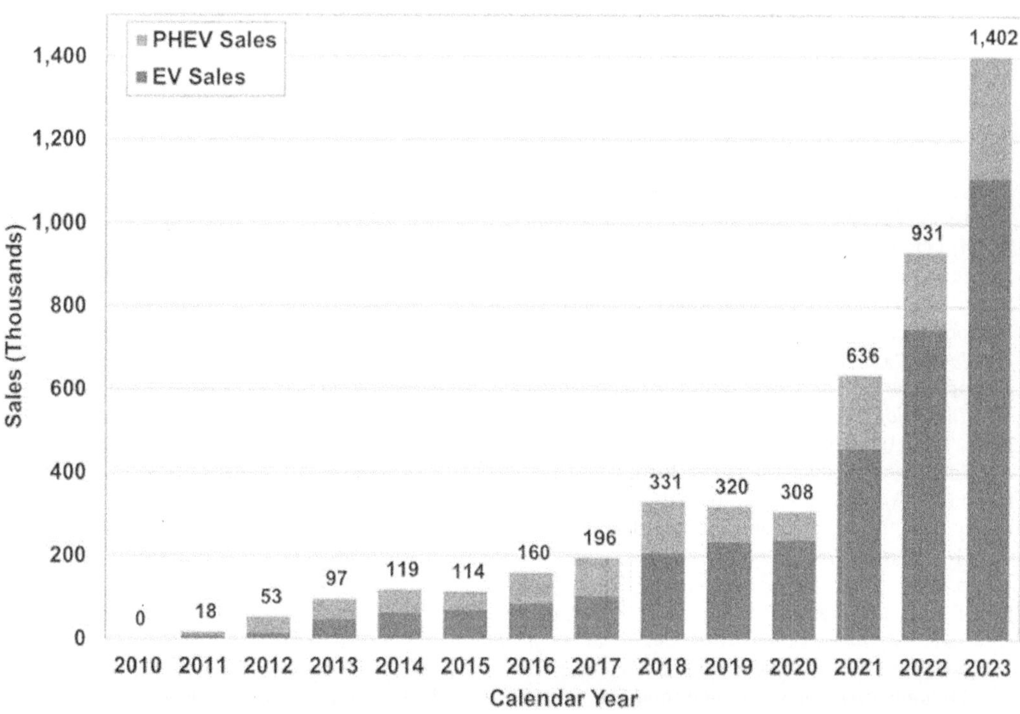

FIGURE 9.8 Annual sales of plug-in hybrid electric vehicles (PHEVs) and all-electric vehicle (EVs) from 2010 to 2023. (From https://www.energy.gov/eere/vehicles/articles/fotw-1327-january-29-2024-annual-new-light-duty-ev-sales-topped-1-million, 2024.)

9.3.3 Electric Vehicles

Electric vehicles, considered to be ZEVs, have been used for years in a variety of utility applications, ranging from golf carts and forklifts to small automobiles. The limitations of electric vehicles have historically been their limited driving range, limited availability of charging stations, and the time required to charge the vehicles, making them more attractive for daily commutes rather than long-distance travel. However, these limitations are being addressed in multiple ways, including the development of home chargers that can be purchased in the \$400–\$1,000 range (2020), installation of public and employee charging stations, and installation of dedicated charging stations by EV manufacturers, such as Tesla. The mileage ranges of EVs have also increased as a result of technological developments in battery energy density and lower-weight materials, and it is now common to see advertised EVs with driving ranges between 200 and 315 miles. In early 2020, there were about 10 large manufacturers who were selling EV models in the United States, and six new models were being introduced for MY2021.

Projections about the growth of EVs, however, vary widely; the US EIA's AEO2020 reference case projects that the growth rate of EVs (ZEVs) with driving ranges of 200–300 miles will be the highest of any vehicle category for the next 30 years, with annual sales growing from 280,000 vehicles/year in 2019 to 1.9 million/year by 2050. Sales of hybrid and plug-in hybrids are projected to grow to nearly 1.0 million/year by 2050, while gasoline (and flex-fuel) powered vehicle sales are projected to remain flat or decline slightly. Based on their projections, gasoline vehicle sales, as a percentage of the overall market, are expected to decrease from 94% (2019) to 81% by 2050 but will still dominate the market in the United States. Figure 9.8 shows PHEV and EV sales from 2010 to 2023. After a sharp growth phase beginning in 2021, annual EV sales quadrupled by 2023, when they topped one million units for the first time. Countries that have made commitments to the Kyoto Protocol and the Paris Agreement are likely to transition much more rapidly than the United States. The transition to electric vehicles will rely mostly on the availability of recharging infrastructure in most countries.

READINGS

Amman, C.A., Alternative fuels and power systems in the long term, *Int. J. Vehicle Design*, 17, 5/6 (Special Issue), 510–549, 1996.

Calvert, J.G., Heywood, J.B., Sawyer, R.F., and Seinfeld, J.H., Achieving acceptable air quality: Some reflections on controlling vehicle emissions, *Science*, 261, 37, 1993.

Chang, T.Y., Alternative transportation fuels and air quality, *Environ. Sci. Technol.*, 25, 1190, 1991.

Charlton, S.J., Control technologies in compression-ignition engines, in *Handbook of Air Pollution from Internal Combustion Engines*, Sher, E., Ed., Academic Press, San Diego, CA, 1998, p. 358.

Dekiep, E., and Patterson, D.J., Emission control and internal combustion engines, in *Handbook of Air Pollution Technology*, Calvert, S., and Englund, H., Eds., John Wiley & Sons, New York, 1984, p. 484.

Garret, T.K., *Automotive Fuels and Fuel Systems*. Vol. I: *Gasoline*, Pentach Press and Society of Automotive Engineers, Washington, DC, 1991.

Harmon, R., Alternative vehicle-propulsion systems, *Mech. Eng.*, 105, 67, 1992.

Hawley, J.G., Brace, C.J., and Wallace, F.J., Combustion-related emissions in CI engines, in *Handbook of Air Pollution from Internal Combustion Engines*, Sher, E., Ed., Academic Press, San Diego, CA, 1998, p. 281.

Heck, R.M., and Farranto, R.J., *Catalytic Air Pollution Control*, Van Nostrand Reinhold, New York, 1995.

Heywood, J.B., *Internal Combustion Engine Fundamentals*, McGraw-Hill, New York, 1988.

Heywood, J.B., Motor vehicle emissions control: Past achievements, future prospects, in *Handbook of Air Pollution from Internal Combustion Engines*, Sher, E., Ed., Academic Press, San Diego, CA, 1998, p. 1.

Hockgreb, S., Combustion-related emissions in SI engines, in *Handbook of Air Pollution from Internal Combustion Engines*, Sher, E., Ed., Academic Press, San Diego, CA, 1998, p. 119.

Medlen, J., MTBE: The headache of cleaner air, *Environ. Health Perspect.*, 103, 666, 1995.

Milton, B.E., Control technologies in spark ignition engines, in *Handbook of Air Pollution from Internal Combustion Engines*, Sher, E., Ed., Academic Press, San Diego, CA, 1998, p. 189.

National Research Council, *Rethinking the Ozone Problem in Urban and Regional Air Pollution*, National Academy Press, Washington, DC, 1991.

Olsen, D.R., The control of motor vehicle emissions, in *Air Pollution*. Vol. I: *Engineering Control of Air Pollution*, 3rd ed., Stern, A.C., Ed., Academic Press, New York, 1977, p. 596.

Peyle, R.J., Reformulated gasoline, *Automot. Engin.*, 99, 29, 1991.

US EIA, *Annual Energy Outlook 2020 (With Projections to 2050)*, January 2020. www.eia.gov/aeo.

US EPA, *The 2019 EPA Automotive Trends Report: Greenhouse Gas Emissions, Fuel Economy, and Technology since 1975*, March 2020. epa/gov/sites/production/files/2020-03/documents/420r20006.pdf.

Watson, A.Y., Bates, R.R., and Kennedy, D., Eds., *Air Pollution, the Automobile and Public Health*, National Academy Press, Washington, DC, 1988.

QUESTIONS

1. Describe the types and sources of pollutant emissions from spark ignition engine-equipped motor vehicles.
2. What is the relationship between air/fuel ratios and emissions of HCs, CO, and NO_x?
3. How is the stoichiometric ratio related to emissions of HCs, CO, and NO_x?
4. How does the compression ratio used in a particular vehicle affect emissions?
5. What is the relationship between spark timing and emissions?
6. How do emissions differ in spark ignition and compression ignition engines?
7. What are blowby gases? How are they controlled?
8. Describe factors that contribute to evaporative emissions. How are they controlled?
9. What is an equivalence ratio? What is its relationship to pollutant emissions?
10. Since both CO and HCs are products of incomplete combustion, why can't they be controlled by application of the same control principles?
11. Decreased compression ratios result in decreased HC emissions. Why?
12. Even with the best pollution control systems, the highest emissions occur under what vehicle operating conditions?
13. Describe the principle of exhaust gas recirculation and the pollutant that it controls.
14. How does a thermal reactor work? What are its limitations in controlling pollutant emissions?
15. Describe how an oxidation catalyst works.
16. What pollutants are controlled using a three-way catalyst?
17. Why are lead, phosphorous, and sulfur levels in gasoline a concern?
18. How are particulate-phase emissions from CI engines reduced?
19. Why do gasolines have a relatively high octane rating?
20. Why was lead used in gasoline? What is being used to replace lead?
21. What are oxygenates? Why are they used in gasoline?
22. From an emissions standpoint, what makes gasoline reformulated?
23. Why use fuels such as CNG, LPG, or methanol in light-duty motor vehicles?
24. Describe how hybrid electric vehicles reduce emissions of HCs, CO, CO_2, and NO_x.
25. While an electric vehicle has no direct emissions of HCs, CO, CO_2, PM, or NO_x, discuss the secondary emissions that might occur due to using electricity to charge the vehicle.
26. What is the ORVR?

10 Control of Emissions from Stationary Sources

Stationary sources, as seen in Tables 2.2 and 2.4, are significant sources of pollutant emissions as well as greenhouse gas emissions. These emissions are quite varied in both their nature and quantity, depending on the industrial and commercial activities involved. Pollutants may be emitted to the atmosphere through specially designed stacks and local exhaust ventilation systems or escape as fugitive emissions.

Emissions from stationary sources can be controlled, or their effects reduced, by the application of one or more control practices. These include the use of tall stacks, changes in fuel use, implementation of pollution prevention programs, fugitive emissions containment, and "end of the pipe" control systems that remove pollutants from waste gas streams before they are emitted into the ambient environment through stacks or vents.

10.1 CONTROL PRACTICES

10.1.1 Tall Stacks

The stacks or vents exiting most stationary sources do not reduce source emissions. Rather, they reflect source efforts to reduce potential pollution problems near the source by elevating emissions above the ground where they may be more effectively dispersed and diluted. When public concern was limited and control devices were not available, a sufficiently sized stack was a relatively effective way of reducing ground-level concentrations and human exposures. The effectiveness of a stack in dispersing pollutants depends on its height, plume exit velocity and temperature, atmospheric conditions such as wind speed and stability, topography, and proximity to other sources (see Chapters 3 and 7).

Utility (EGU) and smelter operations have used tall stacks (200–400 m, ~600 to 1,200 ft) to reduce ground-level concentrations and thus comply with air quality goals or standards for more than five decades. Such stacks were often necessitated by the high sulfur dioxide (SO_2) emissions of these sources. The higher the stack, the greater the plume height, wind speed, and air volume between the plume and ground available for dilution. Although they reduce ground-level concentrations, tall stacks disperse pollutants over larger geographical areas. Ironically, and as emission standards mandated further reductions through the use of wet SO_2 scrubbing systems, many of the electricity-generating coal-fired power plants required the further construction of shorter stacks due to the cooler exit temperatures and reduced buoyancy of the saturated flue gases and the challenges related to exhausting the more dense, saturated plumes through the taller stacks. Figure 10.1 illustrates an excellent example of a plant where one can see the progression from a series of shorter stacks to two very tall stacks (304.8 m or 1000 ft) to one single intermediate height stack as a direct result of increasingly more stringent SO_2 emission standards over the last 40 years.

10.1.2 Fuel Use Policies

Significant reductions in emissions from combustion processes can be achieved by changes in fuel use. Indeed, after the passage of the 1970 Clean Air Act (CAA) Amendments, much of the early progress in emission reductions in the United States was achieved by switching from high-sulfur (S) fuel oil and coal to cleaner-burning fuels such as natural gas and lower-S fuel oil and coal. The cleanest-burning fuel is natural gas.

DOI: 10.1201/9781003664109-10

FIGURE 10.1 Example of a power plant with multiple height stacks in Kingston, TN. (The shorter stacks on the right were used in the 1960s when only PM was controlled with electrostatic precipitators (ESP); the tall stacks were installed as a means of reducing the ground-level effect of sulfur dioxide emissions in the 1970s; the shorter stack on the left was constructed when a high-efficiency limestone SO_2 wet scrubbing system was installed in the late 2000s.)

These fuel use changes have not been problem-free. Supplies of low-S fuels such as oil have always been limited, and therefore, these fuels are more expensive. Low-S oil must be imported, increasing this country's reliance on sometimes unreliable foreign suppliers. Use of low-S western coals decreased markets for high-S eastern coals, raising political and economic concerns. The recent resurgence of the availability of natural gas through fracking technologies and the simultaneous need to reduce carbon (CO_2) emissions due to global warming concerns has resulted in many utilities shifting their long-term power generation strategies away from coal, leading to a number of planned shutdowns of older coal-fired plants in the 2010 to 2020 period in the United States. In 2013, an estimated 73,000 megawatts electric (MwE) of aging coal-fired capacity was retired and plans were under consideration to retire an additional 250,000 MwE by 2015.

Significant improvements in air quality were also achieved in certain regions of the United States as a result of utility decisions to use nuclear reactors (rather than coal) to generate electricity. In 1971, nuclear reactors produced only 2.4% (38,000 MWh) of electricity. By 1988, nuclear reactors were producing about 20% (527,000 MWh), and the percentage has held at about 20% since that time with a total generation of 807,000 MWh in 2018. As a result of accidents at Three Mile Island (Pennsylvania) in the late 1970s and Chernobyl (Ukraine, in the former Soviet Union) in 1986, escalating construction costs, and opposition to nuclear power, no new nuclear power plants were constructed in the United States for nearly 20 years. In the 2005 to 2011 period, nuclear power was projected to see a renaissance as a result of global warming concerns. Although nuclear power has many advantages over fossil fuels from an air emissions and global warming standpoint, the 2011 event in which an earthquake-induced tsunami resulted in a catastrophic failure of the Fukushima Nuclear Power Plant in Japan caused many companies to cancel or delay plans for the initiation of new plants. Reactor safety and radioactive waste disposal remain major environmental concerns. Only one nuclear power plant, Georgia Power's Vogtle 3 and 4 units, was under construction as of early 2020, and the retirement of a number of reactors in the late 2010s may result in a decrease in nuclear power capacity. It is anticipated that the loss of capacity will be replaced by sources that are less carbon intensive than coal e.g., natural gas) or zero carbon (e.g., solar and wind).

10.1.3 Fugitive Emissions

Significant fugitive emissions of vapor-phase and particulate-phase materials occur within industrial and commercial facilities and from outdoor materials-handling activities. Fugitive emissions have, for the most part, gone unregulated. This lack of regulation reflected a dearth of information on the nature and extent of such emissions and a regulatory preoccupation with stack emissions, which were amenable to reduction using existing control technologies.

Fugitive emissions are increasingly being controlled in industrial facilities. Such control reflects both economic considerations and the fact that fugitive emissions are often a significant problem. Under the bubble concept, which treats a single facility as a source, it may be more economical to control fugitive emissions than to apply more expensive technology to control stack gases. A case in point may be a steel mill where fugitive emissions from process and road surface dusts become entrained upon disturbance by vehicle movement. Such dusts can be (and are) controlled by periodic roadway surface wetting. Fugitive emissions control can be achieved by implementing a variety of operating practices, materials handling, and process changes, as well as technology.

Considerable potential exists to reduce fugitive emissions through equipment and process changes, equipment maintenance, and containment strategies. Examples of the latter include the use of hoods and negative pressure within a facility to minimize the escape of hazardous materials and the construction and operation of negative pressure enclosures. Such enclosures are required under occupational safety and health rules for asbestos and lead (Pb) abatement projects. They have the benefit of reducing emissions to the outdoor environment. Under National Emissions Standards for Hazardous Air Pollutants (NESHAP) asbestos abatement requirements, friable asbestos materials must be wetted before and during removal operations to prevent visible emissions.

Fugitive emissions have received considerable attention under Title III in the air toxics provisions of the 1990 CAA Amendments, where maximum achievable control technology (MACT) standards require that plantwide emissions of hazardous pollutants be controlled.

10.1.4 Pollution Prevention

As indicated in Chapter 8, pollution prevention as a waste control strategy was codified in the Pollution Prevention Act of 1990 and given new attention in achieving air quality goals under the CAA Amendments of 1990. Although pollution prevention efforts were originally for reduction of solid waste disposal on land, its application has been broadened to include other media as well.

The goal of all pollution prevention activities is to reduce waste generation at the source. The concept is applied both narrowly (focusing primarily on industrial processes) and broadly (focusing on such things as energy conservation and emissions from finished products).

At the level of an industrial or commercial facility, pollution prevention may include substitution of process chemicals (materials that are less hazardous to human health or the environment than those presently being used); enhanced maintenance of plant equipment; and changes in process equipment, plant operating practices, plant processes, and the like.

10.1.4.1 Substitution

Examples of substitution include the use of (1) *n*-methyl pyrrolidone (NMP) as an alternative for methylene chloride-based paint strippers, (2) reformulated gasoline, and (3) hydrochlorofluorocarbons (HCFCs) and hydrofluorocarbons (HFCs) for chlorofluorocarbons (CFCs). NMP is less toxic and less volatile than methylene chloride. Use of reformulated gasoline to achieve air quality goals was discussed in Chapter 9, and use of HCFCs and HFCs to protect the ozone (O_3) layer was discussed in Chapter 8.

10.1.4.2 Process Equipment Changes

A variety of process equipment changes can be used to reduce pollutant emissions. These include the use of completely enclosed vats in place of open vats where solvent emissions occur; retrofitting or replacing leaky coke oven doors with state-of-the-art seals; and use of enclosed electric arc furnaces instead of reverberatory furnaces in the metals processing industry.

10.1.4.3 Plant Operating Practices

Excess production of pollutants and subsequent emissions may occur as a result of poor equipment and plant operation. This is particularly true in the operation of boilers and other combustion equipment. Good boiler and incinerator operation requires adequate supplies of combustion air. Insufficient combustion air results in incomplete combustion, with the production of elemental carbon (C) particulate matter (PM) and a variety of gas-phase substances. Good operating practices are particularly important in achieving emission reductions for which pollution control equipment are designed.

10.1.4.4 Maintenance

The performance of equipment in terms of producing products and releasing contaminants to the environment may be compromised by inadequate maintenance. It is particularly important to maintain combustion equipment in good operating condition. Maintenance is necessary to reduce leakage of solvents and other chemicals from vats, valves, and transmission lines. It is also necessary to reduce spill-related emissions and maintain adequate seals around coke oven doors in coke oven operation.

10.1.4.5 Process Changes

In many cases, pollutant emissions are related to processes used in product manufacturing. Emissions of solvent vapors, for example, can be reduced in painting operations by dry powder painting. In this process, specially formulated thermoplastic or thermosetting heat-fusible powders are sprayed on metallic surfaces that are subsequently heat cured. Dry powder painting can also be conducted electrostatically. In either case, significant reductions in hydrocarbon (HC) emissions are achieved.

10.1.4.6 Energy Conservation

Reduction in energy use by application of a variety of energy conservation measures can result in significant emission reductions. All fuels, when fully or partially combusted, produce by-products that pollute the atmosphere. Any measure that reduces energy consumption associated with fossil fuel use will result in a reduction in both fuel use and associated emissions.

Energy conservation measures may include the (1) manufacture, sale, and use of fuel-efficient motor vehicles; (2) development and use of energy-efficient combustion and heat-recovery systems in industry and homes; (3) use of mass transit and carpooling; (4) construction of new buildings and retrofitting existing buildings to reduce energy loss; (5) manufacture and use of energy-efficient appliances and equipment; (6) time of day residential energy pricing to encourage using energy during peak periods, (7) use of secondary (as compared with primary) materials in product manufacturing; and (8) recycling and reuse of materials.

Significant reductions in energy usage have occurred in varied economic sectors in the United States since several energy crises were experienced in the 1970s. Most notable have been the improvements in motor vehicle fuel economy and energy-use reductions in manufacturing facilities. Programs with less impact but of some note are the US Environmental Protection Agency's (US EPA) Energy Star Computer and Green Light Programs.

The Energy Star Program is a voluntary partnership between the US EPA and computer manufacturers. Computers identified with the Energy Star logo are designed to power down automatically when not in use. The Green Light Program encourages voluntary reductions in energy use through more efficient lighting in buildings (http://www.energystar.gov/).

10.2 GAS CLEANING TECHNOLOGY

Compliance with emission limits required under state implementation plans (SIPs) often necessitates the application of pollution control systems. Because of the expense and complexity that may be involved, the selection of appropriate equipment requires a careful engineering evaluation of the nature of the emission problem, including physical or chemical characterization of pollutants, process conditions, and determination of gas discharge rates. The selection of a control system will, in good measure, be determined by its performance relative to achieving mandated emission limits. It also involves a consideration of the capital, operating, and maintenance costs.

10.2.1 System Performance

Gas cleaning systems may be designed to remove PM, individual gases, or both. Control system design and performance is determined by specific control requirements. Performance can be quantified by determining collection efficiency, which is defined as a percentage of influent PM or gas collected or removed from the waste gas stream. Collection efficiency (E) can be calculated from Equation 10.1:

$$E = 100\left(1 - B / A\right) \tag{10.1}$$

where A is the concentration of the pollutant entering the control system and B is the emission level achieved or emission reduction required. If the influent to a particle collector were 600 kg PM/h and effluent gas emissions 30 kg PM/h, the calculated collection efficiency would be 95%:

$$E = 100\left[1 - \left(30 / 600\right)\right] = 95\%$$

Emission limits or standards generally do not specify the collection efficiency required of control systems. Therefore, performance must be calculated from emission data and applicable emission limits. For example, if process emissions (A) were 600 kg/h and allowable emissions (B) 60 kg/h, the collection efficiency necessary to comply with the emissions standard would be 90%. To minimize costs, industrial sources may select control equipment that meets only the emission limit mandated with a reasonable safety margin to ensure compliance.

10.2.2 Particle Collection Efficiency

The major determinants of particle collection efficiency are particle size and control systems employed. Some control systems are inherently more efficient than others. One or several of the following collection principles may be utilized for specific control applications: inertial separation, interception, diffusion, and electrostatic precipitation. The overall efficiency and collection techniques for commonly used particle collectors are summarized in Table 10.1 and discussed in the following sections.

Variation in collection efficiency associated with particle size is illustrated for commonly used particle control systems indicated in Figure 10.2. Note that fractional collection efficiency decreases

TABLE 10.1
Collection Efficiencies of Particle Control Devices

Equipment Type	Overall Collection Efficiency (%)
Cyclone	
Medium efficiency	68–85
Multitube collector	70–90
Electrostatic precipitator	70–99.5+
Fabric filter	99.5+
Wet scrubber	
Spray tower	90
Baffle plate	95
Venturi	99.5+

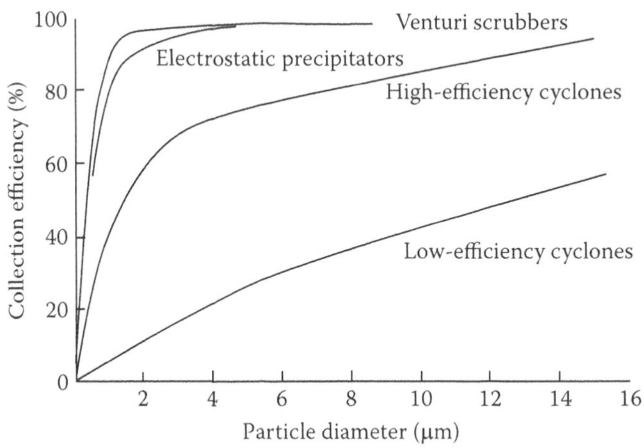

FIGURE 10.2 Comparative fractional collection efficiencies of particle collecting devices.

sharply for small particles, whereas over a broad range, large particles are collected at efficiencies that approach 100%.

Although fractional collection efficiency is important in equipment design, performance-based (i.e., overall) collection efficiency is the principal determinant of compliance with emission limits. As shown in Figure 10.2, particle collection efficiency decreases significantly with decreasing particle size, particularly with particles in the submicrometer range. Paradoxically, it is these small particles that have the greatest human health significance. The efficiency of particles smaller than approximately 0.1 μm increases for some of the devices such as fabric filters and electrostatic precipitators (ESPs), resulting in the minimum collection efficiency according to size being in the 0.1 to 1 μm range—the same range that is responsible for the scattering of visible light and the production of haze.

As shown in Table 10.1, average overall collection efficiencies are dependent on the collection principle used. Under actual operating conditions, performance may vary considerably from the average values of Table 10.1. This variation is due to several factors, for example, gas flow rate, PM loading, and particle properties such as density, shape, viscosity, and resistivity (see Section 10.3.2).

10.2.3 Gas-Phase Contaminant Collection Efficiency

The collection efficiency of control systems designed to control gas-phase substances depends on the gas treatment techniques employed. Commonly used control principles include thermal oxidation, adsorption, and absorption. In general, the collection efficiency of control systems employing these control principles will be influenced by factors such as gas flow rates and pollutant concentrations. Other factors affecting efficiency may be specific to the treatment process or equipment design.

10.3 CONTROL OF PARTICULATE-PHASE POLLUTANTS

10.3.1 Cyclonic Collectors

Cyclonic systems (i.e., cyclones) are widely used particle collectors intended primarily to remove larger particles during materials transfer or to reduce the particle load to more efficient collectors such as fabric filters or electrostatic precipitators. They employ the principle of centrifugal separation to remove particles from waste streams. Collection efficiencies vary from relatively low for large, single cyclones to moderate–high efficiency in multiple smaller diameter cyclone systems where centrifugal force is substantially greater due to the smaller radii. Cyclonic particle collection systems are characterized by low capital, operating, and maintenance costs.

In a simple, single cyclone, particle-laden air enters an inlet near the top of the collector. Cyclonic flow is induced by the design of the entry, which may be tangential, helical, involute, or vane axial. In simple vane axial systems, air enters the top of the collector and a downward-spiraling vortex is produced whose nature is determined by the shape of the system. This curvilinear flow produces centrifugal forces that cause larger suspended particles (particularly those > 10 μm aerodynamic diameter) to move radially outward until they strike the collector wall. They are caught in a thin laminar area of air near the wall and fall downward by gravity into a collection hopper. As pollutants and waste gases continue along their path downward, the narrowing cone-shaped section produces an increase in rotational velocity that confines particles to the wall area and minimizes their reentry. At the bottom of the cone, the vortex reverses direction and cleaned gases move upward and enter the center of the collector. This is a reverse-flow cyclone; other cyclones are designed as straight-through devices.

The particle collection efficiency of cyclones is proportional to the square of the tangential gas velocity and inversely proportional to their radii. Collection efficiency can be optimized by using high inlet velocities, small diameters, and longer cylinders. Smaller diameters, however, increase the pressure drop and operating costs.

Large individual cyclones such as those serving grain elevators (Figure 10.3) accommodate gas flows up to 18.9 m^3/s (40,000 actual cubic ft/min [acfm]). Parallel arrays of cyclone tubes are commonly used for higher gas flows and collection efficiencies. These cyclone tubes vary in diameter, with flow capacities up to 2.1 m^3/s (4,500 acfm).

In multiple-cyclone systems, a large number of cyclone tubes (Figure 10.4) have a common gas inlet and outlet in a larger chamber that houses them. Cyclonic motion is imparted by axial inlet vanes. Overall efficiency is determined by the arrangement of individual tubes and their diameters.

The collection efficiency of all cyclones increases significantly with increasing particle size. In common cyclones, collection efficiency is less than 50% for particles smaller than 5 μm and up to 99% for particles larger than 40 μm. Efficiency also increases as particle density and loading increase.

Cyclonic collection systems are commonly used to collect PM from grinding, crushing, sanding, conveying, machining, and materials-handling operations. Depending on the volume of waste gas and nature of the particulate dusts produced, cyclones may be used singly (e.g., woodshop), as arrays of simple cyclones (e.g., grain handling), or as arrays of many tubular cyclones in a housing (e.g., boiler soot collection, precleaners). Multiple-cyclone systems are often used on coal-fired boilers as precleaners for more efficient downstream systems such as fabric filters or ESPs. They are

FIGURE 10.3 Cyclone collectors on grain silos in Galveston, TX, harbor.

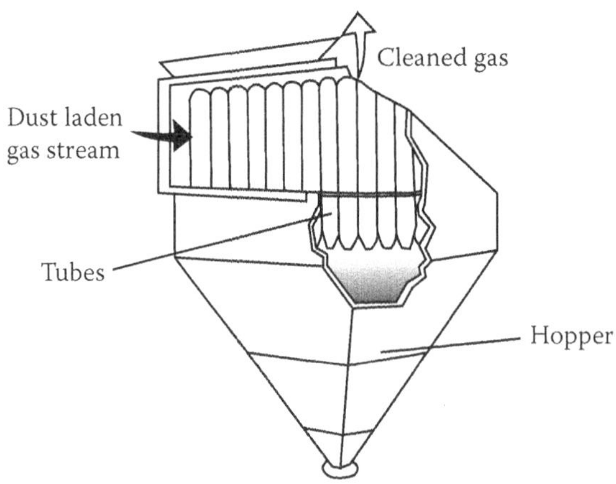

FIGURE 10.4 Multitube high-efficiency cyclone collector.

commonly used downstream of wet scrubbers to collect particle droplets and serve as demisters. In such applications, gas entry occurs near the bottom of the collector.

10.3.2 Electrostatic Precipitation

Electrostatic precipitators are widely used particle collecting systems, particularly where waste gas streams have large, steady volumetric flow rates. They are commonly used to remove fly ash from coal-fired power plants. Historical applications have included the collection of sulfuric acid (H_2SO_4) mist (as early as 1907) and collection of metal oxides and particle dusts from a variety of metallurgical operations, including ferrous and nonferrous metal processes.

Electrostatic precipitators remove solid or liquid particles from waste gases passing through an electrical field where negative ions are produced from high-voltage wires or plates and imparted to

FIGURE 10.5 Electrostatic precipitator.

particles entering the ESP. The negatively charged particles are then collected on positively charged collection plates.

One or more configurations are used in ESP design. These are the (1) plate-wire precipitator, (2) flat-plate precipitator, (3) tubular precipitator, (4) wet precipitator, and (5) two-stage precipitator. The plate-wire precipitator is the most commonly used.

In the plate-wire precipitator (Figure 10.5), effluent gases flow between vertical parallel sheet metal collection plates and high-voltage wire electrodes. Plates are spaced approximately 30 cm (12 in.) apart, and wire electrodes, spaced approximately 5 cm (2 in.) apart, are at the center of the channel between collection plates. Each collection plate (present in series) may be up to a meter (3 ft) in length and 3 to 6 m (10–20 ft) high. The path length through the collector may be approximately 6 m (~20 ft) or more.

A high direct current (dc) voltage with negative polarity is applied to the wire electrodes. This causes gas molecules to break down, producing a corona. Ions generated in this corona follow electrical field lines from the electrodes to the collecting plates. Particles intercept these ions and become negatively charged. As they pass each downstream electrode, they are driven closer and closer to the positively charged plates, where they are collected.

With time, the collecting surface becomes coated with particles that reduce the electrical potential of collection plates. As a consequence, collection plates must be periodically "cleaned" by mechanical rapping (dry ESP) or wet washing (wet ESP). Particles may become re-entrained during rapping, with subsequent loss of collection efficiency. Rapping losses can be minimized in systems that have multiple plates in series as shown in Figure 10.5, as plate rapping is done in sequence, not all at one time.

Although rapping losses are important, collection efficiency is primarily influenced by particle retention time in the electrical field and particle resistivity (resistance to taking a charge). Retention time is determined by gas path length (distance gas travels across the electrical field) and gas velocity. The highest collecting efficiencies are achieved in precipitators using long gas path lengths and relatively low gas velocity.

When particle resistivity is high, as is the case of fly ash from low-S western coals, the charge is not neutralized at the collection plate; as a result, an electrical potential builds up on the collected particles. As the potential increases, the incoming particles receive less than the maximum charge. As a result, collection efficiency decreases. If this condition continues, a corona discharge (sparking) may occur at the collection plate, which may cause the collector to malfunction. Ironically, low-S coals, widely used to comply with SO_2 emission standards in many coal-fired utility boilers, produce fly ash of high resistivity. High-S coals produce fly ash with low resistivity; thus, collection efficiency is high. As a result of the resistivity problem, more plate area is required, increasing the capital and operational cost. In many cases, baghouses (or fabric filtration) are used as an alternative to ESPs in power plants burning low-S coal, or where there is a need to control PM as well as capture part of the sulfur dioxide in alkaline-based fly ash on the filter.

On the other end of the spectrum are particulate dusts with low resistivity (e.g., carbon black). Collection efficiency is decreased because C particles are highly conductive and lose their charge too readily. As collected particles lose their negative charge, they become re-entrained in the waste stream. They become negatively charged and are again attracted to collection plates. The process is repeated until particles reach the precipitator outlet where they are discharged to the atmosphere.

Flat-plate precipitators are used in smaller volumetric flow rate applications and for small (1–2 μm), high-resistivity particles. Plates instead of wires are used for the high-voltage electrodes. They increase the strength of the electrical field and provide an increased surface area for particle collection. The distance between collection plates is approximately 16 cm (6.5 in.).

Electrostatic precipitation is a widely used particulate dust collection method because of its high collection efficiency for a range of particle sizes, including submicron particles (Figure 10.2). Collection efficiencies of more than 99% can be achieved even on very large waste gas flows. Electrostatic precipitators have low operating and power requirements because energy is needed to impart charges onto only the particles collected and not the entire waste stream. Energy costs are therefore low compared with other high-efficiency particle collection systems. Improvements continue to be made using advanced microprocessors, high-frequency rectifier systems, and optimum rapping operations, providing for reduced sparking or short circuiting within the ESPs, and allowing the units to operate at maximum voltage across the electrodes, all of which increase the collection efficiency.

10.3.3 Filtration

In filtration systems, solid particles are physically collected on woven or felted fabrics or mats (commonly called fabric filters) or on pleated filters that are composed of nonwoven fibers. Particle collection occurs as a result of sieve action, inertial impaction, interception, and diffusion onto filter fibers. As the filter becomes soiled, a mat of collected particles is produced that increases the filtration system's collection efficiency, although with an increased pressure drop across the filter.

10.3.3.1 Fabric Filters

Fabric filters are employed to control particulate dust emissions from industrial sources such as cement kilns, foundries, oil-fired and coal-fired boilers, carbon black plants, and electric and oxygen furnaces for steelmaking operations. They are also used to collect fly ash from EGUs burning low- to medium-S coal. Fabric filters are commonly used to control particulate emissions when particle sizes are small and high collection efficiencies are required. Fabric filters are designed to handle gas volumes in the range of 4.7 to 23.6 m^3/s (10,000–50,000 acfm). They are often preceded by cyclonic precleaners in high dust-loading applications to minimize abrasion and pressure drop. It is common to talk about the air to cloth ratio (filtration velocity), which is the total air flow rate divided by the total area of the fabric through which the air flow is measured in either feet per minute or meters per second. Typical values range from 2 to 15 fpm.

Fabric filter collection systems consist of multiple tubular collecting bags suspended inside a housing. A single housing, called a baghouse, may contain several hundred to several thousand bag filters. Bags are made from a variety of fibrous materials including fiberglass, polyester, and several patented materials designed for heat or chemical resistance (Nomex, Teflon, Gore-Tex, etc.). Fabric choice depends on temperature, moisture, and chemical composition of the waste gas, as well as the physical and chemical nature of particles collected. Filter bags are made of both woven and nonwoven (felted) materials. Glass fiber bags are commonly used for high-temperature or high corrosion–potential gas cleaning. In the typical baghouse unit (Figure 10.6), waste gas enters from the side and flows downward toward a hopper or directly into the side of the hopper where the flow is reversed (upward) into the bag array and filtered onto the inside of the bag. Alternatively, the cylindrical filter bags may be hung from the top of the baghouse, in which case the gas stream enters the side

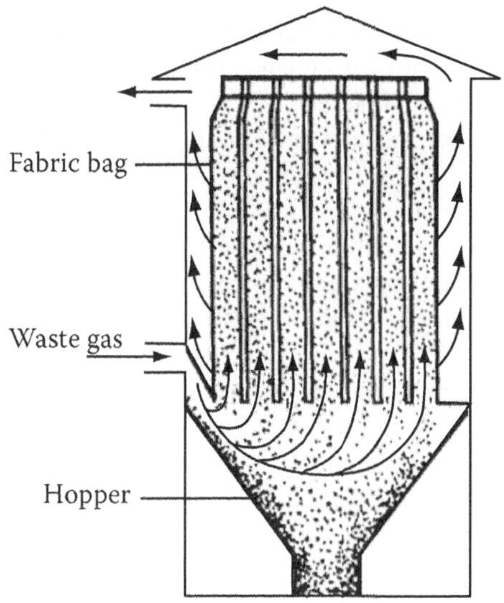

FIGURE 10.6 Fabric filter (baghouse) particle collection system.

of the collector, passes through the outside of the bags into the inside of the bags, and exits through the top of the bags. As the gas stream changes direction, large particles are removed by inertial or centrifugal separation and collected in the hopper below. As gas passes through the tubular bags, particles are collected on the bag surface by a combination of inertial impaction, interception, and diffusion and filtered gas is discharged to the atmosphere.

Fabric filtration is similar to the process employed in a home vacuum cleaner except that positive, rather than negative, air pressure causes dirty air to pass into the collection bag. Like a vacuum cleaner, collected dust must be periodically removed. Bags may be cleaned by automated mechanical shaking or reverse air cleaning for bags where the dust is filtered on the inside of the bags, or by pulsing of air through the top of the bags where the dust is captured on the outside of the bags. In each case, cleaning is conducted to dislodge collected particles into a hopper where they are removed from the collection system. Collection efficiencies are lowest when filters are first installed and immediately after cleaning. Maximum efficiency and pressure drop occur when particle mat buildup has occurred.

Although fabric filters provide high overall collection efficiencies (> 99%) and often represent the best available technology for controlling submicron particles, they have limitations. These include high capital and operating costs, flammability hazards for some particulate dusts, high space requirements, flue gas temperatures limited to 285°C (545°F), and sensitivity to gas moisture content. Operating costs are high because of the high pressure drop that develops as particulate mats form.

Because of the abrasiveness of PM and sensitivity to chemicals, bag wear is a major maintenance concern. Bag life varies with the nature of the application. Bags are commonly changed every 12 to 18 months, although in very low velocity collectors, they may last as long as 3–4 years.

Filtration is also employed in cases where noncleanable filters, composed of pleated fiberglass or nonwoven media, are employed. In these cases, the filter(s) cannot be easily cleaned and are designed to be discarded after use in a proper disposal site. These filters, housed in boxes or cartridges, are used in many applications ranging from sites where radioactive particles are being controlled, to the indoor air cleaner, commonly found in retail stores, as well as in HVAC systems associated with residential use. The filters range in efficiency from those just intended to remove large fibers/dust/dander to certified filters, such as high (HEPA) and ultra-high (ULPA) efficiency particulate air filters. HEPA and ULPA filters (pleated filters typically made from fiberglass fibers) are 99.9% and

99.9999% efficient, respectively, on 0.1 μm particles. In these filters, the fiber diameters are very small (micron size) and/or electrostatically enhanced and therefore collect both larger particles (due to impaction and interception) and smaller particles (due to electrostatics and/or diffusion), resulting in higher collection efficiencies for particles smaller and larger than the certified 0.1 μm size.

10.3.3.2 Wet Scrubbers

Particulate dust collection by wet scrubbing involves the introduction of a liquid into the effluent stream, saturation of the gas stream, and a subsequent transfer of particles into the scrubbing liquid. The saturated gas stream allows the particles to grow in size and simultaneously become impacted into the liquid (film or droplet).

Scrubber designs vary from one manufacturer to another. However, most scrubbers have three basic components: (1) a humidification section, (2) a section where liquid–gas contact occurs, and (3) a separation section where liquid droplets (containing the collected particles) are removed. Particles in scrubbers come in contact with liquid droplets to form a particle–liquid agglomerate. The contact process is achieved by forcing a collision between liquid droplets and particles. Collisions may be promoted by gravity, impingement, and/or mechanical impaction. When contact is made, particles significantly increase in size and mass. The resultant particle–gas agglomerates are removed by inertial devices. Commonly used de-entrainment mechanisms include impaction on extended baffles and centrifugal separation. In most systems, the contact liquid is water (H_2O).

10.3.3.2.1 Spray Towers

In the open tower design, a scrubbing liquid is sprayed downward as preformed drops through low-pressure nozzles. The PM-laden waste gas enters from the side and moves downward. Waste gases then change direction and move countercurrent to the flow of the scrubbing liquid and are discharged at the top. Dust particles are captured by the falling droplets. The liquid–particle agglomerate is collected in the liquid pool at the bottom of the tower. Spray towers are limited to low gas flows to prevent liquid drop entrainment in the scrubbed gas stream. These scrubbers achieve relatively low collection efficiencies (80%–90%) and are usually employed as precleaners to remove particles larger than 5 μm.

10.3.3.2.2 Venturi Scrubber

The Venturi scrubber is used when high collection efficiencies are required. In the system illustrated in Figure 10.7, effluent gases enter the Venturi section, where they flow through the wetted cone and throat. Waste gas velocity increases as it enters the annular orifice of the Venturi. The resulting high velocities produce a shearing action that atomizes the scrubbing liquid into many fine droplets. The high differential velocity between the gas stream and the liquid promotes inertial impaction of dust particles on the droplets. As waste gas leaves the Venturi section, it decelerates, and impaction and agglomeration of particles and liquid continue to occur. Particle–liquid droplets are removed from the waste stream by centrifugal forces in a cyclonic de-entrainment section.

10.3.3.2.3 Cyclonic Scrubbers

Scrubbing liquids can be injected into cyclonic airflows to increase particle collection efficiencies. The rotating gases cause particles entrained in droplets to migrate to scrubber walls where they are removed by impaction.

10.3.3.2.4 Operating Considerations

Because of their high collection efficiencies, wet scrubbers are used to control fine particles. Venturi scrubbers, for example, are commonly used to collect particles from basic oxygen steelmaking processes.

Although capital costs are relatively low, operating and maintenance costs are often high. These high operating costs are primarily a result of the energy requirements needed to atomize the droplets, provide high collection efficiencies for fine particles, and overcome high pressure drops of the

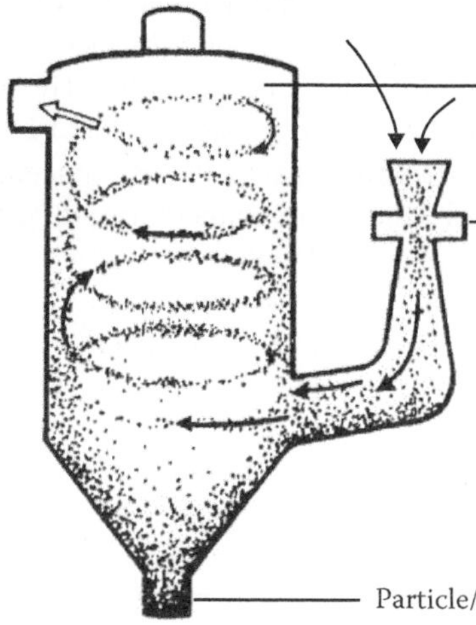

FIGURE 10.7 Venturi scrubber particle collection system.

Venturi section. Additional operating costs are associated with the disposal of the collected slurry of particle-laden waste liquid. High maintenance requirements often result from the erosion and corrosion of scrubber surfaces produced by abrasion and corrosive acids.

10.4 CONTROL OF GAS-PHASE POLLUTANTS

Public and regulatory concern for the control of PM emissions, for the most part, predated concern for gas-phase pollutants by several decades. In the collection of particulate dusts, specific collection principles can be utilized for a variety of collection problems and sources; the major consideration in selecting control systems is the performance required. Application of control technology for gaseous pollutants cannot be generically applied. Gas cleaning systems must be developed or designed to control specific gas-phase pollutants (or pollutant categories) that vary in their chemical and physical properties. Control of some gas-phase pollutants is a relatively complex activity. It is also an inherently more expensive one, as both capital and operating expenditures for control equipment are often high.

Major physical and chemical principles used to remove gas-phase contaminants from effluent gas streams include thermal oxidation (combustion), adsorption, and absorption. Increasingly, biological systems are being used to control emissions of hazardous organic air pollutants.

10.4.1 Thermal Oxidation

Thermal oxidation is widely used to remove combustible vapors from waste gas streams. The term *incineration* is commonly used to describe such processes. However, its use is problematic because it is also associated with the combustion of municipal, medical, and industrial wastes. In the air pollution control sense, incineration is limited to the combustion or thermal oxidation of gas- and particulate-phase substances in waste air streams associated with various processes, including the burning of municipal, medical, and industrial waste.

Thermal oxidation as applied to pollution control is a high-temperature or catalytic process in which oxygen (O_2) is combined with organic pollutants to convert them to simpler substances such as CO_2 and H_2O vapor, releasing heat in the process.

In most cases, thermal oxidation is applied to dilute streams of waste gases containing HCs or carbon monoxide (CO), such as those associated with paint spraying and baking operations, solvent use, and incompletely combusted gases from other combustion processes (e.g., waste incinerators). It is also used to control organic aerosols, malodorants such as hydrogen sulfide (H_2S) and mercaptans, and combustible gases produced in refineries. Dilute HC airstreams below the lower flammable limit (LFL), better known as the lower explosion limit (LEL), cannot propagate a flame. For many HC–air mixtures, the LEL varies from 1% to 5% HC concentration and more than 15% O_2. As such, they must be subject to sufficiently high temperatures (800°C, 1,500°F), or to catalytic materials at lower temperatures, for complete oxidation to occur.

Three types of systems use the principle of thermal oxidation to control pollutants. These are flare systems, conventional thermal oxidizers, and catalytic systems.

10.4.1.1 Flare

Direct flare systems are used to combust waste gas streams that are HC-rich, with concentrations above the LEL and below the upper explosion limit (UEL). Under these conditions, HCs ignite and propagate a flame. Such waste gas streams are commonly found in landfill gases, chemical processing plants, and refineries (Figure 10.8). They pass through stacks that may be 30 to 40 m (100–130 ft) or so in height.

HCs, released intermittently or because of an emergency, are ignited by pilot lights (or some other ignition system). Combustion occurs continuously as long as HCs above the LEL continue to flow through the system. Air jets at the tip of the tower promote turbulent mixing of effluent gases and the immediate atmosphere. Operated properly, flares can achieve HC oxidation efficiencies of 99% or more.

10.4.1.2 Conventional Oxidizers

Conventional oxidizers are used to treat relatively dilute HC effluent gases (often only a few hundred parts per million [ppmv]). Complete oxidation is facilitated by combustion at high temperatures (660°C–1,100°C, 1,200°F–2,000°F), with sufficient residence time (0.5–2 seconds) and turbulent mixing of effluent gases in air.

FIGURE 10.8 Refineries in Galveston, TX, using flare incineration.

Thermal oxidizers are often used as afterburners; that is, they are used downstream of primary combustion chambers. Afterburners are used on municipal and hazardous waste incinerators.

The use of afterburners as a part of hazardous waste incineration systems is standard practice. They are often required to have removal efficiencies of 99.99% of the principal organic hazardous constituents. US EPA operating permits often require the use of high temperatures (~1,100°C, 2,000°F).

10.4.1.3 Catalytic Oxidation

Catalytic oxidizers are used industrially to reduce both the size of control equipment and fuel use. Catalysts reduce temperature requirements for efficient combustion. Catalytic oxidizers operate in the range of 165°C to 440°C (300°F–800°F).

Catalysts promote combustion but are not consumed in the process. Catalytic materials are applied to ceramic or metal substrates that have very high surface areas. Catalysts are usually noble metals such as platinum (Pt) or palladium (Pd), but may include chromium (Cr), manganese (Mn), copper (Cu), cobalt (Co), or nickel (Ni).

The length of the catalytic bed varies from approximately 15 to 60 cm (6–12 in.). Despite this short length, the time required to achieve oxidation of HCs is an order of magnitude less for catalytic oxidizers compared with conventional thermal oxidizers.

Although catalytic oxidizers are used in many industrial applications, they have disadvantages. These include higher capital costs than thermal oxidizers; a tendency to be "poisoned" by contaminants such as chlorine (Cl), S, and Pb; plugging by PM; and damage by high-temperature excursions associated with transiently higher HC concentrations.

10.4.1.4 Operating Considerations

Conventional thermal oxidation systems have high fuel requirements to maintain system temperatures in the optimum range. To reduce fuel usage costs, most systems are equipped with heat recovery units. Such units have thermal extraction efficiencies on the order of 70%. Recovered heat is transferred to influent gases to increase their temperature and reduce energy needs.

Thermal oxidation in both primary combustion chambers and afterburners can produce unwanted by-products. Most notable are Cl compounds such as hydrochloric acid (HCl) and possibly dioxins and furans associated with the combustion of mixed wastes containing chlorinated plastics. The production of such contaminants is common in medical and municipal incinerators.

Any primary or secondary thermal oxidation process will produce nitrogen oxides (NO_x) as an emission product. The higher operating temperatures used in afterburners provide optimum conditions for NO_x production. Lower NO_x production and emissions are associated with catalytic systems because of the lower operating temperatures. Nitrogen oxide emission control practices are described in detail at the end of this chapter.

10.4.2 Adsorption

Adsorption is a process in which pollutants are removed from a waste gas stream by the physical attraction of gas and vapors to a solid sorbent as the gas stream moves through a sorbent bed. This adherence is similar to that of van der Waals forces. Physically adsorbed molecules retain their chemical identity and properties.

Adsorption is primarily used to collect organic vapors from solvent-using operations, including degreasing, dry cleaning, surface coating, rubber processing, and several different printing processes. It is also used to control toxic and malodorous substances emitted from food processing, rendering, and chemical and pharmaceutical manufacturing processes.

Sorbents are characterized by their porosity, high surface area, and ability to collect different substances on their surfaces. Commonly used sorbents include activated carbon, silica gel, activated alumina, and zeolites.

Surface polarity is a major distinguishing feature of sorbents and determines the type of vapors collected. Nonpolar sorbents collect nonpolar vapors; polar sorbents collect polar vapors. Because they sorb H_2O vapor, polar sorbent use in pollution control is very limited.

Activated carbon is the most widely used sorbent, as it has high affinity for nonpolar compounds that have molecular weights of more than 45 mass units. Activated carbon is primarily produced from wood or coal by carbonization at elevated temperatures in the absence of O_2. Such heating drives off volatile compounds. Postcarbonization activation at high temperatures by using steam, air, or CO_2 produces a material of high internal porosity and surface area.

Sorbents are prepared for special applications by impregnating them with reactant substances or catalysts. Examples of impregnants are bromine (Br) to collect ethylene (C_2H_4), iodine (I) for mercury (Hg), lead acetate for H_2S, and sodium silicate for hydrogen fluoride (HF). Catalytic impregnants include metal salts containing Cr, Cu, silver (Ag), Pd, or Pt. These substances catalyze oxidation reactions. In such chemical adsorption (chemisorption), a chemical reaction occurs that converts the sorbate to a new chemical species. As such, the process is not reversible.

Sorbents are produced as granular materials of appropriately small size to increase collection efficiency without excessive resistance to gas flow. They may be used in disposable or rechargeable canisters, fixed regenerable beds, traveling beds, and fluidized beds. Disposable and rechargeable canisters and regenerable beds have received the greatest industrial and commercial applications.

Disposable and rechargeable canisters are commonly used for low-volume exhaust flows and waste streams with low contaminant concentrations; a common use is vapor recovery from storage tanks.

Fixed regenerable beds are typically used when the quantity of gas treated or adsorbate concentration is high. They are particularly attractive for solvent recovery when the cost of regeneration is less than the cost of the sorbent.

Both thin-bed (2.5 cm, 1 in. thick) and thick-bed (0.3–1 m, 1–3 ft thick) regenerable systems are available. The latter, because of their size and capacity, are most appropriate for industrial applications. They are attractive when effluent concentrations of solvent vapors exceed 100 ppmv and exhaust flows exceed 4.7 m^3/s (10,000 acfm).

A two-unit bed regenerable adsorption system of the type used in vapor recovery is shown in Figure 10.9. In this adsorber system, one unit can be used in an on-stream adsorbing mode, the other

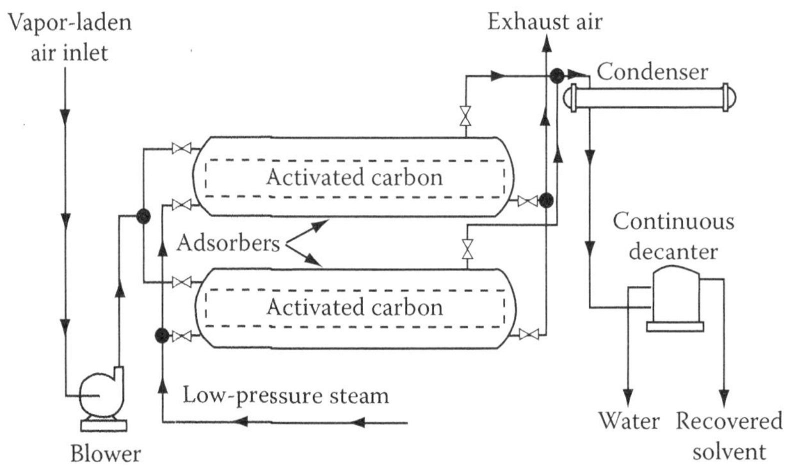

FIGURE 10.9 Dual-bed adsorption system. (With permission from Springer Science+Business Media: *Air Pollution Control Equipment: Selection, Design, Operation and Maintenance*, 1994, Theodore, L. and Buonicore, A.J., Eds.)

in regeneration. They can also be used simultaneously when vapor concentrations are relatively low and regeneration occurs at the end of the work shift. Parallel operation doubles the air-handling capacity of the system. Thick-bed systems can be oriented vertically or horizontally.

In regenerable systems, desorption can be achieved by heating, evacuating, stripping with an inert gas, displacing with a more adsorbable material, or a combination of these. Heating is the most common desorption method; the adsorbate can be subsequently collected and reused. Stripping can be used to concentrate contaminants for incineration. Steam displacement is also common. As H_2O vapor is adsorbed, the adsorbate is displaced; desorbed vapors are condensed along with excess steam. The bed is then regenerated by passing hot air over it to remove H_2O.

A variety of factors may affect the efficient operation of adsorption systems. These include adequate sorbent capacity, ample contact time, low resistance to airflow, and uniform airflow distribution. It may also be desirable to pretreat the waste gas stream to remove PM and competing vapors, and to cool it. Ample contact time can be achieved by utilizing relatively low flow velocities and extended bed depths.

Adsorption is not generally deemed advisable when waste gas contains PM that may clog the sorbent bed or substances that cannot be easily desorbed from the sorbent. It is also not desirable to use adsorption to collect flammable vapors. Because the adsorption process is exothermic, heat is released. If the adsorption bed is large, a substantial increase in temperature may occur and the collected vapor and sorbent may catch fire by autoignition.

10.4.3 Absorption

Pollution control systems that utilize liquid or liquid–solid media to remove selected gases from waste gas are called scrubbers. Gaseous pollutants are absorbed when they come into contact with a medium in which they are soluble or with which they can chemically react. Although H_2O has been the most common scrubbing medium, it is almost always amended with substances such as lime, limestone, sodium carbonate, or ethanolamine to increase collection efficiency and absorptive capacity. Collection efficiency is also affected by factors such as gas solubility, gas and liquid flow rates, contact time, mechanism of contact, and type of collector.

The fixed-bed packed tower scrubbing system is commonly used to control gas-phase pollutants in industrial applications. It can achieve removal efficiencies of 90% to 95%.

Packed tower scrubbers can be operated as countercurrent, cocurrent, and cross-flow systems. Countercurrent systems are the most widely used. The scrubbing medium is typically introduced at the top and flows downward; waste gases flow upward. Countercurrent systems are used for difficult-to-control gases because the packing height is not restricted, thus allowing for higher removal efficiencies. It also has the highest pressure drop and least capacity to handle solids. A schematic cross-section through a countercurrent fixed-bed packed tower can be seen in Figure 10.10.

Cocurrent systems have both the scrubbing media and waste gases flowing in the same direction. They have a finite limit on absorption efficiency but are useful when the scrubbing medium has a high solids content. They can operate at relatively high scrubber liquid flow rates without plugging.

Cross-flow designs are used for moderately to highly soluble gases or when chemical reactions are rapid. Waste gas flows horizontally whereas the scrubbing liquid flows downward. An attractive feature of such systems is that they have lower pressure drops.

Absorption of a gas in a scrubbing medium depends on its solubility or chemical reactivity. Solubility depends on gas type and concentration, system pressure, scrubbing liquid, and temperature.

As with sorbents, a scrubbing medium is limited by the amount of a gas it can absorb. Upon saturation, it cannot absorb additional gases. In most packed tower scrubbers, the liquid is recirculated from a sump. Fresh scrubbing liquid is added and used liquid is withdrawn continuously to prevent saturation.

Collection efficiency in scrubbers is enhanced by using a packing medium. Packing media include steel, ceramic, and thermoplastic materials. These materials come in a variety of shapes

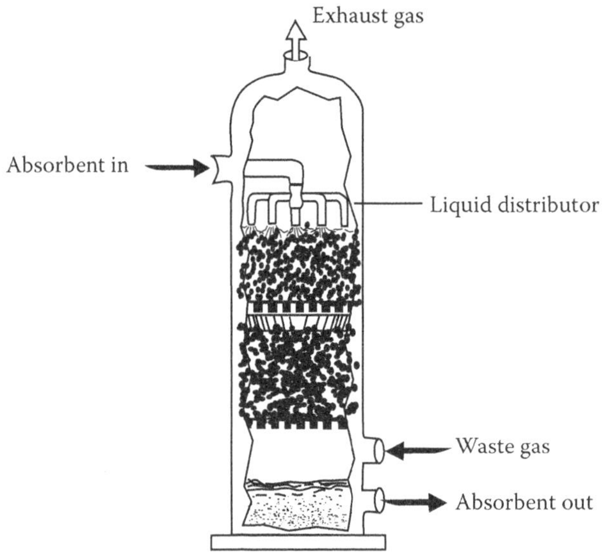

FIGURE 10.10 Countercurrent fixed-bed scrubbing tower.

FIGURE 10.11 Plastic packing material used in a scrubber. (Courtesy of Lantec Products, Inc.)

and have a characteristically large surface area. By providing a large surface area over which the scrubbing liquid is spread, a greater contact area is provided, thereby increasing collection efficiency. Thermosetting plastics are the most widely used packing materials. They are less subject to corrosion and pressure drop than metal or ceramic types. A plastic packing material can be seen in the photograph in Figure 10.11.

Scrubbers experience several operating and maintenance problems. These include corrosion of scrubber surfaces and plugging of the packed beds. To reduce corrosion, internal scrubber surfaces need to be constructed with stainless steel, lined steel, exotic metals, polyvinyl chloride, or fiberglass-reinforced plastic. Fiberglass or lined steel is commonly used to reduce costs. Plugging may occur due to the deposition of undissolved solids or insoluble dusts, or precipitation of dissolved salts or salts produced upon absorption. Scaling often occurs when calcium (Ca) or magnesium (Mg) reacts with CO_2 in the alkaline scrubbing medium to produce carbonate precipitates. Such scaling may require flushing with an acidic solution.

The scrubbing medium must be treated and disposed of after it is removed from the system. Waste scrubbing liquids used for acidic gases may require neutralization. Solids may be removed in settling ponds and disposed of in landfills.

10.4.4 Control of Sulfur Oxides

Sulfur oxides (SO_x) generated in the combustion of coal, high-S fuel oils, and smelting of metallic ores have enormous effects on air quality and the environment. As a result, considerable regulatory attention and allocation of economic resources have focused on the reduction of SO_x emissions to the atmosphere.

When SO_x emission reduction requirements were first promulgated by state and local agencies to achieve air quality standards for SO_x, many sources chose to comply by using low-S fuels. Due to the economic impact of fuel use policies (particularly the use of low-S coal) on the midwestern and eastern high-S coal fields of the United States, Congress, in 1977, mandated that the US EPA promulgate new source performance standards (NSPSs) for large coal-fired power plants that required the use of flue gas desulfurization (FGD) systems. As a consequence, new coal-fired power plants given permits to construct or operate between 1977 and 1990 were required to use FGD.

A variety of SO_x control technologies are now used to reduce SO_x emissions to the atmosphere. Some SO_x technologies, for example, coal beneficiation, coal gasification, and solvent refining, remove S before combustion. Others, for example, fluidized bed combustion, remove S during the combustion process; FGD systems remove it after combustion has been completed and before gases are emitted to the atmosphere.

10.4.4.1 Coal Beneficiation

Coal is a very heterogeneous mineral consisting of both combustible organic and noncombustible inorganic matter. Both fractions contain S, with concentrations ranging from 0.5% to more than 5% by weight. The inorganic fraction that contains pyritic S can be removed by coal washing. In coal washing, inorganic minerals are separated from the organic portion by suspending coal fragments in water; differences in specific gravity result in heavy inorganic mineral matter sinking and lighter organic coal floating. Coal cleaning can, on average, remove 50% of pyritic S and 20% to 30% of total S.

Coal washing is widely used by coal producers to reduce the ash content of coal both to enhance its combustibility and to meet the ash specifications of coal users. Sulfur reduction is a side benefit associated with the enhanced coal quality produced by washing. Sulfur reduction by coal beneficiation or cleaning is usually not adequate to meet emission reduction requirements. However, it may be useful when combined with other control technologies such as FGD.

10.4.4.2 Solvent Refining

Solvent refining, or chemical cleaning, is a technology that has been undergoing development for decades. It is based on the principle that S and other impurities can be removed by solvent extraction. Such a technology has the potential to obviate the need for FGD systems. However, it is not currently economically viable.

10.4.4.3 Coal Gasification

Another technology that has been explored and utilized with limited success in the United States is that of integrated gasification combined cycle (IGCC). In this process, coal or petroleum-coke is gasified by applying heat under pressure in the presence of steam to convert coal or petroleum-based coke to a gaseous fuel. In these processes, the impurities such as sulfur, mercury, metals and ash are removed prior to the synthetic gas, or "syngas" being burned in a more typical natural gas combined cycle type EGU. Considerable development efforts were focused on synfuels in the United States in the late 1970s and 1980s when energy concerns were high, and demonstration projects were developed under DOE's Clean Coal Technologies program, including the Wabash River IGCC project in Indiana. The 262 MW EGU began construction in 1993, was completed in 1995 and was operated until 2016 and achieved a thermal efficiency of about 40%, meeting its DOE goals. Another successful IGCC EGU is the Edwardsport Power Station (618 MW), also in Indiana. The Edwardsport IGCC plant began construction in 2007 and it was placed into commercial operation in 2013. The plant initially experienced operational challenges and was the subject of other economic/political controversies, but, it reached an annual production of about 4 million MWh of energy (74% of its maximum capacity) by 2018. While other IGCC EGUs have been proposed in the United States, and some were even partially built, most of those were halted or shuttered as a result of the success of fracking which produced an alternative abundant source of lower cost natural gas.

Even though IGCC technologies have not been adopted to any large extent by EGUs in the United States, the Global Syngas Technologies Council reported in 2020 that were more than 270 operating gasification plants worldwide with the majority providing syngas to the chemical and fertilizer industries. Worldwide, more than 25% of the production of ammonia and methanol is from gasification from coal or petroleum coke. IGCC is also increasing substantially in other countries where coal is abundant and natural gas is in short supply such as Asia (i.e., China, India, South Korea and Japan), as a means of providing the cleaner synthetic or substitute natural gas (SNG) for use as a fuel.

10.4.4.4 Flue Gas Desulfurization

Flue gas desulfurization systems are widely used to remove SO_x from effluent gases of large, and sometimes small, utility boilers, industrial boilers, and primary and secondary metal smelters. Several hundred FGD systems are currently operated in the United States with thousands more around the world.

A major use for FGD systems has been to control sulfur dioxide emissions from large coal-fired power plants. Depending on S content, the 1979 to 1990 NSPSs for coal-fired power plants required a 70% to 90% reduction of SO_2 emissions using FGD systems. The average design efficiency for new and retrofit FGD systems applied to coal-fired power plants was 82% and 76%, respectively. After the 1990s, new systems were designed with more than 95% efficiency.

FGD systems vary in the types of sorbents used and how flue gas is treated. In most cases, flue gases are treated downstream of the combustion chamber; in fluidized bed control systems, they are treated within the combustion chamber, with further treatment downstream if needed (see Section 10.4.5).

FGD systems can also be described in terms of sorbent reuse. In regenerable systems, expensive sorbents are recovered by stripping SO_2 from the scrubbing medium. Useful by-products may include elemental S, H_2SO_4, liquefied SO_2, and gypsum. In nonregenerable (historically called throwaway) systems, the scrubbing medium is relatively inexpensive and sorbent recovery is not normally economical. In some cases, where the flyash is removed upstream of a calcium-based FGD system, the reacted product containing calcium and sodium can be directly utilized in a gypsum plant for use in sheetrock. The three dominant FGD processes used to meet NSPS and SIP SO_2 emission limits are throwaway systems. Such systems have significant waste treatment and disposal requirements.

FGD systems are of two primary types: wet scrubbing systems and dry scrubbing systems.

10.4.4.4.1 Wet Scrubbing Systems

Wet systems typically use a scrubbing medium that is a mixture of a liquid (typically H_2O) and suspended and partially dissolved solids. Thus, the scrubbing medium is a slurry. Common scrubbing media include lime, limestone, lime–limestone, lime–alkaline fly ash, limestone–alkaline fly ash, sodium carbonate, and dual alkali. A limestone slurry is used in approximately 50% of FGD applications, and lime in 16%.

Limestone is considerably less soluble in H_2O than lime is. It is also less expensive. Sulfur dioxide is removed from calcium-based wet scrubbers in a complex set of chemical equilibrium reactions that involve the absorption of gaseous sulfur oxides into the liquid phase in the wet scrubber; hydrolysis of the sulfur oxides; dissolution of the solid phase limestone ($CaCO_2$) or lime ($Ca[OH]_2$); acid-base neutralization; the stripping of CO_2 in the case of the limestone scrubber; formation of calcium sulfite/sulfate reaction products in the liquid; precipitation of the calcium sulfite/sulfates from the scrubbing solution in the primary scrubber or in a secondary oxidation tank; separation of the reacted products; and final disposal/use of the precipitated products. The overall initial reactions can be summarized by the following overly simplified equations:

Limestone:

$$CaCO_3(s) + SO_2(g) + \tfrac{1}{2}\,H_2O \rightarrow CaSO_3 \cdot \tfrac{1}{2}\,H_2O\ (s) + CO_2(g) \tag{10.2}$$

Lime:

$$2Ca(OH)_2(s) + 2SO_2(g) + 2H_2O \rightarrow 2CaSO_3 \cdot \tfrac{1}{2}\,H_2O\ (s) + 3H_2O(liq) \tag{10.3}$$

Further oxidation of the calcium sulfite hemihydrate, either within the scrubber/reactor or by forced oxidation in a secondary reactor, results in the formation of calcium sulfate dihydrate (gypsum):

$$2CaSO_3 \cdot \tfrac{1}{2}\,H_2O\ (s) + 3H_2O\ (liq) + O_2(g) \rightarrow 2CaSO_4 \cdot 2\,H_2O\ (s) \tag{10.4}$$

As a consequence of these overall reactions, the slurry, after it passes through the scrubber, contains unreacted $CaCO_3$ (or unreacted $Ca[OH]_2$ in lime scrubbers), $2CaSO_3 \cdot \tfrac{1}{2}\,H_2O$, and $CaSO_4 \cdot 2\,H_2O$.

In limestone and lime slurry FGD systems (Figure 10.12), flue gases flow upward through a packed tower countercurrent to the scrubbing medium. After coming into contact with flue gases, the scrubbing medium is pumped to a holding tank where some of the calcium sulfite and sulfate are precipitated. Because considerable sorption capacity remains, scrubbing fluid is recirculated to the scrubber. Scrubbing medium that is not recycled is pumped to a thickening tank and then to a stabilizing pond. Although continuously recirculated, the scrubbing medium is continuously recharged with freshly prepared scrubbing fluid.

Both limestone and lime scrubbing are similar in process flow and equipment used. Lime, however, is much more reactive than limestone, due to being produced by slaking pebble-sized CaO. The slaking process (a highly exothermic reaction of CaO with water) produces a high surface area reactant, whereas limestone is brought from a quarry and ballmilled into a pulverized slurry, but with much less surface area per unit volume. Lime scrubbers can routinely achieve 95% removal of SO_2 (whereas limestone achieves 90% removal).

Although lime and limestone scrubbing systems can achieve high FGD efficiencies, their operation is technically demanding. Several factors reduce performance and reliability. Scaling and plugging of scrubber equipment by insoluble compounds is a common occurrence. Scaling problems can be minimized by proper control of pH.

High desulfurization efficiencies require high scrubbing medium/gas ratios. However, high ratios result in higher operating costs because of the pressure drop in the scrubber and higher energy

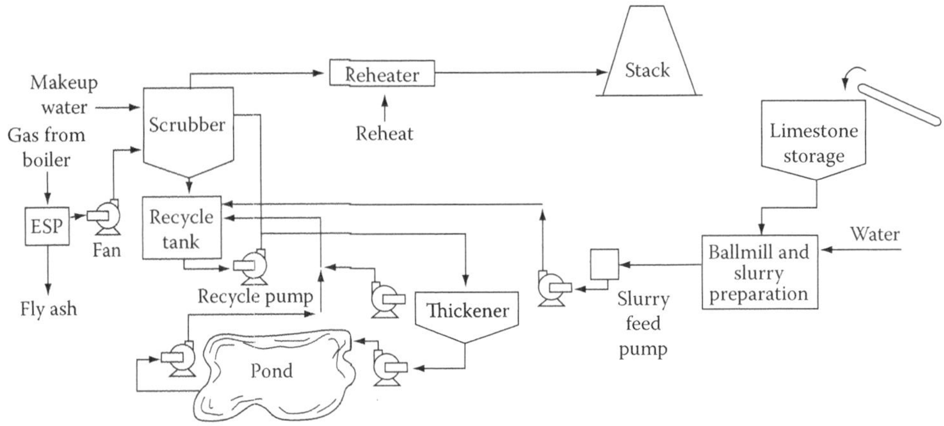

FIGURE 10.12 Flue gas desulfurization system. (With permission from Springer Science+Business Media: *Air Pollution Control Equipment: Selection, Design, Operation and Maintenance*, 1994, Theodore, L. and Buonicore, A.J., Eds.)

needs for pumping. Energy consumption by FGD systems is on the order of 3% to 6% of the power generated by a coal-fired facility. In addition to these high operating costs, FGD systems used on coal-fired utility boilers have high capital costs; on a 1,000 MwE unit, these typically exceed 150 million US dollars (USD).

As an example of the application of limestone scrubbing in a large coal-fired power plant, the 2,400 MwE Cumberland Steam Plant in Tennessee burns 20,000 tons of coal/day, and utilizes selective catalytic removal to reduce NO_x by 90%, followed by high-efficiency electrostatic precipitators for PM removal, and then by a high-efficiency limestone FGD scrubber that removes 95% of the SO_x. Because the fly ash is separated ahead of the wet FGD scrubber, the collected slurry containing primarily calcium sulfate-based materials is transported from the steam plant to an adjacent manufacturing facility where it is used as the raw material for producing gypsum board (sheetrock; https://www.tva.gov/Energy/Our-Power-System/Coal/Cumberland-Fossil-Plant).

10.4.4.4.2 Dry Scrubbers

In dry scrubbing, flue gas is treated by direct injection of a dry-state reagent aerosol into the flue gas upstream of a particulate collector (ESP or fabric filter) or by spraying a liquid slurry into a spray dryer where the flue gas heat dries the reacted droplets into a powder.

In direct injection, the reagents used are either trona (naturally occurring sodium carbonate, Na_2CO_3) or nahcolite (naturally occurring sodium bicarbonate, $NaHCO_3$). These dry aerosol particles react with SO_2, and the fly ash particles are collected on baghouse filters downstream. The PM deposited on the filters serves as a porous reagent bed that converts SO_2 to sodium sulfite (Na_2SO_3) and sodium sulfate (Na_2SO_4).

In spray dryer or semidry scrubbing, the reagent is normally pebble CaO which is slaked to form a $Ca(OH)_2$-based slurry that is atomized into droplets that mix with flue gases in a spray dryer reaction chamber. The SO_2 absorbs into and reacts with the calcium in the atomized droplets. The hot flue gas evaporates the H_2O, leaving suspended dried reactants. The dried particles are on the fabric filter in a large baghouse downstream of the spray dryer. Sulfur dioxide not removed in the reaction chamber may be further removed by sorption/reaction on the filter cake; such removal may be on the order of 15% to 30%.

Dry scrubbing is used in approximately 10% of coal-fired utility FGD systems. Spray drying with lime is the predominant method employed.

10.4.5 Fluidized Beds and FBC

Because of higher thermal efficiencies, fluidized bed control (FBC) systems are sometimes used in a variety of applications, including industrial and utility boilers, incineration of solid wastes, and reclamation of coal refuse piles.

FBC units contain a bed of small particles that are fluidized by air passing through the bed at a sufficient velocity (Figure 10.13). A variety of materials can be introduced into the bed and thermally oxidized. Limestone can be mixed with the fuel to capture SO_2 and convert it to SO_4^{2-}. The ash, SO_2, and SO_4^{2-} can be separated from the circulating mixture as new fuel and limestone are added.

In the primary aluminum industry, a unique application of the fluidized bed reactor is utilized to control fluoride emissions from pot rooms. In the process, a fluidized bed followed by a fabric filter baghouse is used to control both particles and gaseous fluoride emissions from the potrooms. The fluidized bed consisting of powdered alumina (the raw material used to make aluminum) reacts with and captures the fluoride emissions. The alumina from the bed and the alumina captured on the filters are then used to feed the reduction pots where the aluminum is made. The fluorides are a necessary part of the manufacturing process, and the control system allows for the fluoride emissions to be recycled back into the process. The process provides greater than 99% control of fluorides.

10.4.6 Biological Treatment

Biological treatment systems or bioreactors utilizing microorganisms to metabolize pollutants include three types: bioscrubbers, trickle bed reactors, and biofilters. In addition to wastewater treatment, biological treatment systems are used industrially to control odors associated with food processing and fermentation. They have been used to control alcohols, aldehydes, amines, aromatic HCs, esters, ketones, organic acids, and organic sulfides.

Bioscrubbers consist of a packed bed in which microorganisms are suspended in a scrubbing medium. Inert packing materials serve to increase the contact surface to enhance the transfer of

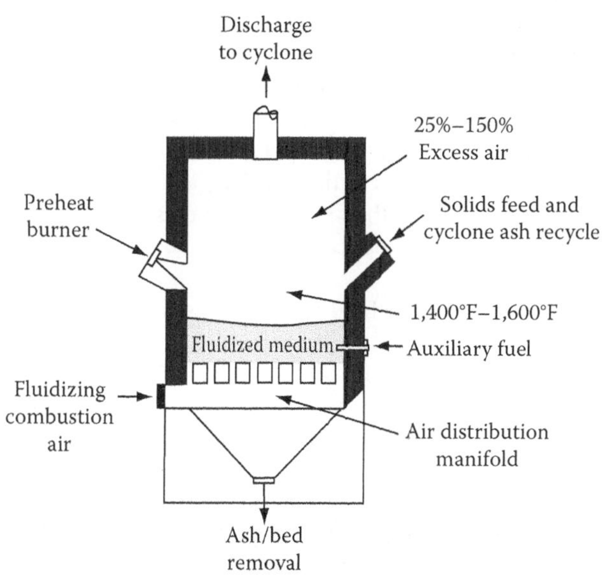

FIGURE 10.13 Fluidized bed combustion. (Davis, W.T., Ed.: *Air Pollution Engineering Manual*. 2000. Copyright Wiley-VCH Verlag GmbH & Co. kgaA. Reproduced with permission.)

pollutants from the gas to the liquid phase. The scrubbing medium is aerated to provide O_2 and stirred to keep organisms in suspension.

Biofilters consist of one or more beds of biologically active material such as compost, municipal waste, and the like, which provides a food base for microorganisms. Bed depths are on the order of 1 m (~3 ft) to minimize pressure drop. Waste gases enter the bottom of the unit and diffuse into the biofilm (biologically active layer) attached to the compost where pollutants are metabolized. If metabolism is complete, the end products are CO_2, H_2O, and microbial mass.

Trickle bed reactors consist of a packed bed of compost particles that have a low surface/volume ratio to provide large voids that allow for biological growth. A recirculating fluid of nutrients flows downward, wetting the biofilm attached to compost particles. Waste gases travel upward countercurrent to the flow of liquid. Water-soluble pollutants are absorbed, diffuse to the biofilm, and are metabolized.

10.4.7 Control of Nitrogen Oxides

Because of concerns associated with the role of NO_x in photochemical oxidant formation and acidic deposition, stationary sources, particularly large fossil fuel–fired power plants, are increasingly being required to reduce NO_x emissions.

A variety of control techniques for reducing emissions from utility and industrial boilers have been developed. One of the most widely used is the low NO_x burner, now a standard part of new utility boiler designs. Low NO_x burners inhibit the formation of NO_x by controlling air and fuel mixing. By using these burners, located in pulverized coal boiler walls, the amount of excess air used to ensure good combustion can be reduced. Reducing excess air from 20% to 14%, for example, has been shown to reduce NO_x production by 20%. It also reduces the amount of O_2 available to combine with nitrogen (N_2). Additional NO_x reductions can be achieved by staged combustion (fuel is combusted in a primary zone that is fuel-rich, followed by secondary and following zones that are fuel-lean). Combustion staging methods reduce NO_x formation by either decreasing available O_2 or providing excess O_2 to cool the combustion process.

A variety of flue gas treatment techniques are available for NO_x emission control. The two primary ones are selective catalytic reduction (SCR) and selective noncatalytic reduction (SNCR). In SCR, ammonia or ammonia-based compounds like urea are injected into the flue gas downstream of the combustion chamber and the air preheater/economizer in an optimum temperature range of 288 °C to 399°C (550°F to 750°F), as shown in Figure 10.14. The flue gas then enters a reaction chamber that houses a catalyst. The catalyst material is coated or impregnated with a precious metal, a base metal oxide such as mixtures of titanium (Ti) and vanadium (V). The following reactions involving NH_3 are illustrative, with the first being the most important since the concentration of NO is much higher than NO_2 in most applications:

$$4\ NO + 4\ NH_3 + O_2 \rightarrow 4\ N_2 + 6\ H_2O \tag{10.5}$$

$$2\ NO_2 + 4\ NH_3 + O_2 \rightarrow 3\ N_2 + 6\ H_2O \tag{10.6}$$

$$NO + NO_2 + 2NH_3\ 2N_2 + 3H_2O \tag{10.7}$$

In 2019, it was reported that there were more than 1,000 SCR units installed in the United States alone. NO_x reduction efficiencies were reported to range from 70% to 90%. Variables affecting the efficiency include the temperature, type of catalyst, residence time, reducing agent and its injection rate, and NO_x and O_2 levels. In SNCR, ammonia or ammonia-based compounds are injected at a much higher temperature of 930°C to 1,090°C (1,700°F to 2,000°F) downstream of the combustion chamber but upstream of the air preheater, eliminating the need of the catalyst. While SNCR is less complicated and less expensive, lower efficiencies are generally achieved when compared to SCR. NO_x reduction

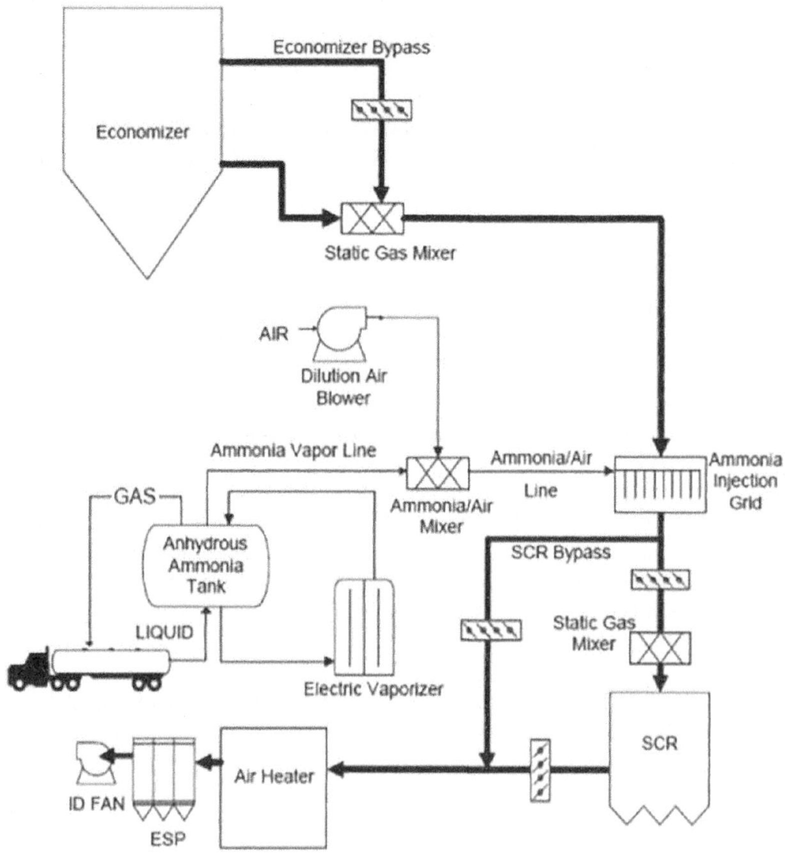

FIGURE 10.14 Schematic of a NO_x selective catalytic reduction system on a large combustion boiler. (US EPA, 2019, https://www.epa.gov/economic-and-cost-analysis-air-pollution-regulations/chapter-2-selective-catalytic-reduction.)

efficiencies of 20%–65% have been reported on large utility boilers. On smaller industrial boilers, efficiencies have ranged from 25% to 80%, again depending on the operating variables.

10.4.8 Control of Carbon Dioxide

The worldwide emissions of carbon dioxide continue to increase as shown in Figure 2.5 in spite of the significant efforts made by countries committed to the Kyoto Protocol and the Paris Agreement; the adoption of CAFE and greenhouse gas emission standards (Sections 8.6.2 and 8.6.3), and the recent declines in emissions in the EU and the United States, among others. For the most part, the reductions that have occurred were achieved by a combination of improvements in the thermal efficiency of combustion systems (both stationary and mobile) and retirement of older existing coal-fired boilers (often in favor of more efficient and less carbon-intensive natural gas units). The recent commitments by automotive manufacturers to increasing the production and percentage of electric vehicles in future model years in the United States, as well as in other countries, has the potential to further reduce carbon emissions, although these are not as zero-carbon as it would first seem due to the transfer of the energy source to EGUs.

The use of carbon taxes (Chapter 8, Section 8.6.3), if adopted, will enhance the reduction or elimination of carbon-intensive sources and could potentially create greater interest and adoption of control technologies that capture CO_2 emissions, particularly from fossil fuel–based EGUs. A number of technologies have been developed to control CO_2 emissions, and while these might be available, their widespread adoption may not occur unless policies are adopted that require reductions in emissions in a time frame that are greater than those that can be accomplished by carbon-intensive source retirements, increased combustion efficiencies, and switching to less carbon-intensive fuels. The following sections, summarizing some of the available control technologies, are included for the first time in this seventh edition of *Air Quality*.

10.4.8.1 Carbon Capture and Storage

Carbon capture and storage (CCS) is a technology whereby potential CO_2 emissions are captured from the source, whether it be an EGU or an industrial process, followed by its being stored or transported for use in another industrial process or by its longer term storage in geological formations as liquid CO_2 under pressure. Typical uses include its use in enhanced oil recovery (EOR), the manufacture of fuels, or building materials. The Global CCS Institute (GCCSI) reported in December 2019,

> There are now 51 large-scale CCS facilities in operation or under development globally in a variety of industries and sectors. These include 19 facilities in operation, four under construction, and 28 in various stages of development. Of all the facilities in operation, 17 are in the industrial sector, and two are power projects. The United States is currently leading the way in CCS development and deployment with 24 large-scale facilities, followed by 12 facilities both in Europe and the Asia Pacific region, and three in the Middle East.

Many of the operating systems come from the petroleum industry. For example, ExxonMobil has operated its Shute Creek Treating Facility in LaBarge, Wyoming, since 1986. The facility is a gas production plant in which the incoming gas stream consists of roughly 21% methane, 65% CO_2, 0.6% He, and 5% H_2S. Although the first three of these were processed for commercial sale, it was necessary to dispose of a concentrated acid gas stream of H_2S and CO_2 by injection back into another area of the reservoir for long-term storage. A new CO_2 capture technology (Controlled Freeze Zone™) was also developed and installed at the plant in 2013, which enhanced carbon dioxide capture by freezing it out of the gas stream followed by converting it to a liquid. The captured CO_2 from the plant is transported by pipeline for injection into oil production wells in Wyoming and Colorado and is the largest carbon capture facility in the world at approximately 6.3 million metric tons annually. To date, most of the commercial carbon dioxide capture facilities around the world have been utilized for EOR. Other options include injecting the captured CO_2 into underground depleted oil/gas reservoirs, non-minable coal beds, or deep ground saline formations for the sole purpose of long-term storage.

As stated earlier, carbon capture has also been applied to fossil fuel–fired combustion systems, but acceptance has been limited in most countries, including the United States, due to the lack of mandatory control of or carbon taxes on greenhouse gas emissions. Most of the systems have historically been on relatively small systems in the 0.5 to 45 MW range, and not on full-scale EGUs. While a carbon tax does not yet exist, the United States passed the Furthering carbon capture, Utilization, Technology, Underground storage and Reduced Emissions (FUTURE) Act in 2017 that was updated in 2018, and IRS guidelines were issued in February 2020. The FUTURE Act, as updated, includes a federal 45Q energy tax credit of \$10/ton, increasing to \$35/ton by 2024, for CO_2 captured and used for EOR. For CO_2 that is captured and stored in geological storage (true CCS), the credit rises from \$20/ton to \$50/ton by 2024. Both tax credits then increase with inflation thereafter. Projects that begin construction within 6 years of the enactment of FUTURE (i.e., before 2024) can claim the annual credit for up to 12 years. The GCCSI reported that there were approximately 10 projects that were ramping up in the United States, motivated in part by the tax credit.

The technologies for CCS are generally separated into those that capture the CO_2 prior to combustion (i.e., pre-combustion IGCC systems—see Section 10.4.4.3), post-combustion carbon capture, and oxyfuel carbon capture. The oxyfuel process is a unique post-combustion process whereby pure oxygen is utilized as the source of oxidation in the combustion process rather than air (which contains 79% nitrogen). This results in a carbon dioxide concentration in the flue gas exceeding 50% by volume, making it more amenable to control; however, it also introduces additional capital and operating costs associated with producing a relatively pure oxygen air supply. This is compared to a typical coal-fired or gas-fired power plant where the carbon dioxide concentration is approximately 14% and 8%, respectively, due to the utilization of air as the source of oxidation. Most post-combustion capture processes involve the scrubbing of the flue gas to capture the CO_2 by utilizing the absorption process (see Section 10.4.3) where the CO_2 is absorbed into an amine- or ammonia-based scrubber. This process of capturing CO_2 in amine-based solvents in scrubbers was first patented almost 90 years ago, with the applications mostly aimed at purifying gas streams ranging from natural gas production to the food and beverage industry. It is anticipated that most large-scale systems will utilize an amine-based solvent. While different amine solvents have been studied, monoethanolamine (MEA) is a likely choice for many scrubbers. The reaction is:

$$C_2H_4OHNH_2^+ + H_2O + CO_2 \rightarrow C_2H_4OHNH_3 + HCO_3^- \tag{10.8}$$

During the absorption process the MEA equilibrium reaction proceeds to the right and, under the appropriate operating conditions, can remove 85%–90% of the CO_2. Figure 10.15 illustrates a typical flowchart for the absorption system. The flue gas is cooled and enters the absorber and then exits through a reheater and the stack. The scrubber solvent absorbs the CO_2. The CO_2-rich solvent containing the captured CO_2 (as HCO_3^-) is pumped to a separate regeneration unit (or stripper) where steam (or heat) is applied. The heat reverses the reaction in Equation 10.8, releasing the more concentrated CO_2 which is then dehydrated, condensed, and compressed in preparation for transport to either an end user or into geologic long-term storage. The CO_2-lean solvent is then further treated and

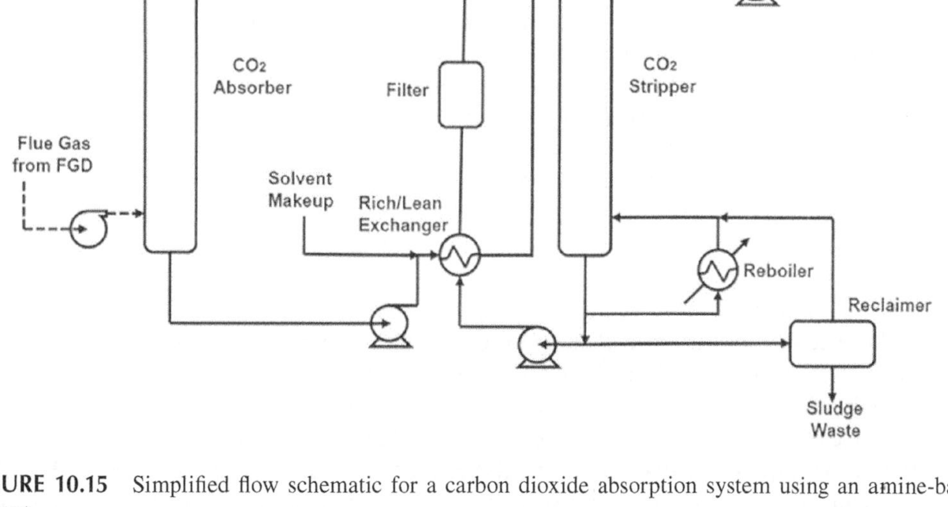

FIGURE 10.15 Simplified flow schematic for a carbon dioxide absorption system using an amine-based solvent.

returned to the absorber. Other systems that have been proposed for CO_2 removal include membrane separation and cryogenic separation.

The two largest coal-fired power plants utilizing carbon capture and a combination of EOR and geologic storage, as reported in 2017, were the Boundary Dam Unit 3 Power Station (115 MW) in Estevan, Saskatchewan, Canada, that began operation in October 2014, and the Petra Nova carbon capture system (240 MW) installed on the W.A. Parish Plant near Houston, Texas, which became operational in 2017. Due to its uniqueness of being the first large-scale utility to adopt CCS, the Boundary Dam unit had significant operational problems in the first several years. In its latest monthly update in March 2020, the EGU reported that it had captured over 3 million metric tons since its startup, was operational 67% of the time over the last year, and averaged the capture of 2,130 metric tons of CO_2 per day. The operational costs of carbon capture systems are high; however, when the captured CO_2 is piped and utilized in EOR, the enhanced production of the oil wells can recover the costs or even provide a positive return. That is not the case for geologic storage, although the 45Q tax credit in the United States provides further financial incentive for both EOR and geologic storage.

10.4.8.2 Other Carbon Dioxide Capture Technologies

In 2017, based on pilot studies, a consortium of international organizations and universities developed an initiative to capture CO_2 at the Hellisheidi Power Plant in Iceland. The geothermal power plant is located on the Hengill volcano and is Iceland's largest geothermal plant. The program, referred to as CarbFix, captures about one-third of the CO_2 emissions from the steam at the plant. It absorbs the CO_2 into water (which is abundant in Iceland) at low concentrations at a ratio of about 4% CO_2 by weight (a process described as being similar to carbonated drinks). The liquid is then pumped into the underlying basalt volcanic rock where it reacts with the calcium, magnesium, and iron to form additional rock. While the CarbFix plant is small and captures about 10,000 metric tons per year, it demonstrates the ability to capture and permanently store carbon in areas where basaltic rock is present. A similar technology, referred to as CarbonCure™ in the United States, diverts CO_2 captured from or present at industrial sources to concrete plants, where it is injected as an aqueous mixture into cementitious materials to form concrete in construction and related applications. In early 2020, it was reported that the technology was continuing to grow and had already captured 62,000 metric tons.

In recent years, technologies have been developed to actual capture CO_2 from the air, rather than from point sources, since it is present worldwide at an average concentration of about 410 ppmv. While these systems, referred to as direct air capture (DAC) systems, would not be categorized as stationary sources of emissions (i.e., the primary focus of this chapter), they are stationary systems that capture ambient CO_2, so DAC is included in this section. DAC has the advantage that it could be installed at any location where there is a need for CO_2 or where carbon storage is desired. There has been a renewed interest in DAC in the United States, under the 45Q tax credit, and in other countries with similar programs. One of the first DAC systems to capture CO_2 at the industrial scale is the Climeworks AG facility near Zurich, Switzerland. The ambient air is pulled in through fans that push the air through a filter system that chemisorbs the CO_2. When the filter system is saturated, the filters are then heated to approximately 100°C, releasing the CO_2 which is then captured for commercial use. In general, DAC systems are far more costly than capturing CO_2 from flue gases due to the low concentrations, but they do provide access to CO_2 at any location for specialty uses and may become more viable in the future.

Another approach being proposed for reducing the concentration of CO_2 in the atmosphere is a natural DAC whereby massive tree-planting programs are being proposed through reforestation or afforestation. According to NASA in 2014, forests and other land vegetation remove up to 30% of the anthropogenic emissions during photosynthesis. In 2018, the UN's IPCC projected that holding the world's average temperature to a 1.5°C rise by 2030 would require an additional one billion hectares (2.4 billion acres) of trees. At a tree density of 200/hectare, this would correspond to about 200 trillion trees, occupying about 8% of the Earth's surface area. Some have argued that such an undertaking would be unrealistic, as it would compete with other essential land uses, including agriculture and food production. In the case of forest fires, such as those that occurred in Australia and United States in recent years, trees can also become a contributor to global CO_2 emissions. However, it is projected that it will take the implementation of many different approaches to address climate change, and tree planting is one of the viable approaches, if properly managed. Several large-scale global and country-specific initiatives have been proposed and/or are in progress as of 2020. In January 2020, world leaders at the World Economic Forum in Davos, Switzerland, committed to the "trillion trees" initiative intended to plant over one trillion trees by 2050. The Priceless Planet Coalition was launched in February 2020 by Mastercard and a host of other partners, including corporations and nonprofits like such as Conservation International and the World Resources Initiative. The goal is to plant 100 million trees within 5 years. As a follow-up to that initiative, a bill was introduced in the United States house, referred to as the Trillion Trees Act, that would commit the United States to the program and incentivize the use of wood products as carbon storage or sequestration devices.

Other emission control technologies for stationary sources involve hydrogen reforming for the production of low-carbon hydrogen, such as blue or green hydrogen, and its use as a cleaner fuel alternative in power plants, industrial facilities, and other fixed emission sources.

READINGS

Allen, D.T., and Rosselot, K.S., *Pollution Prevention for Chemical Processes*, Wiley-Interscience, New York, 1997.

Bubenick, D.V., Control of fugitive emissions, in *Handbook of Air Pollution Technology*, Calvert, S., and Englund, H., Eds., John Wiley & Sons, New York, 1984, p. 745.

Buonicore, A.J., and Davis, W.T., Eds., *Air Pollution Engineering Manual*, Van Nostrand Reinhold, New York, 1992.

Cooper, C.D., and Alley, F.C., *Air Pollution Control: A Design Approach*, 2nd ed., PWS Engineering, Boston, 1994.

Croker, B.B., and Schnelle, K.B., Jr., Control of gases by absorption, adsorption and condensation, in *Handbook of Air Pollution Technology*, Calvert, S., and Englund, H., Eds., John Wiley & Sons, New York, 1984, p. 135.

Davis, W.T., *Air Pollution Engineering Manual*, John Wiley & Sons, New York, 2000.

Dempsey, C.R., and Oppelt, E.T., Incineration of hazardous waste: A critical review update, *JAWMA*, 43, 25, 1993.

DeNevers, N., *Air Pollution Control*, McGraw-Hill, New York, 1995.

Devinny, J., Deshusses, M., and Webster, T., *Biofiltration for Air Pollution Control*, Lewis Publishers/CRC Press, Boca Raton, FL, 1999.

ExxonMobil, *Cleaner Power: Reducing Emissions with Carbon Capture and Storage*, November 20, 2018. https://corporate.exxonmobil.com/Research-and-innovation/Carbon-capture-and-storage/Cleaner-power-reducing-emissions-with-carbon-capture-and-storage#researchOpportunitiesWithCCS.

Flagan, R.C., and Seinfeld, J.H., *Fundamentals of Air Pollution Engineering*, 2nd ed., Prentice Hall, Saddlebrook, NJ, 1988.

Freeman, H., Harten, T., Springer, J., Randall, P., Curran, M.A., and Stone, K., Industrial pollution prevention: A critical review, *JAWMA*, 42, 618, 1992.

Heck, R.M., and Farrato, R.J., *Catalytic Air Pollution Control: Commercial Technology*, John Wiley & Sons, New York, 1995.

Heinsohn, R.J., and Kabel, R.L., *Sources and Control of Air Pollution*, Prentice Hall, Saddlebrook, NJ, 1999.

Henzel, D.S., Laseke, B.A., Smith, E.O., and Swenson, D.O., *Handbook for Flue Gas De-Sulfurization: Scrubbing with Limestone*, Noyes Data Corporation, Park Ridge, NJ, 1982.

Hesketh, H.E., *Air Pollution Control: Traditional and Hazardous Pollutants*, Technomics Publishing Co., Lancaster, PA, 1991.

Heumann, W.L., *Industrial Air Pollution Control Systems*, McGraw-Hill, New York, 1996.

Higgins, T.E., Ed., *Pollution Prevention Handbook*, Lewis Publishers/CRC Press, Boca Raton, FL, 1995.

Hinds, W.C., *Aerosol Technology*, Wiley-Interscience, New York, 1982.

Kilgroe, J.D., Coal cleaning, in *Handbook of Air Pollution Technology*, Calvert, S., and Englund, H., Eds., John Wiley & Sons, New York, 1984, p. 419.

Klaassen, G., and Forsund, F.R., Eds., *Economic Instruments for Air Pollution Control*, Kluwer Academic Publishers, Boston, 1994.
Mycak, J.C., McKenna, J.D., and Theodore, L., *Handbook of Air Pollution Control and Engineering Technology*, Lewis Publishers/CRC Press, Boca Raton, FL, 1995.
Noll, K.E., and Anderson, W.C., *Fundamentals of Air Quality Systems: Design of Air Pollution Control Devices*, American Academy of Environmental Engineers, Annapolis, MD, 1999.
Schifftner, K., and Hesketh, H.E., *Wet Scrubbers*, 2nd ed., Technomics Publishing Co., Lancaster, PA, 1992.
Spencer, D.F., Gluckman, M.J., and Alpert, S.B., Coal gasification for electric power generation, *Science*, 215, 1571, 1982.
Strauss, W., *Industrial Gas Cleaning*, Pergamon Press, Tarrytown, NY, 1996.
Turk, A., Adsorption, in *Air Pollution*. Vol. IV: *Engineering Control of Air Pollution*, Stern, A.C., Ed., Academic Press, New York, 1977, p. 329.
Turner, L.H., and McKenna, J.D., Control of particles by filters, in *Handbook of Air Pollution Technology*, Calvert, S., and Englund, H., Eds., John Wiley & Sons, New York, 1984, p. 249.
US DOE, *Clean Coal Technology: The New Coal Era*, DOE/FE-0193P, DOE, Washington, DC, 1990.
US EPA, *Clearinghouse for Inventories and Emissions Factors (CHIEF)*. https//epa.gov/chief.
US EPA, *Sulfur Emission: Control Technology and Waste Management*, EPA 600/9-79-019, EPA, Washington, DC, 1979.
Warner, C.F., Wark, K., and Davis, W.T., *Air Pollution: Its Origin and Control*, 3rd ed., Addison-Wesley-Longman, Boston, 1998.
White, H.J., Control of particulates by electrostatic precipitation, in *Handbook of Air Pollution Control Technology*, Calvert, S., and Englund, H., Eds., John Wiley & Sons, New York, 1984, p. 283.

QUESTIONS

1. What are the advantages and disadvantages of using tall-stack technology as a pollution control measure?
2. Why is natural gas such a popular fuel for use in industrial boilers and single-cycle gas turbine electrical power production?
3. Describe several applications of pollution prevention principles for controlling air pollutants.
4. What are fugitive emissions? What is their significance in pollution control?
5. Describe the relationship between pollution prevention principles and control of hazardous air pollutant emissions.
6. The collecting efficiency of particle and dust air cleaners decreases as particle size decreases. What is the environmental significance of this phenomenon?
7. Describe the relative fractional collection efficiencies of the following control devices: (a) simple cyclone, (b) multicyclone, (c) baghouse, and (d) electrostatic precipitator.
8. Describe the physical principles used to collect particles in the following: (a) cyclones, (b) baghouse, and (c) electrostatic precipitator.
9. Why does an electrostatic precipitator experience decreased performance when a coal-fired power plant burns low-sulfur coal?
10. When would thermal oxidization be an appropriate method for controlling pollutants in a waste gas stream?
11. What is an LEL? How does it relate to controlling emissions of hydrocarbon gases and vapors?
12. Under what circumstances would catalytic oxidation be used in an industrial facility to control emissions of pollutants? What are its limitations?
13. What is an afterburner?
14. Describe the process of adsorption and its industrial applications.
15. How do physical and chemical adsorption differ?
16. What physical and chemical processes are utilized to remove SO_2 from a waste gas stream?
17. What is coal beneficiation? What are its advantages and limitations in meeting regulatory SO_2 emission limits?

18. Describe the chemical processes involved in the removal of SO_2 from flue gases using a wet scrubber system.
19. What is the role of packing materials in wet scrubbing systems?
20. Describe the techniques used to control emissions of SO_2 using dry scrubbing.
21. What is fluidized bed combustion? How can SO_2 emissions be reduced from such a combustion system?
22. Describe the methodologies used to reduce emissions of NO_x from industrial and utility fossil fuel–fired systems.
23. Describe the difference between DAC and control of CO_2 from a stationary source.

11 Indoor Air Quality

As we have seen thus far, pollution of the ambient atmosphere is an enormous environmental concern. As a consequence, regulatory measures have been implemented to safeguard public health, protect air quality in pristine air regions, regulate or ban ozone (O_3)-destroying chemicals, and reduce acidic deposition; measures are being contemplated and debated to reduce emissions of greenhouse gases to mitigate the anticipated adverse consequences of global warming. The primary focus of air pollution control activities on a day-to-day basis is to protect public health. In the more than four decades that have passed, these health-based pollution control efforts have required an investment of several trillion dollars.

Exposure to polluted ambient air is but one of several air pollution and public health concerns that humans face. Particularly notable are pollutant exposures in the indoor environments where we live, work, learn, and play.

Based on numerous scientific studies and field investigations, the air inside our homes, office buildings, and schools is contaminated by a variety of gas-phase and particulate-phase substances that may be present in sufficient concentrations to cause acute or chronic symptoms or illness in individuals of a relatively large exposed population. Exposures may vary from approximately 8 hours/day in office, commercial, and institutional buildings to 16 to 24 hours/day in homes. This compares, on average, to approximately 2 hours/day in the ambient environment.

Exposure duration, populations at risk, pollutant concentrations, and pollutants themselves are, in many cases, significantly different in indoor and ambient environments. Exposure durations are longer and populations at risk larger in the former. In residential buildings, exposures occur to individuals ranging in age from infants to elderly, and health status from healthy to having a variety of illnesses and infirmities.

Based on studies of illness prevalence, pollutant exposures, and long-term cancer risks associated with exposures to radon, asbestos, pesticides, and combustion by-products, public health concerns associated with indoor air may rival, if not significantly exceed, those associated with the ambient environment.

11.1 PERSONAL, INDOOR, AND OUTDOOR POLLUTANT RELATIONSHIPS

11.1.1 Personal Pollutant Exposures

In National Ambient Air Quality Standards (NAAQSs) promulgated under clean air legislation, there is an implicit (although unrealistic) assumption that exposures occur similarly in both ambient and indoor environments. They do not, as a consequence, take into account actual (personal) exposures. Scientific studies have shown that such personal exposures are often considerably different from those implied in NAAQSs. Differences in outdoor, indoor, and personal exposures to respirable particles (RSPs) in a Topeka, Kansas, study can be seen in Figure 11.1. Note that although all three exposure measures track each other, indoor exposures are greater than outdoor exposures, and for the most part, personal exposures are greater than indoor exposures. Personal exposures include different environments and activities, including motor vehicle travel and hobbies. Research conducted on personal pollution exposures to criteria pollutants indicates that exposures inferred from fixed-site monitoring stations may be significantly less than those experienced by a significant portion of the US population.

11.1.2 Indoor and Outdoor Pollutant Concentrations

Humans are exposed to a variety of pollutants in indoor and ambient air. Although ambient pollutants are present in indoor environments, indoor and outdoor concentrations often differ, depending

DOI: 10.1201/9781003664109-11

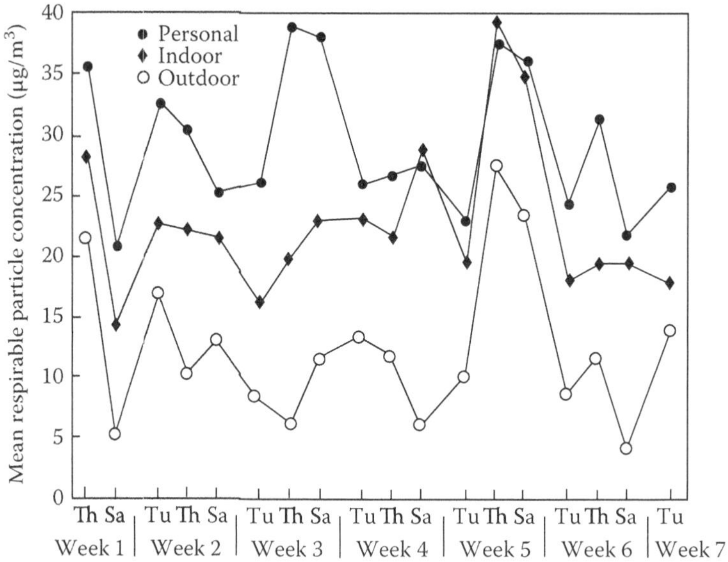

FIGURE 11.1 Indoor, outdoor, and personal exposure of Topeka, KS, residents to RSPs. (Courtesy of Spengler, J.D., School of Public Health, Harvard University, Boston.)

on pollutant reactivity and whether indoor sources are present. Indoor levels of reactive chemical species such as O_3 and sulfur dioxide (SO_2) are significantly lower than they are outdoors. In other cases, significantly higher concentrations are found in indoor environments as a result of indoor sources.

The level of air pollutant concentrations is related to emission rate of sources, dispersion, or/and ventilation rate of pollutants. In indoor environments, if little outdoor air enters indoors, pollutants can accumulate to levels that can pose health and comfort problems. High ventilation rate could significantly decrease the level of indoor air pollutants. Unless buildings are built with special mechanical means of ventilation, those buildings designed to minimize the "infiltration" of outdoor air could have higher indoor pollutant levels.

The indoor/outdoor ratio (I:O) of O_3 is typically in the range of 0.1 to 0.3, but it may be as high as 0.7 in mechanically ventilated spaces. Typically, ambient air is the major source of O_3 observed indoors. However, elevated localized indoor O_3 concentrations and exposures may result from the use of electronic air cleaners or electrostatic photocopy machines.

Sulfur dioxide ratios vary from 0.3 to 0.5 when outdoor levels are moderately high, to 0.7 when they are relatively low. Elevated indoor SO_2 concentrations and exposures may result from kerosene heaters using 0.20% sulfur (S) fuel.

Although ambient air is a source of volatile organic compounds (VOCs), concentrations of individual VOC species indoors are, in most cases, several times to an order of magnitude higher than outdoors. This is because many materials used to manufacture building materials, furnishings, and the like emit VOCs for extended periods of time.

Nitrogen dioxide (NO_2) is a relatively reactive gas found in the ambient atmosphere, and as a consequence, I:O ratios of less than 1.0 should be expected. However, significant sources of NO_2 are often present in residential environments. These include unvented combustion appliances such as gas and kerosene space heaters, gas cooking stoves and ovens, and, increasingly, ventless gas fireplaces.

As carbon monoxide (CO) is a relatively unreactive substance, it passes freely through building envelopes and ventilation systems. In the absence of indoor sources, concentrations in ambient and indoor environments should be the same. However, in many cases, indoor concentrations are higher because of sources such as gas cooking stoves and ovens, ventless gas and kerosene heaters or fireplaces, tobacco smoking, and others.

In a "closed" environment, the building envelope and systems tend to "filter out" ambient particles as they are transported into buildings. In the absence of indoor sources, I:O ratios can be expected to be lower than 0.5. However, the indoor environment is often a major source of particles, particularly when smoking is occurring; in such cases, I:O ratios are higher than 1.0.

Concentrations of asbestos; radon (Rn); formaldehyde (HCHO); tobacco smoke; allergens such as those from dust mites, pets, cockroaches, and the like; and other biological contaminants such as bacterial endotoxins, fungal glucans, certain mold species, and mycotoxins are usually significantly higher indoors than outdoors. Exposure to environmental tobacco smoke (ETS) is, for the most part, unique to indoor environments.

11.2 INDOOR AIR QUALITY CONCERNS

Indoor air quality (IAQ) health and exposure concerns are primarily focused on residential dwellings and nonresidential, nonindustrial buildings such as offices and schools. Although the same pollutants may be found in both building categories, exposures and exposure concerns are often quite different.

Occupants of residences can be exposed to potentially toxic, allergenic, or infectious pollutants upward of 8 to 24 hours/day, with exposed populations including infants/children, healthy adults, the aged, and the infirm. Pollutant exposures that are unique (particularly as a result of elevated concentrations) include Rn, HCHO, ETS, combustion appliance by-products, pesticides, VOCs from personal and home care products and arts-and-crafts activities, and biological contaminants such as bacterial endotoxins and allergens associated with dust mites, cockroaches, pets, and mold. Exposure concerns are increased as a result of limited ventilation in enclosed conditions during hot or cold seasons.

Nonresidential, nonindustrial buildings are subject to relatively unique pollutant exposure, health, and comfort concerns. Such buildings are typically mechanically ventilated and may have high occupant densities (e.g., schools). Populations subject to pollutant exposures depend on the building's function; for example, adults in office buildings, a mixture of children and adults in schools, and the infirm or aged in hospital convalescent or senior citizen facilities.

Pollutant exposure and health concerns in nonresidential, nonindustrial buildings include

1. Elevated bioeffluent (pollutants released by humans and other organisms) levels associated with high occupant densities and inadequate ventilation
2. Emissions from office equipment and materials
3. Cross-contamination from contaminant-generating areas
4. Entrainment of contaminants generated outdoors
5. Reentry of building exhaust gases
6. Contamination of air-handling units (AHUs) by organisms and biological by-products that can cause illness, for example, hypersensitivity pneumonitis, humidifier fever, and Legionnaires' disease
7. Transmission of contagious diseases such as flu, colds, and tuberculosis (TB)
8. Exposure to resuspended surface dusts
9. Exposure to ETS where smoking is not restricted

11.3 MAJOR INDOOR POLLUTANTS

A number of pollutants and pollutant categories have been identified as significant exposure and potential health risks in buildings. These include asbestos, Rn, combustion by-products, aldehydes, VOCs, semivolatile organic compounds (SVOCs), pesticides, and a variety of contaminants of biological origin.

11.3.1 Asbestos

Asbestos became a major IAQ concern in the United States in the late 1970s when public health and environmental protection authorities concluded that friable (hand-crushable) asbestos-containing building materials (ACBM) in school buildings posed a potential cancer risk to children and other building occupants. Such materials were widely used in the construction of schools and other large buildings prior to 1978.

"Asbestos" is a collective term for a number of fibrous mineral silicates. The most widely used asbestos mineral in the United States was chrysotile, or white asbestos. It accounted for more than 90% of the fibrous mass used in the 3,000 or so different asbestos-containing products. Amosite, or brown asbestos, was less commonly used in the United States. Crocidolite, or blue asbestos, has been widely used in Europe and Southeast Asia.

Asbestos fibers are fire and heat resistant, with high tensile strength. Therefore, they were used in sprayed-on fireproofing, acoustical plaster, thermal system insulation, friction products, fibrocement board and pipes, roofing products, spackling compounds, mastics, and vinyl floor tile.

Asbestos has been regulated as a hazardous air pollutant by the US Environmental Protection Agency (US EPA) since 1973. Occupational exposures have been causally associated with asbestosis (a fibrogenic, debilitating lung disease), lung cancer, mesothelioma (a cancer of the lining of the chest and abdominal cavities), and scarring of the pleural cavity (found in the chest) called pleural plaques. Asbestosis is caused by heavy, long-term exposures to asbestos fibers and is unlikely to be a health risk to building occupants. Because there is no apparent threshold or safe level of exposure, public health concern has primarily focused on risks of developing lung cancer or mesothelioma. Because children would be exposed earlier in life, providing for a longer latency period, their risk of developing mesothelioma would be increased. Although relatively rare (an estimated 1,200–1,500 cases/year in the United States), mesothelioma has a mortality rate of 100%.

Asbestos fibers in many products are held in a matrix that varies in its friability. Fiber release occurs when this matrix is broken by hand pressure (hand friable) or mechanical force (mechanically friable). Because hand-friable ACBM poses the greatest risk of fiber release and building air contamination, the US EPA, in 1973, required friable asbestos removal before building demolition or renovation and banned sprayed-on fireproofing materials (Figure 11.2). In 1978, the EPA banned sprayed-on and troweled-on acoustical plaster and premolded asbestos products used in thermal insulation (Figure 11.3). These materials are present in hundreds of thousands of buildings in the United States.

Over time, friable ACBM may become damaged or lose its adhesion to the substrate to which it was applied. Thermal system insulation on steam and hot water lines, boilers, and the like may become friable when damaged by maintenance and service activities. Damage to ACBM increases the likelihood that fibers become airborne, resulting in human exposure. Exposure may be increased

FIGURE 11.2 Sprayed-on asbestos fireproofing.

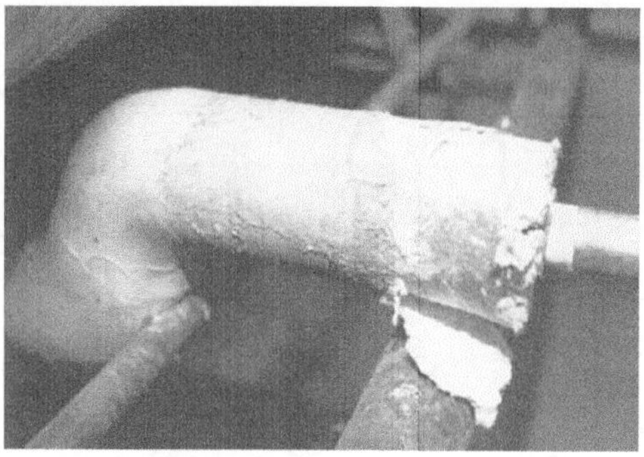

FIGURE 11.3 Partially damaged asbestos thermal system insulation.

by custodial activities such as dusting and cleaning and activities of occupants that may resuspend settled fibers.

The US EPA initiated a guidance and technical assistance program, the Asbestos-in-Schools Program, in 1979 in response to asbestos health concerns in schools. Because of inadequacies in the program, Congress, in 1986, passed the Asbestos Hazard Emergency Response Act (AHERA). This law required all schools to be inspected by accredited inspectors and school corporations and districts to prepare and implement a management plan for each building under their jurisdiction.

In the late 1980s, considerable public health attention was given to potential asbestos fiber exposures and health risks to the general school population, that is, children, faculty, and administrative staff. Subsequently, a consensus evolved in the scientific and regulatory communities that exposure risks to these groups were very small. At the same time, there was increasing scientific evidence that custodial and service staff in buildings containing ACBM were at significant risk of developing asbestos-related disease as a result of asbestos exposures. Custodial workers, sheet metal workers, and carpenters were shown to have significantly elevated prevalence rates of asbestos-related pleural plaques that result in impaired pulmonary function. Consequently, the Occupational Safety and Health Administration (OSHA) promulgated rules designed to protect custodial and service workers in office and other buildings; materials presumed to contain asbestos are to be identified, and workers notified and trained to reduce exposure risks associated with their disturbance.

11.3.2 Radon

Radon is a noble, nontoxic, radioactive gas produced in the decay of radium-226. Radium is found in relatively high concentrations in uranium (U) and phosphate ores and U mill tailings and, to a lesser extent, in common minerals such as granite, schist, limestone, and others. As a consequence, it is ubiquitous in soil and air near the surface of the Earth. As Rn undergoes radioactive decay, it releases an alpha particle, gamma rays, and progeny that quickly decay (Figure 11.4) to release alpha and beta particles and gamma rays. Because Rn decay products (RDPs) are electrically charged, they readily attach to particles, producing radioactive aerosols. These aerosol particles may be inhaled and deposited in the bifurcations of respiratory airways.

11.3.2.1 Concentrations and Sources

Radon concentrations determined for buildings in the United States are calculated and reported as picocuries per liter (pCi/L). A picocurie is equal to a trillionth of a curie, or 2.2 radioactive disintegrations per minute. A working level (WL) is a quantity of Rn that will produce 1.3×10^5 million electron volts (MeV) of potential alpha particle energy per liter of air. One WL is equal to the concentration of RDPs in equilibrium with 100 pCi/L Rn. Because the equilibrium value for

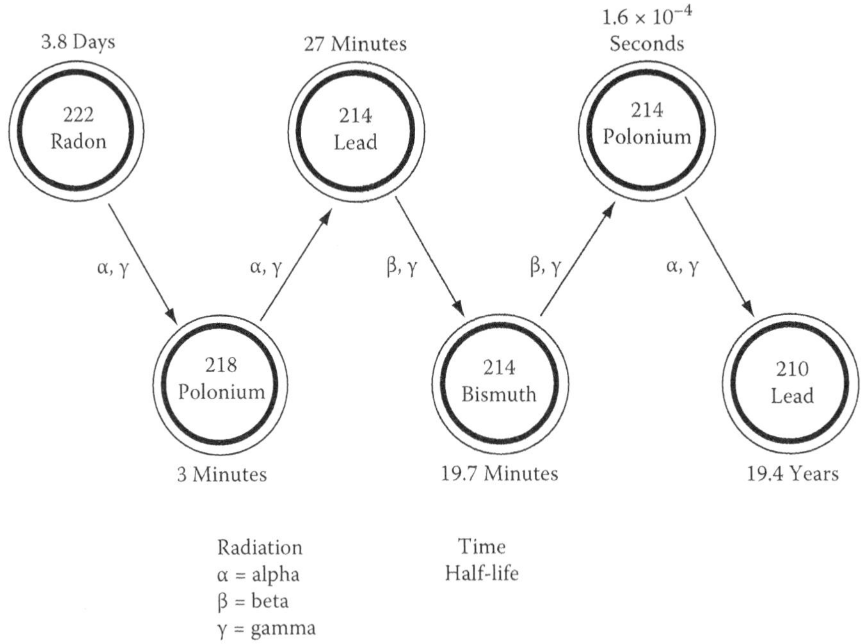

FIGURE 11.4 Radioactive decay of radon and its progeny.

RDPs in indoor environments is typically 0.5, Rn concentrations corresponding to 1 WL would be approximately 200 pCi/L.

Because of the relatively common occurrence of radium-226 in the Earth's crust, Rn is present in ambient air with average concentrations in the range of 0.20 to 0.25 pCi/L. Radon levels in buildings may be several times or orders of magnitude higher than ambient levels. This is particularly true in single-family dwellings. In one of the worst cases ever reported, Rn concentrations as high as 2,600 pCi/L (13 WL) were measured in a southeastern Pennsylvania home.

Significant geographical differences in Rn levels in the United States have been reported. Geometric mean values based on 3-month averages for 30,000 nonrandomly tested houses were reported as 3.43 pCi/L in the Northeast, 2.36 in the Midwest, 0.64 in the Northwest, 2.40 in the mountain states, and 1.43 in the Southeast. The geometric mean for the entire sample population was 1.74 pCi/L; 0.5% were higher than 100 pCi/L. Other studies have reported a lower geometric mean value of approximately 1.0 pCi/L.

The major source of indoor Rn is soil gas transported by pressure-induced convective flows. Other sources include well water and masonry materials. Buildings constructed on soils with high Rn release and transport potential typically have the highest Rn levels; these vary in response to temperature-dependent and wind-dependent pressure differentials and changes in barometric pressure. Radon transport is enhanced when the base of a building is under significant negative pressure.

11.3.2.2 Health Effects and Risks

The major health risk definitively linked to Rn exposures is lung cancer. Extrapolations from U miner Rn exposure and health data indicate that humans exposed to the lower concentrations characteristic of buildings may also be at risk. Risk assessments have been made by the National Council on Radiation Protection (NCRP), the National Academy of Sciences Committee on the Biological Effects of Radiation (BEIR IV), and the US EPA. These assessments predict (with some degree of uncertainty) that approximately 13,000 to 16,000 lung cancer deaths a year can be attributed to Rn

TABLE 11.1
Radon Risk for Smokers

Radon Level	If 1,000 people who smoked were exposed to this level over a lifetime* . . .	The risk of cancer from radon exposure compares to** . . .	What to Do: Stop Smoking and . . .
20 pCi/L	About 260 people could get lung cancer	← 250 times the risk of drowning	Fix your home
10 pCi/L	About 150 people could get lung cancer	← 200 times the risk of dying in a home fire	Fix your home
8 pCi/L	About 120 people could get lung cancer	← 30 times the risk of dying in a fall	Fix your home
4 pCi/L	About 62 people could get lung cancer	← 5 times the risk of dying in a car crash	Fix your home
2 pCi/L	About 32 people could get lung cancer	← 6 times the risk of dying from poison	Consider fixing between 2 and 4 pCi/L
1.3 pCi/L	About 20 people could get lung cancer	(Average indoor radon level)	(Reducing radon levels
0.4 pCi/L		(Average outdoor radon level)	below 2 pCi/L is difficult)

Note: If you are a former smoker, your risk may be lower.
Source: A Citizen's Guide to Radon (EPA 402/K-12/002 | 2016 | www.epa.gov/radon.)

exposure, with interaction effects contributing to a higher risk for smokers. For purposes of communicating with the public, the US EPA published tables (Tables 11.1 and 11.2) wherein lung cancer risks are projected for different average lifetime Rn exposures and compared with equivalent risks from exposures to smoking or nonsmoking people. Note the relatively large lifetime lung cancer risk of 2 and 20 per 1,000 population for outdoor levels of 1.3 pCi/L and 7 to 62 per 1,000 for an average lifetime exposure to 4 pCi/L for nonsmoking and smoking people, respectively, the US EPA action level for remediation. In the latter case, an annual average lifetime exposure to 4 pCi/L would be equivalent to the lung cancer risk associated with smoking five cigarettes per day. Using these risk estimates, or even more conservative ones, Rn seems to be one of the more hazardous substances to which humans are exposed. Because of these risks, the US EPA and the US Surgeon General, in 1988, issued a public health advisory recommending that all homes be tested and remediation be undertaken when long-term (3 months to 1 year) test results are excessive, that is, above the action guideline of 4 pCi/L. In June 2011, a federal action plan was established as a comprehensive collaborative effort between the US EPA, the US Departments of Health and Human Services, Agriculture, Defense, Energy, Housing and Urban Development, Interior, and Veterans Affairs, and the US General Services Administration. The plan represents a multiyear approach to protecting public health by reducing the risk of radon exposure. The goals are (1) to have operating radon mitigation systems in 30% of the 9.2 million homes in the United States that have elevated radon levels and (2) to ensure that 100% of the new homes in high-radon potential areas have radon-reducing features.

There is a limited amount of epidemiolocal evidence to suggest that Rn exposures may be a risk factor for acute myeloid leukemia, cancers of the kidney, melanoma, and certain childhood cancers. In Great Britain, approximately 6% to 12% of acute myeloid leukemia cases are estimated to be associated with Rn exposures; for regions where Rn levels are higher, the predicted range is 23% to 43%. For the world average exposure of 1.3 pCi/L, approximately 13% to 25% of acute myeloid leukemias are estimated to be associated with Rn.

TABLE 11.2
Radon Risk for Nonsmokers

Radon Level	If 1,000 people who smoked were exposed to this level over a lifetime* . . .	The risk of cancer from radon exposure compares to** . . .	What to Do: Stop Smoking and . . .
20 pCi/L	About 36 people could get lung cancer	← 35 times the risk of drowning	Fix your home
10 pCi/L	About 18 people could get lung cancer	← 20 times the risk of dying in a home fire	Fix your home
8 pCi/L	About 15 people could get lung cancer	← 4 times the risk of dying in a fall	Fix your home
4 pCi/L	About 7 people could get lung cancer	← The risk of dying in a car crash	Fix your home
2 pCi/L	About 4 people could get lung cancer	← The risk of dying from poison	Consider fixing between 2 and 4 pCi/L
1.3 pCi/L	About 2 people could get lung cancer	(Average indoor radon level)	(Reducing radon levels
0.4 pCi/L		(Average outdoor radon level)	below 2 pCi/L is difficult)

Note: If you are a former smoker, your risk may be higher. *Lifetime risk of lung cancer deaths from EPA Assessment of Risks from Radon in Homes (EPA 402-R-03-003). **Comparison data calculated using the Centers for Disease Control and Prevention's 1999–2001 National Center for Injury Prevention and Control Reports.

Source: A Citizen's Guide to Radon (EPA 402/K-12/002 I 2016 I www.epa.gov/radon.)

11.3.3 Combustion By-Products

Combustion by-products are released into indoor air from sources such as (1) open cooking fires fueled by wood, charcoal, dung, and others, in developing countries; (2) unvented gas cooking appliances, space heaters, fireplaces, and kerosene space heaters; and, to a lesser extent, (3) vented combustion systems, for example, fireplaces, wood-burning appliances, gas and oil furnaces, and hot water heaters. Miscellaneous sources include tobacco smoking (the most ubiquitous source of combustion products in indoor environments), candle and incense burning, propane-fueled burnishers and forklifts, ice resurfacing machines, arena events, entrainment from outdoor sources, and flue gas reentry.

11.3.3.1 Unvented Combustion Systems

11.3.3.1.1 Cooking Stoves in Developing Countries

According to a study conducted in 2018, almost 3 billion people still rely on biomass (wood, charcoal, crop residues, and dung) and coal as their primary source of domestic energy for cooking (Pattanayak et al., 2019). Biomass accounts for more than one-half of domestic energy in many developing countries and for as much as 95% in some lower-income households. These practices result in what is referred to as a triple treat in that they (1) are high emitters of carbon, (2) result in deforestation due to daily collection of wood for fuel, and (3) expose users to excessively high concentrations of air pollutants. Most cooking fires are simple pits and are not vented to the ambient environment (Figure 11.5). As a consequence, indoor concentrations of combustion by-products are very high. This is evident in particulate matter (PM) concentrations measured in biomass cooking homes in villages in several countries (Table 11.3). What is notable about these total suspended particulate (TSP) matter exposure levels (primarily to adult women and young children) is that they are an order of magnitude or more above the old US TSP standard of 260 μg/m^3 (not to be exceeded more than once per year) and the World Health Organization's (WHO) PM_{10} (PM with an aerodynamic diameter of ≤ 10 μm) guidelines of 100 to 150 μg/m^3 24-hour average.

FIGURE 11.5 Home cooking fire in a dwelling. (Courtesy of Smith, K.R., School of Public Health, University of California-Berkeley.)

TABLE 11.3 INDOOR AIRBORNE PARTICULATE MATTER (TSP) CONCENTRATIONS ASSOCIATED WITH BIOMASS COOKING IN DEVELOPING COUNTRIES

Location/Report Year	Measurement Conditions	PM Concentration (μg/m³)
Papua, New Guinea		
1968	Overnight, floor level	200–4,900
1975	Overnight, sitting level	200–9,000
India		
1982	Cooking with wood	15,000
1988	Cooking with dung	18,000
	Cooking with charcoal	5,500
	Cooking, measured near ceiling	4,000–21,000
Nepal		
1986	Cooking with wood	8,800
2015	Cooking with wood	656–1,580
China, 1987	Cooking with wood	2,600
Gambia, 1988	24 hours	1,000–2,500
Kenya, 1987	24 hours	1,200–1,900

Source: Smith, K.R., School of Public Health, University of California-Berkeley.

High indoor PM concentrations have also been reported in homes in developing countries that use coal for cooking and heating. PM associated with biomass cooking and coal use has high polycyclic aromatic hydrocarbon (PAH) concentrations. Indoor concentrations of these potent carcinogens are also very high.

Indoor levels of gas-phase substances associated with gas cooking are also elevated. These include CO (10–50 ppmv, with averages of 30–40 ppmv common), NO_2 (70–160 ppbv), SO_2 (60–100 ppbv), and aldehydes (0.67–1.2 ppmv).

The WHO estimates that pollutant exposures associated with biomass cooking are responsible for approximately 2.5 million premature deaths worldwide each year. Exposures to biomass cooking smoke significantly increase the risk of acute respiratory infections, chronic obstructive lung disease, cor pulmonale, and lung cancer. Untreated acute respiratory infection commonly progresses to pneumonia, a major cause of children's deaths worldwide.

Numerous improved cookstoves (ICS) have been developed; however, acceptance of these and their diffusion into the market has been very slow as reported as recently as 2019 due to economic conditions, culture, geography, and many other factors.

11.3.3.1.2 Gas Cooking Appliances

Gas cooking stoves and ovens can be significant sources of carbon dioxide (CO_2), NO_2, nitrogen oxides (NO_x), aldehydes, RSPs, and a variety of VOCs. Levels of CO in kitchens with operating gas stoves have been reported to be in the range of 10 to 40 ppmv. Nitrogen dioxide levels in homes using gas stoves and ovens have been reported in the range of 18 to 35 ppbv, compared with 5 to 10 ppbv in non-gas cooking residences.

Exposure to elevated levels of NO_2 associated with gas cooking has been proposed as a potential causal factor for respiratory symptoms and pulmonary function changes in school-age children living in gas-cooking homes. However, a number of epidemiological studies evaluating the relationship between gas cooking (and, to a lesser extent, associated NO_2 levels) and respiratory disease have not shown a consistent pattern between such exposures and respiratory symptoms or lung function changes in either children or adults.

In a recent Australian study, the presence of a gas cooking stove and exposure to NO_2 were observed to be significant risk factors for both asthma (increased threefold) and respiratory symptoms (increased twofold) in children. Although an independent relationship was observed for asthma and NO_2 exposures, statistical analyses revealed that some other aspect of gas stoves posed an additional asthma risk.

11.3.3.1.3 Space Heaters

Unvented gas and kerosene heaters are widely used to heat residential spaces. Gas space heaters are commonly used in the southern United States and Australia where winter temperatures are relatively moderate. Kerosene heaters are widely used in Japan and were once popular (late 1970s to early 1980s) in the United States as a space-heating alternative.

Modern unvented space heaters have low emissions of CO and do not pose the asphyxiation hazard they once did. Nevertheless, they have the potential to significantly contaminate indoor spaces with NO_2, nitric oxide (NO), CO, CO_2, SO_2, aldehydes, VOCs, and RSPs. Emissions depend on heater or burner design, fuel used, and heater maintenance.

Unvented kerosene and gas-fired space heaters are used by homeowners and apartment dwellers under varying home space conditions, ventilation rates, number of heaters used, and daily and seasonal hours of operation. In a large monitoring study, northeastern US homes operating one kerosene heater averaged approximately 20 ppbv of NO_2, those with two heaters averaged 37 ppbv, whereas control homes averaged approximately 4 ppbv. Approximately 8% of kerosene heater-using homes had peak NO_2 levels of more than 255 ppbv. More than 20% had average SO_2 levels that were higher than 240 ppbv. These concentrations are well above the NO_2 and SO_2 1-hour NAAQSs of 100 and 75 ppbv, respectively. Other studies have reported CO levels of 1 to 5 ppmv in kerosene heater-using homes, with RSP increases above the background ranging from 10 to 88 μg/m^3. Elevated SO_2 levels were associated with use of grade 2-K kerosene, which has a S content of approximately 0.2%.

Epidemiological studies of households using kerosene heaters for space heating have shown that children under the age of 7 exposed to average NO_2 levels of more than 16 ppbv had a more than twofold higher risk of lower respiratory symptoms and illness (fever, chest pain, productive cough, wheezing, chest cold, bronchitis, pneumonia, or asthma). They also experienced increased risk of upper respiratory symptoms (sore throat, nasal congestion, dry cough, croup, or head cold).

11.3.3.1.4 Gas Fireplaces

Ventless gas-burning fireplaces have been installed in hundreds of thousands of new homes in North America. Limited studies have indicated that in high-altitude Colorado, such fireplaces could produce CO levels that varied from 40 to more than 100 ppmv, with elevated levels of NO_x as well. Carbon dioxide levels often exceeded the 5,000 ppmv OSHA 8-hour permissible exposure limit. At such levels, CO_2 is a respiratory stimulant; that is, it increases breathing rates.

11.3.3.2 Vented Combustion Systems

In general, vented combustion systems such as gas, oil, and wood furnaces; wood-burning and coal-burning stoves; and wood-burning fireplaces pose limited indoor contamination concerns. However, flue gas venting systems are not perfect and some flue gas spillage can be expected under normal operation, particularly for systems using natural draft. Wood-burning appliances, because of design deficiencies and a requirement for manual recharging, pose unique indoor contamination concerns.

11.3.3.2.1 Flue Gas Spillage

Flue gas spillage commonly occurs in residences using natural draft gas furnaces, gas water heaters, or wood-burning appliances. In most cases, spillage is momentary, resulting in minimal emission of pollutants to the building environment. Serious flue gas spillage does occur, with occasional deaths or, more commonly, sublethal CO poisoning. Such spillage usually occurs in aging or poorly installed or maintained combustion and flue systems.

Flue gas spillage occurs because of chimney blockage or backdrafting. Backdrafting results when the chimney is cold, under some weather conditions, and when the house is depressurized by competing exhaust systems (e.g., fireplaces, exhaust fans, and furnaces). Backdrafting is also a concern in energy-efficient houses that do not have sufficient air for combustion or exhaust.

Flue gas spillage also occurs when furnace or flue components malfunction due to corrosion, improper installation, or, in the case of flue pipes, accidental disconnection. Perforation of heat exchange elements or flue pipes is more common as systems age.

11.3.3.2.2 Wood-Burning Appliances

During the late 1970s to the mid-1980s, wood-burning appliances (stoves and furnaces) became popular as residential space heating alternatives in the United States. An estimated 8 to 10 million units were in use. Although no recent surveys have been conducted, it seems that with fuel price stabilization, wood heating appliance use has dramatically decreased.

Combustion by-products may be emitted from wood-burning appliances because of improper installation; negative air pressure and downdrafts; leaks in appliance parts and flue pipes; and in starting, reloading, stoking, and ash removal operations. Wood-burning appliances have been reported to emit CO, NO_x SO_2, aldehydes, RSPs, and a variety of VOCs. Some of the more than 200 VOCs identified in wood smoke are known carcinogens, for example, PAHs.

Significant contamination of indoor spaces by wood-burning emissions has been reported for conventional stove and furnace units. Airtight appliances, on the other hand, emit only trace quantities of combustion gases and particles into indoor spaces. The effect of heater type on PAH concentrations both indoors and outdoors is illustrated in Figure 11.6.

Because they are open to room air, fireplaces are potentially significant sources of indoor air contamination during use. However, they are used primarily for their aesthetic appeal, and frequency of use is often limited.

In some parts of the United States where wood supplies are abundant and firewood inexpensive, wood-burning appliance use was so extensive that emissions resulted in ambient concentrations that exceeded air quality standards for PM_{10}. This was the case in mountain valley communities such as Missoula, MT; Corvallis, OR; Aspen, CO; and Danbury, VT. As a result, the US EPA promulgated a new source performance standard (NSPS) for woodburning stoves and furnaces; wood stoves and

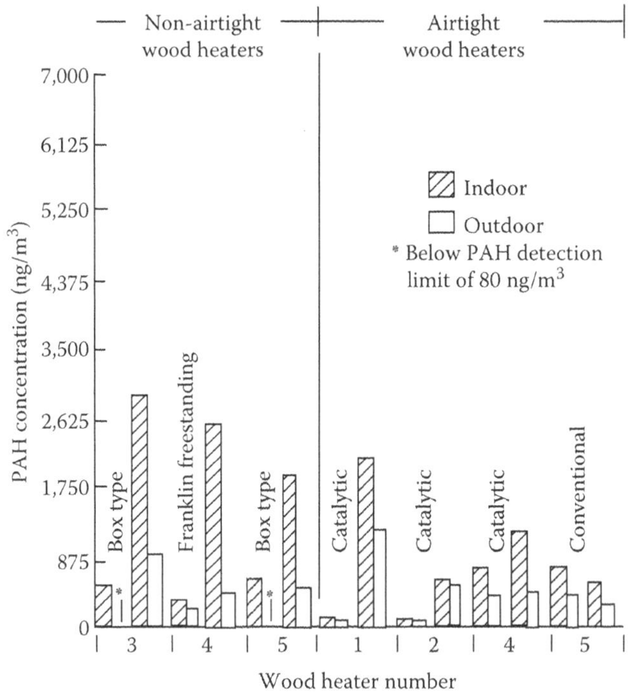

FIGURE 11.6 PAH concentrations associated with different wood-burning heating appliances. (From Knight, C.V. et al., *Proceedings of IAO'86: Managing Indoor Air for Health and Energy Conservation*, ASHRAE, Atlanta, 1986. With permission.)

furnaces subject to the NSPS have significantly decreased emissions of both PM and CO to both ambient and indoor environments.

11.3.3.3 Miscellaneous Sources

A variety of combustion-generated substances cause contamination of indoor spaces as a result of human activities. The most important of these is tobacco smoking. Burning candles and incense in homes has, in the past decade, become an increasingly popular activity. Combustion by-product contamination of indoor spaces may result from the use of propane-fueled floor burnishers and forklifts, propane-fueled and gas-fueled ice-resurfacing machines, arena events involving gasoline-powered vehicles, entrainment of motor vehicle emissions, reentry of flue gases, and, surprisingly, deterioration of duct liner materials.

11.3.3.3.1 Tobacco Smoking

Use of tobacco products by approximately 40 million smokers in the United States results in significant indoor contamination from combustion by-products that poses significant exposures to millions of others who do not smoke but must breathe contaminated indoor air. Several thousand gas-phase and particulate-phase compounds have been identified in tobacco smoke. The more significant of these include RSPs, nicotine, nitrosamines, PAHs, CO, CO_2, NO_x, acrolein, HCHO, and hydrogen cyanide (HCN).

Environmental tobacco smoke is a combination of sidestream (SS) smoke (released between puffs) and exhaled mainstream (MS) smoke. Exposure to ETS depends on the number and type of cigarettes consumed in a space, room volume, ventilation rate, and proximity to the source. Concentrations of major tobacco-related contaminants measured in a variety of building environments are summarized

TABLE 11.4
Tobacco-Related Contaminant Levels in Buildings

Contaminant	Type of Environment	Levels	Nonsmoking Controls
CO	Room (18 smokers) 15 restaurants	50 ppmv	0.0 ppmv
		4 ppmv	2.5 ppmv
	Arena (11,806 people)	9 ppmv	3.0 ppmv
RSPs	Bar and grill	589 μg/m^3	63 μg/m^3
	Bingo hall	1,140 μg/m^3	40 μg/m^3
	Fast food restaurant	109 μg/m^3	24 μg/m^3
NO_2	Restaurant Bar	63 ppbv	50 ppbv
		21 ppbv	48 ppbv
Nicotine	Room (18 smokers)	500 μg/m^3	—
	Restaurant	5.2 μg/m^3	—
Benzo-α-pyrene	Arena	9.9 ng/m^3	0.69 ng/m^3
Benzene	Room (18 smokers)	0.11 mg/m^3	—

Source: Godish, T., *Sick Buildings: Definition, Diagnosis & Mitigation*, Lewis Publishers/CRC Press, Boca Raton, FL, 1995.

in Table 11.4. From these as well as other data, tobacco smoke (where smoking is occurring) seems to be the major source of RSPs in many indoor environments.

Health hazards of MS smoke on smokers are well known. It is the major cause of lung cancer (and causes other cancers as well) and respiratory diseases such as chronic bronchitis and pulmonary emphysema. It is also a major risk factor for cardiovascular disease.

Several gas-phase (and presumably particulate-phase) constituents of MS and SS smoke are potent mucous membrane and upper respiratory system irritants in both smokers and nonsmokers. Most notable of these irritants are the aldehydes, acrolein, and HCHO. Exposure to acrolein in a smoke-filled room is believed to be the major cause of eye irritation. Nonsmoking office workers exposed to ETS have reductions in small airway function comparable to those in smokers consuming 1 to 10 cigarettes per day. Other workplace studies indicate that nonsmokers exposed to ETS experience more cough, phlegm production, eye irritation, chest colds, and days lost from work than those not exposed.

Respiratory symptoms in infants, especially significant when exposure occurs *in utero*, have been associated with parental smoking. Exposure to ETS from parental smoking is a major risk factor for asthma in children. Asthma attack incidence decreases dramatically when parents quit smoking or smoke outdoors.

Involuntary exposure to ETS seems to be a major risk factor for lung cancer. In a review of several epidemiological studies, the US EPA concluded that ETS poses a significant lung cancer risk, with an estimated 3,000 deaths per year associated with such exposure.

ETS may also increase the lung cancer risk associated with exposure to Rn. Tobacco smoke particles serve as foci for RDP deposition, suspension in air, and transport and deposition in lung tissue.

Recent epidemiological studies have indicated that the risk of cardiovascular disease mortality associated with ETS exposure constitutes a more serious public health problem than ETS-associated lung cancer. An estimated 35,000 to 40,000 deaths annually in the United States have been attributed to ischemic heart disease associated with ETS exposure.

11.3.3.3.2 Candles and Incense

Millions of homeowners and apartment dwellers burn candles daily or several times per week. Because of low flame temperatures, candles can produce significant RSPs. The effect of candle

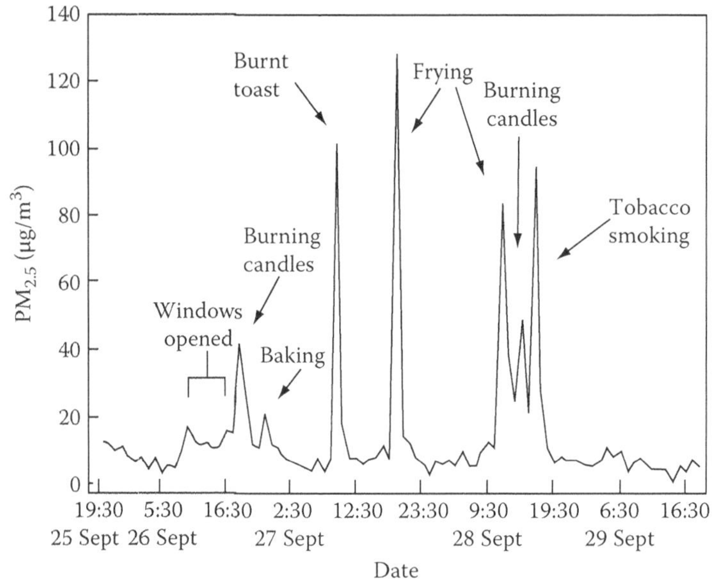

FIGURE 11.7 Effects of various indoor combustion sources and cooking activities on indoor $PM_{2.5}$ concentrations. (Courtesy of Koutraikis, P., School of Public Health, Harvard University, Boston.)

burning and other human activities on indoor PM levels can be seen in Figure 11.7. Soot particles, because of their small size and potential to contain potent carcinogens, may pose a small but unknown health risk.

In epidemiological studies evaluating risk factors for childhood leukemia, a significant increase in childhood leukemia was observed in households where incense was burned once a week or more during pregnancy, with an even greater risk when it was burned regularly. Incense is reported to produce benzo-α-pyrene, several other PAHs, and the nasal carcinogen sinapaldehyde.

The primary environmental concern allegedly caused by candle burning is the soiling of building surfaces, called ghosting. Although candle burning is known to cause soiling of interior building surfaces, ghosting has, paradoxically, been reported in residences where no candle or incense burning has occurred or, if so, at a minimal level.

11.3.3.3.3 Gas and Gasoline-Using Equipment

Several propane-fueled or gas-fueled devices and machines used indoors can cause significant indoor contamination. These include floor burnishers, forklifts, ice resurfacers, and motor vehicles operated in arenas.

Propane-fueled burnishers are used to polish vinyl flooring in small to large buildings. They produce significant emissions of CO and other combustion by-products. Employees using such devices have reportedly been exposed to concentrations of several hundred ppmv CO. In a case investigation of illness in an unventilated building (a childcare center), CO concentrations of more than 500 ppmv and carboxyhemoglobin levels as high as 26% were reported.

Propane-fueled forklifts are widely used in industry and warehouse operations. They pose exposure concerns when used in warehouse operations with attached office facilities. Carbon monoxide levels of 50 to 75 ppmv have been observed in office spaces adjoining warehouse operations.

Skating and hockey rinks use resurfacing machines that are propane or gasoline fueled. Recent reports indicate that CO concentrations in such rinks are in the range of 10 to 30 ppmv. Nitrogen dioxide concentrations are also relatively high. Short-term exposure concentrations of more than 1 ppmv are not uncommon, with peak levels of approximately 3 ppmv. Seven-day average NO_2 concentrations of approximately 225 and 360 ppbv have been reported in rinks in Europe and Japan and in the northeastern United States, respectively.

Exposures to relatively high CO and NO_2 in skating rinks may be of particular concern to those who (1) practice for extended hours; (2) exercise significantly, for example, hockey players; or (3) are full-time employees. For other users, exposures are relatively transient.

Arena events involving monster trucks, tractor pulls, and motocross events are popular in North America. Average CO concentrations in the range of 80 to 140 ppmv have been reported for such events, with peak levels of more than 225 ppmv.

11.3.3.3.4 Entrainment and Reentry

Combustion-generated contaminants from motor vehicles, nearby incinerators and boilers, residential heating appliances (e.g., fireplaces), leaf burning, and the like can enter buildings by infiltration or through open windows or air intakes. Contamination problems arise when major sources are nearby. Motor vehicle pollutants become entrained in building air systems from loading docks and stack effect phenomena.

Indoor air can also be contaminated by combustion by-products as a result of reentry. This may occur in residences when fireplace smoke is drawn indoors after it has been exhausted from a chimney. Factors that cause building depressurization, such as the use of exhaust fans and fireplaces themselves, increase reentry risks. Reentry commonly occurs in large buildings when flue gas discharges are located upwind close to outdoor air intakes.

11.3.3.3.5 Duct Liners

Duct liners are fiberglass insulating materials used to reduce (1) loss of heat and cold through supply ducts and AHUs and (2) sound transmission from supply and return air ducts and AHUs. To reduce friction and dust accumulation, airstream surfaces commonly have been coated with acrylic latex polymers.

These polymers are colorless but are both extended (increased bulk) and pigmented by addition of carbon black (~30% by weight). Because they contain unreacted double bonds, they are subject to cracking, even under low ambient O_3 levels.

Over time, the latex fragments may be blown out of the duct system as large visible fragments or smaller particles of liberated carbon black. This deterioration has recently been suggested to be the cause of the ghosting phenomenon reported in some residences in the United States. Particles are thermophoretically deposited along thermal gradients. Particle deposition is commonly observed to be more intense on cooler wall surfaces. Ghosting can be seen in Figure 11.8.

11.3.4 Aldehydes

Aldehydes are organic substances that belong to a class of compounds called carbonyls, with the carbonyl group in a terminal position. Aldehydes include several substances that differ in molecular structure, solubility, reactivity, and toxicity. Only a relatively few have industrial or commercial applications that may cause significant indoor exposures, are by-products of processes, or have biological activities that have the potential to pose major, or even minor, IAQ concerns. Those known to cause significant indoor air contamination or health effects are HCHO, acrolein, and glutaraldehyde. Most aldehydes are sensory (mucous membrane) irritants, some are skin sensitizers, whereas others may be human carcinogens.

11.3.4.1 Formaldehyde

Widespread contamination of residences and, to a much lesser extent, public access buildings, by HCHO came to the attention of public health authorities in the late 1970s and early 1980s as a result of health and odor complaints associated with urea-formaldehyde foam insulation (UFFI) and

FIGURE 11.8 Ghosting in a residential building.

mobile and conventional homes constructed or furnished with pressed wood products. Like asbestos, HCHO was a widely used industrial and commercial chemical. As a consequence, it was, and continues to be, a ubiquitous contaminant of indoor air. However, changes in product manufacture, which significantly lowered emissions, and changes in materials used in construction have resulted in significantly lower indoor levels than once were common. As such, HCHO is no longer the air quality or public health issue that it once was.

Formaldehyde is a potent mucous membrane irritant causing, on acute exposure, eye and upper respiratory system symptoms. It is also a potent dermal irritant. Upon chronic exposure, it seems to cause neurological symptoms such as headache, fatigue, disturbed sleep, and depression. It is a sensitizing substance causing allergic reactions of the skin, whole-body reactions from renal dialysis exposures, and asthma. Animal studies, as well as epidemiological studies of exposed workers and residents of mobile homes, indicate that it is a potential human carcinogen. A variety of field investigations, as well as systematic epidemiological studies, have demonstrated strong associations between HCHO exposures and health effects commonly found in residences during the early 1980s. Statistically significant relationships have been reported for residential HCHO exposures and eye irritation, runny nose, sinus congestion, phlegm production, coughing, shortness of breath, chest pain, headache, fatigue, unusual thirst, sleeping difficulty, dizziness, diarrhea, rashes, and menstrual irregularities.

11.3.4.1.1 Sources and Levels

Although HCHO has been, and continues to be, used in a variety of products, only a relatively few emit free HCHO at levels that could cause significant indoor air concentrations. Problem products include pressed wood materials such as particleboard (underlayment, paneling, cabinetry, and furniture), medium-density fiberboard (furniture and cabinetry), hardwood plywood (paneling, cabinetry), UFFI, and acid-catalyzed finishes (applied to furniture and cabinetry).

Urea-formaldehyde resins are chemically unstable. They release HCHO from the volatilizable, unreacted HCHO in the resin and from hydrolytic decomposition of the resin copolymer itself. Release of the former was responsible for the initially high HCHO levels associated with new mobile homes, conventional homes with particleboard underlayment, and homes that had been recently insulated with UFFI.

Formaldehyde-emitting products differ in emission potentials. As a consequence, indoor HCHO levels reflect the nature and potency of sources. Indoor levels are also affected by the quantity of HCHO-emitting material present. Buildings with high load factors (surface area of HCHO-emitting products/air volume ratio), such as mobile homes, experienced the highest reported HCHO levels.

Most research on HCHO levels in residences was conducted in mobile homes, UFFI houses, and conventional houses with a variety of HCHO sources in the late 1970s and early to mid-1980s. As a consequence, data on HCHO levels reflect that period, which can be described as historically high for indoor HCHO.

The highest levels of HCHO were reported for mobile or manufactured homes. Concentrations varied as a function of mobile home age, environmental factors at the time of testing, and geographical area. Concentrations varied from 0.05 to more than 1 ppmv, with average concentrations in the range of 0.09 to 0.18 ppmv. In general, concentrations of HCHO in mobile homes manufactured today are much lower. Some new mobile homes may have initial HCHO levels as high as 0.15 ppmv, but most are less than 0.10 ppmv. Residences with particleboard underlayment had HCHO concentrations that, at their extreme, exceeded 0.50 ppmv, but were more typically in the range of 0.05 to 0.20 ppmv. Particleboard underlayment in conventional stick-built houses today is uncommon, as it has been replaced by oriented-strand board, a material with very low HCHO emissions.

Residences with UFFI initially experienced HCHO levels of less than or equal to 0.40 ppmv, but these decreased rapidly to a range of 0.03 to 0.12, with an average of 0.06 ppmv 1 or 2 years after application. Although still widely used in Great Britain, very few US houses are newly insulated with UFFI today. Residences with one or more miscellaneous sources (e.g., pressed wood furniture, hardwood plywood, and pressed wood cabinets) had HCHO concentrations that ranged from 0.03 to 0.15 ppmv, with average concentrations of 0.03 to 0.05 ppmv.

Formaldehyde concentrations decrease significantly with time. As a consequence, the highest concentrations and exposures occur in building environments in which HCHO-emitting products are new.

11.3.4.1.2 Guideline Values and Their Adequacy

In response to building-related problems, several countries adopted air quality guidelines for HCHO exposures in residences. Countries with guideline values of 0.10 to 0.12 ppmv include Australia, Germany, Denmark, the Netherlands, and Italy. Canada and California have adopted guideline values of 0.10 ppmv as an action level (in need of remediation), with a target level of 0.05 ppmv. The WHO recommends a guideline value of 0.082 ppmv.

Compliance with these guideline values may not be sufficient to protect the health of sensitive populations, for example, asthmatic children. Studies conducted in Arizona have shown significant pulmonary function deficits in asthmatic children with average household exposures of 0.06 ppmv. Other studies in Canada observed dose–response relationships between HCHO levels and eight symptoms at average exposure concentrations of 0.045 ppmv. Studies in Australia have shown that HCHO exposures higher than 0.05 ppmv result in increased breath levels of NO, a primary indicator of inflammatory responses.

11.3.4.2 Acrolein

Acrolein is an aldehyde with one carbon–carbon double bond, which makes it highly reactive. It is released into the environment as a combustion oxidation product from oils and fats, wood, tobacco, and automobile or diesel fuels and produced in ambient air as a result of photochemical reactions.

Acrolein is a potent eye irritant causing tearing at relatively low (< 0.10 ppmv) exposure concentrations. It is believed to be the primary cause of eye irritation in smoke-filled rooms and is one of the major eye irritants in photochemical smog or haze.

11.3.4.3 Glutaraldehyde

Glutaraldehyde is a dialdehyde. It is widely used as a biocidal ingredient in medical and dental disinfectants and not uncommonly in disinfectant applications such as duct cleaning. It is also reported to have been used in carbonless copy paper.

Glutaraldehyde, although relatively nonvolatile, is a potent mucous membrane irritant. Irritant effects include nose, throat, and sinus symptoms. Other symptoms associated with exposures include nausea, headache, chest tightening, and asthma.

11.3.5 Volatile and Semivolatile Organic Compounds

11.3.5.1 Volatile Organic Compounds

A large variety of natural and synthetic organic compounds are found in indoor air. These include compounds that have boiling points ranging from (1) less than 0°C to 50°C to 100°C (< 32°F to 122°F–212°F; very volatile organic compounds [VVOCs]), to (2) 50°C to 100°C to 240°C to 260°C (464°F–500°F; VOCs), to (3) 240°C to 260°C to 380°C to 400°C (716°F–752°F; SVOCs). Because of their variety, relative abundance in indoor air, potential to cause sensory irritation and central nervous system symptoms, and vapor-phase presence, VOCs have received significant attention as potential causal factors in building-related illness symptoms, particularly among office workers. In the office context, VOCs also include the more volatile VVOCs.

VOCs include aliphatic hydrocarbons (HCs), which may be straight, branched chained, or cyclic; aromatic HCs; halogenated HCs (primarily by chlorine); and oxygenated HCs such as aldehydes, alcohols, ketones, esters, ethers, and acids. They are emitted by a variety of sources, including building materials and furnishings, consumer products, building maintenance materials, humans, office equipment, and tobacco smoking.

Studies in Danish residences and office buildings have reported the presence of 40 different VOCs, consisting mainly of C_2–C_{13} alkanes, C_6–C_{10} alkylbenzenes, and terpenes. In a study of 42 commonly used building materials, Danish investigators identified emissions of, on average, 22 different compounds associated with each material. The 10 compounds with the highest steady-state concentrations were toluene, *n*-xylene, terpene, *n*-butylacetate, *n*-butanol, *n*-hexane, *o*-xylene, ethoxyethyl-acetate, *n*-heptane, and *p*-xylene. Organic compounds identified in 40 East Tennessee homes included toluene, ethyl benzene, *n*-xylene, *p*-xylene, nonane, cumene, benz-aldehyde, mesitylene, decane, limonene, undecane, naphthalene, dodecane, tridecane, tetradecane, pentadecane, hexadecane, and 2,2-methlynaphthalene.

Concentrations of individual VOCs measured in indoor air are typically several orders of magnitude lower than occupational threshold limit values or OSHA permissible exposure levels. As a consequence, it is not likely that individual VOCs (other than HCHO) are responsible for building-related health complaints.

A potential causal relationship between sick building–type symptoms and exposures to a combination of VOCs at relatively low individual concentrations has been proposed by Danish investigators and supported by human exposure studies. It is described as the total volatile organic compound (TVOC) theory. In the TVOC theory, sensory irritation and possibly neurological symptoms (headache and fatigue) in problem buildings may be due to the combined effect of the many VOCs found in indoor air.

Effects have been reported on exposure to a mixture of 22 different compounds at concentrations as low as 1 ppmv (5 mg/m^3) TVOC (toluene equivalent) as well as submixtures of these compounds. A proposed dose–response relationship between TVOC exposures and discomfort and health effects is indicated in Table 11.5.

Several investigations have been conducted to determine whether building-related illness symptoms are associated with exposures to TVOCs. Although no such relationships were observed in Danish town hall and California office building studies, significant correlations were observed between the logarithmic value of TVOC concentrations and illness symptoms in problem Swedish school buildings (Figure 11.9). Significant loglinear relationships were also observed between symptom prevalence and concentrations of terpenes, C_8–C_{11} *n*-alkanes, and butanols in problem Swedish office buildings.

Several VOCs found in indoor air are mutagenic or carcinogenic (e.g., benzene, styrene, tetrachloroethylene, 1, 1, 1-trichloroethane, trichloroethylene, dichlorobenzene, methylene chloride, and chloroform). This suggests that VOC exposures in indoor air have the potential (albeit at very low risk) to cause cancer in humans. The nature of potential health risks associated with multiple potential carcinogens at very low exposure levels is not known.

TABLE 11.5
Dose–Response Model Relationships between TVOC Exposures and Discomfort/Health Effects

TVOC Concentration (mg/m³)	Response	Exposure Range
< 0.20	No effects	Comfort range
0.20–3.0	Irritation/discomfort possible	Multifactorial exposure range
3.0–25.0	Irritation and discomfort; headache possible	Discomfort range
> 25.0	Neurotoxic effects	Toxic range

Source: Molhave, L., *Proc. 5th Int. Conf. Indoor Air Qual. Climate*, Toronto, 5, 15, 1990.

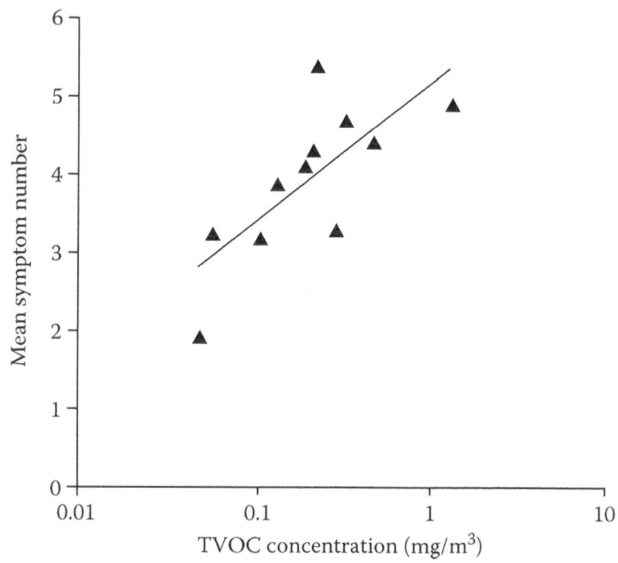

FIGURE 11.9 Dose–response relationships between building TVOC concentrations and symptom prevalence. (From Norback, D. et al., *Scand. J. Work Environ. Health*, 16, 121, 1990. With permission.)

11.3.5.2 Semivolatile Organic Compounds

SVOCs have low vapor pressures (10^{-1} to 10^{-7} mmHg) and exist in gaseous and condensed phases (adsorbed on particles). They include a variety of compound types, including solvents; long-chained alkanes, aldehydes, and acids; pesticides; PAHs; polychlorinated biphenyls (PCBs); and plasticizers. The concentrations of selected SVOCs identified in Danish buildings are summarized in Table 11.6. Many of these are plasticizers, present in relatively low concentrations.

11.3.5.2.1 Plasticizers

Plasticizer compounds such as phthalic acid esters are ubiquitous contaminants of indoor air and dust. They are used in vinyl products to make them soft and relatively flexible. They typically comprise 25% to 50% of the weight of resilient vinyl floor covering. As such materials lose their plasticizers, they become hard and brittle and subsequently crack.

Some scientists have suggested that exposures to phthalic acid ester plasticizers may be partly responsible for the significant increase in asthma observed in developed countries over the past two

TABLE 11.6
Concentrations of Selected SVOCs Measured in Four Danish Office Buildings, a Daycare Center, and School Classroom (ng/m³)

Compound	Office 1	Office 2	Office 3	Office 4	Day Care Center	School Classroom
Tridecane	1,128	890	1,070	980	1,144	4,599
Texanol	2,055	4,378	1,910	3,871	5,492	1,451
Pentadecane	1,003	863	1,173	1,226	1,306	1,488
Dodecanoic acid	1,144	987	2,174	2,146	3,040	2,987
TXIB	957	710	658	1,282	8,341	7,803
Heptadecane	993	718	1,039	965	952	692
Diisobutyl phthalate	1,393	977	891	578	1,240	1,279
Dibutyl phthalate	789	710	1,078	840	1,346	1,195
Di-2-ethylhexyl phthalate	201	278	241	404	1,053	111

Note: TXIB = 2, 2, 4-trimethyl-1, 3-pentadioldiisobutyrate.
Source: Clausen, P. A., Wolkoff, P., and Svensmark, B., *Proc. 8th Int. Conf. Indoor Air Qual. Climate*, Edinburgh, 2, 434, 1999.

decades. An association between the presence of plastic interior surfaces and bronchial obstruction in children has been reported. It has been suggested that the structural similarity between plasticizer compounds and prostaglandins (which mediate inflammatory responses) may be a contributing factor to asthma initiation.

11.3.5.2.2 Polychlorinated Biphenyls

Polychlorinated biphenyls are SVOCs that may pose indoor air exposure and health risks. The primary source of PCBs in indoor air and building surfaces seems to be ballasts in old fluorescent lighting fixtures. Such contamination is particularly significant when a building has experienced a structural fire.

11.3.5.2.3 Pesticides

Pesticides are toxic compounds formulated to control a variety of pest species. Indoor use includes a number of applications, for example, (1) biocides used by manufacturers to prevent biodeterioration of products and by homeowners or tenants to disinfect surfaces such as toilet bowls, and others; (2) insecticides to control cockroaches, flies, ants, spiders, and moths; (3) termiticides; (4) flea and tick sprays and pet shampoos; (5) insecticides for indoor and outdoor plants; and (6) fungicides and herbicides used as lawn and garden treatments (the latter two may enter the home on shoes and clothing).

Because of the deliberate use of toxic substances indoors and their passive transport on shoes and subsequent indoor accumulation, pesticides are unique contaminants of indoor environments. Users have only a limited understanding of the potential toxic hazard to which they and other building occupants may be exposed.

Biocides are used to control the growth of microorganisms such as bacteria and fungi. They are used as disinfectants in cleaning agents, on AHU filters and carpeting, and in paints and duct cleaning. Commonly used biocides include hypochlorites, alcohols, *o*-phenylphenol, and quaternary ammonium compounds. Of these, *o*-phenylphenol has been one of the most common pesticide compounds observed in indoor air and dust samples. Mercury (Hg) is also commonly found, as it was widely used as a biocide in latex-based paint until its use was discontinued in a voluntary agreement between manufacturers and the US EPA in 1990.

In general, fungicides are not used indoors. However, pentachlorophenol (PCP), a pesticide used to prevent wood decay in outdoor timbers, is one of the most common pesticides found in human blood samples. The source of this widespread exposure is unknown. Exposures to PCP have been associated with PCP-treated logs in log houses. Because it is a teratogen and potential human carcinogen, its use has been reduced.

Insects represent major pest control concerns in buildings. Active ingredients in insecticidal formulations used in or around buildings have included diazinon, *p*-dichlorobenzene, chlordane, heptachlor, lindane, aldrin, dieldrin, bendiocarb, methoxychlor, propoxur, methyl demeton, naphthalene, and pyrethrins. Paradichlorobenzene is the active ingredient in mothballs, and naphthalene in moth flakes. Major insecticidal classes include organochlorines, organophosphates, carbamates, and pyrethrins.

Organochlorines, or chlorinated hydrocarbons, such as chlordane, heptachlor, and *p*-dichlorobenzene, have been used indoors—chlordane and heptachlor as termiticides, and *p*-dichlorobenzene as a moth control agent. Despite the fact that use of chlordane and heptachlor has been restricted (not used as termiticidal applications in the United States for > 15 years), these two substances are the most common and abundant pesticidal chemicals found in residential air and dust samples. This is due to their long environmental lifetimes, which are extended indoors because of limited exposure to ultraviolet light. Paradichlorobenzene is also one of the most common and abundant pesticides found indoors. The primary health concern associated with these compounds, and the reason for restrictions on their use, is their potential human carcinogenicity. Organochlorine exposure risks are increased by the fact that they tend to bioaccumulate in fat tissue.

Organophosphates such as dichlorvos, chlorpyrifos, diazinon, and malathion are characterized by their acute neurotoxicity. Dichlorvos has been used in bug bombs, slow-release pest strips, and dog and cat flea collars. Because of health concerns (animal carcinogenicity), their use has dramatically decreased since 1988. Chlorpyrifos was once widely used indoors for most termiticidal treatments and to control cockroaches and fleas. Because of potential exposures to children, its use was phased out in 2000 in a US EPA–industry agreement. Use of diazinon has been restricted for similar reasons.

Because of their lower toxicity and health risks, pyrethrins and their synthetic cousins, pyrethroids, are increasingly being used to control insects indoors. Although persistent, they are rarely observed in the gas phase.

The herbicide 2, 4-d (widely used on lawns to control broad-leaved species) is commonly found in indoor dust samples, apparently due to track-in phenomena. Exposure to 2, 4-d has been suggested to be a risk factor for non-Hodgkin's lymphoma.

Pesticides pose a variety of indoor exposure and health concerns. These include acute symptoms associated with misapplication, their persistence in house dust, bioaccumulation of organochlorine compounds in human fat tissue, suspected human carcinogenicity of several pesticides, children's exposures to house dust while at play and from pets' flea collars, and potential immunological effects.

11.3.6 Biological Contaminants

It is becoming increasingly evident that exposure to particulate-phase and, to a lesser extent, gas-phase biological contaminants is responsible for a significant percentage of illness symptoms and disease reported to be associated with building environments. These contaminants include airborne mold (fungi), bacteria, and viruses that may be viable (living) or nonviable; fragments of bacteria and mold that may contain antigens/allergens, endotoxins, glucans, or mycotoxins; microbially produced VOCs (MVOCs); and antigens/allergens produced by insects, arachnids (mites/spiders), and pets.

Biological contaminants such as bacteria, viruses, and fungi can cause infectious disease as a result of airborne transmission. Exposures to mold spores or fragments and other allergenic substances produced by insects, arachnids, and pets can cause immunological sensitization, which may result in allergic rhinitis and, in more severe cases, asthma. Exposures to bacterial cells or mold spores may cause hypersensitivity pneumonitis. Respiratory system inflammatory responses may be produced because of exposures to bacterial endotoxins, fungal glucans, or MVOCs. Mycotoxin exposures may, depending on the substance, produce a variety of adverse physiological responses.

Of particular public health concern are those biological contaminants that cause major illness syndromes such as chronic allergic rhinitis, asthma, hypersensitivity lung disease, Legionnaires' disease, and TB.

11.3.6.1 Illness Syndromes

11.3.6.1.1 Chronic Allergic Rhinitis

Chronic allergic rhinitis, or common allergy, is a minor illness characterized by inflammatory symptoms of the upper respiratory system, including runny nose, congestion, phlegm production, coughing, and sneezing. These inflammatory responses in sensitized individuals are caused by exposure to substances that have antigenic properties. In indoor environments, major sources of allergens include dust mite and cockroach fecal waste, mold spores and hyphal fragments, and pet danders.

Approximately 40% of the US population has been sensitized to common allergens, with about half this number experiencing clinical symptoms. Approximately 10% of the US population (~25 million people) has experienced chronic allergic rhinitis severe enough to be diagnosed and treated by a physician.

11.3.6.1.2 Asthma

Asthma is an ailment that affects the respiratory airways. It is characterized by episodes of severe bronchial constriction resulting in chest tightness, shortness of breath, coughing, and wheezing.

Asthma prevalence in the United States has been estimated to be 4.3%; that is, approximately 12 million individuals a year experience mild to severe symptoms. Its prevalence is much higher in children, with rates of 8.3% in midwestern children, 11.8% in southern children, and 13.4% in black children. The prevalence rate has increased 50% since 1980, with the annual death rate doubling to approximately 5,000/year.

Asthma is a major medical cause of absenteeism in school-age children and the leading cause of childhood hospital admission and long-term medication use. Approximately 1% of all health care costs are reportedly associated with the treatment of asthmatic symptoms.

The primary cause of asthma is exposure to, and sensitization by, inhalant allergens associated with dust mites, mold, pet danders, or cockroaches.

11.3.6.1.3 Inflammatory Lung Disease

Inflammatory lung diseases resulting from indoor exposures include hypersensitivity pneumonitis and humidifier fever.

Hypersensitivity pneumonitis, or extrinsic allergic alveolitis, is a group of immunologically mediated lung ailments associated with repeated exposures to small-diameter biological aerosols that affect alveoli (site of gas exchange).

Classic symptoms occur after immunological sensitization is induced. They may include fever, chills, nonproductive cough, shortness of breath, myalgia (ache all over), and malaise. Onset occurs within hours of exposure, with spontaneous recovery in 18 to 24 hours. In office and industrial environments, symptoms diminish as the workweek progresses and increase in severity at the beginning of a new week. Chronic exposures cause irreversible lung tissue damage, progressive shortness of breath, coughing, malaise, weakness, and weight loss.

Outbreaks of hypersensitivity pneumonitis in buildings have been associated with contaminated components of heating, ventilating, and air-conditioning (HVAC) systems (including condensate drip pans and ductwork) and a variety of mold-infested materials. Hypersensitivity pneumonitis seems to be caused by exposures to very high air concentrations of bacteria (particularly *Actinomycetes*) or small-diameter mold spores (e.g., those of *Penicillium* and *Aspergillus*).

Humidifier fever is a respiratory ailment not easily distinguished from hypersensitivity pneumonitis. Clinical manifestations include fever, chest tightness, increased respiratory rate, airway hyperreactivity, and bronchitis. Symptoms are produced as a result of inflammatory responses of lung tissue upon exposure to bacterial endotoxins. Gram-negative bacterial contamination of cool-mist humidifiers (commonly used in European office buildings) in mechanically ventilated buildings is believed to be the cause of humidifier fever.

11.3.6.1.4 Infectious Diseases

Several infectious diseases are transmitted in building environments as a result of inhalation exposures. These include viral diseases such as colds and influenza, bacterial diseases such as TB and Legionnaires' disease, and fungal diseases such as aspergillosis. Of major IAQ concern are TB and Legionnaires' disease.

Until the early twentieth century, TB was one of the major causes of death in the United States and in many other countries. As a result of advances in diagnosis, prevention, and treatment, the prevalence of TB dramatically declined. It has experienced a small resurgence as a result of an influx of immigrants from countries where it is endemic, increases in susceptible populations (e.g., those immunocompromised), and the development of antibiotic resistance by the causal bacterium.

Recent outbreaks of TB have occurred in hospitals, nursing homes, homeless shelters, correctional facilities, and residential AIDS care centers. The risk of exposure and subsequent infection among health care workers may be high in health care facilities treating patients with undiagnosed TB.

Legionnaires' disease has been reported in a number of indoor environments. The disease syndrome is characterized by coughing, fever, shortness of breath, chest pain, myalgia, diarrhea, and confusion. It is a progressive pneumonia caused by infection with *Legionella pneumophila* or similar species, with a 15% fatality rate.

Legionnaires' disease may occur as outbreaks (10% of reported cases) or, more commonly, as individual cases.

The causal organism grows well at elevated temperatures (~35°C, 95°F), producing abundant populations in waters of cooling towers and evaporative condensers, hot water heaters, whirlpools, spas, and hot tubs. Exposure results from inhalation of aerosols associated with these sources.

Infections by *Aspergillus fumigatus*, a common mold species found in dust and compost, is a major infection control issue in hospitals where opportunistic infections are a concern in emergency, oncology, and AIDS treatment wards. Aspergillosis is a fungal pneumonia that does not respond to conventional antibiotic therapy.

11.3.6.2 Other Health Concerns

A variety of new health concerns associated with exposures to biological contaminants have developed over the past decade because of increased scientific understanding of both exposures and potential health risks. These include exposures and health responses to bacterial endotoxins, fungal glucans, and mycotoxins.

11.3.6.2.1 Endotoxins

Endotoxins are large molecules produced in the outer membrane of gram-negative bacteria and released into the environment upon the death of cells.

They are common contaminants of settled dusts in indoor environments. Upon respiratory tissue exposure, they induce inflammatory responses, producing a number of common symptoms. They are believed to cause some of the symptoms in building occupants reportedly due to "sick building syndrome" (SBS; see Section 11.4.2) and inflammatory symptoms among children in daycare centers. There is scientific speculation that early childhood endotoxin exposures may subsequently protect individuals from immune system diseases such as chronic allergic rhinitis and asthma. As indicated previously, endotoxin exposure is believed to be the cause of humidifier fever.

11.3.6.2.2 Fungal Glucans

Fungal glucans are large molecules found in the cell walls of most fungal species. They are common in the settled dusts of buildings. Although they can also cause inflammatory responses in human respiratory systems, their potential for causing adverse health effects has not been well documented.

FIGURE 11.10 *Stachybotrys* growing on water-damaged gypsum board.

11.3.6.2.3 Mycotoxins

Mycotoxins are large molecules produced by many fungal species when they are subject to nutrient stress. Of particular note are mycotoxins produced by *Aspergillus* species and the black mold, *Stachybotrys chartarum*. *Aspergillus flavus* produces the liver toxins and human carcinogens aflatoxin A and B, while *S. chartarum* produces the highly toxic trichothecenes saratoxin H and G and verucarin A and B.

Historically, mycotoxin exposures and poisoning have been associated with contaminated livestock feed and human foodstuffs. Recently, however, concerns have been expressed in relation to inhalation exposures to mold spores containing highly toxic mycotoxins. This is particularly the case with *S. chartarum*, widely found in building environments infesting the face paper of gypsum board, ceiling tiles, and other processed wood fiber materials subject to repeated wetting (Figure 11.10).

Building investigations conducted by the US Centers for Disease Control (CDC) in the 1990s linked potential *S. chartarum* mycotoxin exposure to hemosiderosis, a frequently fatal pulmonary hemorrhagic disease of infants. Animal exposure studies show that similar lung hemorrhaging can be induced by exposures to *S. chartarum* spores. However, in 2001, after conducting both internal and external reviews of the data, the CDC concluded that their studies were insufficiently definitive to show a causal relationship between hemosiderosis and exposure to *S. chartarum*. Consequently, the CDC has advised homeowners that for remediation purposes, *S. chartarum* should be treated like other molds. Despite this advice, most *S. chartarum* remediations are conducted using asbestos-type remediation procedures.

11.4 PROBLEM BUILDINGS

Buildings subject to health, comfort, or odor complaints can be best described as problem buildings. They comprise a building population in which some form of dissatisfaction with air quality has been expressed by occupants sufficient to require an IAQ investigation. In many cases, the cause or causes of complaints can be readily identified and remediated. In other cases, there seems to be no single causal factor and mitigation efforts prove ineffective. When a building includes a relatively large percentage of occupants with complaints, it may be described as a sick building.

11.4.1 Building-Related Illness

When the causal factors for health complaints have been identified, the air quality problem is described as building-related illness (BRI). BRI is characterized by a unique set of symptoms accompanied by clinical signs, laboratory test results, and specific pollutants. Included in BRI are nosocomial (hospital-acquired) infections, hypersensitivity diseases (e.g., common allergy, asthma, hypersensitivity pneumonitis, humidifier fever), fiberglass dermatitis, and toxic effects associated with contaminants such as CO, ammonia (NH_3), and HCHO.

11.4.2 Sick Building Syndrome

SBS is used to describe a spectrum of nonspecific symptoms that, on investigation, seem to have no identifiable cause. These commonly include mucous membrane symptoms (irritation of the eyes, nose, throat, sinuses); general symptoms (headache, fatigue, lassitude); skin irritation (dryness, rashes); and, to a lesser extent, respiratory symptoms (coughing, shortness of breath).

11.4.3 Field Investigations and Systematic Building Studies

An apparent relationship between illness complaints and building environments has been reported in many problem building investigations conducted by governmental, occupational, and public health agencies, and private consultants responding to occupant or building management requests. In the United States, more than 1,000 such investigations have been conducted by the National Institute of Occupational Health and Safety (NIOSH) health hazard evaluation teams. Symptoms reported have included eye, nose, throat, and skin irritation; headache and fatigue; respiratory symptoms such as sinus congestion, sneezing, coughing, and shortness of breath; and, less frequently, nausea and dizziness. The reported potential causal and contributory factors to complaints for 529 investigations conducted between 1971 and 1988 are summarized in Table 11.7.

Field investigations have had an important role in initially identifying and defining problem building phenomena. However, because of inherent bias involved in building studies subject to occupant health and comfort complaints and the relatively unsystematic approaches used in many investigations, they are of limited scientific usefulness. Therefore, investigators in several countries have attempted to conduct systematic studies of noncompliant office buildings. These have included 47 office buildings in the United Kingdom, 14 town halls in Denmark, 60 office buildings in the Netherlands, a large office building study in northern Sweden, and 12 buildings in California. Significant prevalence rates of mucous membrane, general (headache, fatigue, lassitude), and skin symptoms were observed in these noncompliant buildings, with complaint rates varying from building to building. Symptom prevalence rates for both males and females in 14 noncompliant Danish town halls are summarized

TABLE 11.7
Problem Types Identified in NIOSH Building Investigations

Problem Type	Number of Buildings Investigated	%
Contamination from indoor sources	80	15
Contamination from outdoor sources	53	10
Building fabric	21	4
Microbial contamination	27	5
Inadequate ventilation	280	53
Unknown	68	13
Total	529	100

Source: Seitz, T.A., *Proceedings Indoor Air Quality International Symposium: The Practitioner's Approach to Indoor Air Quality Investigations*, American Industrial Hygiene Association, Akron, OH, 1989. With permission.

TABLE 11.8
Symptom Prevalence Rates (%) among Occupants of 14 Danish Town Hall Buildings

	Males (n = 1,093–1,115)	Females (n = 2,280–2,345)
Symptoms		
Eye irritation	8.0	15.1
Nasal irritation	12.0	20.0
Blocked, runny nose	4.7	8.3
Throat irritation	10.9	17.9
Sore throat	1.9	2.5
Dry skin	3.6	7.5
Rash	1.2	1.6
Headache	13.0	22.9
Fatigue	20.9	30.8
Malaise	4.9	9.2
Irritability	5.4	6.3
Lack of concentration	3.7	4.7
Symptom Groups (Males and Females)		
Mucous membrane irritation	20.3	
Skin reactions	4.2	
General symptoms	26.1	
Irritability	7.9	

Source: Skov, P. and Valbjorn, O., *Environ. Int.*, 13, 349, 1987. With permission.

in Table 11.8. As a comparison, symptom prevalence rates for 11 Swedish sick (complaint) buildings are summarized in Table 11.9. Note that prevalence rates were relatively high in both cases, with the latter being much higher.

11.4.4 Risk Factors

Investigators conducting systematic building studies have attempted to identify potential risk and contributory factors to building-related and work-related health complaints. These include people-related, work or building environment-related, and contaminant-related concerns.

A variety of people-related risk factors are associated with symptom prevalence. These are gender, allergy, a variety of psychosocial factors, and tobacco smoking. All studies have shown that symptom-reporting rates are several times higher among females than males. Although most studies report self-diagnosed allergy as a risk factor for SBS-type symptoms, objective immunological studies have failed to show such a relationship. Symptom prevalence is also significantly associated with psychosocial variables such as work stress and job satisfaction. It increases with increased workplace stress and decreases with increased job satisfaction. Nonsmoker exposure to ETS increases symptom prevalence rates.

A variety of environmental factors (temperature, humidity, air movement, etc.) have been investigated as potential risk factors for SBS symptoms. Significant relationships have been reported for elevated temperature (21°C–25°C, 70°F–78°F) and low relative humidity (< 30% RH). Low ventilation rates (< 10 L/s per person, < 20 cubic feet per minute [CFM]/person) seem to be a risk factor for SBS-type symptoms.

Various office materials and equipment have been implicated as contributory factors to work-related or building-related health complaints. These include (1) handling carbonless copy paper and

TABLE 11.9
Symptom Prevalence among Occupants of 11 "Sick" Swedish Office Buildings

Symptom	Total Mean Prevalence (%)	Range
Eye irritation	36	13–67
Swollen eyelids	13	0–32
Nasal catarrh	21	7–46
Nasal congestion	33	12–54
Throat dryness	38	13–64
Sore throat	18	8–36
Irritative cough	15	6–27
Headache	36	19–60
Abnormal tiredness	49	19–92
Sensation of getting a cold	42	23–77
Nausea	8	0–23
Facial itch	12	0–31
Facial rash	14	0–38
Itching on hands	12	5–31
Rashes on hands	8	0–23
Eczema	15	5–26

Source: Norback, D., Michel, I., and Winstrom, J., *Scand. J. Work Environ. Health*, 16, 121, 1990. With permission.

other papers; (2) working with office copy machines such as wet process and electrostatic photocopiers, laser printers, diazo-copiers, microfilm copiers, and spirit duplicators; and (3) using video display terminals.

A variety of gas-phase and particulate-phase substances have been evaluated as potential risk factors for occupant health complaints. These have included bioeffluents (e.g., CO_2), HCHO, VOCs, and dust. There is currently little evidence to implicate bioeffluents as a cause of health complaints. Formaldehyde levels in nonresidential, nonindustrial buildings are relatively low and are not considered sufficient to cause building-related health complaints. TVOCs have been proposed as a causal factor for SBS-type symptoms, and airborne and settled dust has been implicated as a risk factor; most notable for the latter has been an association with the macromolecular or organic fraction of floor dust.

11.5 INVESTIGATING INDOOR AIR QUALITY PROBLEMS

As indicated in previous sections, several significant IAQ problems exist in buildings. Except in cases of severe, acute illness, most go unrecognized. The need to conduct an investigation to identify the nature of a suspected problem only occurs when occupant complaints are sufficiently intense to convince building management that the air quality needs to be investigated.

The responsibility for investigating building-related health, and sometimes odor, complaints in nonresidential buildings falls to local, state, and federal public or occupational health agencies or private consultants. A variety of protocols are used to conduct such investigations.

The US EPA, in cooperation with NIOSH, has developed a model protocol for conducting health and comfort investigations in public access buildings. Although designed for in-house personnel, it can be readily adapted for use by public agencies and private consultants.

The US EPA/NIOSH investigative protocol is characterized by systematic information gathering and hypothesis testing (Figure 11.11). It includes an initial walkthrough of problem areas to gather information on building occupants, ventilation systems, contaminant sources, and pollutant pathways. Additional information gathering is recommended if the walkthrough does not identify the cause or causes of the reported problems. Information gathering is intended to provide a basis for developing a hypothesis as to what may be the cause or causes. Because many complaints occur in poorly ventilated buildings and spaces, it is desirable to inspect and evaluate the components of ventilation systems that may serve problem areas and other parts of the building.

This investigative protocol gives limited attention to air sampling, suggesting that such sampling may not be required to solve most problems and may even be misleading. Air sampling is only recommended after all other investigative activities have been employed. Exceptions include CO_2 to determine ventilation adequacy and temperature, humidity, and air movement for thermal comfort.

Relatively speaking, the diagnosis of an IAQ problem in residences is a simple undertaking. Residences are small, contain few individuals, and have simple mechanical systems. Additionally, a few contaminants are responsible for most residential health and odor complaints, and causal connections are easily made. Commonly encountered IAQ problems in residences include exposure to allergens such as mold, dust mite excreta, cockroach excreta and body parts, and animal dander; HCHO that causes irritant-type symptoms; CO that causes central nervous system symptoms; and a variety of other problems, including sewer gas odors, misapplied pesticides, and others. Diagnosis involves information gathering on occupant symptoms (type and onset patterns) and assessment of the building for possible sources. It may also include air testing or collection of surface samples to confirm a hypothesis of what the potential causal factor or factors may be.

11.6 MEASUREMENT OF INDOOR CONTAMINANTS

Pollutant and contaminant measurements are conducted in most IAQ investigations. In addition to sampling airborne concentrations of gases, vapors, particulate-phase substances, and biological contaminants, other environmental samples may be collected. These include surface dust sampling using wipe or vacuum sampling techniques, surface sampling for mold, and bulk sampling for asbestos or mold.

Commonly measured contaminants include CO_2 (to determine ventilation adequacy), CO, HCHO, VOCs, and mold and its by-products. Measurements are often used as a screening tool to determine whether pollutant levels are within acceptable guideline values.

Although in most cases pollutant measurements are made as a screening tool, they are best conducted to confirm a hypothesis that a substance(s) or organism(s) is causally associated with reported illness symptoms. This is especially the case with exposures to CO and biological contaminants such as mold spores.

11.7 PREVENTION AND CONTROL MEASURES

A variety of approaches are available to prevent IAQ problems before they occur and to control and mitigate them once they are manifested. These can be described as source management, exposure control, and contaminant control. Source management includes source exclusion, removal, and treatment. Exposure control focuses on scheduling activities to reduce occupant exposures. Contaminant control focuses on reducing airborne contaminant levels, usually achieved by ventilation or air cleaning.

11.7.1 Source Management

11.7.1.1 Exclusion

In new building construction and remodeling, IAQ problems can be reduced or minimized with the use of low-emission products, which are becoming increasingly available. The carpeting and mastic products that are currently being produced and sold have relatively low emissions of VOCs and

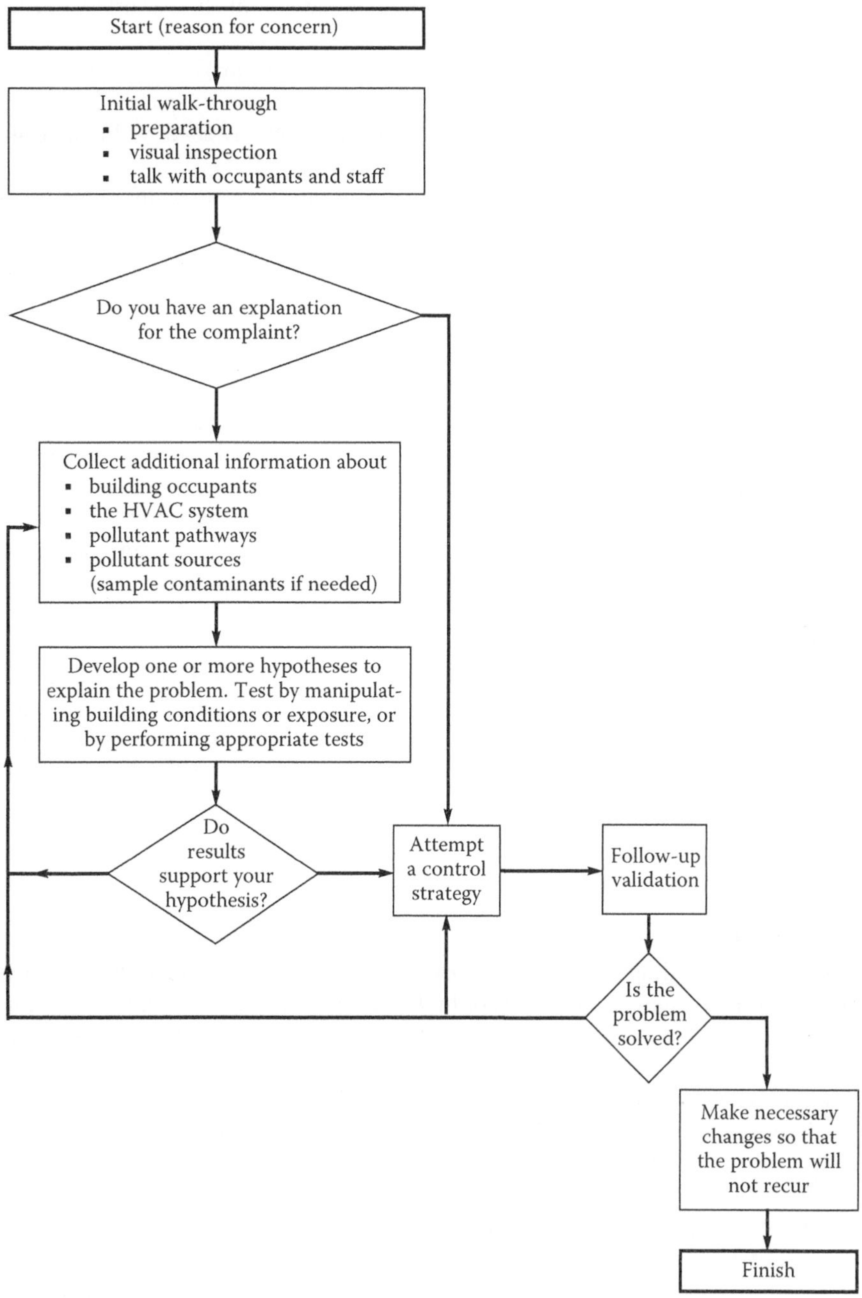

FIGURE 11.11 US EPA/NIOSH building investigation protocol. (From US EPA/NIOSH, EPA/400/1-91/003, DDHS Publication 91–114, EPA, Washington, DC, 1991.)

the odor-producing substance called 4-phenylcyclohexane (4-PC). Formaldehyde emissions from pressed-wood products are much lower than they once were as lower-emission products have been developed. Low levels of HCHO can be attained and maintained by avoiding the use of HCHO-emitting products such as particleboard, hardwood plywood paneling, medium-density fiberboard, and acid-cured finishes. Alternative products include softwood plywood, oriented-strand board, decorative gypsum board and hardboard panels, and, in the case of finishes, a variety of HCHO-free varnishes and lacquers.

Contamination of indoor air with combustion by-products can be avoided by using electric rather than gas appliances for cooking, using only electric or vented gas or oil space heating appliances, and restricting smoking. Indoor pesticide use can also be largely avoided by implementing integrated pest management practices.

High Rn levels can be avoided by selecting homesites that have soils with a high clay content. However, such soils tend to have moisture problems that increase health risks associated with mold and dust mites.

11.7.1.2 Source Removal

The single most effective measure to reduce contaminant levels is to identify and remove the source. In the case of HCHO control, its effectiveness depends on correct identification of the most potent source or sources in a multiple source environment. Because of interaction effects, HCHO levels are determined by the most potent source or sources present; removal of minor sources would not reduce HCHO levels.

In buildings with mold infestation, mitigation of respiratory health problems, for example, chronic allergic rhinitis, asthma, and hypersensitivity pneumonitis, will, in many cases, require the removal and replacement of infested construction materials and furnishings.

11.7.1.3 Source Treatment

Sources may be treated or modified to reduce contaminant emissions. One can, for example, use encapsulants to prevent the release of asbestos fibers from acoustical plaster. Formaldehyde-emitting wood products can be treated with scavenging coatings or encapsulated with vinyl materials. Radon treatment measures include sealing cracks in masonry walls and slabs and gaps around floor-based plumbing. Minor mold infestations can be treated using biocides or encapsulating paints.

11.7.2 Contaminant Control

11.7.2.1 Ventilation

Contaminant levels in building spaces can be reduced by diluting indoor air with less contaminated outdoor air. Such dilution may occur in part or in whole by infiltration or exfiltration, natural ventilation, and mechanical ventilation. Ventilation air exchange in residential buildings occurs by natural ventilation; that is, opening windows and doors during mild to warm seasonal conditions, and by infiltration or exfiltration under enclosed conditions. Infiltration and exfiltration also play a role in building air exchange in mechanically ventilated office, commercial, and institutional buildings.

11.7.2.1.1 Infiltration and Exfiltration

All building structures "leak"; that is, the building envelope contains many avenues whereby air enters and leaves. These include cracks and gaps around windows, doors, electrical and exhaust outlets, ceiling light fixtures, and the building base. Infiltration of outside air and exfiltration of inside air are affected by the tightness of the building's envelope and environmental factors such as indoor–outdoor temperature differentials and wind speed. High air exchange occurs on cold, windy days, and low air exchange occurs on calm, moderate days when temperature differences are small. During the heating season, indoor–outdoor temperature differences result in pressure differences that draw air into the base area of the building (infiltration) and force air out at the top (exfiltration). This is the so-called stack effect. It occurs in all buildings (including single-story) but is most pronounced in tall buildings.

11.7.2.1.2 Natural Ventilation

A building is "naturally" ventilated when occupants open windows and, in some cases, doors. Air exchange can be considerable, as can contaminant dilution. The degree of air exchange, or ventilation, depends on the magnitude of window and door opening, the position of open windows and doors relative to each other, and the same environmental factors that affect infiltration and exfiltration. In the absence of year-round climate control, natural ventilation is the primary means by which residential and some nonresidential building spaces are made more comfortable during warm weather.

11.7.2.1.3 Mechanical Ventilation

Mechanical ventilation is the most widely used contaminant control measure in large office, commercial, and institutional buildings. Mechanical systems are used to provide general dilution and local exhaust ventilation.

General dilution ventilation is used in most large buildings to dilute and remove human bioeffluents that cause odor and comfort complaints. It is also applied as a generic measure to reduce overall building contaminant levels and mitigate SBS-type symptoms and IAQ complaints.

Use of mechanical ventilation for contaminant control is based on the general dilution theory, wherein contaminant levels are expected to decrease by 50% each time air volume or exchange is doubled. It is likely to be relatively effective in contamination problems that are episodic, for example, tobacco smoking and emissions from office copiers, or the relatively constant bioeffluent emissions. It is less effective in reducing the levels of contaminants (e.g., HCHO and VOCs) emitted by diffusion processes.

There have been attempts to evaluate the effectiveness of general dilution ventilation in reducing both contaminant levels and occupant health complaints, but results have been mixed. In general, it seems that increasing ventilation rates to 10 L/s per person (20 CFM/person) results in decreased symptom prevalence, with higher ventilation rates being relatively ineffective in achieving further reductions. The effect of ventilation rate on personal satisfaction or dissatisfaction with air quality can be seen in Figure 11.12.

A consensus has developed that adequate ventilation is essential to maintain comfortable and low-human-odor building environments. Consequently, guidelines have been developed to help building designers and operators provide adequate ventilation rates to meet the needs of occupants. In North America, ventilation guidelines are developed and recommended by the American Society of Heating, Refrigerating and Air-Conditioning Engineers (ASHRAE). In Europe and East Asia, they have been developed by governmental agencies, regional collaboration, and the WHO.

Ventilation guidelines specify a rate of outdoor airflow that can reasonably be expected to provide acceptable air quality relative to human odor and comfort. Ventilation guidelines specify quantities of outdoor air that will maintain building steady-state CO_2 levels (typically 1,000 ppmv [ASHRAE] or 800 ppmv [WHO]) at maximum design capacity. At design occupant capacity, recommended

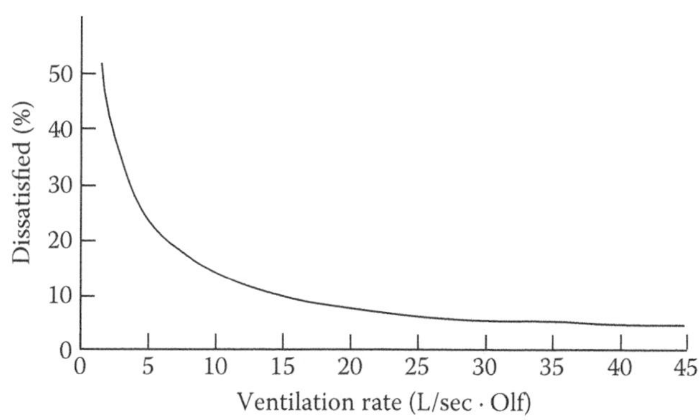

FIGURE 11.12 Relationship between ventilation rate and dissatisfaction with air quality. (From Fanger, P.O., *Proc. 5th Int. Conf. Indoor Air Qual. Climate*, 5, 353, 1990.)

minimum ventilation rates (liters per second per person, CFM per person) would maintain CO_2 levels below guideline values. ASHRAE guideline ventilation values for different building spaces or conditions of use are summarized in Table 11.10.

Local exhaust ventilation is used in special circumstances to minimize the levels of contaminants in buildings. It is particularly well suited to situations in which sources are known, emissions are high, and localized and general dilution ventilation would not be acceptable. Its major application is to control odors from lavatories. It is also used to control combustion by-products and odors from cafeteria kitchens, vapors and gases from institutional laboratories, and ammonia from blueprint machines.

11.7.2.2 Air Cleaning

As with ventilation, air cleaning systems are designed to reduce contaminant levels after contaminants become airborne. Air cleaning systems can be designed to control PM (dust), gas-phase contaminants, or both.

Particle or dust cleaning systems have a long history of use. A minimum level of air cleaning (dust-stop filters) is used in most building HVAC systems to protect mechanical equipment from soiling. Filtration systems are less commonly employed to maintain clean building spaces. A variety of filters and filtration systems are available for building dust control. These vary from low-efficiency dust-stop filters, to medium-efficiency pleated panel filters, to high-efficiency pleated panel filters and electronic air cleaners, to very high-efficiency HEPA (high-efficiency particulate absolute) filters, which have the ability to remove 99.97% of particles at the hardest-to-control diameter of

TABLE 11.10
ASHRAE Ventilation Guidelines

		Outdoor Air Requirements	
Facility Type	**Estimated Occupancy (persons/1,000 ft²)**	**(CFM/person)**	**(L/s per person)**
Food and Beverage			
Restaurant dining rooms	70	7.5	3.8
Bars, cocktail lounges	100	7.5	3.8
Hotels, Motels, Resorts, etc.			
Lobbies/prefunction	30	7.5	3.8
Multipurpose assembly rooms	100	7.5	3.8
Barracks sleeping areas	20	5	2.5
Bedroom/living room	10	5	2.5
Laundry rooms, central	10	5	2.5
Laundry rooms with dwelling units	10	5	2.5
Office Buildings			
Reception areas	30	5	2.5
Break rooms	25	5	2.5
Office space	5	5	2.5
Telephone/data entry	60	5	2.5
Education			
Classrooms (age > 9)	35	10	5.0
Libraries	10	5	2.5

Source: ASHRAE Standard 62.1–2022: Ventilation for Acceptable Air Quality, ASHRAE, Atlanta, 2022.

0.3 μm. Particle filtration systems can be used relatively effectively to reduce indoor particle and mold spore levels.

Gas-phase contaminant control by air cleaning is more difficult, more expensive, and usually less effective than particle control. Such air cleaning typically employs sorbents or specially impregnated sorbents in panel beds or pleated filters. The most widely used sorbent is activated carbon (charcoal). It is effective in removing VOCs that have a relatively high molecular weight (e.g., toluene, benzene, xylene, and methyl chloroform). It is relatively ineffective in removing low-molecular-weight substances such as HCHO, ethylene, acetaldehyde, and others. Such substances may be removed from air by potassium permanganate ($KMnO_4$) on activated alumina and specially impregnated charcoal.

Many air cleaners are commercially available (particularly in the residential market) whose manufacturers claim a high degree of effectiveness in removing a broad range of indoor contaminants. While the claims may be true in a laboratory environment or in small enclosures, they may not have been validated in actual residential environments where infiltration and exfiltration may overshadow the ability of the devices to clean the air.

11.8 INDOOR AIR QUALITY AND PUBLIC POLICY

In the United States, the identification of a potential health-threatening environmental problem has historically been followed by the enactment of federal legislation that provides a regulatory framework for control or promulgation of specific regulations by the US EPA. In the case of IAQ, a variety of federal agencies have limited authority under existing legislation. These include the Consumer Product Safety Commission (CPSC), which regulates dangerous or defective products; the US EPA, which has specific authority relative to Rn and general authority to conduct IAQ research; the Department of Housing and Urban Development (HUD), which sets standards on the habitability of manufactured houses and federally subsidized housing; OSHA, which sets and enforces workplace health standards; and NIOSH, which has the authority to recommend workplace health standards, conduct research, and conduct workplace health hazard evaluations (including IAQ).

Under its statutory authority, the CPSC attempted to ban the use of UFFI. The ban was subsequently overturned by a federal appeals court. The CPSC also evaluated emissions of HCHO from wood products and kerosene and gas space heaters for possible regulatory action. Under its statutory authority, HUD issued product standards for emissions of HCHO from particleboard and hardwood plywood used in manufactured housing. It also required manufacturers to prominently post warning signs in or on their products about the health dangers of HCHO.

The US EPA has been the most active federal agency in actions involving IAQ. It has authority to promulgate and enforce regulations under AHERA, which requires schools to use accredited personnel to conduct asbestos inspections, prepare management plans, and conduct any abatement operations.

The US EPA has maintained a leadership role in the broad area of IAQ, in part because of its expertise related to air quality and the agency's research and public policy interests. Over the past decade, the US EPA has increasingly emphasized voluntary efforts in controlling indoor air pollution. Because IAQ in many instances (e.g., houses) is not in the public domain, it naturally lends itself to voluntary efforts and the leadership experience of an agency such as the US EPA.

OSHA has attempted to deal with IAQ concerns associated with workplaces. First, it identified ETS as a human carcinogen and subject of workplace exposure concern. Although OSHA proposed rules to regulate smoking and ventilation in nonindustrial workplaces, such rules were never promulgated.

There are inherent limitations in the ability of the government to solve IAQ problems through regulatory action. EPA's recent IAQ actions focus on cleaner, safer indoor air. In 2025, the EPA updated its IAQ materials, emphasizing stronger ventilation and filtration to curb airborne disease and improve productivity. In March 2024, the TSCA rule bans ongoing uses of chrysotile asbestos (with April 2024 corrections), retaining EPA's policy baseline despite some litigation activity in 2025. In May 2025, EPA issued a Best Practices Guide to keep wildfire smoke out of public and commercial buildings and schools. Consequently, public policy is focused primarily on public education so that individuals can make informed choices about the quality of the air they breathe indoors.

READINGS

American Conference of Governmental Industrial Hygienists, Evaluating office environmental problems, *Ann. ACGIH*, 10, 21–35, 1984.

American Society of Heating, Refrigerating and Air-Conditioning Engineers, *ASHRAE Standard 62–1989: Ventilation for Acceptable Air Quality*, ASHRAE, Atlanta, GA, 1989.

Bearg, D.W., *Indoor Air Quality and HVAC Systems*, Lewis Publishers/CRC Press, Boca Raton, FL, 1993.

Burge, H.A., Ed., *Bioaerosols*, Lewis Publishers/CRC Press, Boca Raton, FL, 1995.

California Environmental Protection Agency, Air Resources Board, *Proposed Identification of Lead as a Toxic Air Contaminant*, California Environmental Protection Agency, Sacramento, CA, 1997.

Cone, J.E., and Hodgson, M.Q., Eds., Problem buildings: Building-associated illness and the sick building syndrome, in *State of the Art Reviews: Occupational Medicine*, Hanley and Belfus, Inc., Philadelphia, PA, 1989.

Cox, C.S., and Wathes, C.M., Eds., *Bioaerosols Handbook*, Lewis Publishers/CRC Press, Boca Raton, FL, 1995.

Dillon, H.K., Heinsohn, P.A., and Miller, J.D., Eds., *Field Guide for the Determination of Biological Contaminants in Environmental Samples*, American Industrial Hygiene Association, Fairfax, VA, 1996.

Godish, T., *Indoor Air Pollution Control*, Lewis Publishers, Chelsea, MI, 1989.

Godish, T., *Indoor Environmental Quality*, Lewis Publishers/CRC Press, Boca Raton, FL, 2000.

Godish, T., *Sick Buildings: Definition, Diagnosis and Mitigation*, Lewis Publishers/CRC Press, Boca Raton, FL, 1995.

Light, E., and Sundell, J., *General Principles for the Investigation of Complaints, TF11–1998*, International Society of Indoor Air Quality and Climate, Milan, Italy, 1998.

Maroni, M., Siefert, B., and Lindvall, T., *Indoor Air Quality: A Comprehensive Reference Book*, Elsevier, Amsterdam, 1995.

Morey, P.R., Feeley, J.C., and Othen, Q.A., Eds., *Biological Contaminants in Indoor Environments, ASTM STP 1071*, American Society of Testing Materials, Philadelphia, PA, 1990.

National Research Council, *Health Risks of Radon and Other Internally-Dispersed Alpha Emitters*, BEIR IV, National Academy Press, Washington, DC, 1988.

National Research Council, Committee on Indoor Pollutants, *Indoor Air Pollutants*, National Academy Press, Washington, DC, 1981.

Norback, D., and Edling, C., Indoor air quality and personal factors related to the sick building syndrome, *Scand. J. Work Environ. Health*, 16, 121, 1990.

Pattanayak, S.K., Jeuland, M., Lewis, J.J., Usmani, F., Brooks, N., Bhojvaid, V., Kar, A., Lipinski, L., Morrison, L., Patange, O., Ramanathan, N., Rehman, I.H., Thadani, R., Vora, M., and Ramanathan, V., Experimental evidence on promotion of electric and improved biomass cookstoves, *PNAS*, 116(27) 13282–13287, 2019. https://doi.org/10.1073/pnas.1808827116.

Rafferty, P.J., Ed., *The Industrial Hygienists Guide to Indoor Air Quality Investigations*, American Industrial Hygiene Association, Fairfax, VA, 1993.

Salthammer, T., Ed., *Organic Indoor Air Pollution: Occurrence, Measurement, Evaluation*, John Wiley & Sons, New York, 1999.

Samet, J.M., and Spengler, J.D., Eds., *Indoor Air Pollution: A Health Perspective*, Johns Hopkins University Press, Baltimore, 1991.

Seitz, T.A., NIOSH indoor air quality investigations, 1971–1988, in *Proceedings of the International Indoor Air Quality Symposium: The Practitioner's Approach to Indoor Air Quality Investigations*, Weekes, D.M., and Gammage, R.B., Eds., American Industrial Hygiene Association, Akron, OH, 1989.

Smith, K.R., *Biofuels, Air Pollution and Health: A Global Review*, Plenum Publishing Co., New York, 1987.

Spengler, J.D., Samet, J.M., and McCarthy, J.C., Eds., *Indoor Air Quality Handbook*, McGraw-Hill, New York, 2001.

US EPA, *A Citizen's Guide to Radon*, EPA-86-004, EPA, Washington, DC, 1986.

US EPA, *Protecting People and Families from Radon—A Federal Action Plan for Saving Lives*, June 20, 2011. http://www.epa.gov/radon/pdfs/Federal_Radon_Action_Plan.pdf.

US EPA, *Respiratory Health Effects of Passive Smoking: Lung Cancer and Other Disorders*, EPA/600/6-90/006 B, EPA, Washington, DC, 1993.

US EPA/NIOSH, *Building Air Quality: A Guide for Building Owners and Facility Managers*, EPA/400/1-91/1003, DDHS (NIOSH) Publication 91–114, EPA, Washington, DC, 1991.

US Surgeon General, *The Health Consequences of Involuntary Smoking*, DDHS (CDC) 87–8398, Department of Health & Human Services, Washington, DC, 1986.

Weekes, D.M., and Gammage, R.R., Eds., *Proceedings of the International Indoor Air Quality Symposium: The Practitioner's Approach to Indoor Air Quality Investigations*, American Industrial Hygiene Association, Akron, OH, 1989.

World Health Organization, Indoor air quality: Organic pollutants, in *EURO Reports and Studies*, WHO Regional Office for Europe, Copenhagen, 1988, p. 171.

Yocum, J.E., Indoor–outdoor relationships: A critical review, *JAPCA*, 32, 500, 1982.

Yocum, J.E., and McCarthy, S.M., *Measuring Indoor Air Quality: A Practical Guide*, John Wiley & Sons, New York, 1991.

QUESTIONS

1. Indoor and outdoor concentrations of pollutants vary considerably. Which pollutants are higher indoors? Outdoors?
2. How do personal exposures to PM differ from outdoor and indoor exposures?
3. Describe the nature of building air quality concerns associated with asbestos.
4. What is radon? Describe its decay.
5. Because radon is inert, why should it be considered a danger to public health?
6. Describe combustion sources that result in elevated NO_x in building environments.
7. What health risks are or may be associated with the use of (a) unvented appliances, (b) candle and incense burning, and (c) vented combustion appliances?
8. Describe major health concerns associated with exposures to environmental tobacco smoke.
9. What exposures and health risks are associated with biomass cooking?
10. What exposure and health concerns are associated with aldehydes in buildings?
11. What are VOCs? What exposure and health concerns are associated with them?
12. What are SVOCs? Describe major SVOC contaminants in indoor environments.
13. Why are pesticides an indoor environment concern?
14. Describe illness patterns characteristic of chronic allergic rhinitis, asthma, hypersensitivity pneumonitis, and Legionnaires' disease.
15. Identify primary allergenic biological contaminants in indoor environments.
16. What is the health significance of endotoxins and mycotoxins?
17. What is a problem building? Sick building?
18. Distinguish between building-related illness and sick building syndrome.
19. What are some of the major risk factors for developing SBS-type symptoms?
20. How would one go about conducting an indoor air quality investigation?
21. Why are indoor air quality measurements made in problem buildings?
22. Describe differences between source control and pollutant control measures.
23. What is the relationship between indoor air quality and ventilation? Between ventilation and illness symptoms?
24. How is indoor air quality regulated in the United States?

12 Environmental Noise

Environmental noise associated with human activities is a phenomenon that characterizes the acoustical environment and our perceptions of it in the communities in which we live. Noise is sound energy that is objectionable because of its physiological and psychological effects on humans.

Humans vary in their reception and perception of sound. As such, the concept of noise can and does vary from one individual to another. What is an acceptable or unacceptable sound, or level of it, depends on the individual exposed. Nevertheless, there are sound types and exposure levels that most individuals agree are not acceptable and, therefore, noise. Major noise sources include transportation, industry, construction, buildings and households, and humans and pets. In most cases, noise is produced because of human activities.

Noise in industrial, construction, and military environments poses unique exposure concerns because it has the potential to cause hearing loss. Noise exposure in industrial and construction environments is regulated by governmental agencies whose mission is to protect worker health (in the United States, the federal Occupational Safety and Health Administration [OSHA] and a number of state agencies since EPA concluded in 1982 that state and local agencies were best able to handle noise standards). These occupational safety and health concerns differ from noise exposures in the ambient environment, that is, environmental or community noise.

Noise is treated in this book as a form of ambient or atmospheric pollution since the sound energy that is the cause of noise is primarily transmitted through the air environment.

12.1 SOUND AND ITS MEASUREMENT

Sound is a form of energy produced by vibrating objects or aerodynamic disturbances. The former includes our vocal cords and operating motors, automobiles, and airplanes; the latter, thunder, sonic booms, and air movement.

Disturbance of air molecules by sound energy produces variations in normal atmospheric pressure. As these pressure variations reach our ears, they cause the eardrums to vibrate. Transmission of vibrations to the inner ear, and their interpretation by the brain, results in our sensory perception of sound. Sound can only be transmitted through a medium that contains molecules; that is, it cannot move through a vacuum. The speed of sound depends on the density of the transmission medium, with increased speed associated with increased density. The speed of sound in air, water, and steel is 343, 1,482, and 5,030 m/s (1,125, 4,860, and 16,500 ft/s), respectively.

12.1.1 Sound Energy

When objects vibrate, they radiate acoustical energy. The energy produced by a source is described as sound power. In conventional use, sound power (Sp) values are expressed in dimensionless units called decibels (dB), calculated from the following equation:

$$\text{Sp} = 10\log_{10}\left(W_{\text{m}} / W_{\text{r}}\right)\text{dB} \tag{12.1}$$

where W_{m} is the measured sound energy in watts and W_{r} is the reference sound energy, 10^{-12} W. Sound power is the logarithm of the ratio of the measured and reference sound energy multiplied by 10.

DOI: 10.1201/9781003664109-12

When sound energy passes through a given area, its flow is described as sound intensity (SI), expressed as watts per square meter (W/m²). Sound intensity values are also expressed in decibels and are calculated similarly to Sp:

$$\text{SI} = 10\log_{10}\left(I_{\text{m}} / I_{\text{r}}\right)\text{dB} \tag{12.2}$$

where I_{m} is the measured sound energy and I_{r} is the reference sound energy, 10^{-12} W/m².

When sound energy radiating from a source strikes a surface, it induces a pressure. This pressure can be measured by instruments known as sound pressure level meters. Sound pressure values are also expressed in decibels. Calculation of sound pressure differs slightly from those for Sp and SI. Because pressure in sound waves varies sinusoidally, the mean square values of the pressure changes are used to calculate sound pressure levels (Spl):

$$\begin{aligned}\text{Spl} &= 10\log_{10}\left(P_{\text{m}}^2 / P_{\text{r}}^2\right)\text{dB} \\ &= 20\ \log_{10}\left(P_{\text{m}} / P_{\text{r}}\right)\text{dB}\end{aligned} \tag{12.3}$$

In calculations of Spl, the reference pressure (P_{r}) is the threshold of human hearing, 2×10^{-4} µbars or 2×10^{-5} N/m², where N is newtons. P_{m} is the measured sound pressure. For purposes of illustration, let us calculate the Spl when $P_{\text{m}} = 2.0$ µbars.

$$\begin{aligned}\text{Spl} &= 20\log_{10}\left(2.0 / 0.0002\right)\text{dB} \\ &= 20\log_{10}\left(10{,}000\right)\text{dB} \\ &= 20\left(4\right)\text{dB} \\ &= 80\,\text{dB}\end{aligned}$$

As an object vibrates, sound waves radiate outward in an ever-expanding sphere. At increasing distances, Sp remains the same, but SI declines as the area through which sound energy passes increases. In a free field (space where sound is not reflected), SI and Spl are reduced by 6 dB (factor of 4) each time the distance from the source is doubled. This relationship is illustrated in Figure 12.1. At a distance of two radii (2*r*), the area of the sound sphere is four times as large as the area at a distance of one radius (1*r*). The change in sound intensity and pressure follows the inverse square law:

$$\text{SI} = \text{W} / 4\pi r^2 \tag{12.4}$$

where W is watts and $4\pi r^2$ is the area of the sound sphere.

Each time the radius doubles, the area of the sound sphere increases by $4\pi r^2$ of its previous value and SI decreases inversely with this increase in area.

Sound produced by a source is emitted in the form of waves. These waves are characterized by alternating compressions (wave peaks) and rarefactions (wave troughs) and vary in length, frequency, and amplitude. The shorter the wavelength, the more frequent the waves are per unit of time. The higher the wave (increased amplitude), the more energy it has.

Frequency is a major characteristic of sound, and its discrimination by the human auditory organ constitutes the phenomenon of hearing. Frequency is expressed in cycles (waves) per second or hertz. If 1,000 complete sound waves pass a point per second, that sound has a frequency of 1,000 Hz.

A sound source does not usually produce discrete frequencies. Most sounds include a range of frequencies characterized by dominant frequencies. Male human speech, for example, is characterized by dominant frequencies lower than 2,000 Hz; for female speech, the range of dominant frequencies is somewhat higher.

Humans can hear sounds in the frequency range of 20 to 20,000 Hz. However, frequencies in this range are not equally perceptible. As can be seen in Figure 12.2, humans hear those frequencies

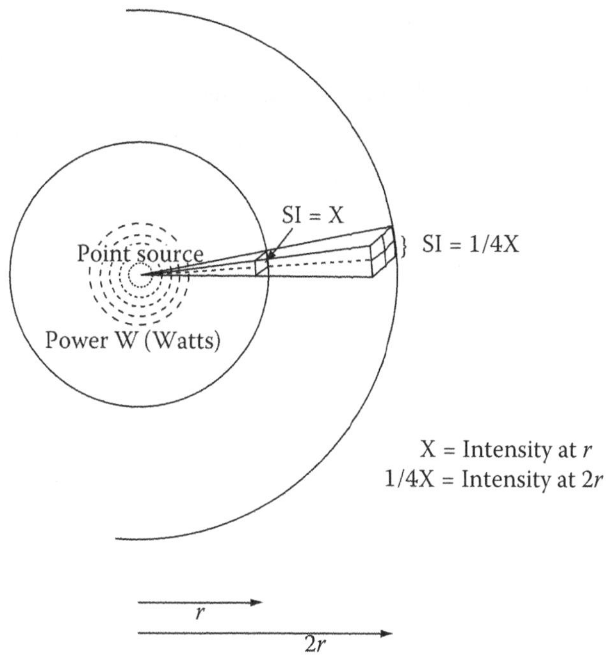

FIGURE 12.1 Sound propagation from a vibrating source.

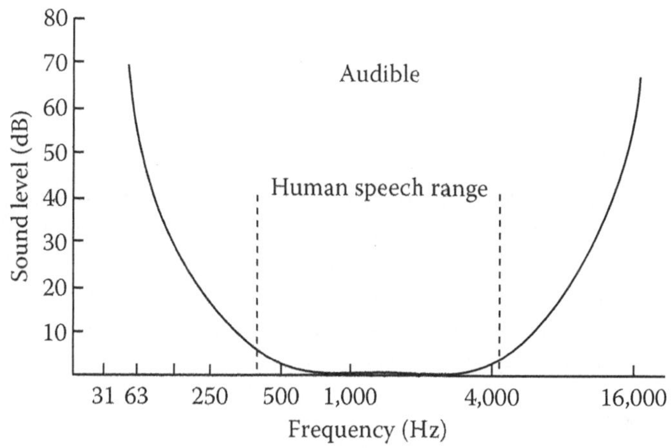

FIGURE 12.2 Frequency range of human hearing.

best that correspond to human speech. Sound frequencies lower than 16 Hz are characterized as infrasound; those higher than 20,000 Hz, as ultrasound. Whether humans can hear a sound depends on its spectrum of frequencies and intensity. Decibel readings for a variety of sound sources experienced in the home, work, and ambient environments are summarized in Table 12.1. Note that 0 dBA corresponds to the threshold of hearing. Sound pressure levels higher than 70 dBA become increasingly objectionable, as they interfere with speech communication. Sounds lower than 0 dBA occur but cannot be heard. Brief exposures to sound levels higher than 140 dBA may be painful, and long-term exposures to sound levels higher than 90 dBA may result in hearing loss.

TABLE 12.1
Sound Pressure Levels, Sources, and Human Responses

Sources	dB	Human Responses
	150	
	140	Painful
	130	
Jet aircraft on ground, 20 ft	120	
Rock and roll band	115	
Loud motorcycle, 20 ft	110	
	100	
Heavy truck, 50 ft	90	Permanent hearing loss
Pneumatic drill, 50 ft	85	
	80	
	75	Annoying
Motor vehicle, 50 ft	70	Interfere with speech communication
	60	
Living room	55	
	50	Quiet
Bedroom	45	
Library	40	
Soft whisper	30	
	20	Very quiet
	10	Just audible
	0	Threshold of hearing

12.1.2 Sound Measurement

Instruments used to measure sound levels in decibels are described as sound pressure meters (Figure 12.3). Such meters consist of a microphone, attenuator, amplifiers, and a digital or mechanical reading. Sound generates a voltage proportional to the acoustical pressure acting on the microphone. This voltage produces a deflection of the meter needle or is converted to a digitally displayed value. An integral part of sound-measuring instruments is weighting scales or networks that emphasize or deemphasize sound in selected frequency ranges.

12.1.2.1 Weighting Scales

Sound level meters are generally equipped with three circuits that correspond to weighting scales. Three weighting scales have been established by the American National Standards Institute (ANSI) for general-purpose use. These are the A, B, and C scales. Other scales (D and E) have been developed and used for special applications. Each scale differs in its ability to measure sound levels across a frequency range (Figure 12.4). The A scale is designed to discriminate against (i.e., not measure) frequencies lower than 600 Hz. Low-frequency discrimination is more moderate on the B scale. The C scale corresponds more closely to the actual sound energy produced; its response is relatively uniform across the frequency spectrum. Readings on all three scales provide an indication of the frequency distribution of a recorded sound. If the level is greater on the C scale, some of the sound is in frequencies lower than 600 or higher than 10,000 Hz. If no differences are observed among the three scales, most of the measured sound is in the middle frequencies.

The sound reading should be expressed as a function of the scale used, for example, 20 dBA, 20 dBB, or 20 dBC. At one time, it was customary to select the weighting scale according to sound

FIGURE 12.3 Sound pressure level meter.

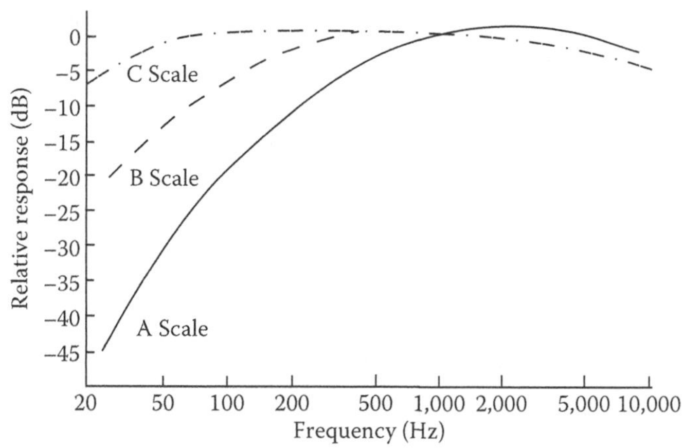

FIGURE 12.4 Frequency discrimination for A, B, and C weighting scales.

level. For levels lower than 55 dB, the A scale was recommended; from 55 to 85 dB, the B scale; and higher than 85 dB, the C scale. It is now common practice to use the A scale for most sound measurements. This dominance of A scale use is due to its application to sound level measurement in occupational environments where compensable hearing impairments are limited to the frequency range of 500 to 2,000 Hz. The A scale approximates noise exposure risk to human hearing relatively well. Because humans show limited response to frequencies lower than 500 Hz, there is little value in measuring such frequencies when controlling worker noise exposure. Similarly, it is the audible middle frequencies that are important in community noise exposures.

Despite its almost universal use for measurements of Spl in industrial and community environments, the A scale has some important limitations. It apparently does not simulate the spectral selectivity of human hearing or its nonlinear relationship to sound energy. If different spectral characteristics are compared, dBA values may not be an accurate indicator of human subjective response. Laboratory and field studies have shown that the A scale predicts perceptions of loudness and annoyance relatively poorly. It is strongly dependent on the exposure pattern with time and underestimates the effect of low-frequency components of sound. It is also unrepresentative of loud sounds that contain a mixture of distinctive tonal components. As such, it is unsuitable for predicting loudness or annoyance.

12.1.2.2 Meter Response and Instrument Accuracy

In addition to weighting scales, most sound pressure level instruments have slow and fast measuring options. The averaging time for the fast response is 0.125 seconds; whereas for the slow response, it is 1 second. Although the fast response corresponds to the response time of the human auditory system, the slow response averaging time is usually used because it is easier to read.

Commonly available sound pressure level meters are rated as Type I, II, or III. These designations reflect their accuracy. Type I meters have an accuracy of ±1 dB; they are primarily research instruments. Type II meters have an accuracy of ±2 dB; they are general-purpose instruments recommended for use in measuring both industrial and community sound levels. Type III meters have an accuracy of ±3 dB and are not recommended for regulatory or technical use.

12.1.2.3 Measurement of Impulse Sound

General-purpose sound pressure level meters cannot be used to measure impulse noises, that is, those produced by a variety of industrial processes such as punch presses and pile drivers. These sounds consist of rapidly rising and falling sound pressures. Instruments must be specially designed to measure both peak and average impulse sound levels. An averaging time of 0.035 seconds is typically used.

12.1.2.4 Spectrum Analysis

The audible frequency range is covered by 10 octave bands. An octave is a range of frequencies in which the upper frequency is, in most cases, twice the lower frequency (Table 12.2). Each octave band is characterized by a center frequency that is the geometric mean of the upper and lower limits. The octave band sound level is the amount of acoustical energy (measured as sound pressure) associated with sound frequencies in the octave band. Noise assessments are often conducted using one-third octave band filters. Each octave is subdivided into thirds to better characterize sound spectra.

12.2 SOUND EXPOSURE DESCRIPTORS

The US Environmental Protection Agency (US EPA) uses a system of four sound descriptors to summarize how people hear sound and how environmental noise affects public health and welfare. These are the A-weighted sound level, A-weighted sound exposure level, equivalent sound level, and day–night sound level.

TABLE 12.2
Octave Band Frequency Ranges and Center Frequencies

Frequency Range (Hz)	Center Frequency (Hz)
18–45	31.5
45–90	63
90–180	125
180–355	250
355–710	500
710–1,400	1,000
1,400–2,800	2,000
2,800–5,600	4,000
5,600–11,200	8,000

12.2.1 A-Weighted Sound Level

A weighting is recommended because of its convenience of use, relative accuracy for most purposes, and widespread acceptance around the world. It is used in the other descriptors described later. When used by itself, an A-weighted decibel value describes a sound value at a given instant, a maximum value, or a steady-state value.

12.2.2 A-Weighted Sound Exposure Level

Sound exposure levels take into account the moment-to-moment variations that occur when measuring environmental noise. The changing magnitude of sound levels can be observed on a strip-chart recording. Changes in sound levels associated with different sound sources can be seen in Figure 12.5. Sound levels vary over a range of 30 dBA. They seem to be characterized by relatively steady lower-level (background, ambient) sound values, over which are superimposed sound levels associated with distinct individual sources. These can be characterized by their maximum values and time patterns. By combining the maximum sound level with the length of time that the sound level is within 10 dB of its maximum value, one can determine the single-event noise exposure level (SENEL).

With the development of both iPhone and Android apps, and while not calibrated, the individual can download a variety of apps (often free) that provide the measurement of sound levels with the ability to save the responses. The author uses two of those that are currently available (Armstrong Ceiling Solutions and NIOSH SLM). Neither can be calibrated, but they provide good indicators of sound levels and can be used for relative comparisons of noise levels at athletic events (where a touchdown or three-pointer can create a seemingly deafening noise), inside of automobiles, inside of classrooms, and for many other useful and fun purposes.

12.2.3 Equivalent Sound Level

The equivalent sound level (L_{eq}) is the average of all sounds measured on the A scale over time. It can be used for any exposure duration. The L_{eq} correlates relatively well with the effects of noise on humans, even for wide variations in sound level and duration. It is used to report average environmental noise levels.

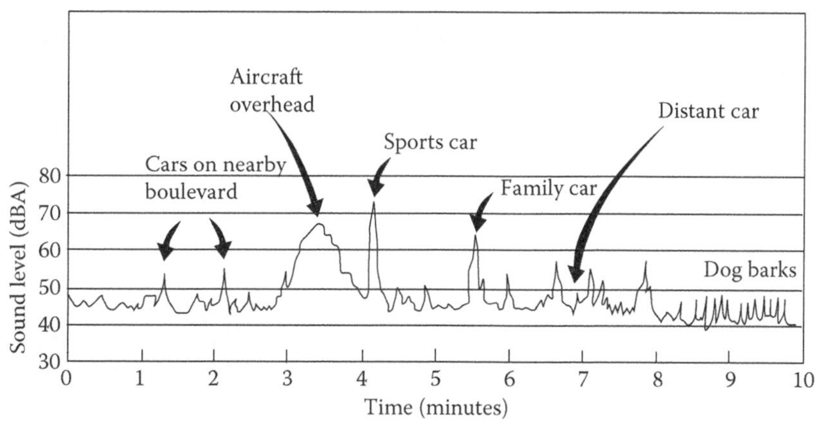

FIGURE 12.5 Variation in environmental noise levels. (From US EPA, *Protective Noise Levels*, http://www.nonoise.org/library/levels/levels.htm, 2000.)

The L_{eq} can be obtained by using dosimeter-type instruments or calculated from individual sound measurements. In the latter case, we use Equation 12.5:

$$L_{eq} = 10\log_{10}\sum_{i=1}^{n}10^{(L_i/10)} / \mathrm{N} \tag{12.5}$$

where L_i corresponds to each measured value in decibels.

Assume that sound measurements were made in a city every 6 minutes during an hour, with the following values: 65, 75, 68, 70, 80, 72, 76, 78, 82, and 65 dBA. Using Equation 12.5:

$$L_{eq} = 10\log_{10}\left[10^{6.5} + 10^{7.5} + 10^{6.8} + 10^{7.0} + 10^{8.0} + 10^{7.2} + 10^{7.6} + 10^{7.8} + 10^{8.2} + 10^{6.5}\right] / 10$$
$$= 76.3\,\mathrm{dBA}$$

A modification of Equation 12.5 can be used to determine sound levels when one is exposed simultaneously to two sounds of equal or different magnitudes. If an individual is exposed to two different sound sources of 80 dB each, then:

$$L_a = 10\log_{10}\left[10^8 + 10^8\right] / 2$$
$$= 83\ \mathrm{dBA}$$

Consequently, a doubling of the sound exposure would be equal to a 3 dBA increase in sound pressure. This relationship is used to construct tables (Table 12.3) to determine equivalent exposures to two or more sounds experienced simultaneously. The exposure value is determined by adding the table value to the highest exposure value. The greatest table value will be 3 dBA. The greater the difference between the two values, the lower the value added to the highest value.

12.2.4 Day–Night Sound Level

The day–night sound level is the equivalent sound level averaged over 24 hours, with 10 dB added to the nighttime (10 p.m to 7 a.m.) equivalent sound level. If the daytime L_{eq} were 60 dBA and the nighttime L_{eq} were 50 dBA, the weighted nighttime sound level would be 50 + 10 dBA = 60 dBA.

TABLE 12.3
Values Used to Determine Actual Sound Exposures Produced by Two Different Sources and Received at the Same Time

Differences in Sound Values	Value to Be Added to Highest Exposure Value
0	3.0
1	2.5
2	2.1
3	1.8
4	1.5
5	1.2
6	1.0
7	0.8
8	0.6
9	0.5
10	0.4

TABLE 12.4
Outdoor L_{dn} Levels by Location

Location	L_{dn} (dBA)
Airport, near freeway	88
0.75 mi from commercial aircraft landing	86
Urban center construction activity	78.5
Urban high-density apartment	78
Urban row housing—major street	68
Old urban residential area	59
Residential—wooded neighborhood	51
Residential—rural	39
Wilderness	35

Source: US EPA, *Protective Noise Levels*, 2000, http://www.nonoise.org/library/levels/levels.htm (condensed version of EPA Levels document).

The day–night sound level (L_{dn}) would therefore be 60 dBA. Table 12.4 indicates L_{dn} values for different outdoor locations. The L_{dn} is commonly used in noise ordinances.

12.3 LOUDNESS

The physical magnitude of sound energy can be described as sound intensity or pressure. The subjective, or perceived, magnitude is called loudness. Perception of loudness depends on sound energy, frequency, and duration. Humans do not hear sound frequencies equally well across the frequency spectrum (Figure 12.2). Additionally, physiological processes in the human auditory system protect it from excessive sound stimulation. Consequently, the relationship between sound energy received and our perception of loudness is not linear; it is closer to being logarithmic.

The sone is the basic unit of loudness. It is defined as the loudness of a pure 1,000 Hz tone at a sound pressure of 40 dBA under specified listening conditions. A 2-sone sound would be perceived as twice as loud as a 1-sone sound.

Loudness is proportional to sound energy over a significant fraction of audible frequencies. This psychophysical "power law" is referred to as Steven's law. In the middle frequencies, the exponent of the power function is such that a twofold increase in perceived loudness corresponds to a 10 dB increase in sound pressure.

The relationship between loudness and sound pressure can be seen from the suite of equal loudness contours in Figure 12.6. Loudness levels are expressed in units called phons, which, at 1,000 Hz, are directly related to the Spl in decibels. This relationship varies with frequency. Although the 40-phon contour corresponds to 40 dBA at 1,000 Hz, it also corresponds to 50 dBA at 100 Hz. Therefore, any tone that is perceived equally well as a 1,000 Hz tone assumes the same loudness or phon value as the 1,000 Hz tone. At 1,000 Hz, phon values are equal to the decibel value.

The relationship between the scale of loudness, sones (S), and loudness level, phons (P), can be expressed for loudness levels of more than 40 P:

$$S = 2(P - 40)/10 \tag{12.6}$$

If the loudness level increases from 40 to 50 P, sound will be perceived as twice as loud; an increase from 40 to 60 P will be four times as loud.

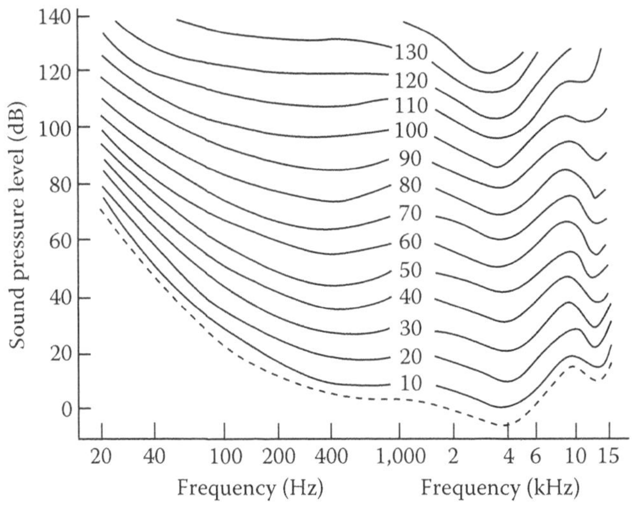

FIGURE 12.6 Equal loudness contours as a function of frequency and sound pressure level.

It would be desirable for measurements of sound pressure to be expressed in loudness readings equal to phons. Due to the complexity of human perceptual processes, this has been difficult to achieve. Such procedures, nevertheless, have been developed and incorporated into modern digital equipment that is not commonly used.

12.4 EFFECTS OF ENVIRONMENTAL NOISE EXPOSURE

Noise exposures experienced daily by an individual depend on how much time one spends in different indoor and outdoor locations and the noise environments present. In the United States, the noise environment outdoors varies over a range of 50 dBA. Day–night sound levels as low as 30 to 40 dBA occur in wilderness areas; in urban areas, they range up to 80 to 90 dBA. Approximately half of the US population lives in urban areas that have L_{dn} values of 48 dBA or higher due to traffic noise alone. More than 62 million people live in areas with urban or freeway traffic noise of L_{dn} values of 60 dBA or higher; 400,000 people in areas where urban or freeway traffic noise results in L_{dn} values of more than 80 dBA. Approximately 10 million people live in areas with L_{dn} values of 60 dBA associated with aircraft noise; 200,000 people where aircraft operations result in L_{dn} values of 80 dBA.

Outdoor noise exposures are significantly diminished by building structures when windows are closed. Despite this reduction, indoor noise levels are usually comparable to, or higher than, those outdoors. This is due to indoor sources such as appliances; mechanical equipment; television, radio, and music playing; and people. Indoor sound sources are often used to mask outdoor noise.

Exposure to excessive sound or noise levels may have a variety of adverse effects. These include hearing impairment; adverse effects on speech communication, performance, and behavior; nonauditory physiological effects; sleep disturbance; and annoyance.

12.4.1 Hearing Impairment

Adverse effects of exposure to excessive noise levels in industrial environments are well known. As a consequence, workplace noise exposures in the United States are regulated by OSHA. There is very little documentation of hearing impairment resulting from community noise exposures. Therefore, hearing impairment risks associated with environmental noise are inferred from evaluations of worker exposures.

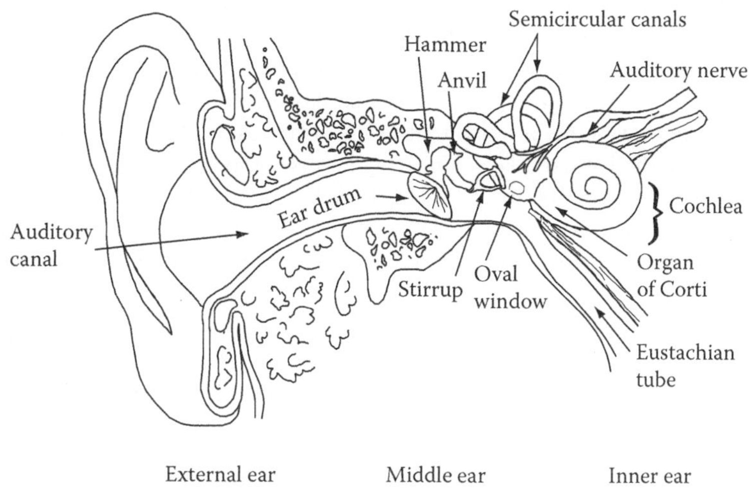

FIGURE 12.7 Human auditory system.

The human auditory system (Figure 12.7) is composed of the outer ear, middle ear, cochlea of the inner ear, and auditory nerve–brain connections.

The outer ear collects sound waves and directs them through the ear canal to the eardrum or tympanic membrane. Collected sound causes a resonance vibration of the eardrum. This vibration is transmitted through the three bones (malleus [hammer], incus [anvil], stapes [stirrup]) of the middle ear into the inner ear through the oval window in the cochlea; alternatively, sound can pass to the inner ear by bone–bone conduction.

The middle ear plays a major role in protecting the auditory system from excessive noise stimulation. The aural reflex is mediated by two small muscles in the middle ear that contract, pulling the stapes and eardrum toward the middle ear cavity, when intense (> 80 dB) sound exposure occurs. This helps protect the middle ear bones from excessive movement and damage and also protects the cochlea. The aural reflex is more responsive to broadband sounds than pure tones, and to lower frequencies than higher ones. It is readily activated and maintained by intermittent, intense sound impulses. As middle ear muscles contract, they impede or attenuate sound energy that would have been received by the cochlea.

The organ of Corti is found between two fluid-filled chambers in the cochlea. Pressure waves produced by sound at the oval window displace the organ of Corti. The basilar membrane of the organ of Corti is associated with two groups of hair cells. The inner hair cells serve as sensory receptors. The outer hair cells serve as an amplification system. It is these outer cells that are commonly damaged by noise.

Acoustic information is transmitted from the organ of Corti to the brain by the auditory nerve. At all steps in this pathway, complex processing enables the brain to interpret sound qualities such as perceived intensity and pitch, speech feature analysis, and noise identification.

As indicated previously, humans can detect sounds in the frequency range of 20 to 20,000 Hz. The wide variation in sensitivity among humans may be affected by different environmental influences.

Hearing sensitivity diminishes significantly with age, particularly at high frequencies (> 10,000 Hz). This age-associated decline in hearing sensitivity may be augmented by the cumulative effects of everyday noise exposure.

In evaluating potential hearing impairment, individuals are tested using audiometry. The concept of hearing level is used to determine whether hearing impairment or loss has occurred. Hearing level is the audiometric threshold of an individual or group in relation to an accepted standard. Hearing threshold levels are auditory thresholds expressed in hearing levels. Threshold levels outside the normal range indicate hearing impairment. The term *hearing loss* is used to describe hearing impairment that poses significant difficulties in leading a normal life, particularly in understanding speech, or a hearing threshold that has deteriorated. It is common to use reference levels to determine whether a noise-induced hearing loss is "compensable" under workers' compensation laws. In the United States, a hearing loss is compensable only for a 25 dB threshold shift in the frequency range of 500 to 2,000 Hz. This is similar to the reference threshold used by the International Standards Organization (ISO). Such a reference standard for hearing disability ignores the fact that sound energy at higher frequencies is important for speech intelligibility and music perception.

Noise may cause a shift in hearing sensitivity that is temporary (temporary threshold shift [TTS]) or permanent (permanent threshold shift [PTS]). One can experience a measurable loss in hearing sensitivity associated with a noisy environment or discrete noise events such as gunshots, explosions, and others. A TTS is reversible and likely to be experienced by a majority of individuals during their lifetime. Recovery, which is exponential, may occur in a matter of minutes, hours, days, or even weeks. Recovery depends on the severity of the threshold shift, type of exposure, and individual sensitivity.

Noise-induced PTS commonly occurs in individuals working in very noisy industrial or construction environments. Hearing impairment occurs gradually, becoming progressively worse with time. Maximal loss in hearing acuity occurs at approximately 4,000 Hz. The frequency pattern of hearing impairment and its progress over time of exposure can be seen for industrial workers in Figure 12.8. As hearing impairment increases, speech communication becomes increasingly more difficult.

There is limited evidence that environmental noise sources, for example, activities such as shooting, motorcycling, snowmobiling, fireworks, and even loud music, can cause hearing impairment. The effects of "background" environmental noise in urban areas on hearing impairment cannot be easily determined. As such, hearing impairment risks have to be extrapolated from industrial exposures. From such extrapolations, an 8-hour continuous equivalent exposure of 70 dB has been identified by the US EPA as the apparently safe level of protection for significant PTS.

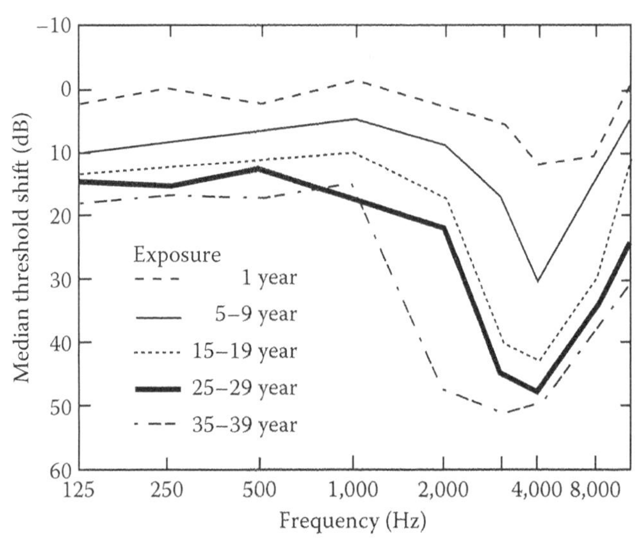

FIGURE 12.8 Hearing losses in industrial workers. (Adapted from Taylor, W. et al., *J. Acoust. Soc. Am*, 38, 113, 1965. With permission.)

12.4.2 Speech Communication

One of the major effects of environmental noise is interference with speech communication. Because speech is an essential element of human interaction, speech interference can significantly degrade the quality of life. It may disturb normal social and work activities and indirectly cause annoyance and stress. Of primary concern are the effects of noise on face-to-face and telephone conversations and use of audio and video devices, both indoors and outdoors.

Speech is a learned motor behavior. It involves feedback between the act of hearing and speech musculature coordinated by the central nervous system. The acoustical energy of speech falls between 100 and 6,000 Hz, with the most important cue-bearing energy between 300 and 3,000 Hz. Speech comprehension under normal circumstances involves distinguishing a finite number of mutually exclusive sounds, and temporal features such as variations in loudness, pitch, and rhythm.

Noise-associated interference with speech communication is a masking process in which noise renders speech sounds unintelligible. The more intense the masking sound and the more energy it contains at speech frequencies, the greater the number of speech sounds that cannot be discerned by the listener.

The degree that noise interferes with speech depends on other factors as well. These include vocal effort, distance between speaker and listener, and surrounding acoustics. It also depends on the speaker's enunciation, conversation topic, and the listener's motivation, hearing acuity, and familiarity with the speaker's vocabulary.

In indoor environments, the highest environmental noise level that allows relaxed conversation with 100% intelligibility is approximately 45 dBA. Conversation is still approximately 95% intelligible at 65 dBA, but it diminishes dramatically thereafter (Figure 12.9). Speakers tend to raise their voice level when environmental sounds are greater than or equal to 60 to 65 dBA. In indoor environments, speech sound levels decrease as an inverse square function of distance up to a distance of 2 m. Further inverse square decreases do not occur because of the effect of reverberations. In outdoor environments, the effect of increasing distance between speakers is a continuum of inverse square decreases.

As can be seen in Table 12.5, if environmental sound levels exceed those at which speech is easily intelligible, speakers must move closer or raise their voices appreciably. For example, if normal

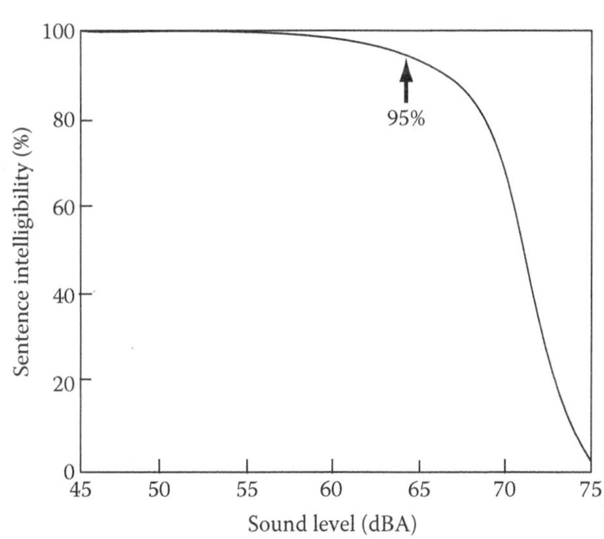

FIGURE 12.9 Decrease in speech intelligibility with increasing environmental sound. (From US EPA, *Protective Noise Levels*, http://www.nonoise.org/library/levels/levels.htm, 2000.)

TABLE 12.5
Environmental Sound Levels That Allow 95% Speech Intelligibility at Various Outdoor Distances and Two Voice Levels

Distance (m)	Normal Voice (dBA)	Raised Voice (dBA)
0.5	72	78
1	66	72
2	60	66
3	56	62
4	54	60
5	52	58

Source: US EPA, *Protective Noise Levels*, 2000, http://www.nonoise.org/library/levels/levels.htm (condensed version of EPA Levels document).

voice levels are intelligible at 3 m in the presence of 56 dBA environmental sound, 66 dBA of environmental sound requires speakers to be only 1 m apart or raise their voices to maintain the same intelligibility.

In a typical residential dwelling with windows closed, sound attenuation is approximately 10 to 15 dBA. Consequently, an outdoor L_{dn} value of 60 dBA would (assuming no major indoor sources are present) permit 100% speech intelligibility indoors.

Such speech intelligibility assumes that most individuals have normal speech perception. However, most individuals belong to sensitive populations, for example, the elderly and those who have hearing impairment. Even individuals as young as 40 years may have difficulty interpreting spoken messages. For sensitive groups or when listening to complex messages (e.g., classrooms or telephone conversations), the signal speech level should be 10 dBA greater than the environmental sound level.

12.4.3 Sleep Disturbance

Sleep disturbance, which includes difficulties falling asleep, alterations in sleep patterns and depth of sleep, and awakenings, is considered to be a major effect of environmental noise. Physiological effects induced by nighttime noise exposure during sleep include increased blood pressure and heart rate, vasoconstriction, change in respiration, and cardiac arrhythmia. Secondary or aftereffects occurring in the morning or day after nighttime exposure may include reduced perceived sleep quality, increased fatigue, and decreased performance and sense of well-being.

The A-weighted equivalent sound level is poorly correlated with sleep disturbance. Better indicators of sleep disturbance risk are the number of sound events exceeding a reference sound level and the difference between maximum and background sound levels.

12.4.4 Psychophysiological Effects

Noise exposure may elicit several kinds of reflex responses, particularly when the noise is sudden, unfamiliar, or of an unwanted nature. These reflect primitive body defense responses and may develop in response to other stimuli as well. Target organs include (1) the heart, blood vessels, intestines, endocrine glands, and others that are enervated by the autonomic nervous system and (2) the hypothalamodiencephalic centers that regulate sleep arousal, endocrine secretions, and other functions.

Environmental noise is known to interfere with several activities such as recreation, sleep, communication, and concentration. Viewed in this context, it can be considered a stress factor that results in physiological responses that are modified in complex ways by individual psychological processes. Most studies have focused on the possibility that noise may be a risk factor for cardiovascular disease. Of particular note are occupational studies and studies of residents near airports that show a tendency for elevated blood pressure, a major risk factor for heart attack and stroke.

High occupational noise level exposures have been associated with the development of neuroses and irritability; high environmental noise exposures have been linked to mental health disorders. In the latter case, noise is not believed to be a direct cause of mental illness; rather, it may accelerate and intensify latent mental health problems.

12.4.5 Performance

Noise exposures may affect human performance. High noise exposures seem to interfere with the conduct of tasks, particularly those (1) that demand attention to multiple cues, (2) for which continuous and detailed attention to frequent signals is required, or (3) that require high levels of working memory. Chronic noise exposures in children seem to affect reading acquisition, which may be related to deficits in auditory discrimination.

12.4.6 Annoyance and Social Behavior

Annoyance resulting from environmental noise exposure is common in urban areas, being experienced in many cases by most inhabitants and visitors. Annoyance can be defined as a feeling of displeasure associated with any agent, condition, or person known or believed to be adversely affecting an individual or group.

Sound-related annoyance may be affected by several physical factors. These include the nature of the sound or noise and its generation and temporal patterns. Sound's tonal qualities (low or high frequency, monotonous), its temporal variation (constant, intermittent, random) and intensity (loudness), the highest levels produced by a source as well as the number of sound events, and the time of day affect annoyance. Airport noise is particularly annoying, due in large part to the number of sound events. Because of the intensity and continuous nature of highway traffic, it is a source of considerable individual and community annoyance.

Annoyance is also affected by psychological factors as well as an individual's living conditions, personal attitude toward the noise source, previous noise exposures, and a variety of socioeconomic factors.

Annoyance with noise seems to have significant effects on social behavior. Noise exposures have been reported to reduce helpfulness and promote aggression when combined with provocation or preexisting anger or hostility. It may affect overt everyday behavior patterns such as opening windows, using balconies, TV and radio use, circulating petitions, and complaining to authorities. It may also be related to increased residential mobility, drug consumption, accident rates, and hospital admissions.

Noise-associated annoyance is, as indicated earlier, affected by a variety of both physical and psychological factors. As such, it is difficult to quantify. In noise and annoyance reduction efforts, the L_{eq} value is commonly used as a reference, although it is only a fair approximation of the annoyance level.

12.5 NONREGULATORY AND REGULATORY CONTROL APPROACHES

12.5.1 Community Exposure Guidelines

The US EPA, in the early 1970s, identified a range of L_{eq} and L_{dn} values that would protect the public from the effects of environmental noise. The values are meant to be viewed as levels below which the general public would not be at risk for noise exposures. They contain a margin of safety to ensure that they are protective. Because the process of identifying noise protection levels did not take into account technical feasibility and economic costs, the US EPA emphasized that they should not be viewed as standards, criteria, or goals.

TABLE 12.6
US EPA Community Exposure Guidelines

Effect	Level	Area
Hearing	L_{eq} (24 h) < 70 dBA	All areas
Outdoor activity—speech interference/annoyance	L_{dn} (24 h) < 55 dBA	Areas where quiet is a basis for use
Outdoor activity—speech interference/annoyance	L_{eq} (24 h) < 55 dBA	Areas where humans spend limited amounts of time
Indoor activity—speech interference/annoyance	L_{dn} (24 h) < 45 dBA	Indoor residential areas
Indoor activity—speech interference/annoyance	L_{eq} (24 h) < 45 dBA	Other indoor areas such as schools

Source: US EPA, *Information on Levels of Environmental Noise Requisite to Protect Public Health and Welfare with an Adequate Margin of Safety*, EPA/ONAC 550/9-74,004, EPA, 1974.

The US EPA's noise protective values are summarized in Table 12.6 (1974). Although based on 24-hour measurements, they are in fact annual average values. For example, to protect against hearing damage, one's 24-hour noise exposure averaged over the whole year should not exceed 70 dBA. Annual yearly outdoor sound levels on the L_{dn} scale of less than 55 dBA in areas such as residences, schools, and hospitals are deemed sufficiently protective of public health and welfare; for inside buildings, an L_{dn} value of less than 45 dBA is deemed sufficiently protective. The US EPA phased out the funding of its Office of Noise Abatement and Control in 1982 as a part of the federal noise control policy to transfer primary responsibility to state and local agencies.

12.5.2 Community Regulatory Efforts

Abatement of community noise requires the identification of noise as a problem of sufficient magnitude to establish a regulatory program for its control. Regulatory programs for noise control exist at local, state, and federal levels. These programs vary in form and scope. Many local noise ordinances are qualitative, prohibiting excessive noise or noise that results in a public nuisance. Because of the subjective nature of such ordinances, they are often difficult to enforce.

Regulation of community noise requires a quantitative approach; that is, noise regulations should be based on well-defined standards, violations of which can be measured. Standards should be based on scientific criteria as to acceptable noise levels. Noise standards used in Minnesota are summarized in Table 12.7. They are based on 1-hour measurement durations and are defined for three noise area classifications: NAC 1 (residential housing, religious activities, camping and picnicking areas, health services, hotels, and educational services); NAC 2 (retail, business and government services, recreational activities, and transit passenger terminals); and NAC 3 (manufacturing, fairgrounds and amusement parks, agricultural and forestry activities). They are specified for daytime (7 a.m.–10 p.m.) and nighttime (10 p.m.–7 a.m.).

The L_{10} value (Table 12.7) indicates that this standard can be exceeded 10% of the time or 6 minutes out of the hour; the L_{50} value, 50% of the time or 30 minutes out of the hour. There is no limit on maximum noise. Note that the most stringent noise standards are applied to residential neighborhoods for nighttime hours. They assume that higher levels are acceptable in commercial, manufacturing, and industrial zones and that sounds of short duration are the most easily controlled.

12.5.3 Federal Noise Control Programs

A variety of federal agencies have authority to regulate and control noise. These include the Federal Aviation Administration (FAA), Federal Highway Administration (FHA), Department of Housing and Urban Development (HUD), OSHA, and the US EPA. The FAA has authority to impose and

TABLE 12.7
Minnesota Noise Pollution Rules (Minn. R. 7030.0040), 2015

	Daytime		Nighttime	
Noise Area Classification	L_{10}	L_{50}	L_{10}	L_{50}
1	65	60	55	50
2	70	65	70	65
3	80	75	80	75

based on 1-hour duration

enforce noise emission standards for commercial aircraft operation. The FHA is responsible for issuing standards for new and improved federal highway construction. OSHA has significant noise control responsibilities that are designed to protect workers from noise-induced hearing losses. HUD's authority involves specification of noise control standards and techniques for HUD-assisted new housing construction.

The principal responsibility for environmental or community noise is vested in the US EPA. Its role in noise abatement was authorized by Congress in the federal Noise Control Act of 1972 and the Quiet Communities Act of 1978. In the 1972 law, the US EPA was required to develop and publish noise criteria and scientific information on noise effects and identify levels of environmental noise, the attainment and maintenance of which are necessary to protect public health and welfare. The US EPA was also required to identify major noise sources that needed regulation. Under this authority, it has promulgated noise emission standards for interstate motor carriers and buses, medium- and heavy-duty trucks, and railroad carriers. Other noise emission standards apply to construction equipment. The Noise Control Act of 1972 also authorized regulation of the manufacture of a variety of equipment and appliances, requiring labeling as to sound levels produced. As noted earlier, the Office of Noise Abatement and Control was defunded in 1982.

The Quiet Communities Act of 1978 was designed to provide financial assistance to state and local governments and regional planning agencies to investigate noise problems, plan and develop a noise control capability, and purchase equipment. It also provides federal assistance to facilitate community development and enforcement of community noise control measures, as well as development of abatement plans for areas around major transportation facilities, including airports, highways, and rail yards.

12.6 CONTROL MEASURES

Achievement of noise abatement objectives requires that noise producers be regulated. Noise control is best achieved at the source. Noise emission standards can be specified and designed into new products such as aircraft engines, motor vehicle tires, industrial machinery, and home appliances. Control of aerodynamically produced noise may require changing airflow to reduce turbulence or placing sound-absorbent materials along its path. Control of noise from vibrating surfaces may require mounts to limit vibration, use of flexible materials in joints, or materials of sufficient mass that are not easily subject to vibration.

For many existing sources, noise reduction may only be achieved by the retrofit application of engineering modifications, including the use of sound-dampening materials. The economics of retrofitting are such that it is far less expensive to initially purchase less-noisy equipment than pay for its control after installation and use. Some existing noise sources can also be isolated, and sound-absorbing enclosures placed around them.

An alternative to noise control at the source is modification of the noise path or its reception. For example, operational changes in commercial airline and flight patterns during takeoff and landing can significantly reduce environmental noise exposures. Such operational requirements are already in effect at the direction of the FAA. At least initially, the location of major highways and airports can minimize noise exposures in residential neighborhoods. Where major highways are close to residential neighborhoods, sound-absorbing walls, earthen mounds, and vegetation may be used to attenuate sound. Inside buildings, sound-absorbing furnishings (carpets and drapery) or construction materials (acoustical tile) can significantly reduce sound.

READINGS

Berglund, B., and Linvall, T., Community noise, *Arch. Center Sensory Res.*, 2, 1, 1995.

Bragdon, C.R., *Municipal Noise Legislation*, Fairmont Press, Atlanta, GA, 1980.

National Bureau of Standards, *Fundamentals of Noise: Measurement, Rating Schemes and Standards*, NTID Publication 300.15. National Bureau of Standards, Gaithersburg, MD, 1971.

Pelton, H.K., *Noise Control Management*, Van Nostrand Reinhold, New York, 1993.

Peterson, A.P., *Handbook of Noise Measurement*, 8th ed., GenRad, Concord, MA, 1978.

Ropes, J.M., and Williamson, D.L., EPA's implementation of the Quiet Communities Act of 1978, *Sound and Vibration*, December 10, 1979.

Tyler, J.M., Hinton, L.V., and Olin, J.G., State standards, regulations and responsibilities in noise pollution control, *JAPCA*, 24, 130, 1974.

US EPA, *Information on Levels of Environmental Noise Requisite to Protect Public Health and Welfare with an Adequate Margin of Safety*, EPA/ONAC 550/9-74,004, EPA, Washington, DC, 1974.

US EPA, *Noise Effects Handbook: A Desk Reference to Health and Welfare Effects of Noise*, National Association of Noise Control Offices, Fort Walton Beach, FL, 1981.

US EPA, *Protective Noise Levels*, 2000. http://www.nonoise.org/library/levels/levels.htm (condensed version of EPA Levels document).

QUESTIONS

1. Describe differences between sound power, sound intensity, and sound pressure.
2. How do sound and noise differ?
3. What range of frequencies can humans hear? What range of frequencies do humans hear best?
4. Identify the primary sources of environmental noise.
5. What reference pressure is used in determining sound pressure levels?
6. Describe the relationship between sound pressure and distance from a source. If the distance is doubled, what is the difference in sound pressure levels in decibels?
7. What are weighting scales? Why are they used?
8. Describe the limitations of the A weighting scale for assessments of loudness and annoyance.
9. Most environmental noise measurements are made using a Type II meter. What is a Type II meter?
10. Why would one measure sound using an octave band analyzer?
11. Describe sound exposure level and equivalent sound level.
12. Calculate the equivalent sound level (L_{eq}) for the following sound values: 65, 70, 60, 55, and 75 dBA.
13. If you are exposed to two sound sources of 83 dBA each, what is your actual exposure in decibels?
14. Describe how the day–night sound level (L_{dn}) is determined and its relevance to community noise exposure.
15. How is loudness related to sound energy?
16. Describe the relationship between sones, phons, and decibels.
17. How does noise exposure affect human health?

18. Describe the relationship between environmental noise and speech communication.
19. How does exposure to noise cause annoyance?
20. The US EPA has identified noise protective levels for community noise exposures. Describe these.
21. If your community wishes to establish an effective noise ordinance, what noise standards would you recommend?
22. Which federal governmental agencies have the authority to regulate noise production and noise exposures? Identify their respective noise-related responsibilities.
23. How could one reduce noise levels in a residential neighborhood adjacent to a freeway?
24. How could exposures to aircraft noise be reduced?

Glossary

Absorption: Process by which pollutants are collected in a sampling or control device by reaction with a collection medium.

Accumulation mode: Particles in the aerodynamic diameter range of 0.1 to 1.0 μm; formed from the agglomeration of nuclei mode particles.

Accuracy: Closeness of a measured value to a true or reference value; measure expressed as ± a percentage of the reference value when precision is taken into account.

Acid-neutralizing capacity (ANC): Measure of the ability of soil and water to neutralize strong acids.

Acid rain: Deposition of H^+ on land or water surfaces by rain droplets.

Acidic deposition: Wet and dry deposition of H^+ on land or water surfaces.

Acidification: Increase of H^+ concentration in water or soil.

Acrolein: Aldehyde produced photochemically in the atmosphere; characterized by its potent eye irritancy.

Activated carbon: Sorbent used to collect nonpolar organic compounds.

Acute exposure: Exposure of humans, plants, and other organisms to high concentrations or doses over a short period; physiological responses occur immediately or within hours of exposure.

Adiabatic lapse rate: Theoretical change of temperature with height when a warm, dry parcel of air is released into a cool, dry atmosphere.

Adsorption: Collection of gas-phase pollutants on solid surfaces by physical forces; collection principle used in air sampling and pollution control.

Aerodynamic equivalent diameter: Particle diameter standardized to a sphere with unit density and settling velocity.

Aerometric Information Retrieval System (AIRS): Repository of pollutant monitoring information maintained by the US EPA.

Aerosol: Solid or liquid particles dispersed in air.

Agglomeration: Process by which atmospheric particles grow in size as a result of collisions and subsequent adherence and associated processes.

Air pollution: Contamination of atmosphere by gas-phase and particulate-phase substances that results in adverse or undesirable environmental effects.

Air pollution episode: A period of abnormally elevated pollutant levels; typically associated with subsidence inversions.

Air quality control region (AQCR): Political jurisdiction or geographical area where NAAQSs must be achieved and maintained.

Air quality criteria: Compilation of scientific facts and their evaluation, which are used to support the development and promulgation of NAAQSs.

Air Quality Index: US EPA program that relates measured air pollutants to their health-affecting potential; used to communicate daily air quality risks to the public.

Air quality management: Pollution control strategy based on the use of air quality standards.

Air quality modeling: The use of mathematical models to predict the effect of a pollutant source, atmospheric reactions, and others, on ground-level concentrations.

Air quality standards: Permissible levels of regulated pollutants in a control region.

Air toxics: Noncriteria pollutants regulated under hazardous pollutant standards.

Air/fuel ratio: Mixture of air to fuel used in motor vehicle engines to enhance engine performance and reduce exhaust emissions.

Aitken mode: Particles produced in the atmosphere by gas-to-particle conversion as well as in combustion processes; same as nuclei mode.

Albedo: The ability of a body to reflect light; relative to the Earth, 30% of sunlight is reflected back to space.

Aldehydes: Organic chemical compounds characterized by relatively high chemical reactivities, irritancy to humans, and role as by-products of hydrocarbon oxidation processes.

Allowance system: Pollution control approach that limits sources to specific quantities of emissions to achieve emission reductions economically.

Allowances: Quantity of SO_2 emissions permitted sources under acidic deposition provisions of the 1990 CAA Amendments.

Ambient air: The free-flowing air outside of buildings.

American Society of Heating, Refrigerating and Air-Conditioning Engineers (ASHRAE): Organization that develops and publishes ventilation guidelines.

Ammonia: Reduced nitrogen species that serves to neutralize strong acids in the atmosphere.

Annoyance: Unwanted effects of sound or noise that reduce one's sense of personal comfort.

Antagonism: Interaction effect in which one pollutant inhibits the effect of another, thus resulting in reduced physiological harm.

Anticyclones: Clockwise airflow in high-pressure systems in the northern hemisphere.

AP-42: Emission factor document compiled and continuously updated by the US EPA.

Arctic haze: Reduced visibility in portions of the Arctic in spring months associated with pollutant transport from Europe and Russia.

Area source: Collective emissions from several small mobile and stationary sources in an urban area.

Aromatic hydrocarbons: Hydrocarbon species that contain one or more benzene rings; simple aromatic hydrocarbons are characterized by moderate photochemical reactivity.

Asbestos: Fibrous silicate minerals used in fireproofing and insulation products in buildings; hazardous air pollutant.

Asbestos Hazard Emergency Response Act (AHERA): Law that requires asbestos inspections and management in school buildings.

Asthma: Respiratory disease characterized by severe episodes of shortness of breath.

Atmosphere: Gas-phase environmental medium that surrounds the Earth.

Atmosphere–ocean general circulation models: Models used by atmospheric scientists to predict the effect of changes in greenhouse gases and associated factors on the Earth's climate.

Atmospheric density: Weight of atmospheric molecules per unit volume of air; density decreases exponentially with increasing height.

Atmospheric deposition networks: Monitoring networks created by the US EPA to collect deposition samples and measure concentrations of target pollutants.

Atmospheric lifetime: Average time a pollutant remains in the atmosphere.

Atmospheric pressure: Force of air molecules on a surface; decreases exponentially with height as density decreases.

Atmospheric stability: Relative measure of air motion associated with vertical temperature gradients.

Atmospheric transport: Movement of pollutants over long distances.

Attainment area: An air quality control region or portion thereof that is in compliance with specific NAAQSs.

Attenuation: The reduction of image-forming light as it passes from an object to an observer.

Audiometric sound threshold: Lowest sound pressure that an individual can hear at various frequencies in the auditory sound spectrum.

Audit: An evaluation of procedures and activities to ensure high-quality and reliable air quality monitoring data.

Automated surface observing systems (ASOS): Automated, real-time visibility monitoring systems at US airports.

Averaging time: Period used to integrate pollution concentrations for data summary, regulatory considerations, and sampling limitations.

A-weighted sound level: Sound pressure levels recorded on the A scale; commonly used measure of sound.

Backscattering coefficient: Measure of light scattering associated with particles and atmospheric gases; major factor contributing to visibility reduction.

Baghouse: System used to control particle emissions from sources by using fabric filtration.

Benzo-α-pyrene: A highly carcinogenic polycyclic aromatic compound commonly found in combustion-derived atmospheric particles.

Best available control technology (BACT): Control technology demonstrated to achieve the highest level of control; necessary to meet emission reduction requirements in nonattainment areas.

Best available retrofit technology (BART): Control measures applied to existing sources designed to provide emission limits associated with PSD programs.

Bias: Property of a data set wherein systematic error is occurring; typically associated with calibration errors.

Bioaccumulation: Uptake of pollutants by plants and animals from the environment.

Biocides: Substances used to control pest organisms; commonly used to control microbial growth.

Biogenic pollutants: Pollutants emitted to the atmosphere from biological sources.

Biological contaminants: Indoor pollutants that have a biological origin.

Biomagnification: Increase of pollutant concentrations as pollutants pass upward through food chains.

Biomass cooking: Use of wood, charcoal, or dung for cooking in developing countries; significant source of combustion by-product exposures indoors.

Bioscrubbers/filters: Pollution control devices that use biological processes to remove contaminants from waste gas streams.

Blowby gases: Uncombusted gasoline vapors that slip by piston rings and accumulate in the crankcase.

Box plot: A graphical representation that shows the distribution of continuous data values around the median.

Brightness: Measure of wavelength-adjusted light received by the human eye from an object.

Bromine monoxide: Chemical species found in the stratosphere that destroys O_3 more effectively than chlorine monoxide.

BTEX: A mixture of aromatic compounds (benzene, toluene, ethylbenzene, and xylene) used as an additive to boost the octane rating of gasoline.

Bubble concept: Source control option wherein emissions are reduced plantwide rather than just from stacks.

Buoyancy flux: Measure of buoyancy of a plume calculated for plume rise equations.

Calibration: Process by which measured values of a pollutant or airflow are compared with standard or reference values; procedure designed to maintain data quality.

Carbon dioxide (CO_2): Natural constituent of the atmosphere that serves biological processes as a source of carbon and the atmosphere as a greenhouse gas.

Carbon monoxide (CO): Colorless, odorless, toxic air pollutant produced by incomplete combustion.

Carboxyhemoglobin: Molecular complex in the blood resulting from the reaction of CO with hemoglobin; measure of CO exposure.

Carcinogen: A substance that can cause cancer.

Carcinogenicity: Ability of a substance to cause cancer.

Cascade impactor: Particle-collecting device that utilizes a series of plates to fractionate particles into different size ranges.

Catalyst: Substance that reduces the energy requirement of a chemical reaction.

Catalytic converter: Control device that removes pollutants from automobile exhaust by catalytic processes.

Catalytic incinerator: Control device that uses catalytic processes to reduce emissions of oxidizable compounds from an industrial source.

Chapman cycle: Chemical reactions that describe O_3 production and destruction in the stratosphere.

Chemiluminescence: Principle used to measure concentrations of NO_x and O_3 by chemical reactions that release light.

Chlorine monoxide: Primary chemical species involved in the catalytic destruction of stratospheric O_3.

Chlorine nitrate: Chemical species that serves as both a sink and reservoir for stratospheric chlorine.

Chlorofluorocarbons (CFCs): Once widely used chemicals characterized by their long atmospheric lifetime, O_3-destroying potential, and thermal sorption properties.

Chlorosis: Loss of chlorophyll in plant leaves associated with chronic exposures to phytotoxic pollutants.

Chronic acidification: Long-term acidification of aquatic and terrestrial ecosystems.

Chronic allergic rhinitis: Relatively mild, persistent illness of the upper respiratory system associated with sensitization and exposure to common allergens.

Chronic bronchitis: Disease of respiratory airways caused by exposure to tobacco smoke or ambient air pollution.

Chronic disease: Disease associated with low-level pollutant exposures over an extended period.

Chronic exposure: Low-level exposure that occurs to humans, plants, or animals over an extended period (weeks, months, years) resulting in characteristic adverse effects.

Citizen suits: Lawsuits that can be brought by citizens against a source or the EPA; legal standing for citizen suits is provided by the 1970 CAA Amendments.

Class I area: Under the CAA Amendments of 1977, areas around national parks, wilderness areas, and the like that are to receive the greatest degree of protection from new, potential, and existing visibility-reducing sources.

Class II area: Areas that include national forests that are subject to lesser visibility protection limits on emissions than Class I areas.

Clean Air Act (CAA) Amendments: Laws enacted by Congress that grant authority to federal agencies to control air pollution according to specific provisions; basis for all federal air pollution regulation.

Climate change: Change in global surface temperature over time.

Climate models: Models used to describe climatic patterns and variations as a result of changes in Earth, water, and atmospheric systems.

Cloud condensation nuclei: Small particles in the atmosphere that provide surfaces for the condensation of H_2O vapor and formation of cloud droplets.

CO_2e: The global warming potential of a specific gas in comparison to the same mass of carbon dioxide, i.e., 1 kg of methane has the equivalent warming potential of 28 kg of carbon dioxide.

Coagulation: Growth of particles associated with the collision and subsequent adherence of particles to each other.

Coal beneficiation: Separation of inorganic sulfur from coal by flotation and settling processes.

Coarse particles: Particles with aerodynamic diameters of ≤ 2.5 µm; produced by fragmentation of matter or atomization of liquids.

Collection efficiency: Relative ability of a sampling train or control device to remove or control pollutants from waste gases.

Combustion: Process by which compounds containing hydrogen and carbon are oxidized to CO_2 and H_2O vapor with the release of heat.

Common law: Legal system based on the evolution of accepted principles of human behavior in the matter of disputes concerning property, contracts, and related matters; legal principles under which a property owner can sue a source for damages and injunctive relief.

Compression ignition engines: Engines that utilize high compression ratios and autoignition of air–fuel mixtures; diesel engines.

Compression ratio: Ratio of the air volume at top dead center of the combustion stroke to that at the bottom of the stroke; factor that influences combustion efficiency, power, and emissions.

Concentration: The relative amount of a substance in a medium such as air, water, or soil.

Condensation: Gas-phase to liquid-phase transformation of matter; in the atmosphere, results in the formation and growth of particles.

Coning plume: Behavior of a plume under neutral to isothermal lapse rate conditions.

Continuous emission monitor (CEM): Monitoring device that provides real-time measurement of stack gases/particles.

Continuous monitor: Sampling device that provides real-time or near real-time pollutant measurements continuously.

Contrast: Difference in light coming from an object relative to its surroundings; indicator of atmospheric clarity.

Convection: Upward movement of warm air and concomitant subsidence of cooler air to replace it.

Coriolis effect: Apparent deflection of wind associated with the Earth's rotation.

Cost–benefit analysis: A quantitative evaluation of control costs relative to benefits to determine economically efficient emission reduction.

Criteria pollutant: Pollutant regulated under NAAQS provisions of clean air legislation.

Cross-contamination: Building air quality problem associated with contaminant transport from a source to an exposure location.

Cyclone collector: Control device used to collect large particles by inertial separation.

Cyclones: Airflow associated with low-pressure systems wherein air moves counterclockwise in the northern hemisphere and clockwise in the southern hemisphere.

Day–night sound level: Average 24-hour sound level weighted by 10 dB for nighttime hours.

Decibels: Unitless expression of sound power, intensity, and pressure.

Deciview: Measure of visibility that is proportional to the logarithm of atmospheric extinction.

Deciview index: Index that expresses uniform changes in haziness in increments from pristine to very impaired visibility.

Deliquescence: Rapid uptake of water vapor from the atmosphere by particles.

Deposition velocity: Rate at which pollutants are removed from the atmosphere by dry deposition.

Dicarboxylic acids: Organic acids produced in photochemical reactions that condense to produce light-scattering particles.

Dichotomous sampler: Particle sampler that collects and separates atmospheric particles into coarse and fine fractions.

Diesel engines: See *Compression ignition engines.*

Diffraction: Physical process by which light passing near, and parallel to, the surface of a particle is scattered from its optical path.

Diffusion: Random movement of gas-phase substances and very small particles from areas of higher to lower concentration.

Dioxins: Family of long-lived, highly toxic, and carcinogenic chlorinated compounds emitted to the environment from the combustion of solid wastes.

Direct air capture: Technology which captures air pollutants or carbon dioxide directly from the ambient air, rather than from a process.

Dispersion: Pollutants mixing with air resulting in lower ground-level concentrations.

Dispersion coefficients: Coefficients used in Gaussian models that are based on standard deviations of concentrations of pollutants as a plume spreads under different atmospheric stability conditions.

Dobson units: Measure of column O_3 concentration using a spectrometric technique.

Dose–response: Toxic responses to pollutant exposures that increase with increasing dose; measure of the relationship between adverse health effects and pollutant exposure.

Dry deposition: Deposition of gases and particles on land and water surfaces in the absence of precipitation.

Dry scrubber: Removal of pollutants by injecting dry sorbents into the pollutant-generating process; flue gas desulfurization technique.

Dust dome: A mass of particle-polluted air observed over a city.

Eccentricity: Time-dependent variation of the Earth's orbit around the sun.

Ecosystem changes: The wide-ranging effects on plant and animal communities associated with environmental pollution.

Effective stack height: The combination of both physical stack height and plume rise; used to predict downwind concentrations in air quality models.

Electrostatic precipitator: Control device that collects particles from waste gases by imparting electrical charges, with subsequent collection of charged particles on plates of opposite electrical charge.

Elemental carbon: Inorganic carbon found in particulate matter samples.

Emission factors: Numerical values compiled by the US EPA for pollutants and sources to estimate emissions.

Emission rate: Source pollutant emissions per unit time.

Emission standard: Legal limit on the quantity of a regulated pollutant that can be released from a source; used as a control strategy; used as a control tactic in achieving NAAQSs.

Emission tax: Pollution control strategy that attempts to reduce emissions from a source by the imposition of a tax; pollution charges.

Emission test cycles: System of testing required of motor vehicle manufacturers to determine compliance with exhaust emission standards under specified driving conditions.

Emission trading: Pollution control tactic in which sources receive credits for control of pollutants in excess of regulatory requirements that may be marketed.

Emissions assessment: Determination of types and quantities of pollutants released from a source per unit time.

Endotoxins: Substances produced in the cell wall of gram-negative bacteria that, on exposure, can cause inflammatory respiratory symptoms; cause of humidifier fever.

Enforcement: Regulatory practice in which legal action is taken against a source for violating air pollution control requirements.

Entrainment: Process whereby outdoor pollutants are drawn into building air.

Environmental noise: Undesirable sound in the ambient environment.

Environmental tobacco smoke (ETS): By-products of tobacco combustion in indoor air that result from sidestream and exhaled mainstream smoke.

Epidemiology: Study of the statistical relationship between disease in a population and potential causal factors.

Episodes: Periods of high ground-level pollution concentrations commonly associated with subsidence inversions.

Episodic acidification: Short-lived acidification of lakes and rivers associated with spring snowmelt.

Equal loudness contours: Graphic representation of the relationship between sound pressure and loudness.

Equivalence ratio: Ratio of the actual/stoichiometric air/fuel ratios; manipulation used to reduce exhaust emissions.

Equivalent method: Any method of sampling and analysis that has been approved by the US EPA as being equivalent to a federal reference method.

Equivalent sound level: Sound averaged over a given time interval.

Ethylene: Pollutant produced in automobile exhaust; natural growth hormone of plants that produces phytotoxic effects upon exposure.

Evaporative emissions: NMHC emissions from the fuel system and in refueling.

Exhaust gas recirculation: Technology used in motor vehicles to control emissions of NO_x by using exhaust gases as a heat sink.
Extinction coefficient: Measure of the reduction of light transmission due to scattering and absorption by particles.
Extratropical pump: Mechanism by which air molecules are transported from the troposphere into the stratosphere.
Fabric filtration: Collection of particles by moving polluted air through a fabric medium.
Fanning plume: Plume behavior under inversion lapse rate conditions.
Federal reference methods (FRMs): Primary methods of air sampling and monitoring approved by the US EPA to determine compliance with NAAQSs.
Filtration: Collection of particles on a fibrous medium.
Fine particles: Particles that have an aerodynamic diameter of ≤ 2.5 μm.
Flare oxidation: Pollution control system wherein combustible gases above the lower explosion limit are ignited and burned.
Flecking: Upper surface leaf injury associated with exposure to O_3.
Flue gas desulfurization (FGD): Control technology designed to remove SO_x from combustion gases.
Flue gas spillage: Combustion by-products that fail to be exhausted by flue gas systems.
Fluidized bed combustion: Combustion process in which a bed of granular material and solid fuel (e.g., coal) are suspended to effect heat transfer and pollutant collection.
Fluorides: Pollutants that can cause injury to plants and animals and etch glass.
Fly ash: Uncombusted inorganic solids released in particle form in the combustion of coal.
Forest decline: Progressive death of a forest species initially characterized by chlorosis and dieback of portions of trees; primary cause is reported to be exposure to atmospheric pollutants.
Formaldehyde: Reactive pollutant emitted in motor vehicle exhaust or produced secondarily in the atmosphere; participates in atmospheric reactions and is a potent mucous membrane irritant.
Four-stroke engines: Engines that have a cycle of intake, compression, power, and exhaust strokes.
Frequency: Property of sound characterized by the number of sound waves produced per unit of time.
Friction: Force that reduces wind speed near the Earth's surface and affects macroscale airflow patterns.
FT-IR: Fourier transform infrared spectrometric technique used to measure pollutants in a portion of the atmosphere.
Fuel additives: Substances added to gasoline; provision in clean air legislation that grants the US EPA authority to regulate these substances.
Fuel use policies: Pollution control approach wherein a source uses fuels that release fewer pollutants.
Fugitive emissions: Emissions not associated with stacks; may be internal or external to a facility.
Fume: An aerosol produced from the heating and vaporization of metals.
Fungal glucans: Substances produced in cell walls of fungi that, upon release and exposure, cause inflammatory respiratory responses.
Gas cleaning: Removal of gas- or particulate-phase substances from a waste gas stream.
Gasolines: Liquid hydrocarbon fuels used to power vehicles equipped with spark ignition engines.
Gaussian models: Models use to determine concentrations downwind of a source based on the assumption that plume spread is primarily due to diffusion.
Geometric mean: The mean of lognormally distributed data.
Geostropic wind: Wind flow that is parallel to isobars (lines of equal pressure) at altitudes above 700 m.
Global warming: Climatic phenomenon that may be caused by increases in atmospheric greenhouse gases due to human activities.

Global warming potential (GWP): The amount of heat trapped by a gas (i.e., methane) compared to the amount of heat trapped by the same mass of carbon dioxide. Methane has a GWP of 28.

Grab sampling: Sample collection over a period of seconds to minutes for purposes of subsequent laboratory analysis.

Gravity: Force of attraction between two bodies; holds atmospheric molecules close to the Earth's surface.

Greenhouse effect: Warming of the Earth's atmosphere resulting from the absorption of thermal energy emitted from the Earth's surface.

Greenhouse gas (GHG): A gas that traps heat in the Earth's atmosphere, contributing to the greenhouse effect and climate change.

Hadley cell: Macroscale circulation pattern that occurs in both tropical and polar regions.

Halogenated hydrocarbons: Widely used industrial and commercial chemicals characterized by their relatively long atmospheric and environmental lifetimes and potential to destroy O_3.

Hazardous air pollutants: Noncriteria pollutants deemed to be especially toxic and regulated under hazardous pollutant provisions of clean air legislation.

Haze: Reduced visibility associated with moderate levels of atmospheric pollutants.

Hearing impairment: Loss of ability to hear sound; usually in the context of exposure to noise.

Heat island: Atmospheric phenomenon of elevated temperatures in urban areas associated with thermal emissions, changes in solar absorption and thermal emission patterns, etc.

Hertz: Quantitative measure of sound frequency; sound waves per second.

High-volume sampler (Hi-Vol): Particle sampler that collects a broad size range of particles (0.3–100 μm) on a glass fiber filter by filtration; concentrations are reported as TSPs.

Histogram: Graphical representation of continuous data using frequency classes.

Humidifier fever: Inflammatory lung ailment caused by exposure to bacterial endotoxins.

Humidity: Concentration of water vapor in a volume of air expressed as weight per unit volume (absolute humidity) or percentage of water vapor that air can hold at a given temperature (relative humidity).

Hybrid electric vehicles: Low-emission vehicles powered by a combination of gasoline and electrical systems.

Hydrocarbons: Substances that contain only hydrogen and carbon in their chemical structure; major pollutants emitted to the atmosphere.

Hydrochloric acid (HCl): Sink and reservoir chemical species in stratospheric O_3-destroying chemical reactions.

Hydrogen sulfide (H_2S): Reduced sulfur gas that is a major malodorous pollutant.

Hydroxyl radical (OH•): Principal atmospheric sink substance that oxidizes pollutants emitted to the atmosphere to more stable chemical forms.

Hygroscopicity: Property of particles that results in their uptake of water vapor from the atmosphere.

Hypersensitivity pneumonitis: Inflammatory lung ailment caused by exposure to high concentrations of bacterial cells or small-spored fungi.

Indirect source: A facility that does not emit pollutants to the atmosphere but attracts sources like motor vehicles that do; may include shopping malls, stadia, arenas, etc.

Indoor air: Air inside nonindustrial buildings.

Indoor air pollution: Contamination of indoor air by gaseous, particulate-phase, and biological pollutants.

Infiltration/exfiltration: Natural movement of air into and out of buildings due to differences in pressure.

Inhalable particulate matter (IPM): Particles with aerodynamic diameters of ≤ 30 μm that enter the upper human respiratory airways of the head.

Inspections and maintenance (I&M): Vehicle inspection and maintenance programs required for motor vehicles in nonattainment areas for specific motor vehicle-related pollutants.

Integrated pest management: Pest control practices that minimize pesticide use.

Interagency Monitoring of Protected Visual Environments (IMPROVE): Cooperative monitoring program of visibility-impairing chemical species and visibility impairment conducted by the US EPA and the National Park Service.

Intergovernmental Panel on Climate Change (IPCC): International organization that conducts scientific assessments and policy reviews on climate change.

Intermittent/integrated sampling: Sampling process wherein samples are collected for a given period (e.g., 24 hours) and schedule (e.g., every 6 days); sample concentrations are averaged over the sampling period.

Interveinal necrosis: Death of tissue between the veins of broad-leaved species associated with exposures to SO_2.

Inverse square rule: Decrease in sound intensity or electromagnetic energy as an inverse square of distance from the source.

Inversion: Atmospheric condition in which temperature increases with height; characterized by limited vertical mixing and pollutant dispersion.

Investigative protocols: Methodologies developed to conduct indoor air quality investigations.

Ionizing radiation: Electromagnetic energy and particles emitted by radionuclides that remove electrons from nearby atoms and molecules, thereby causing ionization.

Ionosphere: Region of the lower thermosphere where the ionization of air molecules takes place.

Isobars: Lines of equal atmospheric pressure used on weather maps.

Isokinetic sampling: Stack sampling technique for particles wherein the sample flow rate is equal to the stack flow rate.

Jet stream: Relatively fast-moving air in the upper troposphere associated with fronts and discontinuities between macroscale circulation cells.

Knock: Sharp, metallic sounds produced by a motor vehicle engine when fuels do not have desired octane ratings.

Koschmeider equation: Mathematical relationship between visual range and the extinction coefficient.

Kyoto Protocol: International treaty developed by the United Nations Framework Convention on Climate Change (UNFCCC) and adopted by parties in Kyoto, Japan, in 1997 that sets binding obligations on industrialized countries to reduce emissions of greenhouse gases.

Land breeze: Lake or seaward movement of air from land areas on calm, clear nights.

Lapse rate: Rate of temperature change with height.

LD50/LC50: Dose of an ingested or injected toxicant, or concentration of an inhaled substance, that will kill 50% of the animals being tested.

Leaching: Loss of plant nutrients from forest trees and soils associated with acidic deposition.

Lead: Criteria pollutant that poses adverse exposure risks to humans and animals.

Lead alkyls: Organic lead additives used in gasolines to reduce knock and increase engine performance.

Lean-burn combustion: Motor vehicle operation that utilizes air/fuel ratios higher than the stoichiometric ratio.

Legionnaires' disease: Infectious pneumonia-type disease associated with *Legionella pneumophila*–contaminated cooling towers, whirlpool spas, hot tubs, and hot water heaters.

Limestone scrubbing: Use of a limestone slurry to remove SO_2 from flue gases.

Limit of detection (LOD): Lowest concentration of a substance that can be accurately quantified.

Lognormal distribution: Environmental data that have been transformed into logarithmic values to form a normal distribution when plotted on semilogarithmic graph paper.

London-type smog: Smog produced when pollutants produced from coal burning are mixed with fog.

Long-range transport: Movement of pollutants tens to hundreds of kilometers downwind from a source.

Looping plume: Behavior of a plume under superadiabatic lapse rate conditions.

Los Angeles–type smog: Smog produced as a result of photochemical reactions in the atmosphere.

Loudness: Human perception of sound intensity.

Low-emission vehicles: Motor vehicles with significantly lower emissions than conventional vehicles; to be phased in under the 1990 CAA Amendments in areas subject to persistently high O_3 levels.

Lower explosion limit (LEL): Lowest concentration of a flammable gas and oxygen that can propagate a flame.

Lowest achievable emission rate: Most stringent emission rate that can, in practice, be achieved for a new source in nonattainment areas.

Macroscale: Air motions that transport long-lived pollutants planet-wide.

Malodors: Odors that smell "bad"; air pollution problem associated with odor-producing pollutants.

Manual methods: Sampling methods in which air is drawn through a sample medium and later analyzed in a laboratory.

Marginal necrosis: Tissue death observed on the margins of leaves associated with fluoride exposures.

Market strategies: Regulatory approaches that allow sources to choose the most cost-effective control methods to achieve emission limits.

Maximum achievable control technology (MACT): Maximum reduction requirements for a hazardous pollutant source considering cost and feasibility; standard that is not less than the average of the top 12% performing existing sources within a source category or utility.

Mechanical turbulence: Random motion of air induced by nonthermal processes.

Median: Middle value in a data set.

Melanoma: A serious form of skin cancer associated with severe sunburns.

Mercury: Toxic heavy metal regulated as a hazardous air pollutant; major environmental concern associated with atmospheric deposition.

Mesoscale: Medium-scale air motions associated with coastal areas and river valleys.

Mesosphere: Layer of the atmosphere extending from 55 to 80–90 km characterized by decreasing temperature with height.

Metal corrosion: Metal oxidation caused by exposure to acids or acidic gases.

Methane: Simplest hydrocarbon species characterized by its low chemical reactivity and strong thermal absorption properties.

Methanol: Liquid fuel used as an alternative to gasoline to achieve lower NMHC emissions.

Methylmercury: Organic mercury (Hg) compound produced by bacterial action; Hg compound that is bioaccumulated.

Microscale: Air motion that occurs in the near vicinity (meters) of a pollution source.

Mie scattering: Scattering of light by particles in the size range of wavelengths of visible light (0.4–0.7 nm).

Milankovitch cycles: Patterns of orbital changes whose periodicity has been linked to major glacial advances.

Mist: An aerosol produced by the atomization of a liquid or condensation of gas-phase substances in the atmosphere.

Mixing height: Vertical volume of air readily available for pollutant dispersion.

Mobile source: Moving pollutant source; for example, cars, trucks, buses, trains, motorcycles, and airplanes.

Mode: Most frequent value in a data set; describes fine particle size range.

Monitoring: Measurement of atmospheric pollutant(s) over time.

Montreal Protocol: International treaty signed in 1987 that began the international effort to phase out and ban the use of O_3-destroying chemicals.

Morbidity: Illness.

Mortality: Death.

Motor vehicle emission standards: Emission limitations placed on major pollutants emitted from motor vehicles.

Mottling: Injury on leaves and conifer needles characterized by large patches of dead cells caused by O_3 exposures.

MTBE: An ether compound (methyl-*t*-butyl ether) used as an oxygenate additive in reformulated gasoline.

Multiport fuel injection: Fuel injection system used on most late-model light-duty motor vehicles whose accurate control of air/fuel ratios allows for the use of three-way catalytic converters.

Mutagen: Substance that can cause mutations.

Mutagenicity: Ability of a substance to cause heritable changes in a gene.

Mycotoxins: Toxins produced by fungi subject to nutritional stress.

National Acid Precipitation Assessment Program (NAPAP): Federal interagency program designed to investigate acidic deposition and its effects on the environment.

National Air Monitoring Stations (NAMS): Monitoring networks administered by the US EPA for purposes of evaluating long-term trends in atmospheric levels of criteria air pollutants in urban areas.

National Ambient Air Quality Standards (NAAQSs): Permissible levels of criteria pollutants in an air quality control region.

National Council on Radiation Protection (NCRP): Federal entity that recommends radiation safety guidelines.

National Emissions Standards for Hazardous Air Pollutants (NESHAP): Provision of the 1970 CAA Amendments that regulated the emission of seven hazardous air pollutants.

Necrosis: Plant tissue death caused by acute or chronic exposures to phytotoxic pollutants.

New source performance standards (NSPSs): Pollutant emission limits on new or significantly modified existing sources.

New source review: Federal requirement that new or significantly modified existing sources undergo a review for compliance with NSPSs, NAAQSs, and PSD requirements before a permit to operate is authorized.

Nitrates: Atmospheric pollutants that serve as plant nutrients and increase nitrogen loadings in aquatic and terrestrial ecosystems.

Nitric acid (HNO_3): Strong acid responsible for approximately 35% of acidity in acidic deposition in the eastern United States and approximately 90% in the western United States.

Nitric oxide (NO): A relatively nontoxic gas produced in high-temperature combustion.

Nitrogen deposition: Deposition of nitrogen to land and water sources associated with atmospheric pollutants.

Nitrogen dioxide (NO_2): A brownish, relatively toxic gas that is a natural constituent of the atmosphere and whose concentrations increase significantly as a result of human activities; gas that absorbs sunlight and initiates photochemical reactions.

Nitrogen oxides (NO_x): Collective term for gas-phase nitrogen compounds that contain one or more oxygen atoms; used to express concentrations of $NO + NO_2$ in the atmosphere.

Nitrogen saturation: Excess nitrogen deposition or use on land that causes nitrate nitrogen to enter standing or moving water.

Nitrous oxide (N_2O): A low-reactivity gas emitted as a result of natural processes and human activities; participates in stratospheric O_3 chemistry and serves as a greenhouse gas.

Noise: Unwanted sound; sound that may cause hearing impairment, speech interference, annoyance, and other effects.

Noise Control Act of 1972: Law that gives the US EPA the authority to promulgate noise emission standards for various devices/equipment.

Noise guidelines: Environmental noise exposure guidelines published by the US EPA.

Nonattainment area: An air quality control region, or portion thereof, in which one or more air quality standards has not been attained.

Nondispersive infrared (NDIR) photometry: Measurement technique that employs sorption of infrared energy to quantify CO concentrations; FRM for CO.

Nonmethane hydrocarbons (NMHCs): Collective term for hydrocarbon species regulated for the purposes of controlling the production of photochemical oxidants; O_3 precursor molecules.

Nonmethane organic gases (NMOGs): Hydrocarbons and their derivatives, excluding methane, that may serve as precursors for O_3 production.

Normal distribution: Environmental data set that, when plotted as a histogram, forms a bell-shaped curve.

Normal lapse rate: Average lapse rate conditions observed in the atmosphere.

Nucleation: Process by which gas-phase substances form liquid particles.

Nuclei mode: Particles in the 0.01 to 0.08 μm range produced from the nucleation of gas-phase substances.

Nuisance suit: Legal action taken against a source by a property owner alleging an unreasonable interference with one's right to the enjoyment of one's property.

Obliquity: Variation in the tilting of the Earth's axis.

Octane rating: Measure of the antiknock quality of gasoline.

Octave band: A range of frequencies in which the upper frequency is twice the lower frequency; characterized by a center frequency.

Odd hydrogen species: Highly reactive oxidized hydrogen species such as hydroxyl radical, hydroperoxy radical, and hydrogen peroxide.

Odor: Olfactory sensation produced by an odiferous substance.

Odor threshold: Lowest concentration of an odoriferous substance that can be detected by the human olfactory system.

Offset policy: Regulatory policy that allows new sources to reduce emissions from existing sources to obtain a permit to construct and operate in a nonattainment area.

Olefins: Hydrocarbons characterized by the presence of one or more double bonds and high photochemical reactivity.

Onboard controls: Control systems on motor vehicles that capture gasoline vapors during refueling and route them to the engine during operation.

Onboard refueling vapor recovery (ORVR): Onboard refueling vapor recovery during outdoor diurnal parking, operation, and running.

Opacity: Ability of a plume to absorb light; measure of plume darkness.

Orbital variations: Changes in the Earth's orbit characterized by obliquity, precession, and eccentricity.

Organic carbon: Particulate matter carbon associated with organic compounds.

Organic nitrate compounds: Chemical substances produced in the atmosphere as a result of reactions between hydrocarbon radicals and NO_2.

Organochlorine compounds: Semivolatile chlorinated compounds, such as pesticides, PCBs, dioxins, and furans, which move through the environment and are found in atmospheric deposition samples.

Organochlorines: Chlorinated compounds used to control insects; persistent, often carcinogenic compounds that are contaminants of ambient and indoor environments.

Organophosphates: Phosphate hydrocarbons used to control insects; potent, relatively nonpersistent compounds known for their neurotoxicity.

Oxidation: Chemical reactions that often include O_2 and result in the production of heat.

Oxidation catalysts: Catalytic systems on motor vehicles that reduce emissions of CO and NMHCs.

Oxygenates: O_2-containing gasoline additives used to increase both octane ratings and combustion efficiency.

Oxyhydrocarbons: Chemical species produced by the oxidation of hydrocarbons; major atmospheric pollutants.

Ozone (O_3): Highly reactive gas found in both the troposphere and stratosphere; elevated tropospheric levels occur as a result of photochemical reactions involving anthropogenic pollutants.

Ozone depletion: Destruction of stratospheric O_3 by long-lived chemical species released as a result of human activities.

Ozone hole: Significant decline in column O_3 levels over the Antarctic continent in the austral spring; human-caused stratospheric O_3 depletion over the South Pole.

Packed tower: Pollution control component of a scrubber system.

Packing material: Small plastic or ceramic high-surface area structures used in scrubbing systems to increase contact time between sorbent and waste gases.

Paraffins: Hydrocarbons characterized by the presence of single covalent bonds and relatively low photochemical reactivity.

Particles: Solid- or liquid-phase substances suspended in the atmosphere or deposited on surfaces.

Particulate matter: Collective term for particles in the atmosphere; more commonly a term used to express atmospheric particle concentrations based on sample collection on a filter.

Passive sampling: Collection of pollutants for quantification using samplers with no moving parts; typically, pollutants are collected by diffusion or sedimentation.

Per- and polyfluoroalkyl substances (PFAS): Human-made chemicals in products such as non-stick cookware, food packaging, and firefighting foams that break down very slowly.

Perfluorooctanoic acid (PFOA): A synthetic chemical from PFAS.

Perfluorooctane sulfonate (PFOS): A long-lasting PFAS chemical.

Permanent threshold shift (PTS): Permanent loss in hearing acuity associated with chronic noise exposures.

Permit: Right extended to a source to emit pollutants to the atmosphere at levels that do not violate regulatory requirements; legal right granted by regulatory agency to construct and operate a pollution source; legal right to operate pollution control equipment.

Peroxy radicals: Highly reactive hydrocarbon species produced as a result of photochemical reactions; key chemical species in the production of elevated tropospheric O_3 levels.

Peroxyacyl nitrate (PAN): An organic nitrate substance produced as a result of photochemical reactions; phytotoxic air pollutant; reservoir chemical for nitrogen transport in the atmosphere.

Personal pollution exposure: Pollutant exposure that includes ambient, indoor, and activity-related human exposures.

Pesticides: Toxic substances used to control a variety of pest species.

pH: A measure of H^+ concentration.

Phons: Measure of sound loudness equal to 1 dB at 1,000 Hz.

Photochemical Assessment Monitoring Stations (PAMS): Federal sampling network in targeted urban locations that measures photochemical pollutants and their precursors.

Photochemical oxidants: Oxidizing chemicals produced in the atmosphere as a result of photochemical reactions; includes such substances as O_3, NO_2, and peroxy radicals.

Photochemical smog: Air pollution caused by reactions between primary pollutants and sunlight as well as subsequent secondary pollutants.

Photochemistry: Chemical reactions in the atmosphere that involve the absorption of sunlight and the production of new oxidized species.

Planetary boundary layer: Depth of the atmosphere where air motion is affected by the Earth's surface.

Plant injury: Visible phytotoxic pollutant-induced effects.

Plume: Visible manifestation of pollutants being emitted by a source.

Plume blight: Impairment of the visual quality of a scene by the plume of an identifiable source.

Plume rise: Vertical motion of pollutant emissions from a source characterized by the centerline of the plume where it levels off.

PM10: Particulate matter fraction with a 50% cutoff aerodynamic equivalent diameter of $\leq 10\ \mu m$.

PM2.5: Particulate matter fraction with a 50% cutoff aerodynamic equivalent diameter of $\leq 2.5\ \mu m$.

Point source: Stationary source of pollutant emissions.

Polar stratospheric clouds: Supercold clouds consisting of ice or nitrate hydrates that, through heterogeneous-phase chemistry, participate in the formation of the Antarctic O_3 hole.

Polar vortex: Unique atmospheric circulation pattern in the Antarctic that indirectly contributes to the formation of the O_3 hole.

Pollution prevention: Pollution control policies that focus on (1) reducing emissions by changing process materials and operating practices and (2) increasing energy efficiency.

Polychlorinated biphenyls (PCBs): Persistent organochlorine compounds that are observed in atmospheric deposition samples, move through food chains, and are found in elevated levels in fish.

Polycyclic aromatic hydrocarbons (PAHs): Hydrocarbon species characterized by multiple benzene rings, presence in the particulate phase, and potent carcinogenicity.

Positive crankcase ventilation: System used on motor vehicles to control emissions of blowby gases by returning them to the combustion chamber.

Potentiation: Ability of one pollutant to enhance the toxic effect of another.

ppmv: Parts per million by volume; mixing ratio used to express air pollutant concentrations on a volume/volume basis (μL/L).

Precession: Top-like orbital variation of the Earth.

Precision: Reproducibility of a measurement; characterized by the variability of data around the mean of repeated measurements of the same concentration.

Pressure gradient force: Force acting on atmospheric molecules due to differences in pressure; perpendicular to isobars.

Prevention of significant deterioration (PSD): Regulatory program that limits visibility-impairing emissions in areas of high air quality.

Primary pollutant: Pollutant released into the atmosphere from a mobile or stationary source.

Primary standard: NAAQS designed to protect public health; in the measurement of pollutant concentration and volume flows, directly traceable to a NIST standard.

Problem building: Building in which indoor air quality complaints are reported and requests made to conduct an investigation.

Quality assurance: Overall management program to assure that monitoring data are accurate and reliable.

Quality control: Routine technical actions taken to ensure that collected data are of high quality; actions include calibration, use of blank and spiked samples, and performance checks.

Quenching: Cooling; process by which the flame is extinguished in motor vehicle combustion chambers due to cooler cylinder walls resulting in NMHC emissions.

Radiational inversion: Ground-based or elevated inversion associated with the radiational cooling of the ground at night.

Radiative forcing: Changes in net radiative energy available to the Earth and its atmosphere.

Radon: Inert gas produced by the decay of radium^{226}, which subsequently undergoes radioactive decay; building air contaminant.

Rainout: An in-cloud process in which pollutants participate in rain making and are subsequently removed from the atmosphere.

Rayleigh scattering: Clear-sky light-scattering caused by very small particles or clusters of molecules.

Reactive hydrocarbons: Hydrocarbon species with one or more double bonds that react readily with other pollutants in atmospheric photochemistry.

Real-time monitoring: Procedure by which concentration data are recorded instantaneously or near instantaneously.

Reasonably available control technology (RACT): Pollutant control technology that provides significant pollutant reduction at relatively reasonable capital and operating costs; applications include compliance with NSPSs and SIPs.

Reduced sulfur: Sulfur compounds produced as a result of reduction reactions, for example, H_2S, CS_2, and COS.

Reduction catalysts: Catalytic systems used on motor vehicles to control emissions of NO_x by reduction reactions.
Reentry: Return of pollutants that had been exhausted from a building.
Reference conditions: Standard conditions of temperature and pressure used to "correct" air volume in pollutant samples.
Reflection: Backward scattering of light.
Reformulated gasoline: Gasoline specially formulated to reduce emissions of reactive NMHCs, aromatic hydrocarbons, and CO.
Refraction: Scattering of light into different wavelengths by atmospheric particles; change in the direction of a light ray as it passes from one medium to another.
Regulatory strategies: Control approaches based on specific air pollution control principles.
Regulatory tactics: Control approaches used in achieving goals of regulatory strategies.
Reid vapor pressure: Measure of the volatility of gasolines.
Resistivity: Relative inability of fly ash to accept an electrical charge in an electrostatic precipitator system.
Respirable particles (RPM): Particles that can be easily deposited in lung tissue.
Ringlemann chart: A series of shaded charts that are used to visually quantify the opacity of a plume.
Risk: Measure of the probability that exposure to a pollutant or other factor will cause harm.
Risk assessment: Qualitative and quantitative process by which policy makers determine the relative harm associated with a pollutant exposure.
Rowland–Molina hypothesis: Theory that described the potential adverse effects of chlorofluorocarbon pollution of the atmosphere on stratospheric O_3.
Rural ozone: Elevated O_3 levels in rural areas associated with the long-range transport of O_3 and its precursor molecules.
Sampling: The act of collecting an atmospheric or stack pollutant for purposes of determining its concentration.
Saturation vapor pressure: Water vapor concentration at which condensation typically occurs.
Scrubber: A pollution control device used to control particles or soluble gases by absorption in a liquid or slurry.
Sea breeze: Landward movement of air from the oceans during calm, clear days.
Sea ice thinning: Reduced thickness of sea ice believed to be due to global warming.
Secondary pollutant: Pollutant produced in the atmosphere as a result of chemical reactions.
Secondary standard: NAAQS designed to protect public welfare; calibration standard derived from a NIST traceable primary standard.
Settling velocity: The rate (cm/s) at which particles settle on a surface.
Sick building syndrome: Characteristic illness complex experienced by building occupants wherein a causal factor or factors cannot be identified.
Single-port fuel ignition: System used in early fuel injection systems that limited the ability to control air/fuel ratios.
Sinks: Processes by which substances emitted to the atmosphere are removed.
Skin cancer: Malignancies of the skin caused primarily by exposure to UV light.
Slurry: Mixture of water and solid particles used in flue gas desulfurization systems.
Smog: Atmospheric condition characterized by significantly reduced visibility associated with elevated pollutant levels.
Smokestack: Physical structure that elevates pollutants above a source to reduce ground-level concentrations.
Solar constant: Solar energy received by the Earth at an average rate of 1.92 cal/cm^2/s.
Solar spectrum: Electromagnetic energy emitted by the sun and received on the Earth's atmosphere and at the ground.
Sone: Unit of loudness based on 40 phons at 1,000 Hz.

Soot: Elemental carbon particles produced as a result of incomplete combustion of a fuel.

Sorbents: Solid materials that collect contaminants by physical attraction to their surface; liquid substances that absorb pollutants.

Sound: Form of energy produced by a vibrating object or aerodynamic disturbance.

Sound intensity: Measure of sound energy at some distance from a source.

Sound power: Sound energy produced by a source.

Sound pressure level: Sound pressure measured at some distance from a source.

Sound pressure level meter: Instrument used to measure sound to characterize exposures.

Spark ignition engines: Conventional light-duty motor vehicle engines that require an electrically generated spark for ignition.

Spark retardation: Modification of engine operation to maintain high exhaust temperatures to continue CO and HC oxidation.

Spatial scales: Used to locate air monitoring sites to better interpret the results of air monitoring in the context of the area affected.

Spectrum analysis: Characterizing sound by its dominant frequencies; conducted with octave band analyzers.

Spray tower: Pollution control system used to control particles by spraying waste gas with water droplets.

Stability classes: Relative atmospheric stability conditions identified by the letters A to F used to calculate and use dispersion coefficients for air quality modeling.

Stable air: Air mass with little vertical mixing.

Stack sampling: Collection and quantification of pollutants from a stack using either manual or automated (continuous) methods.

Standard deviation (σ): Measure of the variability of measurement data around the mean; mean $\pm$ 1 σ = 67% of measured values.

State and local air monitoring stations (SLAMS): Network of air monitoring stations that states operate to determine compliance with NAAQSs.

State implementation plan (SIP): Specific control actions formulated by a state to meet NAAQSs in all control regions under its jurisdiction.

Stationary source: Fixed-site source of pollutant emissions.

Stem-and-leaf diagram: Graphical method in which all values are presented to illustrate the distribution of measured data.

Stoichiometric ratio: An air/fuel ratio in which there is an exact balance between the amounts of fuel and O_2 present for complete combustion.

Stratosphere: Layer of the atmosphere extending from approximately 15 to 50 km characterized by increasing temperature with height and elevated O_3 levels.

Stratosphere–troposphere exchange: Mechanisms by which air from the troposphere is transported into the stratosphere and from the stratosphere into the troposphere.

Stratospheric turbidity: Scattering of incoming solar radiation by volcanically derived sulfate aerosols in the stratosphere.

Subsidence inversion: Inverted lapse rate conditions that result from the compression of air aloft; associated with the sinking of cool air in a high-pressure system.

Sulfates: Collective term for chemical species that contain SO_4^{2-} as part of the molecule.

Sulfur dioxide (SO_2): Pollutant produced by the oxidation of sulfur-containing fuels and metal ores.

Sulfur oxides (SO_x): Collective term for sulfur dioxide and sulfur trioxide.

Sulfuric acid (H_2SO4): Strong acid responsible for ~60% of the H^+ in acidic deposition in the eastern United States.

Sunspots: Cool, dark spots on the sun's surface believed to affect solar emissions to the Earth and subsequent weather and climatic conditions.

Superadiabatic lapse rate: Lapse rate conditions in which temperature changes with height more rapidly than that which would occur adiabatically.

SVOCs: Semivolatile organic compounds with boiling points in the range of 240°C–260°C to 380°C–400°C; chemical compounds characterized by low volatility, low atmospheric concentrations, and long environmental lifetimes and mobility.

Synergism: Pollutant interactions that produce biological responses in exposed organisms that are significantly greater than those resulting from the sum of the individual exposures.

Synoptic scale: Air motions associated with the movement of high-pressure and low-pressure systems.

Tall-stack technology: Pollution control principle used to reduce ground-level concentrations by elevating emissions to significant heights above the ground.

Tampering: Illegal practice of disabling motor vehicle emission control systems.

Tapered element oscillating monitor (TEOM): Continuous particulate air monitoring device that uses a tapered element oscillating balance to determine particle concentrations on a filter.

Technology-based standards: Emission limits based on available technology; standards that do not consider health effects directly.

Temporary threshold shift (TTS): Temporary loss in hearing acuity associated with excessive noise exposure.

Thermal emission spectra: Range of infrared energies radiated to space.

Thermal oxidation: Pollution control technique wherein combustible gases below their lower explosion limit are combusted under elevated temperature conditions.

Thermosphere: Outer region of the atmosphere beginning at an elevation of 90 to 100 km where temperature increases significantly with height.

Thoracic particles (TPM): Particles 10 μm that enter respiratory airways and lung tissue.

Three-way catalysts: Catalytic system used on most late-model motor vehicles that, by precise control of the equivalence ratio, can achieve an 80% reduction in CO, MNHC, and NO_x.

Threshold: Dose or exposure level below which no significant adverse health effects are expected.

Threshold of hearing: The lowest sound pressure a good human ear can hear; 0 dB.

Tip necrosis: Death of the tip of leaves and coniferous needles associated with pollutant exposures.

Topography: Landforms characteristic of an area; topographical features influence air movement and, as a consequence, pollutant dispersion.

Total suspended particulates (TSPs): Concentration of atmospheric particles determined by using a high-volume sampler.

Total volatile organic compound (TVOC) theory: Theory that building illness symptoms are caused by additive and synergistic effects of low concentrations of multiple VOCs.

Toxicant: Substance that may cause biological harm to an exposed organism.

Toxicity: Relative potential of a toxicant to cause biological harm upon exposure; toxicity includes the potency of the substance, exposure dose, and genetic susceptibility.

Transmissometer: Instrument that measures light attenuation over a specified path.

Trespass: Common law principle applied to suits that involve one's exclusive right to the use of property; used when pollutants such as particulate matter "trespass" or violate one's property rights.

Tropopause: Layer of the atmosphere between the troposphere and stratosphere characterized by isothermal lapse rate conditions.

Troposphere: That portion of the atmosphere near the ground where temperature decreases with height.

Turbidity: Reduced ability of air to transmit light; usually considered in the vertical context.

Turbulence: Unstable air movements (eddies) caused by mechanical or thermal factors.

Ultrafine particles: Particles with diameters of < 0.01 μm produced in the early phases of gas-to-particle conversion processes.

Ultraviolet (UV) light: Region of the solar spectrum that can cause sunburn and skin cancer; light energy below the visible light spectrum and mostly absorbed by stratospheric O_3.

Urban plume: Pollutants transported downwind from an urban area.

UV photometry: Monitoring technique wherein O_3 concentrations are determined by the absorption of UV light by O_3.

UV-A: Region of the UV spectrum (320–400 nm) that causes melanin formation and vitamin D production in human skin; regulates plant growth.

UV-B: Region of the UV spectrum (280–320 nm) that causes sunburn and skin cancer.

Ventilation: Natural or mechanical process by which outdoor air is brought into buildings to reduce pollutant concentrations.

Ventilation guidelines: Recommended quantity of ventilation air needed to provide acceptable health and comfort conditions in buildings.

Venturi scrubber: Particle control device that achieves high collection efficiency by using large quantities of small water droplets.

Virtual impactor: Sampling device that fractionates particles into coarse and fine modes by impacting them on a "virtual surface" of a slow-moving column of air; see *Dichotomous sampler.*

Visibility: Visual quality of the atmosphere relative to color and contrast; typically described as a function of distance.

Visual range: Distance that a dark object can be clearly seen against its surroundings.

Volatile organic compounds (VOCs): Compounds with boiling points in the range of 50°C–100°C to 240°C–260°C.

Washout: Process by which particles and gases are removed from the atmosphere by falling precipitation.

Wet deposition: Removal of pollutants from the atmosphere by precipitation-related processes.

Wet scrubbers: Pollution control systems wherein pollutants are removed by bringing them into contact with water or a watery media.

Wind: Physical characterization of atmospheric motion relative to speed and direction.

Working level: Measure of energy associated with radon decay products.

Index

Page locators in **bold** indicate a table
Page locators in *italics* indicate a figure

D

G

H

N

T